Plant Pathology: Disease Detection and Identification

Plant Pathology: Disease Detection and Identification

Editor: Ian Brock

CALLISTO REFERENCE

www.callistoreference.com

Callisto Reference,
118-35 Queens Blvd., Suite 400,
Forest Hills, NY 11375, USA

Visit us on the World Wide Web at:
www.callistoreference.com

ISBN: 978-1-64116-172-5 (Hardback)

Cataloging-in-Publication Data

Plant pathology : disease detection and identification / edited by Ian Brock.
   p. cm.
Includes bibliographical references and index.
ISBN 978-1-64116-172-5
1. Plant diseases. 2. Plant diseases--Diagnosis. 3. Phytopathogenic microorganisms.
I. Brock, Ian.
SB601 .P53 2019
632--dc23

Table of Contents

Preface

This book was inspired by the evolution of our times; to answer the curiosity of inquisitive minds. Many developments have occurred across the globe in the recent past which has transformed the progress in the field.

Plant pathology is the study of diseases in plants that are caused by pathogens. It encompasses the studies of pathogen identification, disease etiology, plant disease epidemiology, economic impact, etc. Pathogens that cause diseases in plants are fungi, viruses, bacteria, protozoa, etc. Effector proteins, cell wall-degrading enzymes and toxins are the prominent methods of pathogenic infection. Some of the severe plant diseases include citrus canker, rice blast, soybean cyst nematode, etc. This book discusses the fundamentals as well as modern approaches of plant pathology. It strives to provide a fair idea about this discipline and to help develop a better understanding of the latest advances within this field, particularly with respect to disease detection and identification. Students, researchers, experts and all associated with botany and agriculture science will benefit alike from this book.

This book was developed from a mere concept to drafts to chapters and finally compiled together as a complete text to benefit the readers across all nations. To ensure the quality of the content we instilled two significant steps in our procedure. The first was to appoint an editorial team that would verify the data and statistics provided in the book and also select the most appropriate and valuable contributions from the plentiful contributions we received from authors worldwide. The next step was to appoint an expert of the topic as the Editor-in-Chief, who would head the project and finally make the necessary amendments and modifications to make the text reader-friendly. I was then commissioned to examine all the material to present the topics in the most comprehensible and productive format.

I would like to take this opportunity to thank all the contributing authors who were supportive enough to contribute their time and knowledge to this project. I also wish to convey my regards to my family who have been extremely supportive during the entire project.

Editor

A Simple, Rapid and Efficient Method of *Pepino mosaic* Virus RNA Isolation from Tomato Fruit

Amal Souiri[1-3], Mustapha Zemzami[3], Khadija Khataby[1], Hayat Laatiris[3], Saaid Amzazi[2] and Moulay Mustapha Ennaji[1]*

[1]*Laboratory of Virology, Microbiology, Quality, and Biotechnologies/Ecotoxicology and Biodiversity, Faculty of Science, and Techniques Mohammedia, University Hassan II of Casablanca, Morocco*
[2]*Laboratory of Biochemistry and Immunology, Faculty of Sciences, University of Mohammed V, Rabat, Morocco*
[3]*Laboratory of Sanitary Control, Control Unit of Plants, Domaines Agricoles Maâmora, Salé, Morocco*

Abstract

The main concern in molecular detection of RNA viral pathogens in plants is the achievement of good quality of the extracted RNA. Various methods of isolating RNAs from both polysaccharide-rich and poor tissues and other recalcitrant plants are available. However, the use of time and reagent consuming methods and those involving hazardous chemicals is somewhat cumbersome and problematic, especially when it is not necessary for specific purposes like isolating viral RNA from tomato fruit, hence the objective of this paper. We describe an alternative, simple and rapid method for preparing viral RNA from tomato fruit without RNA extraction and purification steps, case of *Pepino mosaic* virus (*PepMV*).

The method employs mechanical treatment and suspension in water. The quality of RNA obtained was judged by spectrometric readings and validated in RT-PCR assays. The used protocol was compared with the usual TRIzol method. The results showed that the yield and the quality of RNA obtained using the proposed method are efficient and highly yielded in comparison with TRIzol method. Moreover, the developed method successfully allowed a sensitive and reproducible detection of *PepMV* predicted bands in RT-PCR. Thus, molecular detection of *PepMV* from tomato fruit can be performed routinely without fastidious RNA isolation. As well, this will make the diagnosis of other RNA viruses infecting tomato crops easier and less time-consuming, in comparison with the other methods performed with expensive commercial kits and those involving toxic chemicals. Finally, the described established method will contribute effectively in strategies of phytosanitary and certification programs of tomato crops worldwide.

Keywords: Tomato; RNA isolation; *Pepino mosaic* virus; Molecular diagnosis

Introduction

Tomato (*Solanum lycopersicum*) is one of the most widely grown vegetable crops. In many worldwide regions where tomato is cultivated, viral diseases have become one of the main limiting factors in tomato production. Over the past few years, *Pepino mosaic* virus (*PepMV*), genus Potexvirus, Family Flexiviridae, has caused significant economic losses in tomato production areas in Europe [1-4], North America [5], South America [6], and Asia [7]. Detection and prevention are the main control measures. Diagnosis based on disease symptoms is not reliable, because symptoms can vary according to the *PepMV* isolates [8].

Several methods have been developed for the detection of *PepMV* in plants. Serological techniques like ELISA have been developed and used successfully for a number of years for the detection of plant viruses. Recently, the trend toward molecular biology techniques such as RT-PCR and nucleotide sequencing has risen in last years. The use of RT-PCR in diagnosing *PepMV* is described in Hasiów et al. [9]. An immunocapture-retrotranscription-PCR (IC-RT-PCR) approach is detailed in Mansilla et al. [10].

Prior to RT-PCR, RNA virus isolation is a critical step. For the majority of plant material containing high levels of polysaccharides and polyphenols, pigments and *RNase* is a challenging starting material for high-quality RNA isolation in reasonable amounts because of different amounts of those substances in diverse tissues.

A large number of methods have been developed or widely modified [11,12] for RNA isolation and purification from plant tissues and adopted by researchers and laboratories. The RNA isolation method developed by Chomcynski and Sacchi [11] employ a single extraction with acid guanidinium thiocyanate-phenol-chloroform mixture [11]. This method incorporated TRIzol, a ready-to-use reagent. Since its introduction, this method has become widely used for isolating total

RNA from biological samples of different sources, during last few decades [13]. Thompson et al. [14] used the Plant RNeasy Kit produced by QIAGEN to quickly extract high-quality total RNA from strawberry leaves, but the kit is expensive [14]. In addition, protocols designed for plant are time consuming, require hazardous products and are tissue specific [12,15,16].

In the case of tomato, the level of polysaccharides and other interfering components varies in dependence of fruit ripening stages and nature of tissue (leaf, stem, and root, flower, fruit). Consequently, the yield of total RNA extracted differs from a tomato tissue to another [15]. The method described by Wang et al. [15] for isolation of total RNA from tomato fruit was used in gene expression studies at the microarray-based level [15].

However, isolation of viral RNA from infected tomato fruit is not as difficult and requiring as isolation of functional RNA and small RNAs [17]. Optimizing protocols of RNA virus isolation step, is important in order to meet sensitivity, specificity and rapid means of detecting RNA plant viruses, including *PepMV*, in tomato tissues.

*Corresponding author: Moulay Mustapha Ennaji, M.Sc. Ph. D RSM (CCM) Laboratory of Virology Microbiology, Quality and Biotechnologies /Ecotoxicology and Biodiversity, Faculty of Science and Techniques Mohammedia, University Hassan II of Casablanca, Mohammedia 20650, Morocco
E-mail: m.ennaji@yahoo.fr

For this purpose, we attempted to change the time- and reagent-consuming RNA extraction procedures by a simple isolation method that yield a good quality of RNA from tomato fruit, when compared with TRIzol method, cost effective, and not require ultra-centrifugation or toxic chemicals. The developed method was validated in assays of repeatability and reproducibility performed in RT-PCR. In addition, we evaluate the integrity of the viral RNA obtained and the broad applicability for both high and low titers. The RT-PCR assays were conducted for *PepMV*, targeting various regions of the viral RNA genome; so, the obtained viral RNA from tomato samples would be appropriate in downstream RT-PCR detection of *PepMV*. This is the first study to report detection of plant virus infecting tomato crops without chemical reagents for RNA isolation.

Materials and Methods

Virus isolates and screening of positive samples

Tomato fruit showing characteristic symptoms of *PepMV* infection were collected from fresh vegetables markets in Rabat and other production fields in Morocco. This study was conducted on a total of 20 tomato samples. The experiments were carried out at Laboratory of Sanitary Control, Control Unit of Plants, Domaines Agricoles Maâmora, Sale, Morocco. Samples were analyzed for determination of *PepMV* presence using a double antibody sandwich enzyme-linked immunosorbent assay (DAS-ELISA) employing monoclonal antibodies, 1b11-G10 and 5A1-G5, which were produced in a previous study. The samples were rated positive if the OD exceeded the mean value of two negative control wells by three times [18].

Design of experimental study

In order to develop and optimize the simple and rapid method of viral RNA extraction from tomato fruit, an experimental study was performed following four major phases: the first one was to extract samples (n=10) by phenol-guanidine isothiocyanate (TRIzol®) method, the second was to prepare suspension from the duplicate of each sample using Milli-Q water, and this is considered as an alternative to RNA extraction. The third phase was to perform RNA qualification and quantification and compare the yield between both methods. Finally, the last phase was to evaluate the RNA obtained from both methods by testing in one step RT-PCR the detection of RNA genomic targets of *PepMV*. Reproducibility and repeatability tests were performed in RT-PCR analysis to evaluate the reliability of the developed RNA preparation method. Also, to investigate the sensitivity of viral RNA detection, for both high and low titer of the virus, a serial dilution of RNA extracts was prepared and used in RT-PCR assays.

TRIzol-based RNA extraction

The total RNA from each of the pulp and skin samples was extracted from the infected tomato (90 mg of homogenized fruit sample) by using the phenol-guanidine isothiocyanate procedure according to the manufacturer instructions (TRIzol® reagent; Invitrogen). RNAs thus obtained are stored at -80°C [11].

RNA preparation method

A suspension was prepared by grinding 100 mg of pulp, skin and juice of tomato with 100 µl of Milli-Q water (RNase Free) in a 1.5 ml microtube using a pestle, then total volume was made up to 1 ml with Milli-Q water. After precipitation for 1 hr at 4°C, the supernatant was then recovered in a new microtube and stored at -80°C until use in downstream application.

Quantity and quality control

RNA concentration and purity were assessed using Spectrophotometer (NanoDrop Technologies Inc.). The water was used as blank. Sample optical density was measured at wavelengths of 260 and 280 nm, and the 260/280 ratio was used to assess RNA purity. RNA purity was considered adequate when the 260/280 ratio was between 1.8 and 2, as a lower ratio could indicate the presence of proteins, phenol, or other contaminants that typically show strong absorbance at 280 nm [19].

PepMV RNA detection by RT-PCR

RNA preparations were used to perform RT-PCR analysis to detect *PepMV*-RNA using and optimized one step protocol. For gene-specific amplification primers *Pep3* (5'-ATGAGGTTGTCTGGTGAA-3') and *Pep4* (5'-AATTCCGTGCACAACTAT-3') specific for a part of RNA-dependent RNA polymerase (*RdRp*), primers *PepRecB-D* (5'-GAACTAAATGCCAGGTCT-3') and *PepRec-R* (5'-GTGACTCCATCGAAGAAGT-3') specific for half of triple gene block (TGB) and half of capsid protein, primers *PepUSTGB-D1* (5'-TCACAAACTCCATCAAGG-3') and *PepUSTGB-R* (5'-TTAGAAGCTGTAGGTTGGTTTT-3') specific for TGB, and primers *PepCP-D* (5'-CACACCAGAAGTGCTTAAAGCA-3') and *PepCP-R* (5'-CTCTGATTAAGTTTCGAGTG-3') specific for CP were synthesized to amplify overlapping reading frames of *RdRp* (624 bp), TGB-CP (1028 bp), TGB (1317 bp) and CP (845 pb). RNAs were reverse transcribed using *M-MLV* reverse transcriptase (Promega) and polymerase chain reaction PCR amplified using Taq polymerase (Promega). Briefly, 25 µl of reaction was carried out in 0.2 ml tube containing 1 µl of prepared RNA and 5 µl of M-MLV buffer 5x, 2 µl of MgCl$_2$ (25 mM), 1 U M-MLV reverse transcriptase, 1000 U RNasin (RNasin® Ribonuclease Inhibitor, promega), 0.2 mM each dNTPs, 0.4 µM each primer and 0.05 U Taq polymerase. Ultra-pure water was used as negative control. The cycling parameter were reverse transcribed for 30 min at 50°C, denaturated for 2 min at 94°C, followed by 40 cycles of denaturation at 94°C for 15 sec, annealing at 45°C to 52°C depending on the specific primers, for 30 sec and extension cycle 60°C for 45 sec with a final extent cycle of 7 min at 68°C. An aliquot of RT-PCR preparation (6 µl) from each reaction was applied onto 1.5% agarose gel for electrophoresis and the RT-PCR product were visualized under UV.

Repeatability and reproducibility tests

To assess the repeatability of the obtained RNA extracts in downstream RT-PCR experiments, RNA extraction method was repeated 3 times for 10 samples from infected and non-infected tomato fruits, in the same conditions and same operator.

For the reproducibility assays, we work with the same triplicate of RNA extract samples (n=10) under the previous conditions but we varied the parameter of time by separating the triplicates by one day between each RT-PCR assay, also the experiment was tested by multiple users.

Limit of detection

To study the limit of detection (LOD) of *PepMV* RNA in RT-PCR assays, total RNAs from tomato fruit was prepared in two-fold serial dilutions in water. This experiment was conducted for RNA obtained with the proposal method in comparison with Trizol method following the same conditions. The initial concentration of undiluted RNA was 140 ng/µl, and all amplifications were carried out with 1 µl in a final volume of 25 µl.

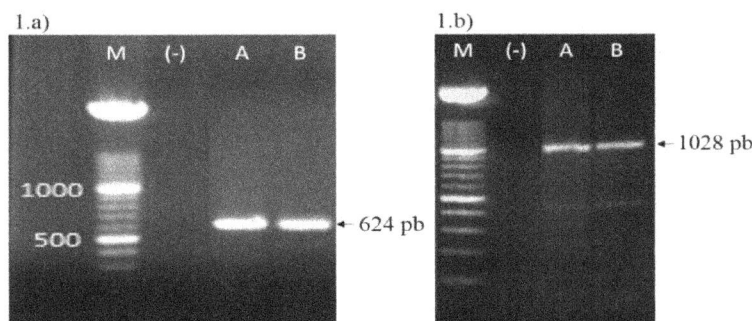

Figure 1: Agarose electrophoretic profile of RT-PCR product of PepMV. (1a) amplification of RdRp gene, (1b) amplification of TGB-CP gene, M: DNA marker 100 pb, (-): Negative control, A: Trizol-based RNA extraction, B: RNA preparation method.

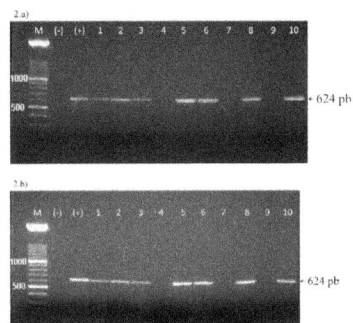

Figure 2: Representative results of repeatability and reproducibility tests for detection of PepMV viral RNA in tomato extract. (2a): Repeatability, (2b): Reproducibility, Target: *PepMV RdRp* gene (624 bp); (-): Negative control, (+): positive control, 1 to 10: RNA template from infected and non-infected tomato, M: DNA marker 100 pb.

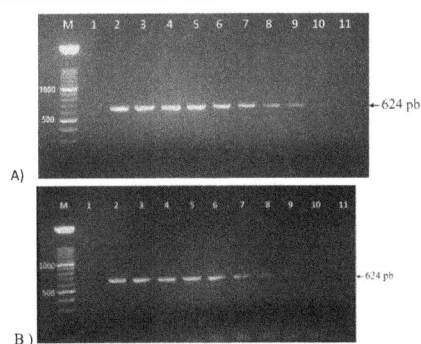

Figure 3: Detection of PepMV viral RNA. Two-fold serial dilution of extracted RNA were prepared in water and used as template in RT-PCR reactions. 3a) developed method, 3b) Trizol method, *Target: PepMV RdRp* gene (624 bp); Template quantity (25 µl of reaction): (1) no template, (2) 140 ng, (3) 70 ng, (4) 35 ng, (5) 17.5 ng, (6) 8.75 ng, (7) 4.37 ng, (8) 2.19 ng, (9) 1.09 ng, (10) 0.55 ng, (11) 0.27 ng.

Sensitivity

In order to evaluate the viral RNA detection at low titer of virus infection, the neat extract from the infected tomato was diluted in extract from healthy tomato using the developed method. Thereby, the background level of inhibitor present in tomato remains the same, but the amount of RNA decreases.

Results

Yield comparison of two total RNA isolation method

In order to compare the yield of total RNA extracted with both methods, quantitative measurements were conducted to determine the nucleic acid concentration and the ratio OD260 nm/OD280 nm. The new developed method shows a high RNA concentration of 140 ng/µl and a good purity of 1.72. Both methods led to slightly identical results. However, the eluted volume is considered more important in the developed method (800 µl from 100 mg of tissue), whereas only 60 µl from 90 mg of tissue in case of TRIzol method.

Detection of *PepMV* viral RNA

Results related to RNA *PepMV* detection from total RNA extracted by both described methods was checked using RT-PCR. 100 ng/µl of nucleic acid was used as template in this experiment. Through electrophoresis migration of RT-PCR amplified products, reliability of the proposal RNA isolation was evaluated by comparing to the TRIzol method. As shown in the Figure 1a and 1b, the correctly sized RT-PCR product (624 bp) and (1028 bp) were obtained for each of the RNA method after amplification with specific *PepMV* primers.

Validation in RT-PCR assays

Repeatability and reproducibility tests

The obtained results of both repeatability and reproducibility assays were satisfactory. Non-infected samples were negative and the positive samples remained positive in each triplicate tests. A full success of the evaluated criteria was rated at 100% for each sample. The representative results are shown in Figure 2 of one of each triplicate, a repeatability assay (Figure 2a) and a reproducibility assay (Figure 2b).

Limit of detection

LOD was investigated including a comparison using RNA from the Trizol method compared to the new method. This approach aims to evaluate how much RNA is available for the PCR amplification in the new method compared with an established method. Two-fold dilutions were achieved for both cases with an equal amount of starting RNA templates.

The results of this part of study indicated that PCR products derived from viral RNA obtained with the proposed method were generated at lower template dilution in comparison with the established method TRIzol.

Indeed, the specific band (624 pb) was visible until the dilution corresponding to a quantity of 0.55 ng of total RNA (Figure 3A), as a low titer, while it became undetectable beyond 1.09 ng for TRIzol method (Figure 3B).

Sensitivity

Sensitivity of viral RNA detection was assessed by simulating

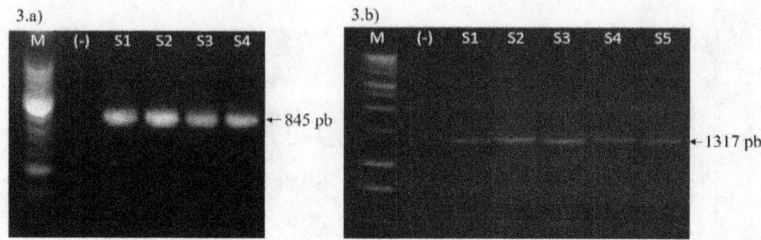

Figure 4: Agarose electrophoretic profile of RT-PCR product of PepMV. (3a) amplification of CP gene. M: DNA marker 100 pb, (-): Negative control, S1 to S4: Tomato RNA samples, (3b) amplification of *TGB* gene, M: DNA marker 1Kb, (-): Negative control, S1 to S5: Tomato RNA samples.

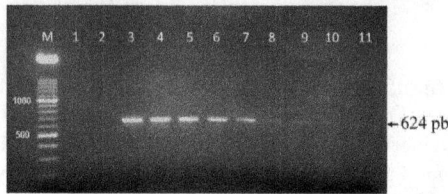

Figure 5: Detection of PepMV viral RNA in tomato extract. Two-fold serial dilution of extracted RNA were prepared in healthy plant extract and used as template in RT-PCR reactions. Target: PepMV RdRp gene (624 bp); Template quantity (25 μl of reaction): (1) No template, (2) Healthy plant extract, (3) 140 ng, (4) 70 ng, (5) 35 ng, (6) 17.5 ng, (7) 8.75 ng, (8) 4.37 ng, (9) 2.19 ng, (10) 1.09 ng, (11) 0.55 ng.

higher and lower virus titer by a serial dilution of RNA template in healthy tomato extracts (Figure 5).

Gel migration reveals that the optimal quantity of total RNA that gives a good aspect of the specific band corresponds to 8.75 ng per 25 μl RT-PCR reaction. This amount of nucleic acid is greater than that required for the previous experiment, because the dilution was performed in tomato extract rather than in water, the amplification may be affected by the level of inhibitor present in tomato Figure 5.

Amplification of other regions of the *PepMV* genome

Further, to evaluate the integrity of viral RNA and the performance of the simplified method for RNA isolation developed in this study, total RNA from each of the tomato samples (n=10) was subjected to other RT-PCR reactions using the set of *PepMV* specific primers cited above. Examples of viral RNA detection in electrophoresis agarose gel 1.5% are shown in Figure 4. The results of these assays were successful in all RNA samples tested. All RT-PCR reactions yielded a DNA fragment at the molecular weight expected, ranged from 624 pb to 1317 pb, from *PepMV*-RNA for the chosen genomic regions.

Discussion

In the present work, we attempted a novel method of viral RNA isolation from tomato fruit that is of good quality and suitable for downstream molecular analysis, namely RT-PCR. This is the first study to report reliable and efficient detection of plant virus infecting tomato crops without chemical and toxic reagents. A comparison with the reference TRIzol method, repeatability and reproducibility were performed to assess the validation process of this procedure using RT-PCR targeting viral RNA of *PepMV*.

The results of the comparison with the reference TRIzol method showed that, the purity of tomato fruit RNA prepared using mechanical treatment and suspended in water was comparable to that extracted with TRIzol. The ratio of A_{260}/A_{280} was approximately 1.8 in all samples, this indicate high purity of RNA and the absence of protein, phenol or other contaminants that absorb strongly at or near 280 nm.

Besides, for RNA yield per gram tomato fresh weight, the results of the developed method were found higher than the obtained yields using TRIzol protocol, and those reported yields in other studies using protocol developed for tomato fruit [20]. This difference is due to the elution volume in the final step, which is important (800 μl) in our method that do not use ultra-centrifugation and purification steps, thus all the RNAs are recovered including DNA traces. Moreover, we noticed that standard deviation of yield from RNA preparations is little significant. This can be explained by the non-homogeneous fruit maturation stage between the tomato samples, which affect the yield of total RNA, as shown by Wang et al. [15], the RNA yield (μg/g) vary between tomato green fruit, orange fruit, red fruit [15]. Nevertheless, the concentration and purity remained comparable for both protocols. For another part of this comparison, all RNA samples achieved with both methods have been amplified successfully for the specific *PepMV* targets and exhibited good gel migration.

In addition, through the number of RT-PCR performed in this study, the results confirm the high repeatability and reproducibility of the viral RNA detection using the extracts obtained with our developed method, making it a reliable technique for RNA preparation from tomato fruit. The use of a set of primers targeting variable viral RNA regions and different length, between 624 pb and 1317 pb, have allowed to evaluate the integrity of the viral RNA, which was successful.

The only limit of the developed method, is the possible degradation of RNA due to the presence of *RNase* and other unpurified components in the RNA suspension after a long storage. In preference, tomato extracts should be used in the same day of their preparation or in the next few days in RT-PCR assays to ensure good quality of results and rapidity of the test. The *RNases* that may remain in an RNA sample are trapped by using protein *RNase* inhibitors (RNasin) in enzyme reactions of RT-PCR.

In the other hand, knowing that the concentration of *PepMV* particles in tomato fruit is high, the broad applicability for both high and low titer of the technique has been investigated. For this purpose, we determined the sensitivity of RNA detection using a serial dilution of total RNA extracted in healthy tomato extract. Also, we noted a very low interference by the constituents of tomato fruit with RT-PCR reactions.

Moreover, the limit of detection of viral RNA study, conducted for both the proposal method and the established TRIzol RNA extraction, give significantly satisfactory results. High dilutions lead to early disappearance of the bands when TRIzol is used due to its PCR inhibitors like ethanol and phenol components. These findings validate the use of relatively small amount of total RNA template for *PepMV* viral detection in tomato fruit using the described method in this paper.

Conclusion

In summary, we developed a simple and reliable method based on mechanical extraction and ultrapure water as the only additive. This method has allowed the isolation of the viral RNA from tomato fruits which is correctly detectable by RT-PCR. This method provides a significant advantage to any laboratory, including those of limited-resource, interested in implementing procedures for viral RNA preparation from tomato fruits without the use of hazardous and noxious chemicals for the manipulator and the environment. Also, the potential of a one-step RT-PCR for rapid detection of *PepMV* would be helpful for both epidemiological studies and genetic characterization.

Acknowledgment

The authors would like to thank Domaines Agricoles-UCP Maâmora and Ministry of Higher Education for project financial support.

References

1. Van der Vlugt RAA, Stijger CCMM, Verhoeven JTJ, Lesemann DE (2000) First report of Pepino mosaic virus in tomato. Plant Dis 84: 103.

2. Cotillon AC, Girard M, Ducouret S (2002) Complete nucleotide sequence of the Genomic RNA of a French isolate of Pepino mosaic virus. Arch Virol 147: 2231–2238.

3. Pospieszny H, Borodynko N (2006) New Polish isolate of Pepino mosaic virus highly distinct from European Tomato, Peruvian, and US2 strains. Plant Dis 90: 1106.

4. Hanssen IM, Paeleman A, Wittemans L, Goen K, Lievens B, et al. (2008) Genetic characterization of Pepino mosaic virus isolates from Belgian greenhouse tomatoes reveals genetic recombination. Eur J Plant Pathol 12: 131–146

5. French CJ, Bouthiller M, Bernardy M, Ferguson G, Sabourin M, et al. (2001) First report of Pepino mosaic virus in Canada and the United States. Plant Dis 85: 1121.

6. Soler S, Prohens J, Díez MJ, Nuez F (2002) Natural occurrence of Pepino mosaic virus in Lycopersicon species in Central and Southern Peru. J Phytopathol 150: 49-53

7. Zhang Y, Shen ZJ, Zhong J, Lu XL, Cheng G, et al. (2003) Preliminary characterization of Pepino mosaic virus Shanghai isolate (PepMV-Sh) and its detection with ELISA. Acta Agric Shanghai 19: 90-92.

8. Hanssen IM, Paeleman A, Vandewoestijne E, Van Bergen L, Bragard C, et al. (2009) Pepino mosaic virus isolates and differential symptomatology in tomato. Plant Pathol 58: 450-460

9. Hasiow B, Borodynko N, Pospieszny H (2008) Complete genomic RNA sequence of the Polish Pepino mosaic virus isolate belonging to the US2 strain. Virus Genes 36: 209-214.

10. Mansilla C, Sánchez F, Ponz F (2003) The diagnosis of the tomato variant of pepino mosaic virus: an IC-RT-PCR approach. Eur J Plant Pathol 109: 139-146.

11. Chomczynski P, Sacchi N (1987) Single-step method of RNA isolation by acid guanidinium thiocyanate-phenol-chloroform extraction. Anal Biochem 162: 156-159.

12. Ding LW, Sun QY, Wang ZY, Sun YB, Xu ZF (2008) Using silica particles to isolate total RNA from plant tissues recalcitrant to extraction in guanidine thiocyanate. Anal Biochem 374: 426-428.

13. Chomczynski P, Sacchi N (2006) The single-step method of RNA isolation by acid guanidinium thiocyanate-phenol-chloroform extraction: twenty-something years on. Nat Protoc 1(2): 581-5.

14. Thompson JR, Wetzel S, Klerks MM, Vasková D, Schoen CD, et al. (2003) Multiplex RT-PCR detection of four aphid-borne strawberry viruses in *Fragaria* spp. in combination with a plant mRNA specific internal control. J Virol Methods 111: 85-93.

15. Wang X, Tian W, Li Y (2008) Development of an efficient protocol of RNA isolation from recalcitrant tree tissues. Mol Biotechnol 38: 57-64.

16. Tong Z, Qu S, Zhang J, Wang F, Tao J, et al. (2012) A modified protocol for RNA extraction from different peach tissues suitable for gene isolation and real-time PCR analysis. Mol Biotechnol 50: 229-236.

17. Rosas-Cárdenas F, Durán-Figueroa N, Vielle-Calzada JP, Cruz-Hernández A, Marsch-Martínez N, et al. (2011) A simple and efficient method for isolating small RNAs from different plant species. Plant Methods 7:4.

18. Souiri A, Amzazi S, Laatiris H, Ennaji MM, Zemzami M (2013) Serological detection of tomato Pepino mosaic virus in Morocco. J Agric Sci Technol B 3: 847-852.

19. Thermo Scientific, "NanoDrop 1000 Spectrophotometer V3.7 User's Manual," 2008.

20. Rodrigues-Pousada R, Van Montagu M, Van der Straeten D (1990) A protocol for preparation of total RNA from fruit. Technique 2: 292-294.

Identification and Validation of a Microsatellite Marker for the Seedling Resistance Gene *Lr24* in Bread Wheat

Pallavi JK[1], Anupam Singh[2], Usha Rao I[1] and Prabhu KV[2]*

[1]Department of Botany, University of Delhi-110007, New Delhi, India
[2]National Phytotron Facility, Indian Agricultural Research Institute, New Delhi-110012, India

Abstract

The background of PBW343, the high yielding and widely cultivated bread wheat cultivar of the Indian sub-continent was utilized. We were able to identify specific microsatellite markers for *Agropyron elongatum* derived seedling resistance gene *Lr24*. The two markers, Xgwm114 and Xbarc71 were mapped at a distance of 2.4 cM from *Lr24* locus. They can be unquestionably utilized as landmarks for identification of these genes. An F_2 population segregating for *Lr24* and *Lr48* in the background of PBW343 was utilized for this study. Though phenotypic reaction of the plants of the progeny populations to leaf rust infection was recorded in the seedling stage, it was difficult to perform the same in the adult plant stage as more than one gene effective against the same pathogen act mutually thus making it difficult to interpret and differentiate the resistance reaction of each of the two different genes. This is a major aspect of concern for many plant breeders in various gene pyramiding experiments since differentiating virulences of pathogens for each and every gene utilized cannot be available within all geographic locations. Molecular markers play a significant role in all such cases.

Keywords: Microsatellite markers; *Lr24*; Seedling resistance; Bread wheat

Introduction

Puccinia triticina, the causative of leaf rust, is a considerable pathogen in wheat which results in substantial amount of losses by decreasing the yield in almost all wheat growing areas of the world. Deployment of rust resistance genes into the cultivar is being used to provide resistance against the locally prevalent pathogen races as an economical, enduring and eco-friendly measure [1]. Diversity for resistance to leaf rust is available in the germplasm of related wheat genera and there are many affirmative reports which assure the effectiveness of genes originating from wild relatives of the cultivated wheat in conferring long lasting rust resistance [2]. So far more than 60 *Lr* genes have been identified in various wheat backgrounds [3]. *Lr24* is one such resistance gene transferred into bread wheat from *Agropyron elongatum* which confers resistance right from the seedling stage all through the life of the plant (seedling resistance). *Lr24* is being used in major wheat breeding and pyramiding programmes as a means to provide resistance to otherwise susceptible cultivars [4,5]. However, many of the seedling resistance genes when incorporated singly tend to become ineffective due the constantly evolving physiological races of the pathogen. To suppress such reviving pathogenesis, an approach to stack more than one gene into the same background has been suggested and is pursued in most of the rust resistance initiatives [6]. In this study, we have employed one such F_2 population which segregates for two *Lr* genes. One of them is *Lr24* and the other is a hypersensitive recessive adult plant resistance (APR) gene, *Lr48* which confers resistance to the plant only from the time the plant reaches its booting stage and in a way decreases the selection pressure on the pathogen thus inhibiting the development of new races [2]. Differentiating the phenotypic resistance reaction of two discrete *Lr* genes existing in the same cultivar is practically impossible in the absence of individual *Lr* gene specific pathogen virulences. In such cases, the presence of exclusive DNA based markers which act as indices for each *Lr* gene will be valuable. Molecular markers are utilized on a huge scale to reduce cumbersomeness and enable rapid detection of specific *Lr* genes. Codominant molecular markers are useful in breeding programmes as only they are efficient in differentiating the

heterozygous and homozygous status in plants exhibiting resistance to the pathogen infection since only the latter are significant to forward for further generations. To enable the early selection of homozygosity at the adult plant rust resistance locus, two RAPD markers $S3_{450}$ and $S336_{775}$ have been utilized as a co-dominant marker system [7]. The SSR marker polymorphic for *Lr24* identified in our lab will be useful in wheat breeding populations and can help in fixing the genes by the F_2 population level itself without any further investment till F_5/F_6 generations.

The findings presented in this paper are a result of the work performed in N.P.F., I.A.R.I., New Delhi, India during the period 2007 to 2010.

Materials and Methods

Plant material

An F_2 population developed from the cross between the most widely cultivated and successful Indian wheat cultivar PBW343 carrying the gene *Lr24* (PBW343-*Lr24*) developed at IARI, India and the Australian cultivar Condor derived CSP44 line (with WW80/2*WW1511/ Kalyansona parentage) carrying the gene *Lr48* (CSP44-*Lr48*) was used for the study. *Lr24* is a seedling resistance gene thus conferring resistance in all stages of the plant and *Lr48* is an adult plant resistance gene, effective only from the time the plant reaches booting stage. The zygosity of each of the F_2 individual plants was established both by F_3 progeny testing and co-dominant molecular marker analysis. A set of

*****Corresponding author:** Pallavi JK, National Phytotron Facility, Indian Agricultural Research Institute, New Delhi-110012, India, E-mail: jd_research@iari.res.in

30 plants per each F_2 family were sown to erect the F_3 population. The experiments were conducted in the controlled conditions of National Phytotron Facility, IARI, New Delhi.

Pathotype of the fungal pathogen

The inoculum of the most virulent *Puccinia recondita* pathotype, 77-5 (121R63-1) was obtained from the Directorate of Wheat Research, Regional Station, Flowerdale, Shimla. Inoculation of the spores of the pathotype was done by spraying inoculum suspended in water fortified with Tween-20˙ (0.75 μl/ml) at an average concentration of 20 urediospores/microscopic field (10x × 10x).

DNA extraction

Young leaves from parents and individuals of the segregating population were collected, lyophilized and ground in liquid nitrogen using a pestle and mortar. DNA extraction was performed by the micro-extraction method described by Prabhu et al. [8]. Final concentration of DNA samples was maintained at 10 μg/μl for PCR reactions.

Seedling test

After sampling for DNA extraction, seedlings 8-10 days old at decimal code DC 11 stage were inoculated during the evening hours [9]. Prior to inoculation, the plants were sprayed with water to provide a uniform layer of moisture on the leaf surface. After inoculation, the seedlings were incubated for 36 h in humid glass chambers at a temperature of 23 ± 2°C and more than 85% relative humidity after which, the pots were shifted to muslin cloth chambers in the same green house. The disease reaction was recorded 12-14 days after inoculation, using the scoring method described by Stakman et al. [10].

PCR amplification using molecular markers

Twenty-four SSR markers specific to the 3D chromosome were selected from published data [11,12].The SSR markers (custom synthesized at Biobasic Inc, Canada) were used to screen the parents (PBW343-*Lr24* and CSP44-*Lr48*), F_2 population (comprising homozygous resistant, homozygous susceptible and heterozygous plants) and bulks (resistant and susceptible).

The PCR reactions with SSR markers were performed in a 20 μl volume which consisted of 10 mM Tris HCl (pH 8.3), 50 mM KCl, 2 mM $MgCl_2$, 200 μM of each dNTP (MBI Fermentas, Germany), 40 ng of each of the forward and reverse primers, 0.75 U Taq DNA polymerase (Banglore Genei Pvt. Ltd., India) and 50 ng template DNA. PCR amplifications for RAPD markers were done following the protocol developed by Williams et al. [13] in 20 μl reaction volume containing 10 mM Tris-HCl (pH 8.3), 50 mM KCl, 2 mM $MgCl_2$, 200 μM of each dNTP (MBI Fermentas, Germany), 0.2 μM of primer, 0.75 U Taq DNA Polymerase (Bangalore Genei Pvt. Ltd., India) and 10-15 ng of genomic DNA. The amplification reactions were carried in a PTC-200 thermal cycler (MJ Research, Las Vegas, NV, USA) with the following thermal profile – initial denaturation of 94°C for 10 min followed by 44 cycles of 94°C for 1 min (denaturation), 60°C, 55°C and 36°C (for Xgwm114, Xbarc71 and RAPD markers respectively) for 1 min (annealing), 72°C (extension) and a final extension step of 72°C for 10 min. This was followed by 4°C for 10 min. The amplified products from SSR markers and RAPD markers were separated on a 3% Metaphor® agarose gel and 2% Agarose gel respectively, in 1X TAE buffer at 80 V for 3 hrs to separate the fragments. The gels were later stained with 10 mg/ml ethidium bromide and viewed in a digital gel documentation system (Alpha Innotech, San Leandro, CA, USA).

Bulked segregant analyses were done to identify the markers' linkage to the dominant resistance gene. Ten randomly selected plants from the homozygous resistant and homozygous susceptible F_2 plants were used to prepare bulks [14]. The bulks differentiated for the presence and absence of the leaf rust resistance gene *Lr24* (Figure 1).

Statistical analysis

Segregation ratios were analyzed using a chi-square test. The individuals from the crosses that were scored as resistant and susceptible in the progeny populations were subjected to chi-square test for goodness of fit to test the deviation from the theoretically expected Mendelian segregation ratios. The linkage analysis was carried out using Mapmaker version 3.0 [15].

Results

Phenotypic reaction

The parent PBW343+*Lr24* showed resistance to rust infection and recorded infection type (IT) of ; to 1 while the other parent CSP44 showed high level of susceptibility (IT of 33⁺) at seedling stage. At adult plant stage the parent PBW343+*Lr24* remained resistant while the other seedling susceptible parent showed resistance by recording a ; reaction type. All the F_1 plants remained resistant to the rust infection recording an infection type of ; to 1.

46 seedlings of the F_2 population showed susceptibility to the leaf rust infection while the remaining 136 plants remained resistant by expressing the seedling resistance conferred by the dominant resistance allele of the *Lr24* locus and the population followed a monogenic segregation ratio. All the susceptible F_2 derived F_3 families remained susceptible whereas only 41 out of the 136 resistant F_2 derived F_3 families were homozygous for resistance. The remaining 95 families were heterozygous thus distributing the F_2 genotypes into 1R:2R:1S monogenic segregation ratio (Table 1). The phenotypic expression of adult plant resistance could not be examined due to the interference of the dominant seedling resistance gene *Lr24* in the same genetic background.

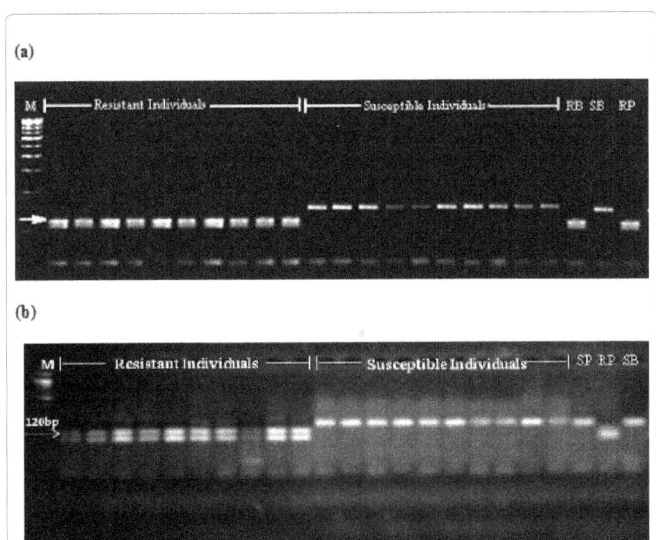

Figure 1: Screening of the SSR markers (a) Xgwm114 and (b) Xbarc71 on the bulked DNA constituent F_2 plants of the cross PBW343 + *Lr24* x CSP44*Lr48* for genetic linkage analysis.

F_1	F_2 reaction*				
R	Total tested	R	S	$X^2_{3:1}$	P
16	182	136	46	0.0072	0.9323

No. of F_2 families for F_3 testing		R in F_2				S in F_2	
R	S	NSeg	Seg	$X^2_{1:2}$	P	NSeg	Seg
136	46	41	95	0.6208	0.7331	46	0

'R: Leaf Rust Resistant; S: leaf rust susceptible, #NSeg: Non-Segregating Family for Leaf Rust; Seg: Segregating Family for Leaf Rust.

Table 1: Reaction of wheat plants in F_1, F_2 and F_3 generations of the cross Pbw343+*Lr24* X CSP44+*Lr48* at seedling stage (DC 11) to infection by the leaf rust pathotype 77-5 under controlled conditions.

Molecular marker study

Out of twenty-four SSR markers specific to the 4AL chromosome, only two markers, Xgwm114 (F: 5' ACAAACAGAAAATCAAAACCCG 3' R: 5'ATCCATCGCCATTGGAGTG3') with annealing temperature of 60°C and Xbarc71 (F:5'GCGCTTGTTCCTCACCTGCTCATA3' R: 5'GCGTATATTCTCTCGTCTTCTTGTTGGTT3') with annealing temperature of 55°C were identified to be polymorphic between the parents. 10 randomly selected samples were taken from the resistant and susceptible plants to prepare bulks for bulk segregant analysis (Figure 1). The markers were found to be putatively linked to the *Lr24* locus. The polymorphic SSR markers were analyzed on the 182 F_2 plants for linkage analysis with the *Lr24* locus. Both the markers were associated with the *Lr24* locus and located at a distance of 2.4 cM from it (Table 2). The PBW343-*Lr24* resistance allele linked SSR marker allele amplified a 120 bp fragment and the CSP44 susceptibility allele linked marker allele amplified a 151 bp fragment. The marker Xgwm114 differentiated the population into 45 homozygous resistant, 89 heterozygous resistant and 48 homozygous susceptible plants. The 120 bp fragment was specific to the *Lr24* resistance allele and did not amplify in other *Lr* genes carrying lines from other native and alien sources (Figure 2).

By employing the flanking RAPD markers $S3_{450}$ (5'CATCCCCTG3') and $S336_{775}$ (5'TCCCCATCAC3') linked respectively to the recessive resistance allele and dominant susceptible allele of the *Lr48* locus; plants which were homozygous for recessive APR gene *Lr48* were identified, as these two markers served as one co-dominant marker system capable of identifying both dominant and recessive alleles of heterozygous plants. 10 F_2 plants were found to be the homozygous at both the dominant seedling resistance locus of *Lr24* and the recessive adult plant resistance locus of *Lr48* (Table 3).

Discussion

The *Lr24* gene transferred from *Agropyron elongatum* is important to wheat since there are reports of its locus being linked to the stem rust resistance gene *Sr24* [16]. The *Agropyron* chromosome segment is located on the satellite of chromosome 1B and the translocation chromosome designated as T1BL·1BS-3Ae#1L. T1BL·1BS-3Ae#1L was inherited from Teewon wheat and carries resistance genes to stem rust (*Sr24*) and leaf rust (*Lr24*). *Sr24* is highly effective against TTKS (Ug99), a recently emerged race with virulence to *Sr31* that is considered to be a serious threat to wheat crop produce all over the world [17]. Though Xgwm114 has not been testified in populations segregating for stem rust resistance in this experiment, an assumption can be made that the identification of presence of *Lr24* through this marker also suggests the existence of stem rust resistance. Such a marker will thus be economically important in wheat breeding programmes. Pathotypes virulent on *Lr24* have been reported from North America , South America and South Africa [5,18-21]. This requires that *Lr24* should be used only in combinations with other *Lr* genes. Worldwide, no virulence has been reported on the combination *Lr9* and *Lr24* [6].

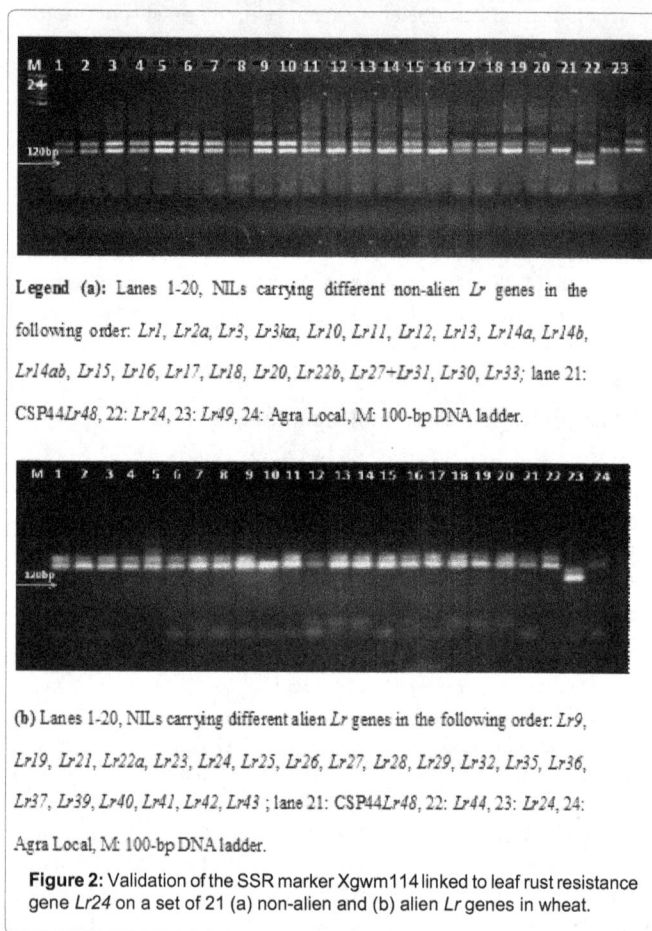

Legend (a): Lanes 1-20, NILs carrying different non-alien *Lr* genes in the following order: *Lr1, Lr2a, Lr3, Lr3ka, Lr10, Lr11, Lr12, Lr13, Lr14a, Lr14b, Lr14ab, Lr15, Lr16, Lr17, Lr18, Lr20, Lr22b, Lr27+Lr31, Lr30, Lr33*; lane 21: CSP44*Lr48*, 22: *Lr24*, 23: *Lr49*, 24: Agra Local, M: 100-bp DNA ladder.

(b) Lanes 1-20, NILs carrying different alien *Lr* genes in the following order: *Lr9, Lr19, Lr21, Lr22a, Lr23, Lr24, Lr25, Lr26, Lr27, Lr28, Lr29, Lr32, Lr35, Lr36, Lr37, Lr39, Lr40, Lr41, Lr42, Lr43* ; lane 21: CSP44*Lr48*, 22: *Lr44*, 23: *Lr24*, 24: Agra Local, M: 100-bp DNA ladder.

Figure 2: Validation of the SSR marker Xgwm114 linked to leaf rust resistance gene *Lr24* on a set of 21 (a) non-alien and (b) alien *Lr* genes in wheat.

Loci	Segregants									Marker Fragment		Lr24 gene & marker fragment		Linkage	
	++/R	-+/R	--/R	++/H	-+/H	--/H	++/S	-+/S	--/S	$X^2_{1:2:1}$	P	$X^2_{1:2:1:2:4:2:1:2:1}$	P	X^2	P
Lr24-Xgwm11	41	0	0	4	89	2	0	0	46	0.1866	0.9109	326.87	0.00	329.68	0.00
Lr24-Xbarc71	41	0	0	2	93	0	2	0	44	0.0987	0.9518	326.69	0.00	336.70	0.00

R: Homozygous Resistant; H: Heterozygous; S: Homozygous Susceptible
Marker Fragment
++: Homozygous Resistant
+-: Heterozygous
--: Homozygous Susceptible

Table 2: Test of linkage between the leaf rust resistance gene *Lr24* and SSR markers (Xgwm114 and Xbarc71) in the F_2 population of the cross PBW343+*Lr24* X CSP44-*Lr48* of wheat.

F_2 plant No.	$S3_{450}$ SCAR	$S336_{775}$ SCAR	Genes Carried (Lr)	F_2 No.	$S3_{450}$ SCAR	$S336_{775}$ SCAR	Genes Carried (Lr)
1	-	+	24,	31	-	+	24
2	-	+	24	32*	+	-	24,48
3 *	+	-	24,48	33	-	+	24
4	-	+	24	34	-	+	24
5	+	+	24,48	35*	+	-	24,48
6	-	+	24,	36	-	+	24
7 *	+	-	24,48	37	+	+	24,48
8	-	+	24	38	-	+	24
9	-	+	24	39*	+	-	24,48
10*	+	-	24,48	40	-	+	24
11	-	+	24	41	-	+	24
12	-	+	24				
13	+	+	24,48				
14	-	+	24				
15	-	+	24				
16	+	+	24,48				
17	-	+	24				
18*	+	-	24,48				
19	-	+	24				
20*	+	-	24,48				
21	-	+	24				
22*	+	-	24,48				
23	-	+	24				
24	-	+	24				
25	+	+	24,48				
26	-	+	24				
27	-	+	24				
28*	+	-	24,48				
29	-	+	24				
30	-	+	24				

Table 3: Marker assisted selection in segregating F_2 progeny of a cross Lr24 and Lr48 for Lr48 in PBW343 background. Only 41 plants with homozygous bands for Lr24 locus were screened with $S3_{450}$ and $S336_{775}$ RAPD markers. *homozygous for Lr48.

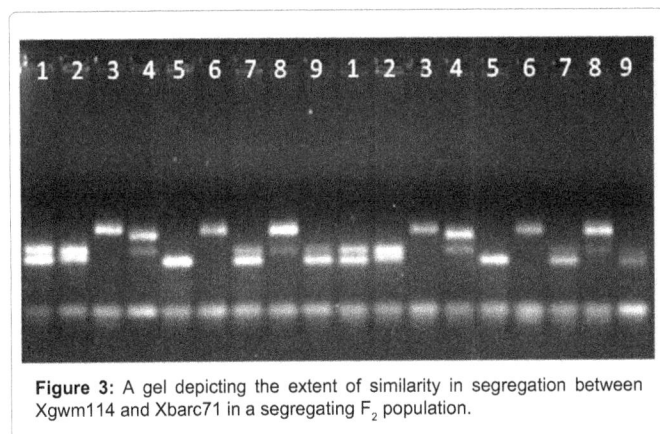

Figure 3: A gel depicting the extent of similarity in segregation between Xgwm114 and Xbarc71 in a segregating F_2 population.

Lr24 still continues to be highly effective in India and three cultivars Vidisha, Vaishali (DL784-3) and HW2004 carrying both Lr24/Sr24 have been released for commercial cultivation in India.

Several markers showing a dominant inheritance pattern have been reported to be linked to the Lr24 resistance locus. A SCAR marker developed by Cherukuri et al., [22] in the same laboratory is currently being successfully employed to track the transfer and establishment of this gene in different genetic background. A PCR-based DNA-STS

marker, six RFLP markers completely linked to Lr24 - one inherited as a codominant marker (PSR1205), one in coupling phase (PSR1203), and four in repulsion phase (PSR388, PSR904, PSR931, PSR1067) were reported to be inherited with Lr24. A RAPD marker, OPJ-09 also was shown to be in complete linkage to the Lr24 resistance gene [23]. The markers have been used in wheat breeding experiments employing MAS [24]. There are other reports of plymorphic RAPD and SCAR markers for Lr24 by Dedryver et al., [25].

Simple sequence repeats DNA called microsatellites are ubiquitously distributed within the eukaryotic genome, and SSR is more polymorphic than other marker systems [12,26]. The genetic map constructed by Roder [26] uses microsatellite markers located on seven chromosome groups and Xgwm114 was located on chromosome arm 3B and 3D. Xgwm114 is reported to be associated with powdery mildew resistance in wheat [27]. Three microsatellite markers, Xgwm247, Xgwm181 and Xgwm114 located on chromosome 3BL, were shown to be associated with the stem solidness locus and with sawfly cutting in durum wheat [28]. McIntosh [29] reports the location of Lr24 on the long arm of 3D and the current experiment shows the linkage of Xgwm114 with the locus of Lr24.

Xbarc71 is reported to be sharing the same position on the long arm of 3D chromosome along with Xgwm114 in the chromosome map developed by Torada et al., [11]. This was reconfirmed by the pattern of segregation shown by both the markers in the F_2 population (Figure 3). However, the same markers are placed considerably far apart in the chromosome map developed by Somers et al., [30,31]. In the present experiment, both the markers were able to differentiate the presence of Lr24 in segregating populations almost with equal precision and here we report that they can be used interchangeably to identify homozygous Lr24 locus.

Such codominant SSR markers will be extremely useful in large scale wheat breeding programmes where selection of homozygous resistant plants which potentially carry the resistance genes will be achieved at very early generations. A segregating F_2 population will suffice to select plants in which the gene is fixed.

In this experiment, the pair of the RAPD markers also was valuable only because they could be employed as a codominant marker system. $S3_{450}$ was linked to the recessive adult plant resistance allele and $S336_{775}$ was linked to the dominant susceptibility linked allele of the Lr48 locus. Since both the alleles are easily differentiated, we could select those plants homozygous for the recessive resistance allele linked adult plant resistance at the Lr48 locus.

References

1. Kolmer JA (1996) Genetics of resistance to leaf rust. Anuual Review of Phytopathology 34: 435-455.

2. McIntosh RA, Wellings CR, Park RF (1995) Wheat rusts: an atlas of resistance genes. Kluwer Academic Publishers, Dordrecht.

3. McIntosh RA, Yamazaki Y, Dubcovsky J, Rogers J, Morris C, et al. (2008) Catalog of gene symbols for wheat. 11th International Wheat Genetics Symposium, Brisbane Qld Australia.

4. Tomar SMS, Menon MK (1998) Introgression of alien genes for leaf rust (Puccinia recondita) resistance in to bread wheat (Triticum aestivum L) cultivars. Indian J Agric Sci 68: 675-681.

5. Long DL, Roelfs AP, Leonard KJ (1994) Virulence and diversity of Puccinia recondite f. sp. tritici in United States in 1992. Plant Dis 78: 901-906.

6. Roelfs AP, Simmonds NW, Ajaram S (1988) Resistance to leaf rust and stem rust in wheat. Breeding Strategies for Resistance to Rust of Wheat 10-22.

7. Samsampour D, Zanjani BM, Singh A, Pallavi JK, Prabhu KV (2009) Marker

assisted selection to pyramid seedling resistance gene Lr24 and adult plant resistance gene Lr48 for leaf rust resistance in wheat. Indian J Genet Pl Breed 69: 1-9.

8. Prabhu KV, Somers DJ, Rakow G, Gugel RK (1998) Molecular markers linked to white rust resistance in mustard Brassica juncea. Theor. App Genet 97: 865-870.

9. Zadoks JC, Chang TT, Konzak CF (1974) A decimal code for the growth stages of cereals. Weed Res 14: 415-421.

10. Stakman EC, Stewart DM, Loegering WQ (1962) Identification of physiological races of Puccinia graminis var. tritici. Agricultural Research Service E617.

11. Torada A, Koike M, Mochida K, Ogihara Y (2006) SSR-based linkage map with new markers using an intraspecific population of common wheat. Genet 112: 1042-105.

12. Roder MS, Korzun V, Wendehake K, Plaschke J, Tixier MH, et al. (1998) A microsatellite map of wheat. Genetics 149: 2007-2023.

13. Williams JGK, Kubelik AR, Livak KJ, Rafalski JA, Tingey SV (1990) DNA polymorphisms amplified by arbitrary primers are useful as genetic markers. Nucleic Acids Res 18: 6531-6535.

14. Michelmore RW, Paran I, Kesseli RV (1991) Identification of markers linked to disease resistance genes by bulked segregant analysis: A rapid method to detect markers in specific genomic regions by using segregating populations. Proc Natl Acad Sci 88: 9829-9832.

15. Lander ES, Green P, Abrahamson J, Barlow A, Daly MJ, et al. (1987) An interactive computer package for constructing primary genetic maps of experimental and natural populations. Genomics 1: 174-181.

16. Menon MK, Tomar SMS (2001) Transfer of Agropyron elongatum-derived rust resistance genes Sr24 and Lr24 into some Indian bread wheat cultivars. Wheat Info Serv 92: 20-24.

17. Singh RP, Hodson DP, Jin Y, Huerta-Espino J, Kinyua MG, et al. (2006) Current status, likely migration and strategies to mitigate the threat to wheat production from race Ug99 (TTKS) of stem rust pathogen. CAB Reviews: Perspectives in Agriculture, Veterinary Science, Nutrition and Natural Resources 1: 1-13.

18. Bowder LE (1973) Probable genotypes of some Triticum aestivum 'Agent' derivatives for reaction to Puccinia recondite f. sp. tritici. Crop Sci 13: 203-206.

19. Kolmer JA, Dyck PL, Roelfs AP (1991) An appraisal of stem and leaf rust resistance in North American hard red spring wheats and the probability of multiple mutations in populations of cereal rust fungi. Phytopathology 81: 237-239.

20. Singh RP, Rajaram S (1991) Resistance to Puccinia recondita f. sp. tritici in 50 Mexican bread wheat cultivars. Crop Sci 31: 1472-1479.

21. Pretorius ZA, Kemp GH (1990) Effect of growth stage and temperature on components of resistance to leaf rust in wheat genotypes with Lr26. Plant Disease 74: 631-635.

22. Cherukuri DP, Gupta SK, Charpe A, Koul S, Prabhu KV, et al. (2005) Molecular mapping of Aegilops speltoides derived leaf rust resistance gene Lr28 in wheat. Euphytica 143: 19-26.

23. Schachermayr G, Messmer M, Feuillet C, Winzeler H, Keller B (1995) Identification of molecular markers linked to the Agropyron elongatum-derived leaf rust resistance gene Lr24 in wheat. Genet 90: 982-990.

24. Slikova S, Gregova E, Bartos P, Hanzalova A, Hudcovicova M, et al. (2004) Development of wheat genotypes possessing a combination of leaf rust resistance genes Lr19 and Lr24. Plant Soil Environ 50: 434-438.

25. Dedryver F, Jubier MF, Thouvenin J, Goyeau H (1996) Molecular markers linked to the leaf rust resistance gene Lr24 in different wheat cultivars. Genome 39: 830-835.

26. Roder MS, Plaschke J, Konig SU, Borner A, Sorrells ME, et al. (1995) Abundance, variability and chromosomal location of microsatellites in wheat. Molecular and General Genetics 246: 327-333.

27. Wang Z, Qi Z, Yongkang P, Chaojie X, Qixin S, et al. (2004) Identification of Random Amplified Polymorphic DNA and Simple Sequence Repeat markers Linked to Powdery Mildew Resistance in Common Wheat Cultivar Brock. Plant Prod Sci 7: 319-323.

28. Houshmand S, Ronald EK, Fran RC, John MC (2007) Microsatellite markers flanking a stem solidness gene on chromosome 3BL in durum wheat. Mol Breeding 20: 261-270.

29. McIntosh RA, Hart GE, Devos KM, Gale MD, Rogers WJ (1998) Catalogue of gene symbols for wheat. Proceedings of the Ninth International Wheat Genetics Symposium, University Extension Press, University of Saskatchewan, Saskatoon, Canada.

30. Somers DJ, Isaac P, Edwards K (2004) A high-density microsatellite consensus map for bread wheat (Triticum aestivum L). Genet 109: 1105-1114.

31. Tomar SMS, Menon MK (2001) Improvement of WH542 a Petkus rye derivative of bread wheat with additional genes for rust resistance. Annals Agri Res 3: 303-308.

3

Effectiveness of *Trichoderma* Biotic Applications in Regulating the Related Defense Genes Affecting Tomato Early Blight Disease

Mohamed E. Selim*

Agricultural Botany Department, Faculty of Agriculture, Menoufiya University, Egypt

Abstract

Early blight disease caused by *Alternaria solani* fungus is one of the most important diseases attacking tomato especially in humid regions with high temperatures (24-29°C). Controlling the early blight disease using fungicides has become unfavorable in the last years due to their environmental and human health concerns. Biotic and abiotic induction of host plants defense mechanisms could be applied as an alternative management strategy against the disease. Nowadays, different *Trichoderma* species could be used as one of promising bio-control agents affecting development and disease incidence of *Alternaria solani* on tomato plants. In present study, the direct and indirect effects of different *Trichoderma* isolates application on *Alternaria solani* infection as well as on the gene expression levels of some related genes to defense mechanisms in tomato plants was investigated. Results indicated that *Trichoderma* species reduced either the mycelial growth or the disease incidence of *Alternaria solani*. Treating tomato roots with *Trichoderma harzianum*-T10 isolate affected the relative expression levels of eight different genes within tomato leaves. Three genes of them i.e., Les.21895, Les.19403 and Les.1097, which involved in auxin, ethylene and lignin pathway respectively, were up regulated while the other three genes i.e., Les.20348, Les.3129 and Les.9833, which related to pyruvate kinase pathways, were down regulated. In addition, treating tomato plants with *Trichoderma harzianum* T10 regulated the expression level of some Pr-protein genes i.e., Pr-1 and Pr-5. These findings suggested that induction of systemic defense mechanisms using the mutualistic *Trichoderma* isolates such as T10 is candidate to be among the mechanisms that can play a crucial role in controlling tomato early blight disease caused by *Alternaria solani*.

Keywords: *Trichoderma* species; *Alternaria solani*; rt-PCR; Gene expression; Defense related genes in tomato; Pr-proteins

Introduction

Tomato (*Solanum lycopersicum* Mill.) is one of the most important vegetables, ranking first, based on production levels, accounting for 14% of the total fruit and vegetable production worldwide [1]. The global tomato production area covers approximately 4 million hectares of arable lands, annually yielding approximately 100 million tons with an estimated value of 5 to 6 billion US$ [2]. Noteworthy, Egypt is among the top 10 tomato producing countries worldwide covering an estimated 181.000 ha and yielding 6, 4 million tons in annual production [1,2].

The fungus *Alternaria solani* (Ellis & Martin) Sorauer, the causal agent of early blight disease, is a major pathogen of tomato [3] causing considerable yield losses all over the world, particularly in humid regions with fairly high temperatures (24-29°C). Epidemics can however also occur in semi-arid climates, like those of Egypt, where frequent and prolonged nocturnal damp periods occur [4]. Within the species, *Alternaria solani* is divided into races, which show significant variability in morphology. These races can also show variations in pathogenicity on various crops, like tomato and potato, and their cultivars [5]. When infecting, all above ground parts, including leaves, stems collars and fruits, can be attacked resulting in stem lesions on the adult plant, collar and fruit rot [6]. On leaves, *A. solani* causes circular concentric rings with yellow halo, which is known as early blight (EB). Earlier it was reported that EB infections in tomato, causes up to 79% of yield losses [7] whereas collar rot only is responsible for about 50% of these losses on tomato seedlings [8].

Control of early blight disease using chemical fungicides has become increasingly difficult due to the limited number of effective fungicides, some of which were recently withdrawn from the market due to the environmental and human health concerns [9]. Moreover, the use of fungicides during fruiting is discouraged, making the chemical control of *A. solani* virtually impossible. The search for alternative control methods is thus quite important and necessary. Although the use of resistant varieties which may be considered as a good alternative especially within integrated pest management strategies, unfortunately has also significant drawbacks due to the lack of suitable resistant tomato germplasm against *A. solani* [10,11]. Alternatively, no suitable biological control agents are available as well [12,13].

Research has characterized several mutualistic endophytic fungi with biological control activity against fungal pathogens or plant parasitic nematodes as well [14,15]. These mutualistic bioagents are considered eco-friendly and have no reported negative effects on non-targeted organisms, including humans, the useful microflora and host plants. In such tripartite interactions between the host plant, fungal pathogen and mutualistic endophyte agent, different mechanisms of action were reported and considered responsible for protecting the host plants from pathogens and parasites [16-18]. In some cases, the endophyte has the ability to trigger systemic resistance responses in different host plants. Different hormones such as auxin and ethylene were found to be related to complex defense signaling pathways against pathogens and parasites [19,20]. Furthermore, the

***Corresponding author:** Mohamed E Selim, Agricultural Botany Department, Faculty of Agriculture, Menoufiya University, Egypt
E-mail: m_elwy76@yahoo.com

studies of Hamberger and Hahlbrock [21] on *Arabidopsis thaliana* proved the direct or indirect association of CoA Ligase isoenzymes in the biosynthesis of lignin and cell wall-bound phenylpropanoid derivatives. Nicholson and Hammerschmitt [22]; Bhuiyan et al. [23] mentioned also that lignification process is involved in plant defenses against pathogen infection. Moreover, pyruvate kinase (PK) enzymes have been known to affect numerous physiological and biological processes in plants [24,25]. Different studies suggest that PK activity can be used as a remarkable physiological indicator of K^+, Mg^{2+}, and Ca^{2+} contents in plant tissues [26,27].

In this study, the potential of various isolates of *Trichoderma* species as promising biocontrol agent against *A. solani*, the causal pathogen of early blight disease, was evaluated under *in vitro* and *in vivo* conditions.

Moreover, in order to study the influence of *Trichoderma* application on eliciting the systemic defenses pathways in tomato, the relative expression levels of related defense genes which involved in the auxin biosynthesis, ethylene signaling, lignin biosynthesis, pyruvate kinase biosynthesis and Pr-proteins production were determined using qRT-PCR technique.

Materials and Methods

Endophyte and pathogen cultures

Pure culture of 5 bio-agents i.e., *Trichoderma hamatum (Th)*, *T. viride (Tv)*, *T. harzianum (Tz22)*, *Trichoderma* species (T32) and *Trichoderma koningii (Tkg)* were obtained from Institute of Crop Science and Resource Conservation (INRES), Bonn University, Lab of Prof. Dr. Richard Sikora. In addition, *Trichoderma harzianum*-T10 isolate and the pathogenic isolate of *Alternaria solani* were obtained from Plant Pathology Department, Faculty of Agriculture, Menoufiya University, Egypt. Pure cultures of all fungi were maintained in stored micro-bank tubes at -80°C. Conidial spores of tested *Trichoderma* isolates and *Alternaria solani* were obtained from pure cultures grown on PDA plates, supplemented with 150 mg l^{-1} of chloramphenicol at 25°C in the dark for two weeks. Five ml of ml sterile distilled water were added to each plate then the spores were scraped far away the mycelial mat using sterilized glass rod. The spore suspension was filtered through three layers of fine sterile cheesecloth. The prepared spore suspensions of tested *Trichoderma* isolates were adjusted at 1×10^7 spores/mL and at 2×10^3 spores/ml for the pathogen [28].

Plant growth

Tomato (*Solanum lycopersicum*) cv. Hellfrucht/Frühstamm that consider as susceptible cultivar to infection with *A. solani*, was used in all experiments of this study. Seeds were surface-sterilized by submersing them in a 75% ethanol solution for 1 min and 1.5% sodium hypochloride (NaOCl) solution for 3 min, respectively. The seeds were then thoroughly rinsed three times with sterile water. The seeds were then sown in plastic trays containing sandy clay soil (1:1, v/v), which was previously autoclaved at 121°C for 1 h. The plants were grown in a growth chamber at 25 ± 3°C with 16 h diurnal light, 60 to 70% humidity and fertilized weekly with 5 ml of 2 g/l N:P:K: (14:10:14; Aglukon, Düsseldorf, Germany). After 2 weeks, when reaching 10-15 cm in height, the seedlings were transplanted into separate pots filled with 300g of a mixture of sand and sandy loam soil in a ratio of 2:1 (v/v), and transferred to a greenhouse, maintained at 27 ± 5°C and 16 h diurnal light.

In vitro antagonism assay

The bio-control activities of tested *Trichoderma* isolates against the tested pathogenic isolate of *A. solani* was studied *in vitro* by the dual culture technique on plate [29]. Five millimeters diameter plugs of both two weeks old pathogen and biocontrol agent cultures were taken with a sterile cork borer. These plugs were placed equidistantly (60 mm) apart on PDA, amended with 150 mg l^{-1} of chloramphenicol, in a 90 mm Petri-dish. As a control, a plug with the pathogen was placed on PDA, amended with 150 mg l^{-1} of chloramphenicol, alone. Each treatment was replicated three times. The Petri-dishes were incubated at 23 ± 2°C. The radial growth of both tested bio-agents and the pathogen were recorded after 10 days and the relative mycelial growth inhibition was calculated according to Vincent [30].

In vivo antagonism assay

The antagonistic capabilities of tested *Trichoderma* isolates against *A. solani* were studied *in vivo*. In this respect, 3 ml of spore suspension of each individual *Trichoderma* species was inoculated into soil at transplanting stage of tomato seedlings to pots (10 cm diameter contains 300 g soil). One week after the inoculation of the *Trichoderma* isolates, spores of *A. solani* was sprayed over the true leaves of tomato plants [28]. Each treatment was replicated three times. The non-treated plants with *Trichoderma* spore suspension were served as control. Potential influence of the tested *Trichoderma* spp. on early blight disease severity was also compared with a Propamocarp HCl (72,2%) fungicide that is generally recommended (purchased from Syngenta company, Egypt). Therefore, three plants were sprayed with 2.5 ml/L of this fungicide at the same time with *Trichoderma* application. The early blight severity on each leaf of the sprayed plants was recorded three weeks after inoculation using disease index described by Vakalounakis [31] based on a scale of 0 to 5, where 0=no visible lesions on leaf; 1=up to 10% leaf area affected; 2=11%– 25%; 3=26%–50%; 4=51%–75%; and 5=more than 75% leaf area affected or leaf abscised. Percentage of disease severity was calculated according to the following formula [32]:

$$\text{Disease severity}(\%) = \frac{\text{sum of all ratings}}{\text{no. of leaves sampled - maximum disease scale}} \cdot 100$$

Gene expression analysis by semi-quantitative real-time qRT-PCR

The effect of *Trichoderma harzianum*-T10 isolate on relative gene expression of eight selected genes coding for IAA6 (Les.21895), ER69 (Les.271), 4-coumarate--CoA ligase 2 (Les.1097), pyruvate dehydrogenase kinase (Les.20348) and two pyruvate kinases (Les.3129 and Les.9833) in addition to two pathogen related genes (PR-1 and PR-5) which were associated with either development signaling pathways or defense responses in tomato was assessed using quantitative PCR (Table 1).

During the transfer of the seedlings to the pots (as described in plant growth paragraph above), roots of three individual plants were dip-inoculated with spore suspension of *T. harzianum* -T10. Control plants were treated only with tap water. Four weeks after T10 inoculation, shoots plus leaves of inoculated and control plants were separately collected in 50 ml falcon tubes and frozen in liquid nitrogen. The leaves were ground into a fine powder using liquid nitrogen and a pre-cooled mortar and pestle. The ground plant tissue powder was kept frozen using liquid nitrogen, transferred into fresh 50 ml falcon tubes and stored at -80°C until further use. Total RNA was extracted from 100 mg tissue powder using the Macherey-Nagel total RNA extraction kit and further purified using the Machey-Nagel RNA clean up kit, both according to the manufacturer's recommendations. The RNA integrity was assessed by gel electrophoresis and the RNA concentration was measured photometrically at 260 nm using the Nanodrop 2000 C (Fisher Scientific, Germany). To assess RNA integrity, 6 µl of the total

Oligo- Name	Putative function according to NCBI	Sequence (5'- 3')	
		Forward- primer	Reverse- primer
Les.21895	auxin-regulated (IAA6)	TGGTCAGTGTGCCAGTGATAAGA	AGAACCATTGAGAAGATCCATCAAG
Les.271	Ethylene-responsive methionine synthse (ER69)	GGTATCGGCCCTGGTGTGTA	TTGTTAACTCTGACGGCAATCTCT
Les.1097	4-coumarate--CoA ligase 2	CGTTACACTCGTATTGTTTCGAAAAT	TTCGCCCCGTCGATTAAA
Les.20348	pyruvate dehydrogenase Kinas (PK)	CTCCTGATTGTGTGGGCTATATACA	ATTCGCGCAAGCAAATAGAAC
Les.3129	pyruvate kinase (PK)	CCTGTTTTGACTACAGACTCTTTCGA	CAAGCCCCTATATACCAAACTGTGT
Les.9833	pyruvate kinase (PK)	CAGTTCCCATGAGTCCTTTGG	ACTTCCTCCCCTGGTTAACACA
Les.20078	PR-1 (pathogenesis-related protein 1)	CACTCGTATCATGAGTCTTC	CCCTATATACCCTGGTTA
(PR-5)	PR-5 (pathogenesis-related protein 5)	CCCATGAGTCCTTGTGGGCTA	ACTCTGACGGCATATATACC

Table1: Genes selected for conducting qRT-PCR within inoculated and non-inoculated tomato plants with *Trichoderma harzianum* -T10.

RNA was mixed with 2 µl 6 x gel loading dye and loaded on a 1% agarose gel supplemented with 0.5 µg/ml ethidium bromide, (Figure 1). Of this total RNA sample, 1 µg was used to synthesis complementary DNA (cDNA) via the reverse transcription kit, according to the manufacturer's recommendations. The pipetting scheme and the PCR program is indicate in the Tables 2 and 3, respectively.

The cDNA was used for qPCR analysis by measuring the relative accumulation of the eight selected genes. For this, the cDNA was diluted ten times with DNA/RNA free water and 1 µl transferred into a well of a 96-well qPCR plate. For each well 19 µl master mix was added to 1 µl of the sample cDNA (the pipetting scheme is indicated in Table 4). For each sample, three biological replicates were prepared. The plate was covered with a qPCR foil, the reaction components were mixed and the plate was briefly centrifuged for 10 sec at 4000 × g. The thermal cycling parameters are indicated in Table 5. After qPCR, 6 µl of the amplification product was mixed with 2 µl 6 x gel loading dye and loaded on a 1% agarose gel, supplemented with 0.5 µg/ml ethidium bromide, for a gel electrophoresis at 60 mV to assess amplification. The accumulation of the genes of interest was normalized to actin transcript accumulation and the fold change in transcript accumulation relative to the control sample was calculated via the $2^{-\Delta\Delta Ct}$ – method described by Livak and Schmittgen [33].

Statistical analysis

The data were analyzed according to the standard analysis of variance procedure with SPSS 14 for Windows. Differences among treatments were tested using one way analysis of variance (ANOVA) followed by the Tukey test for mean comparison in case the F-value was significant [34].

Results

The antagonistic potentialities of the six tested *Trichoderma* isolates viz., *Trichoderma hamatum (Th)*, *T. viride (Tv)*, *T. harzianum (Tz22)*, *T. harzianum* (isolate T10), *Trichoderma* species (T.32) and *Trichoderma koningii* (Tkg) were investigated under *in vitro* conditions onto PDA plates for determining their abilities in inhibiting the mycelial growth of *Alternaria solani*. The results indicate that all tested bio-agents significantly inhibited the mycelial growth of the pathogen compared with the control (Figure 2) where the inhibition % were ranged between 66% with Tz22 to 75% with T10. Conversely, the radial growth of tested *Trichoderma* isolates was not impaired by *Alternaria solani*. Thus, it could be concluded that T10 was the best performing tested isolate followed by Tkg with regard to inhibiting the mycelial growth of *Alternaria solani*.

The influence of the six tested *Trichoderma* isolates i.e. *Th, Tv, Tz22, T10, T.32* and Tkg on the incidence and the severity of early blight disease caused by *Alternaria solani* in tomato was determined

under greenhouse conditions. As a reference for efficacy, treatment with the fungicide (Propamocarp Hcl) was included in the experiment. Comparing with control treatment (infected plants with *Alternaria solani* only), the results indicate that all tested bio-agents significantly reduced disease severity (Figure 3) which ranged from 50% in case of using *T. harzianum* Tz22 to 70% with *T. harzianum*-T10. The second best disease reduction was obtained with *Trichoderma koningii* (Tkg). On the other hand, *T. harzianum*-T10 came therefore close to the disease reduction % obtained with fungicide treatment, which averaged 77.5% (Figure 3).

The indirect reaction of tomato plants against *Alternaria solani*, the causal of early blight disease in response to treating the plants with *T. harzianum*-T10 as the best performing bio-agent was evaluated through determining the relative expression of eight different genes, six genes related to vital and essential bioprocess and two genes involved in defence mechanisms pathways, in tomato plants. The changes in these genes expression, Les.21895, Les.271, Les.1097, Les.20348, Les.3129 and Les.9833 in addition to two genes related to Pr-proteins i.e., pr-1 and pr-5, was analyzed through using the real-time PCR analysis. The obtained results (Figure 4) show that treating tomato plants with *T. harzianum*-T10 led to induction and accumulation of three genes i.e., Les.21895, Les.271 and Les.1097 which involved and related with synthesis of Indol acetic acid, Ethylene and coumarate Ligase pathways respectively, where the fold changes in gene expressions of these genes were increased 2.5, 7.1 and 3.7 folds respectively, comparing to control plants (un-treated with T10 isolate). Also, the relative expression levels of the other three genes i.e., Les.20348, Les.3129 and Les.9833 were down regulated up to two fold changes comparing to un-treated plants with T10. It is remarkable that all these three genes were found to be related to pyruvate kinase metabolisms. With regard to the two pathogenesis related proteins genes i.e., Pr-1 and Pr-5, results indicate that T10 isolate increased the relative expression of Pr-1 while it down regulated the expression of Pr-5. Moreover, the fold change was reached up to three times with Pr-1 and two times with Pr-5 (Figure 4).

Discussion

Using mutualistic bio-agents is a natural way to improve agricultural production and decreasing yield losses either directly, by mechanisms such as mycoparasitism or indirectly, by competing with pathogens for nutrients and space, modifying the environmental conditions, or promoting plant growth and plant defense mechanisms and antibiosis, [35].

The present investigation aimed to study the direct and indirect effects of mutualistic *Trichoderma* application as a promising bio-agent against *Alternaria solani*, the causal pathogen of early blight disease on tomato. In this respect, six different isolates belonging to genus *Trichoderma* i.e., *Trichoderma hamatum (Th)*, *T. viride (Tv)*, *T. harzianum (Tz22)*, *T. harzianum* (isolate T10), *T.* species (T.32) and *T.*

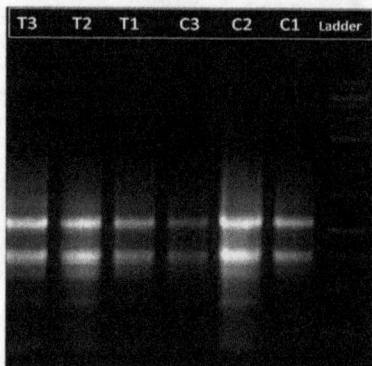

Figure 1: RNA isolated from leaves of three biological replicates (T1, T2 and T3) treated with *T. harzianum*-T10 isolate and three un-treated (control) replicates (C1, C2 and C3). Ladder is 1kbp ladder.

Components	Volume per reaction
10 x RT buffer	2 µ
dNTP mix (10 mM)	2 µl
Random Primers (10 mM)	2 µl
RNAase inhibitor (1 U/µl)	1 µl
Reverse transcriptase (5 U/µl)	1 µl
RNA (1 µg)	x µl
H$_2$Odest	0 - x µl
Total volume	**20 µl**

Table 2: Pipetting scheme containing reverse transcription reagents.

Step	Time	Temperature	Cycles
Anealling of Random primers	10 min	25°C	Hold
Reverse transcriptase activity	120 min	37°C	Hold
Reaction termination	5 min	85°C Hold	Hold

Table 3: Thermal cycling parameter for the cDNA synthesis/reverse transcription.

Components	Volume per reaction
Fast SYBR®- Green Master Mix	10 µl
Forward primer (10 mM)	0,5 µl
Reverse primer (10 mM)	0,5 µl
DNA (1:10 diluted)	1 µl
H$_2$Odest	8 µl
Total Volume	20 µl

Table 4: Pipetting scheme containing Real time PCR Fast SYBR®- Green Master Mix.

Step	Time	Temperature	Cycles
Polymerase activation	20 sec	95°C	Hold
Denature	3 sec	95°	40
Anneal/Extend	30 sec	60°C	40
Melt curve	15 sec	95°C	Hold
Melt curve	1 h	60°C	Hold

Table 5: Thermal cycling parameter for the semi-quantitative RT-qPCR.

koningii (Tkg) were screened either *in vitro* or *in vivo* for this purpose. Also, the ability of *Trichoderma* to affect the expression levels of some related genes to defense mechanisms in tomato was evaluated using qRT-PCR technique.

The direct and indirect influence of all tested biocontrol isolates on *Alternaria solani* the causal of early blight disease was determined either under *in vitro* or in vivo conditions.

Moreover, the potential of the tested biocontrol *Trichoderma* isolates on incidence of early blight disease under *in vivo* conditions was compared with standard and recommended chemical fungicide, propamocarp Hcl, as well as with *Alternaria solani* infected tomato plants, which were treated with tap water only (control).

The obtained results from *in vitro* bioassay revealed that all tested isolates significantly reduced the mycelia growth of *A. solani* in dual cultures under laboratory conditions. The best results were recorded with *Trichoderma harzianum*-T10. Similar finding were reported by Rudresh et al. [36] who noticed that *T. harzianum* inhibited growth of *Rhizoctonia solani* and *Fusarium oxysporum* under *in vitro* conditions with 72.1% and 77.0%, respectively. The *in vitro* antagonistic activity of *Trichoderma* species against six fungal plant pathogens including *A. solani* was proved also by Bell et al. [37].

On the other hand, the *in vivo* bioassay of bio-agents illustrated that *Trichoderma harzianum*-T10 was the superior and the most promising isolate among the six tested bio-control isolates where it exhibited the maximum reduction % of disease severity comparing to control treatment. The realized reduction % of Disease severity as a result of treating tomato plants with *Trichoderma harzianum*-T10 came close to those obtained with fungicide treatment. These results are in agreement with those of Elad et al. [38]; Tu and Vartaja [39] who showed that *Trichoderma* species affected the growth and establishment of different plant pathogens including *Alternaria solani* on their hosts.

It is remarkable that mode of action in the tri-interaction between plant pathogen, *Trichoderma* and host plant still unclear. Several biocontrol agents are known to produce antimicrobial substances to achieve control over various plant ailments [40]. Furthermore, it is known that the genus *Trichoderma* comprises a great number of fungal strains exert biocontrol activity against fungal phyto-pathogens either directly or indirectly. These antagonistic properties are based on the activation of multiple mechanisms [35,40]. Recently, the ability of mutualistic bioagents to induce host plant systemic resistance pathways has been successfully tested. Results showed that some bio-agents including *Trichoderma* were capable to trigger host plant systematic resistance against hazardous plant parasites i.e. *Meloidogyne incognita* through altering the gene expression of some involved genes in different systemic resistance pathway [41].

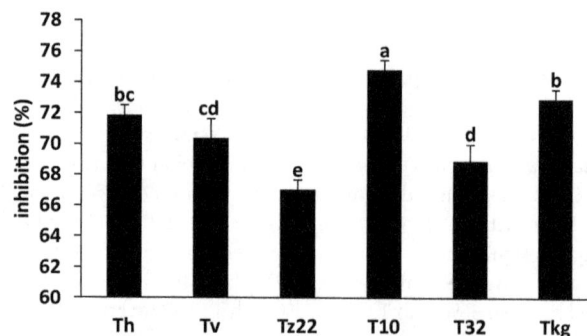

Figure 2: *In vitro* relative radial growth inhibition of *Alternaria solani* mycelial growth on PDA plates by the different *Trichoderma* isolates i.e., *T. hamatum* (Th), *T. viride* (Tv), *T. harzianum* (Tz22), *T. harzianum*-T10, *Trichoderma* sp. (T32) and *T. koningii* (Tkg). The relative growth inhibition was calculated by subtracting the radial growth of *Alternaria solani* in the direction of the biocontrol agent from the radial growth of *Alternaria solani* in the absence of a biocontrol agent (control) divided on radial growth of control , (n=3).

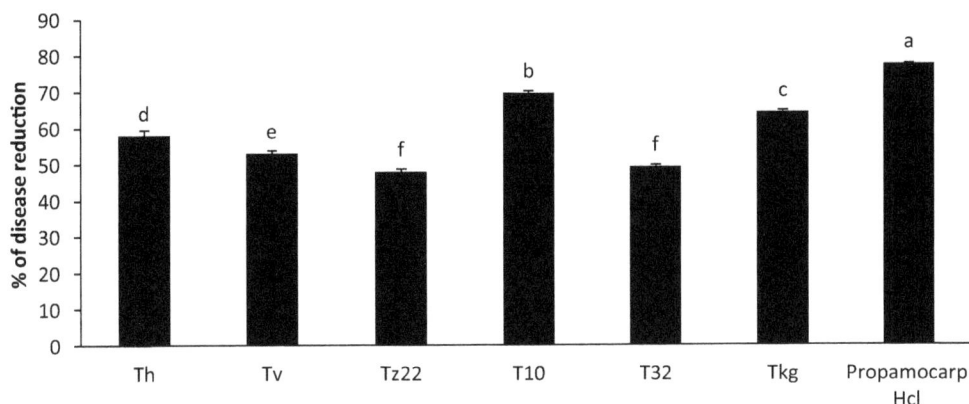

Figure 3: Reduction % of early blight disease recorded on tomato plants treated with the six individual isolates of *Trichoderma* species i.e., *T. hamatum* (Th), *T. viride* (Tv), *T. harzianum* (Tz22), *T. harzianum*- T10, *Trichoderma* sp. (T32) and *T. koningii* (Tkg) compared with treated plants with 2.5 ml/L of fungicide (Propamocarp Hcl).

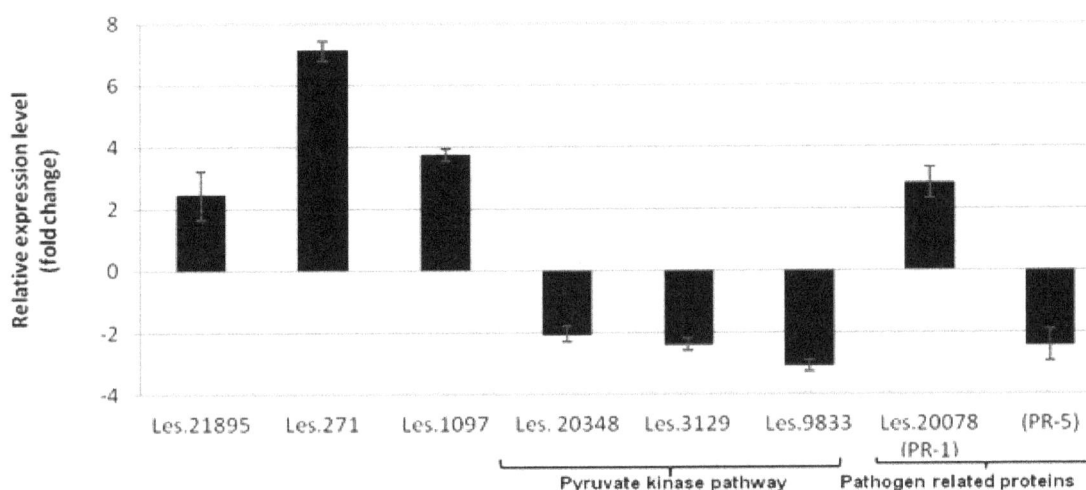

Figure 4: Relative expression levels of eight different related genes to essential bioprocess and defence mechanisms in tomato plants inoculated with bio-agent *Trichoderma harzianum*-T10. Bars indicating the means of fold changes in gene expression as generated by qRT-PCR. The target gene expression changes were normalized to the changes of the Actin gene as a reference gene. Data for three biological replicates.

In the present study, the influence of *Trichoderma* application on the expression levels of 8 different genes in tomato was investigated through quantitative real time PCR, qRT-PCR. The results showed that three genes, Les.21895, Les.271, and Les.1097, which related to auxin (IAA6), ethylene (ER69) and lignin (CL2) production pathways, respectively were up regulated. On the other hand, three genes i.e., Les.20348, Les.3129 and Les.9833 which coding for pyruvate kinase (PK) hormone were down regulated. Noteworthy that T10 affected also the expression levels of two pathogenesis related proteins including Pr-1 and Pr-5. Results demonstrated that *Trichoderma harzianum*-T10, induced the relative gene expression of Pr-1 while it decreased the relative expression of Pr-5 when compared with un-treated control plants. Davies [42] reported that the plant hormone auxin, typified by indole-3-acetic acid (IAA), regulates a variety of physiological processes, including apical dominance, tropic responses, lateral root formation, vascular differentiation, embryo patterning, and shoot elongation. The central role of the Aux/IAA gene family members in auxin signaling has been suggested by molecular genetics and biochemical studies of Woodward and Bartel [43]. Recently,

studies of Overvoorde et al. [44] suggested that auxin hormone can regulate various aspects of plant growth and development. Their investigation showed also that the Aux/IAA proteins regulate auxin-mediated gene expression by interacting with members of the auxin response factor protein family. In the present study, the results showed that when tomato roots were treated with *Trichoderma* isolate (T10), there was an increase in the relative expression of auxin-regulated (IAA6) gene. This indicates that the tested *Trichoderma* isolate can trigger systemically the auxin signaling pathways, which involved in essential physiological and biological processes important for growth and health of tomato plants.

Furthermore, the results obtained from this present study illustrated that *Trichoderma* application increased the expression level of gene Les.271 which involved in ethylene signaling pathway. The change in gene expression was greatly high compared to control plants. Ethylene signaling pathway was found to play a distinct role in the response of host plant to pathogen infection as reported by Dowd et al. [45]. They detected differences between healthy and infected cotton plants with pathogenic *F. oxysporum f.sp.vasinfectum* in ethylene and auxin hormons concentrations. Moreover, they confirmed the role of those two hormons in disease process by direct measurements of those hormone levels inside infected and non-infected plants. Nahar et al. [20] reported also that complex defense signaling pathways were found to be controlled by different hormones such as auxin and ethylene. Their results obtained from q-PCR revealed that Ethylene strongly activates jasmonic acid biosynthesis and signaling genes within roots of rice plants, which led to stimulate the systemic induced resistance (ISR) pathway against root knot nematodes. Based on these results, induction of ethylene pathway using *Trichoderma harzianum*-T10 is candidate to be one of the mechanisms that can be involved in management of early blight disease on tomato plants.

Moreover, the increase in expression level of gene Les.1097 which coding for 4-coumarate-CoA ligase 2 was observed on inoculated tomato plants with *Trichoderma harzianum*-T10 compared to control plants. According to studies of Hamberger and Hahlbrock [21] on *Arabidopsis thaliana*, two classes of CoA Ligase isoenzymes were identified. They reported that class I has been directly or indirectly associated with the biosynthesis of lignin and structurally related soluble or cell wall-bound phenylpropanoid derivatives, whereas class II isoenzymes have been associated with flavonoid biosynthesis. Furthermore, Nicholson and Hammerschmitt [22]; Bhuiyan et al. [23] mentioned that the lignifications, in partial related to CoA ligase isoforms activity, have the potential to act in several ways in plant defenses against pathogen infection. It can establish mechanical barriers to pathogen invasion, chemically modify cell walls to be more resistant to cell wall-degrading enzymes, increase the resistance of walls to the diffusion of toxins from the pathogen to the host and of nutrients from the host to the pathogen, produce toxic precursors and free radicals, and lignify and entrap the pathogen. According to these studies, it seems that the ability of isolate T10 to induce lignifications and/or secondary metabolites like phenylpropanoids in tomato plants can play a curtail role against *A. solani* on tomato.

In contrary, results revealed that inoculating tomato plants with *Trichoderma harzianum*-T10 reduced the relative expression level of three genes, Les.20348, Les.3129 and Les.9833. These three genes are involved in pyruvate kinase biosynthesis (PK). The enzyme pyruvate kinase (PK) had been shown to play a cretin role in various physiological processes in plants [24,25]. Different studies suggest that Pk activity can be used as a remarkable physiological indicator of K^+, Mg^{2+}, and Ca^{2+} contents in plant tissues [26,27]. Ruiz et al. [46] reported that increasing Ca^{2+} concentration in tobacco leaves resulted in decreasing the activity of pyruvate kinase. Moreover, they illustrated also that PK activities were inversely proportional to the leaf concentration of the free forms of K^+, Mg^{2+}.

In conclusion, *Trichoderma* species affected the growth and the establishment of *Alternaria solani*, the causal pathogen of tomato early blight disease, either *in vitro or in vivo*. Moreover, different mechanisms of action including stimulation of auxin, ethylene, lignin and pathogen related proteins can be involved in management of early blight disease on tomato plants using bio-control elicitors as *Trichoderma* species. On the other hand, *Trichoderma* application inhibited the expression level of three related genes to pyruvate kinase bio-synthesis pathway which may point to the role of this hormone in the interaction between host plants and bio–control elicitors with presence of *Alternaria solani* pathogen.

References

1. FAO (2003) Agricultural statistics.

2. Costa JM, Heuvelink E (2005) Introduction: The tomato crop and industry. In: Heuvelink, E. (Eds.). Tomatoes. CAB International, UK, pp. 1-19.

3. Peralta IE, Knapp S, Spooner DM (2005) New species of wild tomatoes (Solanum section Lycopersicon: Solanaceae) from Northern Peru. Syst Bot 30: 424-34.

4. Rotem J, Reichert I (1964) Dew - a principal moisture factor enabling early blight epidemics in a semiarid region of Israel. Plant Dis Rep 48: 211-15.

5. Kemmitt G (2002) Early blight of potato and tomato. The Plant Health Instructor.

6. Walker JC (1952) Diseases of Vegetable Crops. McGraw-Hill Book Co, New York, NY, 699 p.

7. Basu PK (1974) Measuring early blight, its progress and inluence on fruit losses in nine tomato cultivars. Canadian Pl Dis Survey 54: 45-51.

8. Sherf AF, MacNab AA (1986) Vegetable Diseases and their Control. John Wiley and Sons, New York, pp. 634-640.

9. Singh PC, Kumar R, Singh M, Rai A, Singh MC, et al. (2011) Identification of resistant sources against early blight disease of tomato. Indian J Hort 68: 516-521

10. Sikora RA, Schäfer K, Dababat AA (2007) Modes of action associated with microbially induced in planta suppression of plant-parasitic nematodes. Australasian Plant Pathology 36: 124-134.

11. Davies K, Spiegel Y (2011) Root Patho-Systems Nematology and Biological Control. In: Davies K. & Spiegel, Y. (Eds). Biological Control of Plant Parasitic Nematodes: Building Coherence between Microbial Ecology and Molecular Mechanisms, Chapter 12, Springer, The Netherlands.

12. Fageria MS (1997) Genetic analysis of resistance to *Alternaria* leaf spot in tomato. J Mycol Pl Pathol 27: 286-89.

13. Ganie SA, Ghani MY, Nissar Q, U-Rehman S (2013) Bioefficacy of plant extracts and biocontrol agents against *Alternaria solani*. African Journal of Microbiology Research 7: 4397-4402.

14. Alabouvette C, Lemanceau P, Steinberg C (1993) Recent advances in the biological control of Fusarium wilts. Pesticide Science 37: 365-373.

15. Olivain C, Alabouvette C (1997) Colonization of tomato root by a non-pathogenic strain of *Fusarium oxysporum*. New Phytologist 137: 481-494.

16. Alabouvette C, Edel V, Lemanceau P, Olivain C, Recorbet G, et al. (2001) Diversity and interactions among strains of *Fusarium oxysporum*: Application and biological control. In: M.J. Jeger and N.J. Spence (Eds.): Biotic interactions in plant-pathogen associations, pp 131-157. CAB International, London, England.

17. Fravel D, Olivain C, Alabouvette C (2003) *Fusarium oxysporum* and its biocontrol. New Phytol 157: 493-502.

18. Fuchs JG, Moënne-Loccoz Y, Défago G (1997) Nonpathogenic *Fusarium oxysporum* strain Fo47 induces resistance to Fusarium wilt in tomato. Plant Dis 81: 492-496.

19. Abel S, Theologis A (1996) Early genes and auxin action. Plant Physiol 111: 9-17.

20. Nahar K, Kyndt T, De Vleesschauwer D, Höfte M, Gheysen G (2011) The jasmonate pathway is a key player in systemically induced defense against root knot nematodes in rice. Plant Physiology 157: 305-316.

21. Hamberger B, Hahlbrock K (2004) The 4-coumarate:CoA ligase gene family in *Arabidopsis thaliana* comprises one rare, sinapate-activating and three commonly occurring isoenzymes. Proc Natl Acad Sci USA 101: 2209-2214.

22. Nicholson RL, Hammerschmidt R (1992) Phenolic Compounds and Their Role in Disease Resistance. Annual Review of Phytopathology 30: 369-389.

23. Bhuiyan NH, Selvaraj G, Wei Y, King J (2009) Gene expression profiling and silencing reveal that monolignol biosynthesis plays a critical role in penetration defence in wheat against powdery mildew invasion. Journal of Experimental Botany 60: 509-521.

24. Lin M, Turpin DH, Plaxton WC (1989) Pyruvate kinase isozymes from the green alga, *Selenastrum minutum*. II. Kinetic and regulatory properties. Arch Biochem Biophys 269: 228-238.

25. Schuller KA, Plaxton WC, Turpin DH (1990) Regulation of phosphoenolpyruvate carboxylase from the green alga *Selenastrum minutum*: properties associated with the replenishment of tricarboxylic acid cycle intermediates during ammonium assimilation. Plant Physiol 93: 1303-1311.

26. Lavon R, Goldschmidt EE (1999) Enzymatic methods for detection of mineral element defeciencies in citrus leaves. a mini-review. J Plant Nutr 22: 139-150.

27. Ruiz JM, Moreno DA, Romero L (1999) Pyruvate kinase activity as an indicator of the level of $K^{(+)}$, $Mg^{(2+)}$, and $Ca^{(2+)}$ in leaves and fruits of the cucumber: the role of potassium fertilization. J Agric Food Chem 47: 845-849.

28. Giri P, Taj G, Kumar A (2013) Comparison of artificial inoculation methods for pathogenesis studies of *Alternaria brassicae* (Berk.) Sacc. on *Brassica juncea* (L.) Czern. (Indian mustard). Afr J Biotechnol 12: 2422-2426.

29. Utkhede RS, Rahe JE (1983) Interaction of antagonist and pathogens in biological control of onion white rot. Phytopathology 73: 890-893.

30. Vincent JM (1947) Distortion of fungal hyphae in the presence of certain inhibitors. Nature 159: 850.

31. Vakalounakis DJ (1983) Evaluation of tomato cultivars for resistance to *Alternaria* blight. Ann Appl Biol 102: 138-139.

32. Pandey KK, Pandey PK, Kalloo G, Banerjee MK (2003) Resistance to early blight of tomato with respect to various parameters of disease epidemics. J Gen Pl Pathol 69: 364-71.

33. Livak KJ, Schmittgen TD (2001) Analysis of Relative Gene Expression Data Using Real-Time Quantitative PCR and the $2^{-\Delta\Delta CT}$ Method. Elsevier methods 25: 402-408.

34. Plake BS, Kemmerer BE (1987) Statgraphics: Neil W. Polhemus and the Statistical Graphics Corporation. Rockville, MD: STSC. Computers in human behavior. 3: 289-298.

35. Benítez T, Rincón AM, Limón MC, Codón AC (2004) Biocontrol mechanisms of *Trichoderma* strains. Int Microbiol 7: 249-260.

36. Rudresh DL, Shivaprakash MK, Prasad RD (2005) Potential of *Trichoderma* spp. as bio control agents of pathogens involved in wilt complex of chickpea. J Biol Control 19: 157-166.

37. Bell DK, Wells HD, Markham CR (1982) In vitro antagonism of *trichoderma* species against six fungal plant pathogens. Phytopathology 72: 379-382.

38. Elad Y, Chet I, Katan J (1980) *Trichoderma* harzianum, a biocontrol agent effective against Sclerotium rolfsii and Rhizoctonia solani. Phytopathology 70: 119-121.

39. Tu JC, Vartaja V (1981) The effect of hypoparasite (*Gliocladium virens*) on *Rhizoctonia solani* on *Rhizoctonia* root rot of white beans. Canad J Bot 59: 22-27.

40. Yang Z, Yu Z, Lei L, Xia Z, Shao L, et al. (2012) Nematicidal effect of volatiles produced by *Trichoderma* sp. Journal of Asia-Pacific entomology 15: 647-650.

41. Selim ME, Mahdy ME, Sorial ME, Dababat AA, Sikora RA (2014) Biological and chemical – dependant systemic resistance and their significance for the control of root-knot nematodes. Nematology 16: 917-927.

42. Davies PJ (1995) Plant Hormones: Physiology, Biochemistry and Molecular Biology, 2nd ed. (Dordrecht, The Netherlands: Kluwer Academic Publishers).

43. Woodward AW, Bartel B (2005) Auxin: regulation, action, and interaction. Ann Bot 95: 707-735.

44. Overvoorde PJ, Okushima Y, Alonso JM, Chan A, Chang C, et al. (2005) Functional genomic analysis of the auxin/indole-3-acetic acid gene family members in *arabidopsis thaliana*. The Plant Cell 17: 3282-3300.

45. Dowd C, Wilson IW, McFadden H (2004) Gene expression profile changes in cotton root and hypocotyls tissues in response to infection with *Fusarium oxysporum* f.sp. vasinfectum. Molecular plant-microbe interactions 17: 654-667.

46. Ruiz JM, Lopez-Cantarero I, Romero L (2000) Relationship between calcium and pyruvate kinase. Biological Plantarum 43: 359-362.

4

Field Evaluation of New Fungicide, Victory 72 WP for Management of Potato and Tomato Late Blight (Phytophthora infestans (Mont) de Bary) in West Shewa Highland, Oromia, Ethiopia

Mohammed Amin*, Negeri Mulugeta and Thangavel Selvaraj

Department of Plant Science and Horticulture, College of Agriculture and Veterinary Science, Ambo University, Post Box No: 19, Ethiopia

Abstract

The field experiment was carried out to evaluate the management of tomato (Roma-VF) and potato (Guddeni) late blight using a new fungicide, Victory 72 WP and Ridomil gold to select the more effective dosage of the new fungicide option against potato and tomato late blight under field conditions at Toke Kutaye district of West Shoa, Ethiopia, during the main cropping season in 2012. The experiment was arranged in randomized complete block design with three replications. Two types of fungicide applications (Ridomil gold and Victory 72 WP) in weekly intervals were established and unsprayed plot was used as a control. Late blight infection was prevalent in the experimental year, and a significant amount of disease was detected ($P < 0.05$). Application of fungicide treatments considerably reduced late blight progress, with a corresponding increase in tuber and fruit yields of potato and tomato, respectively. Significant differences were observed among treatments in potato and tomato plants in terms of disease severity (DS), area under disease progressive curve (AUDPC), disease progressive rate (r). Among the different treatments, Victory 72 WP fungicide treated potato and tomato plants recorded the lowest DS, AUDPC, disease progressive rate. Based on late blight disease occurrence, application of Victory 72 WP fungicide significantly reduced disease development and increased tuber and fruit yield in both potato and tomato crops, respectively as compared to the Ridomil gold fungicide applications. This is an indication of the reliability and promise as well as the exhibition of great potential of the Victory 72 WP is the effective control of the late blight of potato and tomato in Toke Kutaye district of West Shoa, Ethiopia.

Keywords: Fungicides; Late blight; Potato; Ridomil gold; Tomato; Victory 72 WP

Introduction

Like many other countries in the world, potato (*Solanum tuberosum* L.) and tomato (*Lycopersicon esculentum* Mill.) crops are very important food and cash crops especially on the highland and mid altitude areas of Ethiopia [1]. Ethiopia's potato area had grown to 160,000 ha, with average yields around 9 tons/ha. About 152,956,115 million tons of tomatoes were produced in the world in 2011, and with a yield potential of up to 48.1 tons/ha [2]. In Ethiopia, the total area under production reaches 51,698 hectares and annual production is estimated to be more than 230,000 tons [3]. The national average of tomato fruit yield in Ethiopia is very low (7 tons/ha) compared even to the neighboring African countries like Kenya (16.4 tons/ha) and less than 50% of the current world average yield of 27 tons/ha [3]. Current productivity under farmers' condition in Ethiopia is 9 tons/ha whereas yield up to 40 tons/ha recorded on research plots [4].

Farmers get lower yield mainly due to diseases, pests and sub-optimal fertilization. The most important factors responsible for the low productivity of potato and tomato are diseases and insect pests. Among those diseases, early and late blight fungal diseases of solanaceous vegetables are the most destructive and widespread in tropical, subtropical and temperate regions of the world [5]. Late blight caused by *Phytophthora infestans* (Mont) de Bary is one of the most significant constraints to potato and tomato productions up to 90% of crop losses in cool and wet weather conditions in the country [6]. Yield losses due to the disease are attributed to both premature death of foliage and diseased tubers in potato and foliage, stems and fruits of tomato. The disease is more severe in humid and high rainfall areas and it occurs at a low intensity in dry areas [7]. It causes serious loss in yield and quality as well as reduces its marketability values [8]. Nonetheless,

loss due to the disease was estimated to range between 65-70% and complete crop failures are frequently reported [9].

The management of potato and tomato crops against this pathogen is important to maximize the crops' yield. The disease occurs throughout the major potato and tomato production areas and it is difficult to produce both the crops during the main rainy season without chemical protection measures [1,5,10,11]. Fungicides are among the most efficient control options available to the growers in Ethiopia. This is particularly important in developing countries such as Ethiopia, where the setup of efficient control programs for potatoes and tomatoes are inadequate. The newly introduced fungicide could be treated against late blight of potato and tomato cultivars, which could reduce the need of application of high fungicide, decrease the risk to human health, environmental contamination and increase the economic benefit of farmers. Therefore, the present verification test research work was carried out to evaluate the management of potato (Local var.) and tomato (Roma-VF) against late blight using a new fungicide, Victory 72 WP and Ridomil gold to select the more effective dosage of the new fungicide option against the potato late blight.

---

*Corresponding author: Mohammed Amin, Department of Plant Science and Horticulture, College of Agriculture and Veterinary Science, Ambo University, Post Box No: 19, Ethiopia, E-mail: aminmahammed@gmail.com

Materials and Methods

Description of the study area

Experimental trail was conducted in previous potato and tomato cultivated fields at Guder of Toke Kutaye district, West Showa, Oromia, Ethiopia during the main cropping season on 2012. The altitude of the study areas was between 1900 and 3100 m. a. s. l, geographical positions of N 08° 43.423-N 10° 12.082 and E 037° 28.902-040° 62.590. Heavy rain observed from onset of July to the end of August. The annual rainfall ranges from 1000-1588.06 mm and the temperature of the district ranged between 9.44°C and 21.86°C with average of 15.65°C. The soil of the experimental site is light red in color, clay loam in texture and pH value of 6.8. The late blight pressure is generally high in Guder location during the rainy season.

Materials used

A seed of improved susceptible tomato variety (Roma-VF) was used and obtained from Melkasa Agricultural Research Center, Ethiopia. Seed of improved potato variety (Local) was obtained from Holleta Agricultural Research Center, Ethiopia. Two fungicides, Victory 72 WP with active ingredient of Mancozeb 640 g/kg+Metalaxyl 80 g/kg and Ridomil gold with active ingredient of Metalaxyl 40 g/kg+Mancozeb 640 g/kg) were obtained from Axum Greenline Trading Private Limited Company, Ethiopia.

Experimental design and treatments

The experiments were arranged in randomized complete block design with three replications. Three fungicide treatments (Victory 72 WP, Ridomil gold and no treatment) were used. In this experiment, Ridomil gold was used as a standard check. Likewise, Victory 72 WP was a new fungicide chemical which have not been used before for the controls of late blight disease in Ethiopia. Seedlings of improved potato local variety (Guddeni) which is moderately susceptible to late blight were planted into each plot size of 4.8×3.2 m. Spacing between plants and rows were maintained as 40 cm and 80 cm, respectively. There were 8 plants per row and the four central rows were harvested for determining tuber yield. Also, a seed of tomato variety, Roma-VF was shown on a standard seed bed with a size of 1.8×2.4 m. Apparently healthy seeds were sown into the experimental field with each plot sizes of 3.6×2.7 m, and distance between rows and between plants were 45 and 60 cm, respectively. There were 6 plants per row and the four central rows were harvested for determining fruit yield. Each potato and tomato plots consisted of a total six rows. Each plot and blocks were separated by a buffer zone of 1.5 and 2.0 m, respectively to prevent fungicide drift or cross contamination both for potato and tomato field trials. First spray of fungicides was started soon after the initial appearance of disease symptoms using knap-sack sprayer. The fungicides were applied at the rate of 2.5 kg/ha at an interval of seven days. All agronomic practices such as weeding, cultivation were kept uniform for all treatments in each plot.

Disease assessment

Incidence and severity: Natural inoculation was relied upon in all experimental plots. Disease incidence and severity were assessed on the central two rows every week. Incidence of late blight was assessed by counting the number of plants on the middle four rows and were expressed as percentage of total plants. Five plants were selected randomly from each replicate per treatment, and then five leaves of each plant were used to determine the disease severity [12].

The per cent incidence was calculated as:

$$Disease\ Incidence = \frac{Nunmer\ of\ diseased\ plant}{Total\ number\ of\ plant\ inspected} \times 100$$

Severity of late blight was recorded on the basis of 1-6 rating scales as described by Gwary and Nahunnaro [13]. where scale 1=trace to 20% leaf infection, 2=21-40% leaf infection, 3=41-60% infection, 4=61-80 infection, 5=81-99% infection, 6=100% leaf infection or the entire plant defoliation and then the rating scales were converted into percentage severity index (PSI) for the analysis of disease severity using the following formula:

$$Percentage\ Severity\ Index = \frac{Sum\ of\ Individual\ numerical\ rating}{Total\ Number\ of\ assessed \times Maximum\ score\ in\ scale} \times 100$$

Disease progression analysis

Area under the disease progress curve (AUDPC) and growth curve models were developed for the disease progress data. AUDPC values were calculated for each plot using the following equation [14].

$$AUDPC = \sum_{i=1}^{n-1} 0.5\left(x_{i+1} + x_i\right)\left(t_{i+1} - t_i\right)$$

Where, X_i is the cumulative disease severity expressed as a proportion at the ith observation, t_i is the time (days after planting) at the ith observation and n is total number of observations. AUDPC values were then used in analysis of variance to compare amount of disease among plots with different treatments. Logistic, in [(Y/1-Y)] and Gompertz, -in [-in(Y)] [15]. Models were compared for estimation of disease progression parameters from each treatment. The goodness of fit of the models was tested based on the magnitude of the coefficient of determination (R^2). The appropriate model was used to determine the apparent rate of disease increase (r) and the intercept of the curve.

Assessment of yield data

Data related to yields were recorded from the central four rows on each plot for each treatment. Tubers and fruits were considered ready for picking, when 50% of tomato fruits turned yellow or red. Mean yield of tubers and fruits were assessed on each plot of four central rows.

Data analysis

Data on disease parameters (disease incidence, disease severity, PSI, AUDPC and disease progression rate (r), yield were subjected to analysis of variance (ANOVA) using Statistical Analysis System (SAS) version 9.1 software. Fisher's protected Least Significant Difference (LSD) values were used to separate differences among treatment means (P<0.05) for the field evaluation of potato and tomato late blight disease.

Results

Disease incidence in tomato and potato

Data on the disease incidence revealed that no significant differences among treatments in their initial and final percent disease incidence. During the first disease assessment (53 DAP), treatments did not show uniform initial disease incidence. Minimum initial percent disease incidence was observed in Ridomil gold and Victory 72 WP treated potato plants with the mean values of 18 and 20.0% and in tomato plants with the mean values of 20 and 20%, respectively. At the end of disease assessment (113 DAP), all the experimental plots had recorded 100% disease incidence.

Disease severity in tomato and potato

Among the experimental treatments, percent final disease severity was highly significant (P<0.05). All the treatments showed different level of reaction to late blight of tomato compared to untreated control. The least percent disease severity was recorded in Victory 72 WP and Ridomil gold treatments with mean values of 15.67 and 18.90% in tomato treated plants, respectively. Therefore, in comparisons, the highest percent final disease severity (67.7%) was recorded from untreated plots. Likewise in potato field, the least percent disease severity was recorded in Victory 72 WP treatments with mean values of 40.74% in study area. Similarly in Ridomil gold treatments, the percent severity was recorded 40.74% in Guder field trial of potato, Therefore, in comparisons, the highest percent final disease severities (69.72%) was recorded from untreated control plots (Table 1).

Area under disease progressive curve (AUDPC) in tomato and potato field trail

AUDPC values on tomato and potato late blight exhibited significant differences (P<0.05) within treatments. The minimum AUDPC 711.7%-day was observed on Victory 72 WP treated tomato plots followed by Ridomil gold (894.6%-days) and the maximum AUDPC values were recorded from control with a value of 1923.4%-day. In the same way, the lowest AUDPC 743.7%-day was observed on Victory 72 WP treated potato plots followed by Ridomil gold (842.6%-days) and the maximum AUDPC values were recorded from control with a value of 1973.4%-day (Table 1).

Disease progressive rate (r) of tomato and potato late blight

Disease progressive rates were significantly different at (P<0.05) as compared with the standard control (Tables 2 and 3). In treatments, the acceptable regression equation with coefficient of determination (R^2) ranging from 0.43 to 0.68 in tomato fields and 0.25 to 0.62 in potato

Treatments	Tomato field		Potato field	
	Severity (%)	AUDPC (%-days)	Severity (%)	AUDPC (%-days)
Victory 72 WP	15.67[c]	711.70[c]	40.74[b]	743.70[c]
Ridomil gold	18.90[b]	894.60[b]	40.74[b]	842.60[b]
Untreated control	67.70[a]	1923.40[a]	69.72[a]	1973.40[a]
Mean	34.09	1176.57	50.4	1186.57
CV (%)	15.26	10.79	14.26	12.79
LSD (5%)	20.26	452.2	18.26	432.2

Means followed by the same letter in each column are not significantly different using Least Significant Difference (p=0.05), CV=Coefficient of variation, LSD=least significant difference, AUDPC=area under disease progress curve. Means in column followed by the same letter(s) are not significantly different.

Table 1: Effect of fungicide treatments on late blight of tomato and potato with disease severity percentage and mean AUDPC in field trail.

Treatments	Intercept	Disease progressive Rate (r)	R^2	SEE	Significance (P<0.05)
Victory72WP		0.65	0.43	0.368	0.0001
Ridomil-gold		0.68	0.47	0.297	0.0001
Untreated control		0.83	0.68	0.476	0.0001

SEE: Standard Error of Estimate
The highest disease progressive rate was observed from untreated plot tomato and potato with values of 0.83 and 0.79, respectively which is exceeds Victory 72 WP by 0.29 units per day. Although, disease progressive rate of late blight increased more rapidly in untreated control compared to other treatments.

Table 2: Disease progressive rate (r) of tomato late blight severity on various treatments.

Treatment	Intercept	Disease progressive Rate (r)	R2	SEE	Significance (P<0.05)
Victory72WP		0.50	0.25	0.383	0.0001
Ridomil-gold		0.63	0.40	0.234	0.0001
Untreated control		0.79	0.62	0.456	0.0001

SSE=Standard Error Estimate

Table 3: Disease progressive rate (r) of potato late blight severity on three different treatments from August to November at Guder during 2012 main growing season.

Treatment	Tomato field		Potato field	
	Mean fruits yield yield (tons/ha)	Yield loss (%)	Mean tuber yield yield (tons/ha)	Yield loss (%)
Victory 72 WP	21.1	0	65.21	0
Ridomil gold	18.5	12.35	63.08	3.27
Untreated control	4.3	79.81	47.67	26.89

Table 4: Mean yield and yield loss estimation due to late blight of tomato and potato under various treatments at study area during 2012 main growing season.

treated fields were produced with the linearized form of final late blight severity regressed over time in days after planting. Disease progressive rate in the untreated plot was increased at the rate of 0. 83 units per day which was higher than the best performed treatment so called Victory 72 WP.

Mean yield of tomato fruit and potato tuber

There were significant differences (P<0.05) within treatments on fruit and tuber yield at study location. All treatments were exceeds the untreated plots. The maximum mean fruit yield (21.1tons/ha) was recorded from the Victory 72WP followed by Ridomil gold (18.5 tons/ha). While, the minimum fruit yield was recorded from control (4.3 tons/ha). Moreover, the highest mean tuber yield was harvested from plots treated with Victory 72WP followed by Ridomil gold with mean values of 65.21and 63.08 tons/ha (Table 4).

Yield loss estimation

The variation in tuber and fruit yield losses was observed among different treatments. In comparison, in untreated control plots, tuber and fruit yield losses was notably higher than protected plots. Yield losses were significantly reduced by fungicide chemical compared to other experimental treatments. The highest fruit yield loss was recorded from untreated plots (79.81%) compared to the most protected plots (Victory 72 WP) with a value of (12.35%) at spray frequency of weekly interval (Table 4). Similarly, the maximum tuber loss was observed from control plots.

Discussion

Previously, the fungicides were screened for the control of late blight of tomato and potato in the Central Rift Valley area of Ethiopia. Potential fungicides were also verified on farmers' fields around Melkassa, Ziway and Wondo Genet. Three fungicides (Metalaxyl-M4%+Mancozeb 64% (Ridomil gold 68 WP) 350 g/100 liters, Fungomil 250 g/100 liters, and Mancozeb+Metalaxyl (Mancolaxyl 72%) 250 g/100 liters) were found effective in controlling the disease on tomato and consequently increased marketable fruit yield by 40-66% [16]. Moreover, compared to the rest of treatments, all fungicide treatments provided significantly better foliar late blight control, and significantly gave higher total yield. The minimum AUDPC value also observed on Victory 72 WP followed by Ridomil gold and gave 711.7 and 894.6%-days, respectively.

In this study, Victory 72 WP consistently retarded late blight development and the highest yields were obtained from plots treated with Victory 72 WP followed by Ridomil gold. The study has determined that an application of Victory 72 WP and also Ridomil controls late blight and increases yields markedly. The efficacy of these fungicides can be enhanced by increasing the dosage and frequency of application [17,18]. The use of protectant and systemic fungicides for managing late blight has perhaps been the most studied aspect of late blights management in temperate countries [19]. Olanya et al. [19] also reported that, with the exception of optimum or scheduled fungicide applications based on favourable weather conditions; the most economical option for disease management is the use of host-plant resistance. The use of cultivars with durable resistance combined with scheduled applications of Protective fungicides has been reported as useful for managing late blight [20] as well as other diseases [15]. The performance of Ridomil gold in controlling late blight under present investigation has been supported by many researchers throughout the world [17,18,21,22]. In tropical Africa, the contact fungicide Dithane M 45 (Mancozeb 80% WP) and the systemic fungicide Ridomil MZ are widely used to control late blight [19,23]. Among the fungicides, Filthane M-45, Secure, Melody Duo, Ridomil gold and Metaril are highly effective to minimize late blight and to increase yield of potato. In the present study, a new fungicide, Victory 72 WP was first used in controlling late blight of potato and tomato in West Shoa of Ethiopia. This is because of an excellent potato and tomato late blight control and also reasonable and acceptable price to be invested in chemical control. This study provides new possible alternatives for the management of late blight to both small and large scale potato and tomato producers of Ethiopia.

Conclusions

Victory 72 WP fungicide treatments significantly reduced the late blight severity of tomato and potato as compared with Ridomil gold treatment and the untreated control in the tested location. This is an indication of the reliability and promise as well as the exhibition of the great potential of the Victory 72 WP to effective control of the late blight of potato and tomato in West Showa, Ethiopia. Therefore, due to reasonable reduction in disease severity, easy application made Victory 72 WP superior in controlling tomato and potato late blight. Based on the results of the this location test carried out in West Showa, Ethiopia on the effectiveness of the test fungicide, Victory 72 WP, and consideration of the wide range of late blight disease in potato and tomato controlled, the use of Victory 72 WP by the farming community (both individually and or private sector) will no doubt improve potato and tomato production in West Showa as well as in similar agro-ecology.

References

1. Borgal HB, Arend C, Jacobi, Kanyarukis S, Kulazia A, et al. (1980) Production, marketing and consumption of potato in the Ethiopia highlands' (Holleta, Awassa, and Alemaya). Center of Advanced training in agricultural development technology, University of Berlin, Berlin.

2. FAOSTAT (2011) Statistical database of the Food and Agriculture of the United Nations. FAO, Rome, Italy.

3. Central Statistics Agency (CSA) (2010) Agricultural sample survey, 2008/2007. Report on area and production of crops (Private peasant holdings, main season). Statistical Authority, Addis Ababa, Ethiopia.

4. Mutitu EW, Muiru WM, Mukunya DM (2008) Evaluation of antibiotic metabolites from actimnomycete isolates for the control of late blight of tomatoes under greenhouse conditions. Asian Journal of Plant Sciences 7: 284-290.

5. Hijmans RJ, Forbes GA, Walker TS (2000) Estimating the global severity of potato late blight with GIS-linked disease forecast models. Plant Pathol 49: 697-705.

6. Denitsa N, Naidenova M (2005) Screening the antimicrobial activity of actinomycetes strains isolated from Antarctica. J Cult Collections 4: 29-35.

7. Srivastava A, Handa AK (2010) Hormonal regulation of tomato fruit development: a molecular perspective. Journal of Plant Growth Regulation 24: 67-82.

8. Bekele Kassa, Yaynu Hiskias (1996) Tuber yield loss assessment of potato cultivars with different level of resistance to late blight.pp. 149-152. In: Eshetu, Bekele, Abdurrahman Abdullah and Aynekulu Yemane (Eds). Proceedings of the 3rd annual conference of crop protection Society of Ethiopia 18-19 May, Addis Ababa Ethiopia.

9. Kasa B, Woldegiorgis G (2000) Effect of planting date on late blight severity and tuber yield on different potato varieties. Pest Management Journal of Ethiopia 4: 51-63.

10. Mesfin (2009) Review of research on diseases of root and tuber crops in Ethiopia. In: Increasing Crop Production through Improved Plant Protection Vol II (eds.) Abraham Tadesse). Plant Protection Society of Ethiopia, Addis Ababa, Ethiopia 169-230.

11. Abd-El-Khair, Karima HE, Haggag E, GEI- Gamal N (2004) Biological control of wilt disease caused by Fusarium oxysporum in fennel under organic farming system. Phytopathology 34: 56-65.

12. Gwary DM, Nahunnaro H (1998) Epiphytotics of early blight of tomatoes in Northeastern Nigeria. Crop Port 17: 619-624.

13. Campbell CL, Madden LV (1990) Introduction to Plant Disease Epidemiology. John Wiley, New York.

14. Van der Plank JE (1963) Plant Diseases: Epidemics and Control, Academic Press, New York, London.

15. Abraham Tadesse (2009) Increasing Crop Production through Improved Plant Protection, Vol. II, Plant Protection Society of Ethiopia (PPSE), PPSE and EIAR, Addis Ababa, Ethiopia.

16. Islam MR, Dey TK, Rahman MM, Hossain MA, Ali MS (2002) Efficacy of some fungicides in controlling late blight of potato. Bangladesh Journal of Agriculture Research 27: 257-261.

17. Tsakiris E, Karafyllidis DI, Mansfield J, Paraussi G, Voyiatazis D, et al. (2002) Management of potato late blight by fungicides. Proceedings of the Second Balkan Symposium on Vegetables and Potatoes, Thessaloniki, Greece. Acta Horticulture 579: 567-570.

18. Olanya OM, Adipala E, Hakiza JJ, Kedera JC, Ojiambo P, et al. (2001) Epidemiology and population dynamics of Phytophthora infestans in sub-Saharan Africa: progress and constraints. Africa Crop Science Journal 9: 181-193.

19. Simons MD (1972) Polygenic resistance to plant disease and its use in breeding resistant cultivars. J Environ Qual 1: 232-240.

20. Singh BP, Shekhawat GS (1999) Potato late blight in India. Tech. Bull. No. 27 (revised), India.

21. Singh BP, Singh P, Jhililmil-Gupta H, Lokendra S, Gupta J, et al. (2001) Integrated management of late blight under Shimla hills. National Symposium on Sustainability of Potato Revolution in India, Shimla. Journal of the Indian Potato Association 28: 84-85.

22. Fontem DA (2001) Influence of rate and frequency of Ridomil plus applications on late blight severity and potato yields in Cameroon. Afr Crop Sci J 9: 235-243.

5

Genetic and Phenotypic characterization of *Phytophthora colocasiae* in Taro Growing Areas of India

Vishnu Sukumari Nath*, Shyni Basheer, Muthulekshmi Lajapathy Jeeva and Syamala Swayamvaran Veena

Division of Crop Protection, Central Tuber Crops Research Institute, Thiruvananthapuram, Kerala, India

Abstract

Phenotypic and molecular methods were used for characterizing 40 *Phytophthora colocasiae* isolates obtained from Andhra Pradesh, Assam, Kerala, and Odisha regions of India over a period of five years. Phenotypic parameters such as virulence, colony morphology and mating type varied among isolates collected from different regions over the years. No correlation was observed between phenotypic parameters of the isolates and their geographical origins. Considerable inter and intra specific variation were detected by random amplified microsatellites (RAMS) analysis with 100% polymorphism among the isolates. Dendrogram constructed based on RAMS data using the unweighted pair group method with arithmetic mean (UPGMA) grouped the *P. colocasiae* isolates into two major clusters. No relationship was obtained between RAMS groups of the isolates and phenotypic characters/geographical origin. Population genetic analysis showed that *P. colocasiae* isolates were highly diverse among different regions. Analysis of molecular variance (AMOVA) showed that most of the genetic variability in *P. colocasiae* was confined to within a population (93.21%). These results indicate that *P. colocasiae* populations in India are highly diverse and care should be taken in developing disease management programmes or in breeding resistant cultivars.

Keywords: Disease management; Molecular markers; Pathogen characterization; Population structure; Taro leaf blight

Introduction

Taro (*Colocasia esculenta* (L.) Schott) which belongs to Araceae family is a major root crop with wide distribution in the tropics. Taro leaf blight caused by *Phytophthora colocasiae* is a major bottleneck for taro production worldwide, including India causing yield loss of up to 50% [1-4]. *Phytophthora colocasiae* can infect at any stage of the plant resulting in extensive damage of the foliage. Initial symptoms appear as small, water-soaked circular spots on the edges of the leaves. As the disease progresses, these spots enlarge, coalesce, and become dark brown in color with yellow margins and finally the entire leaf is destroyed. Epidemics are common in temperatures close to 20°C to 25°C with a relative humidity of 90% to 100% [2,4,5]. The disease is more severe in northern and eastern parts of the country which are major areas of taro production. While, in South India, the disease appears periodically in serious proportions [3].

At present, metalaxyl based fungicides are used to manage taro leaf blight. However, the presence of waxy coating on leaf lamina and the occurrence of disease during rainy season make this approach ineffective leading to rapid epidemics and crop loss. Besides this development of resistance to fungicide is another major concern. Build up of resistance to metalaxyl has already been demonstrated in field isolates of *P. colocasiae* [6]. Few cultivars are resistant to leaf blight and resistance breeding offers great potential, but the durability of resistance is largely challenged by the emergence of new virulence strains of pathogen [7].

P. colocasiae is usually diploid and requires the presence of A1 and A2 mating types for production of sexual oospores. The oospore not only serves as the overwintering propagules in the soil and as a source of initial inoculum for subsequent crops, it also contributes to genetic variability through potential new gene combination [8]. In the absence of sexual spores, the pathogen survives as asexual clones in infected plant or in harvested tubers. Mycelium from such infections produces numerous sporangia that are disseminated by wind or rain-splashes to a new host where they germinate directly or release multiple motile

zoospores and thus the cycle continues. The abundant production of sporangia or zoospores, and ability to infect and colonize host tissue combined with the efficient dissemination makes *P. colocasiae* a devastating plant pathogen [4].

Effective management of taro leaf blight is only possible by understanding the characteristics of pathogen population. Molecular markers provide useful tools to track individual genotypes and also to study population diversity. *P. colocasiae* genetic diversity has been assessed using diverse genetic markers. Random amplified polymorphic DNA (RAPD) and Isozyme markers were employed to analyze the genetic diversity in *P. colocasiae* isolates from Southeast Asia and Pacific regions [9] and as well as from India [4]. Recently, amplified fragment length polymorphism (AFLP) and RAPD have proved effective in projecting genetic diversity in *P. colocasiae* isolates [10,11]. However, till date no studies have been reported concerning genetic diversity analysis of *P. colocasiae* using random amplified microsatellite markers (RAMS). RAMS technique was originally described by Zietkiewicz et al. [12] in which the DNA between the distal ends of two closely located microsatellites is amplified and the resulting PCR products are separated electrophoretically. RAMS can be performed easily with sufficient reproducibility and thereby offers advantages over RAPD and laborious techniques like AFLP.

This research was conducted to assess phenotypic and genotypic diversity of the *P. colocasiae* isolates collected from Andhra Pradesh,

*Corresponding author: Vishnu Sukumari Nath, Division of Crop Protection, Central Tuber Crops Research Institute, Thiruvananthapuram 695017, Kerala, India E-mail: vishnu4you007@gmail.com

Assam, Kerala, and Odisha regions of India over a period of five years (2007-2012).

Materials and Methods

Isolates

A total of 40 *Phytophthora colocasiae* isolates were obtained from leaf blight infected samples collected primarily from Andhra Pradesh, Assam, Kerala, and Odisha regions from 2007-2012 which represents the major taro growing regions of India. Isolation and maintenance of the isolates were carried out according to Nath et al. [10]. All isolates were confirmed to the species level using species-specific PCR assay as described earlier [11]. Details of the isolates used in this study are given in Table 1.

Colony morphology

Colony morphology was studied on potato dextrose agar medium (PDA; 250 g/L potato, 20 g/L dextrose and 20 g/L agar). A 5 mm disc excised from the periphery of actively growing colony of *P. colocasiae* was placed at the centre of petri dishes containing PDA. Plates were incubated at 28°C in the dark for two weeks. Following incubation, morphology of *P. colocasiae* was characterized based on the colony texture. Three replicates were used for each isolate to confirm the colony characteristics at same incubation conditions mentioned above.

Virulence assay

All *P. colocasiae* isolates were assessed for their virulence using a floating leaf disc method. Five leaf discs (5 × 5 cm) of taro (cv Sree

Isolate code	Location	District/sampling site	Year of collection	Colony on PDA medium*	Mating type#	Lesion diameter (cm)#
P1	Kerala	Block 2, CTCRI field	2010	Group H	A1	1.26 ± 0.05[c]
P3	Kerala	Block 1, CTCRI field	2010	Group H	A1	2.20 ± 0.10[f]
P21	Kerala	Farm, CTCRI	2011	Group H	A1	3.20 ± 0.10[i]
P42	Kerala	Thiruvananthapuram	2012	Group E	A1	3.30 ± 0.20[i]
P9	Kerala	Thiruvananthapuram	2008	Group E	A1	0.00 ± 0.00[a]
P4	Kerala	Aleppy	2011	Group F	A1	3.86 ± 0.05[j]
P7	Kerala	Pathanamthitta	2011	Group F	A1	3.10 ± 0.10[h]
P35	Kerala	Pathanamthitta	2011	Group F	A1	3.00 ± 0.10[h]
P6	Kerala	Kottayam	2011	Group F	A1	3.86 ± 0.05[j]
P23	Kerala	Kollam	2010	Group E	A1	1.46 ± 0.05[d]
P15	Kerala	Haripad	2012	Group F	A1	4.53 ± 0.05[k]
P28	Kerala	Idukki	2010	Group I	A1	1.26 ± 0.05[c]
P11	Kerala	Calicut	2007	Group B	A2	0.00 ± 0.00[a]
P32	Kerala	Block 2, CTCRI	2012	Group D	A1	3.30 ± 0.20[i]
P33	Kerala	Block 2, CTCRI	2012	Group D	A1	3.28 ± 0.10[i]
P22	Kerala	Calicut	2008	Group D	A1	0.00 ± 0.00[a]
P16	Andhra Pradesh	Veerwada	2010	Group D	A1	1.26 ± 0.05[c]
P34	Andhra Pradesh	Veerwada	2010	Group D	A2	1.24 ± 0.05[c]
P26	Andhra Pradesh	East Godavari	2011	Group D	A1	2.60 ± 0.10[g]
P29	Andhra Pradesh	Parudin pallam	2010	Group D	A1	1.76 ± 0.05[e]
P36	Andhra Pradesh	Parudin pallam	2011	Group D	A1	2.50 ± 0.10[g]
P27	Andhra Pradesh	Veerwada	2011	Group D	A1	2.53 ± 0.05[g]
P5	Odisha	Nayagarh	2007	Group A	A1	0.00 ± 0.00[a]
P12	Odisha	Khandapara	2007	Group A	A1	0.00 ± 0.00[a]
P2	Odisha	RC, CTCRI	2008	Group C	A1	0.00 ± 0.00[a]
P13	Odisha	Salepur	2008	Group A	A1	0.83 ± 0.05[b]
P24	Odisha	Puri	2007	Group A	A1	0.00 ± 0.00[a]
P25	Odisha	Puri	2007	Group A	A1	0.00 ± 0.00[a]
P14	Uttar Pradesh	Malikpur	2007	Group A	A1	0.00 ± 0.00[a]
P39	Uttar Pradesh	Malikpur	2008	Group A	A1	0.80 ± 0.05[b]
P43	Uttar Pradesh	Malikpur	2009	Group A	A1	1.24 ± 0.05[c]
P17	Delhi	New Delhi	2010	Group G	A1	0.83 ± 0.05[b]
P19	Assam	Nellie Road	2007	Group B	A1	0.00 ± 0.00[a]
P46	Assam	Nellie Road	2007	Group C	A2	0.00 ± 0.00[a]
P30	Assam	Nellie Road	2010	Group B	A2	1.26 ± 0.05[c]
P8	Meghalaya	Ribhoi	2009	Group C	A1	0.83 ± 0.05[b]
P20	Meghalaya	Nongpoh	2010	Group G	A1	1.46 ± 0.05[d]
P10	West Bengal	Nadia	2009	Group A	A1	1.76 ± 0.05[e]
P18	Tripura	West Tripura	2010	Group A	A1	2.20 ± 0.10[f]
P38	Tripura	West Tripura	2010	Group A	A1	2.00 ± 0.00[f]

some of the results have been previously published [10,11]. *Group A: Cottony; Group B: Stellate; Group C: Cottony with concentric rings; Group D: Plain with irregular concentric rings; Group E: Irregular pattern; Group F: Plain; Group G: Uniform with concentric rings; Group H: Uniform without pattern; Group I: Flat with concentric rings

Table 1: Characteristics of *Phytophthora colocasiae* isolates used in this study.

Kiran, leaf blight susceptible) were floated in sterile distilled water in 200 mm glass petri plates and inoculated with a mycelial disc excised from the margins of actively growing cultures of *P. colocasiae*. Leaf discs with sterile agar plugs served as control experiment. Following inoculation, the leaf discs were incubated at 25°C in dark and daily examined for disease symptoms. The lesion diameter was recorded 4 days after inoculation (d.a.i.). Isolations were established from resulting lesions in order to confirm the association of the pathogen with the observed symptom. There were five replicates for each isolate and the assay was repeated twice.

Mating type determination

The mating type of isolates was determined by paring each unknown isolate with the isolate of a known A1 (98-111) and A2 (98-35a) mating type on carrot agar (CA) medium at 5 cm apart. After incubation at 28°C in darkness for 4 weeks, agar blocks were examined microscopically. The presence of oospores at the interface between colonies indicated opposite mating type while absence of oospores indicated the same mating type. The single culture of each isolate was paired to examine for oospore formation as a control. The positive control was a cross between two tester isolates of opposite mating types. Three replicates were used for each isolate.

DNA isolation

For DNA isolation, *P. colocasiae* isolates were grown in potato dextrose broth medium (PDB; 250 g l^{-1} potato, 20 g l^{-1} dextrose) at 28°C with 50 rpm. After achieving sufficient growth, DNA was extracted from mycelium using a Genomic DNA purification kit (Fermentas, EU) according to manufacturer's protocol. The nucleic acid obtained was dissolved in TE buffer (100 µl; pH=8.0). The quality and integrity of DNA were assessed by agarose gel electrophoresis and stored at -20°C until further use.

RAMS analysis

The list of primers used for the study is presented in Table 2. To optimize the method, a preliminary RAMS assay was performed using different reagent concentrations and PCR reaction conditions on a random sample of 5 isolates representing different geographical origins (data not shown). Each 25 µl of PCR reaction consisted of 50 ng of template DNA, 100 µM each dNTPs, 20 ng of primer (Integrated DNA Technologies, Coralville, USA), 1.5 mM MgCl$_2$, 2.5 µl Taq buffer (10 mM Tris-HCl pH 9.0, 50 mM KCl, 0.01% gelatin), 1 U of Taq DNA polymerase (Merck Genei, India). The samples were denatured by 10 min incubation at 95°C after which 35 (CGA- and GT-primers) or 37 (CCA-primer) cycles of amplification were carried out (30 s denaturation at 95°C, 45 s annealing at a temperature depending on the primer, 2 min primer extension at 72°C). The annealing temperatures for the primers were as follows: CCA-primer 55°C, CGA-primer 59°C, and GT-primer 46.6°C. After the cycles, the reaction was ended with a 7 min extension at 72°C. Amplified products were separated on a 1.8% agarose gel containing 0.5 µg ml^{-1} ethidium bromide and visualized under UV light. Gel photographs were acquired using a Gel

*Primer code	Sequence#	Total no. of bands	No. of polymorphic bands	Mean no. of bands
CGA	5'DHB(CGA)$_5$	11	11	1.60
GT	5'YHY(GT)$_7$G	10	10	3.0
CCA	5'DDB(CCA)$_5$	10	10	0.97

*Source Hantula et al. [28]. #The following designations are used for degenerate sites: B (G, T, or C), D (G, A, or T), H (A, T, or C), and Y (A, C or G).

Table 2: Summary statistics of RAMS analysis of *Phytophthora colocasiae*.

Doc System (Alpha Imager, Alpha Innotech, California, USA). The size of the amplification products was estimated by comparison with 1 kb plus DNA ladder (Fermentas). The assay was repeated two times with template DNA from two different DNA extractions to ensure the consistency of each band.

Data analysis

All clearly detectable and distinct RAMS bands were scored for their presence (1) or absence (0) by visual inspection. Based on the scores, a genetic similarity matrix was constructed and a dendrogram was deduced to display relationships between isolates using the Nei and Li distance [13] according to the unweighted pair group mean algorithm using the TREECON software package version 1.3 [14]. The reliability of the clustering was assessed by bootstrap analysis (2000 replicates). A cophenetic correlation coefficient was calculated to assess the statistical support for the dendrogram obtained, and Mantel's test [15] was performed to check the goodness-of-fit of the cluster analysis (1000 permutations). The data within a cluster are most likely to be highly reliable when the value of cophenetic correlation coefficient was ≥0.8 [16].

The similarity matrix generated was also used for the analysis of molecular variance (AMOVA) [17] by using FAMD Software version 1.25 [18]. This analysis helps to understand how the genetic diversity is partitioned among populations of *P. colocasiae* (called Phi statistics).

Population genetic parameters such as percentage of polymorphic loci (P), observed number of alleles (N_A), effective number of alleles (N_E), Nei's gene diversity (H), and Shannon index (I) of *P. colocasiae* isolates based on RAMS markers were analyzed using the computer program POPGENE 32 [19]. Loci were considered polymorphic if more than one allele was detected.

Results

Isolation of pathogen

A total of 40 isolates of *P. colocasiae* were isolated from leaf blight infected samples collected from Andhra Pradesh, Assam, Kerala, and Odisha regions of India (Table 1). All isolates were confirmed as *P. colocasiae* using species specific PCR assay and produced 206 bp amplicon when amplified using PCSP RL-F and PCSP RL-R primers [11].

Colony morphology

The isolates exhibited diverse colony morphology when grown on PDA medium (Table 1 and Figure 1). There was no general trend for the morphology exhibited by the isolates from different geographical regions. The isolates collected from the same location but different sampling sites had different morphology. Also, isolates collected from same sampling site in consecutive years showed difference in their morphology (e.g. Block 2, CTCRI field). Based on the colony morphology the isolates were classified into different morphology groups as described earlier [10].

Virulence assay

The majority of the isolates was able to infect leaf disc and produced typical symptoms of leaf blight. A few isolates that were collected before 2008 failed to initiate disease symptoms. The isolates initiated lesion development 2 d.a.i. Lesions appeared yellow in the beginning, which turned dark brown with progression of the disease. There was a significant difference in the lesion diameter among the isolates collected from different regions with recently obtained isolate (P15) from kerala state being more aggressive (4.53 ± 0.05; P ≤ 0.5). The *P.*

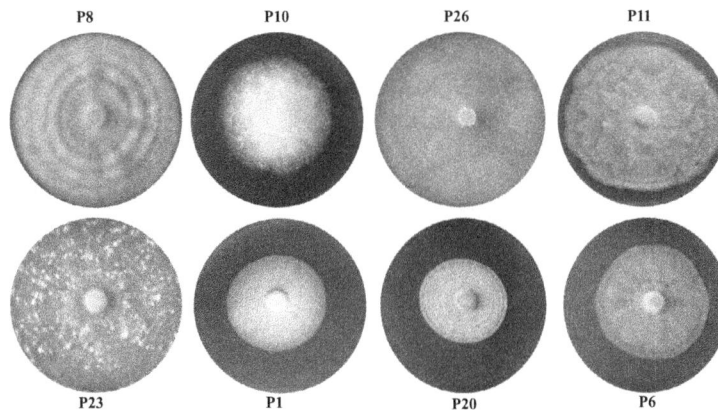

Figure 1: Colony morphology exhibited by *P. colocasiae* isolates collected from different regions of India on PDA medium.

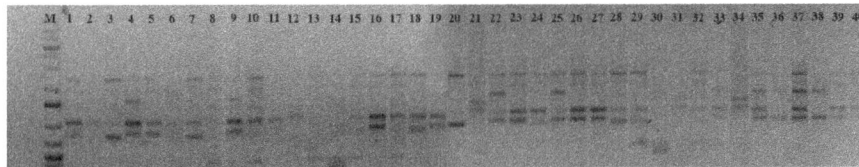

Figure 2: Example of RAMS amplification of *P. colocasiae* isolates by GT primer.

colocasiae isolates collected earlier (2009-2010) were found to be less virulent (0.83 ± 0.05 to 2.20 ± 0.10; P ≤ 0.5) (Table 1). No lesions were produced on the control leaf discs. The pathogen was successfully re-isolated from the lesions completing Koch's postulate.

Mating type determination

All the isolates were heterothallic and produced oospores when paired with known A1 or A2 mating types on carrot agar. Out of the 40 isolates analyzed, 34 (90%) isolates were of A1 mating type while the remaining 4 (10%) isolates were of A2 mating type (Table 1). Isolates collected from different regions showed A1 and A2 mating types. No isolate produced homothallic oospores (A0, or A1/A2).

RAMS analysis

Forty isolates of *P. colocasiae* from various geographical origins of India were analyzed to estimate the level of genetic diversity by using RAMS markers. The bands were distinct, reproducible and easy to score. To ensure reliability in scoring, all markers were scored at least twice. The three primers amplified 31 reproducible fragments ranging in size from 500 to 2000 bp, of which 31 (100%) were polymorphic. The highest number of amplification products (11) was obtained with the CGA primer, while GT and CCA primer had 10 bands each. A summary of the RAMS data is presented in Table 2. Examples of RAMS DNA fingerprints are shown in Figure 2.

The UPGMA dendrogram grouped the isolates into two major clusters (Figure 3) with high bootstrap values. Cluster I had 45 isolates and formed the major group, while cluster II had only 5 isolates. The clustering of isolates in the dendrogram was not correlated with geographical origin or phenotypic characters. The cophenetic correlation coefficient between dendrogram and the original similarity matrix was significant for RAMS marker (r=0.898).

Analysis of genetic diversity

Genetic parameters varied among populations with the percentage of polymorphic bands (*PPB*) values ranging from 45.45% (Kerala) to 72.73% (Andhra Pradesh), with an average of 61.365%. The average Nei's gene diversity (*H*) was estimated to be 0.10 within populations and 0.11 for the pooled populations. The observed number of alleles (*NA*) and the effective number of alleles (*NE*) varied among populations (Table 3).

Analysis of molecular variance (AMOVA) revealed that a high percentage (93.62%) of the *P. colocasiae* genetic diversity in this study was distributed within populations and only 6.37% among populations (Table 4). The coefficient of genetic differentiation among populations (*GST*) was 0.049, which supports the AMOVA analysis indicating only limited genetic diversity among populations and high diversity within populations. The estimate of gene flow (*Nm*) among populations was 9.68 migrants per generation, obtained from the GST value.

Discussion

Taro leaf blight caused by *P. colocasiae* leads to significant economic loss in taro cultivation globally. A better understanding of *P. colocasiae* population dynamics will contribute to more durable disease management strategies. Here, we analyzed *P. colocasiae* isolates collected over a period of five years (2007-2012), to understand the overall population structure with respect to phenotypic and genotypic diversity. Our results indicate that considerable intra specific diversity exists among *P. colocasiae* isolates used in this study.

Analysis of colony morphology on PDA revealed that *P. colocasiae* isolates have highly diverse morphology with isolates collected from different regions showing different morphologies. Our results were consistent with a previous study by Misra et al. [20], however the present study revealed more morphological groups, probably due to the greater number of isolates used in this study, which provides a better coverage of different geographical regions of India. One of the striking results of the study was the finding that *P. colocasiae* isolates collected from the same sampling site in different years had different

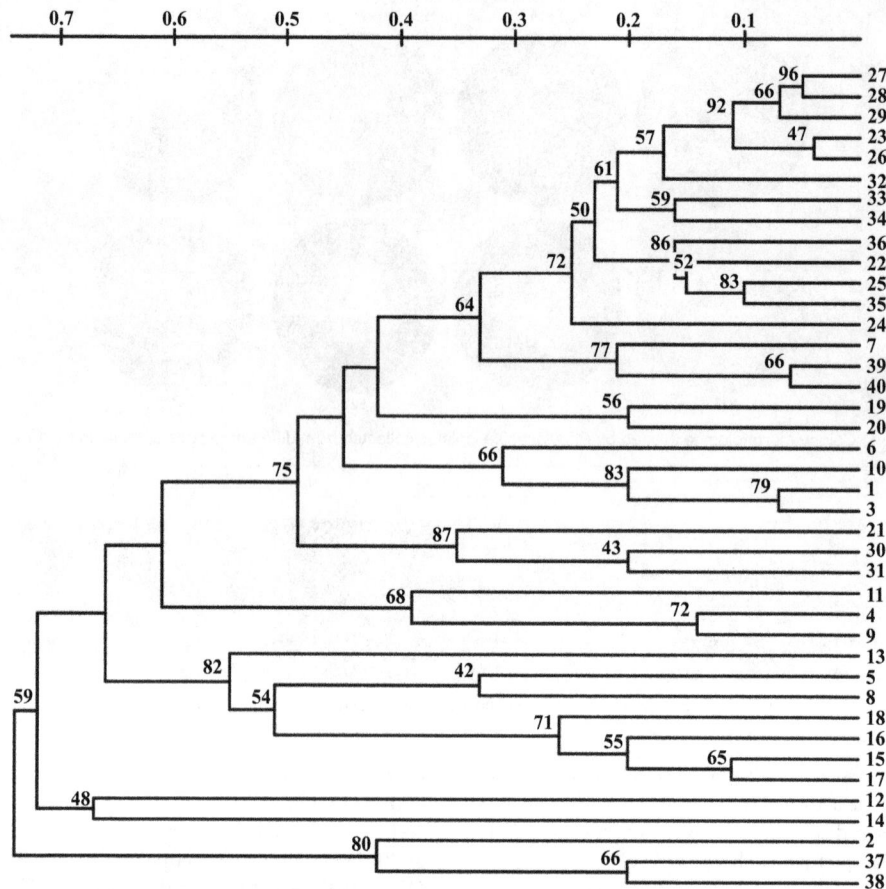

Figure 3: UPGMA dendrogram of *Phytophthora colocasiae* isolates based on RAMS analysis. Numbers at nodes indicate bootstrap values (2000 replications).

Marker	Population Code	Polymorphic bands	PPB[a](%)	$N_A{}^b$	$N_E{}^c$	H^d	I^e
	Kerala	15	45.45	1.45 ± 0.50	1.14 ± 0.19	0.10 ± 0.13	0.17 ± 0.21
	Andhra Pradesh	24	72.73	1.72 ± 0.45	1.14 ± 0.14	0.11 ± 0.10	0.21 ± 0.16
RAMS	Assam	19	57.58	1.57 ± 0.50	1.14 ± 0.18	0.10 ± 0.12	0.18 ± 0.19
	Odisha	23	69.70	1.69 ± 0.46	1.12 ± 0.10	0.10 ± 0.08	0.18 ± 0.14
	Total	33	100	2.00 ± 0.00	1.13 ± 0.09	0.11 ± 0.07	0.21 ± 0.11

[b] Observed number of alleles (NA), [c] Effective number of alleles (NE); [d] Nei's gene diversity (H); [e] Shannon's information index (I)

Table 3: Population genetic analysis of *Phytophthora colocasiae* isolates.

Marker	Source	df	SSD	Φ statistics	Variance components	Proportion of variation components (%)
	Among Populations	3	2.25	0.067	0.032	6.78
RAMS	Within populations	36	16.07		0.446	93.21
	Total	40	18.33		0.479	

Table 4: AMOVA analysis of *Phytophthora colocasiae* isolates.

morphology. This observation clearly suggests that *P. colocasiae* populations are constantly evolving in nature.

P. colocasiae is a soil borne pathogen and a heterothallic species, requiring the presence of A1 and A2 mating types for sexual reproduction. In our study, the presence of both mating types was detected, with the A1 type occurring at a higher frequency than the A2 type. However, the A1 and A2 type were not found in close proximity (e.g. in a single field), which facilitates sexual reproduction under favourable conditions. Similar results were reported by Misra et al. [20], who also observed the lack of compatible mating types (A1 and A2) in India. Recently, a study by Tyson and Fullerton [21] found only

one mating type of *P. colocasiae* (A2) throughout the Pacific region, including Guam, Hawaii, Indonesia, the Philippines, Papua New Guinea, and Samoa. Therefore, from the results it can be commented that sexual recombination may not be playing a major role in the high diversity seen among the Indian *P. colocasiae* isolates.

The results of the virulence tests showed a significant difference in the mean lesion diameter of all the isolates studied. Majority of the isolates was highly virulent and were able to cause serious infection on taro leaf discs irrespective of the geographical origin. The variation in aggressiveness of the isolates projects their high genetic diversity. It is also possible that the variation could be the outcome of the differences in resistance levels of various cultivars from which these isolates were obtained. Difference in virulence property as observed in this study has also been reported in other *Phytophthora* spp. [22,23] and in other plant pathogens [24,25]. No correlation could be elucidated from the phenotypic or genotypic characteristics and geographical origin of the isolates.

This is the first time that RAMS markers have been used for assessing the genetic diversity of Indian *P. colocasiae* isolates. In our study, these markers revealed an intraspecific diversity of *P. colocasiae* isolates from different regions. Profound genetic diversity was evident with 100% polymorphism among the isolates. UPGMA dendrogram grouped the isolates into two clusters with high bootstrap values. Nei's gene diversity examination revealed that the *P. colocasiae* were highly different within a close population, which was further confirmed by AMOVA analysis. The isolates were not grouped according to their geographical origin or phenotypic characters, which was in agreement with previous reports on genetic diversity analysis in *P. colocasiae* [4,7,9,10]. This observation reinforces the fact that *P. colocasiae* frequently move within the country, contributing to the increased genetic diversity of the pathogen.

Several reasons could be attributed to the high intrazonal diversity detected in the present investigation. It is an accepted fact that sexual recombination increases genotype diversity in populations by creating novel recombinants. Mechanisms such as mutation, translocations, chromosomal deletions and duplications are common in *Phytophthora* species [26], which may also be responsible for the genetic variation observed in the *P. colocasiae* populations. According to Goodwin [26], mutation is regarded as the primary source of genetic variation in oomycetes. These mutations in most cases can be neutral and may not cause any observable changes in phenotype, but it is possible that at least a part of the genotype variation might have been the result of spontaneous mutation [27,28]. In *Phytophthora sojae*, mitotic gene conversion was observed to occur at remarkably high frequencies leading to rapid generation of variation.

Conclusion

This study represents a comparative analysis of phenotypic and genotypic diversities present among isolates of *P. colocasiae* from India. Although the number of isolates used is limited, the study nevertheless projects the extend of genetic diversity among *P. colocasiae* isolates from different regions of India. The presence of high phenotypic and genetic variation within the *P. colocasiae* population in India may prove detrimental to the development of sustainable management strategies to curb the disease. More precise conclusions regarding gene flow and genetic recombination would require comprehensive studies on the *P. colocasiae* population with more representative isolates using codominant DNA markers.

Acknowledgements

The authors thank the Indian Council of Agricultural Research for the funding under the IISR Outreach programme.

References

1. Jackson GVH, Gollifer DE, Newhook FJ (1980) Studies on the taro leaf blight fungus *Phytophthora colocasiae* in the Solomon Islands: control by fungicides and spacing. Ann Applied Biol 96: 1-10.

2. Thankappan M (1985) Leaf blight of taro-a review. J Root Crops 11: 1-8.

3. Misra RS, Chowdhury SR (1997) Phytophthora Leaf Blight Disease of Taro, CTCRI Technical Bulletin Series 21, Central Tuber Crops Research Institute, and St Joseph Press, Trivandrum p: 32.

4. Mishra AK, Sharma K, Misra RS (2010) Isozyme and PCR-based genotyping of epidemic *Phytophthora colocasiae* associated with taro leaf blight. Arch Phytopathol Plant Protect 43: 1367-1380.

5. Trujillo EE (1965) The effects of humidity and temperature on Phytophthora blight of taro 55: 183-188.

6. Nath VS, Senthil M, Hegde VM, Jeeva ML, Misra RS, et al. (2012) Evaluation of fungicides on Indian isolates of Phytophthora colocasiae causing leaf blight of taro. Arch Phytopathol Plant Protect. 46(5): 548-555

7. Nath VS, Senthil M, Hegde VM, Jeeva ML, Misra RS, et al. (2013) Molecular evidence supports hypervariability in *Phytophthora colocasiae* associated with leaf blight of taro. Eur J Plant Pathol 136: 483.

8. McDonald BA, Linde CC (2002) Pathogen population genetics, evolutionary potential and durable resistance. Ann Review Phytopathol 40: 349-379.

9. Lebot V, Herail C, Gunua T, Pardales J, Prana M, et al. (2003) Isozyme and RAPD variation among *Phytophthora colocasiae* isolates from South East Asia and the Pacific. Plant Pathol 52: 303-313.

10. Nath VS, Senthil M, Hegde VM, Jeeva ML, Misra RS, et al. (2014a) Analysis of genetic diversity in *Phytophthora colocasiae* causing leaf blight of taro (*Colocasia esculenta*) using AFLP and RAPD markers. Ann Microbiol 64: 185-197.

11. Nath VS, Hegde VM, Jeeva ML, Misra RS, Veena SS, et al. (2014b) Rapid and sensitive detection of *Phytophthora colocasiae* responsible for the taro leaf blight using conventional and real-time PCR assay. FEMS microbiology letters 352: 174-183.

12. Zietkiewicz E, Rafalski A, Labuda D (1994) Genome fingerprinting by simple sequence repeat (SSR)-anchored polymerase chain reaction amplication. Genomics 20: 176-183.

13. Nei M, Li WH (1979) Mathematical model for studying genetic variation in terms of restriction endonucleases. Proc Natl Acad Sci U S A 76: 5269-5273.

14. Vandepeer Y, Dewachter R (1994) Treecon for Windows-A software package for the construction and drawing of evolutionary trees for the Microsoft Windows environment. Computation and Applied Bioscience 10: 569-570.

15. Mantel N (1967) The detection of disease clustering and a generalized regression approach. Cancer Res 27: 209-220.

16. Rohlf FJ (1993) Relative warp analysis and an example of its application to mosquito wings. In: Marcus LF, Bello E, Garcia-Valdecasas A (Eds.) Contributions to morphometrics, vol 8. Museo Nacional de Ciencias Naturales Madrid pp: 131-159.

17. Excoffier L, Smouse PE, Quattro JM (1992) Analysis of molecular variance inferred from metric distances among DNA haplotypes: application to human mitochondrial DNA restriction data. Genetics 131: 479-491.

18. Schluter PM, Harris SA (2006) Analysis of multilocus fingerprinting data sets containing missing data. Mol Ecol Notes 6: 569-572.

19. Yeh FC, Yang R (1999) Microsoft window-based freeware for population genetic analysis (POPGENE version 1.31). University of Alberta, Canada.

20. Misra RS, Mishra AK, Sharma K, Jeeva ML, Hegde V (2011) Characterisation of *Phytophthora colocasiae* isolates associated with leaf blight of taro in India. Arch Phytopathol Plant Protect 44: 581-591.

21. Tyson JL, Fullerton RA (2007) Mating types of *Phytophthora colocasiae* from the Pacific region, India and South-east Asia. Australas Plant Dis Notes 2: 111-112.

22. Costamilan LM, Clebsch CC, Soares RM, Seixas CDS, Godoy CV, et al. (2012) Pathogenic diversity of *Phytophthora sojae* pathotypes from Brazil. Eur J Plant Pathol 135: 845.

23. Granke LL, Quesada-Ocampo LM, Hausbeck MK (2011) Variation in phenotypic characteristics of *Phytophthora capsici* isolates from a worldwide collection. Plant Disease 95: 1080-1088.

24. Baskarathevan J, Jaspers MV, Jones EE, Cruickshank RH, Ridgway HJ (2012) Genetic and pathogenic diversity of *Neofusicoccum parvum* in New Zealand vineyards. Fungal Biol 116: 276-288.

25. Mahto NB, Gurung S, Nepal A, Adhikari TB (2012) Morphological, pathological and genetic variations among isolates of *Cochliobolus sativus* from Nepal. Eur J Plant Pathol 133: 405-417.

26. Goodwin SB (1997) The population genetics of phytophthora. Phytopathology 87: 462-473.

27. Silvar C, Merino F, Diaz J (2006) Diversity of *Phytophthora capsici* in Northwest Spain: analysis of virulence, metalaxyl response, and molecular characterization. Plant Disease 90: 1135-1142.

28. Hantula J, Dusabenyagasani M, Hamelin RC (1996) Random amplified microsatellites (RAMS) - a novel method for characterizing genetic variation within fungi. Eur J Forest Pathol 26:159-166.

6

Biological Control of Potato Brown Leaf Spot Disease Caused by *Alternaria alternata* Using *Brevibacillus formosus* Strain DSM 9885 and *Brevibacillus brevis* Strain NBRC 15304

Ahmed IS Ahmed*

Plant Pathology Unit, Department of Plant Protection, Desert Research Center, Cairo, Egypt

Abstract

Brown leaf spot is one of the prevalent diseases caused by *Alternaria alternata* in different growing areas of potato worldwide. Eight *A. alternata* isolates were screened from forty-two isolates collected from different potato growing regions in four Egyptian governorates *viz*, North Sinai (Baloza), Beheira (El-Nubaria and Wadi El-Natrun), Ismailia (Abu Suweir, Fayed and Tell El-Kebir), Sharqia (New Salheya and El-Husseiniya). The virulence of the isolates was tested based on the Per cent of Disease Index (PDI) which ranged from 28.2% to 70.3% PDI by *Alternaria* isolates of Baloza and Fayed respectively. Two bacterial strains "*Brevibacillus formosus* strain DSM 9885, and *Brevibacillus brevis* strain NBRC 15304 were selected to control of *A. alternata*. The bacterial strains have a higher inhibitory effect on mycelial development and spore germination of *A. alternata*. To determine the effects of the bacterial strains on disease index and severity, the most virulent of *A. alternata* isolates were selected for greenhouse experiments where the potato plants were sprayed with bacterial strains individually and mixture treatments. Superior effect of treatments in disease reduction was observed when the two bacterial strains were combined. The effect of leaf age was studied where the leaf position has significant effect on disease progress. The changes of soluble protein in potato leaves due to *Brevibacillus* strains application were studied. Protein profiling by sodium dodecyl sulfate polyacrylamide gel electrophoresis (SDS-PAGE) revealed that the plant treated with mixture of biocontrol agents able to synthesize some new proteins with maximum number of bands followed by treatment by *B. formosus* strain. The presence or absence of the bands in protein profiling might be responsible for resistance response against *A. alternata* in potato. The present work suggests that use of *B. formosus* strain DSM 9885, and *B. brevis* strain NBRC 15304 could be considered as potential management tools for reducing the impact of *A. alternata* causing brown leaf spot disease on potato.

Keywords: Biocontrol agents; Spore germination; Antagonism; Biochemical; Protein profiling

Introduction

Potato (*Solanum tuberosum* L.) is one of the major agricultural crops worldwide and plays an important role in Egyptian agriculture. In 2014 approximately 381.7 million tons of potatoes were produced worldwide. In Egypt, potato has an important position among all vegetable crops where potato production in 2014 exceeded 4.6 million tons, produced on approximately 172.000 ha [1]. Potato crop is vulnerable to infect by several pathogenic fungi. Along with the devastating potato diseases brown spot is caused by *A. alternata* (Fries) Keissler. It is distributed over a wide range of climatic conditions so it can be found in many potatoes growing regions of the world [2]. *Alternaria* diseases on potato cause yield losses and reduce the quality of the crop and very difficult to control [3]. *A. alternata* is one of the prevalent pathogens causing potato brown leaf spot in different parts worldwide [4]. *A. alternata* mainly affects the potato leaves and leads to brown leaf spots. This disease causes a risk to crop production and significant yield losses especially in case of severe infection where losses result from reduced photosynthetic area, loss of weakened leaves plant and increases its susceptibility to infection, subsequently increases the imbalance between nutrient demand in the tubers and nutrient supply from the leaves, subsequently leading to reduced yields [5,6]. The quality and quantity of potato yield may be reduced by infections caused by pathogens that attack both the aboveground parts of potato plants [3,7,8]. Crop losses due to *A. alternata* are around 20 percent; but there have been cases of 70% to 80% losses in case of severe infections or when the disease is combined with other disease such as early blight [9,10]. Also, there is no major resistance gene for *A. alternata* is known. Genetic sources for partial resistance have been determined within some potato wild species [11,12]. So, this disease is one of the destructive diseases in most potato growing areas [13]. To suppress *Alternaria* spot disease causal agents and to prevent the losses it causes, potato fields are intensively sprayed with fungicides [14,15]. Fungicides of various chemical groups are currently used worldwide to control *Alternaria* spp. on potato. Optimization of fungicide use for the control of *Alternaria* diseases is still a considerable challenge due to the capacity of pathogen to produce huge amounts of inoculum [15], so there is a challenge of selecting fungicide resistance in target populations of *Alternaria* spp. [16]. The high efficiency of these chemical pesticides can result in environmental contamination and the effect of pesticide residues on food, in addition to social and economic impacts. Several investigations have been carried out to improve *Alternaria* disease management and to reduce the number of sprays [17]. Eco-friendly methods using biocontrol agents and induce resistance agents to suppress plant disease provides a useful alternative tool to use these evaluated agents with similar targets [3]. Biological control and use of antagonistic microorganisms such as bacteria has considered as a promising alternative strategy. Indeed, these bio-pesticides provide many advantages in term of ecofriendly disease control methods. Antagonistic bacterial isolates are widely

Corresponding author: Ahmed ISAhmed, Plant Pathology Unit, Department of Plant Protection, Desert Research Center, Cairo, Egypt
E-mail: ahmed_drc@yahoo.com

used for the biocontrol of fungal plant diseases [18]. Rhizo-Bacteria are one of the important groups of biological control agents which have revolutionized the field of biological control of several plant pathogens [19-22]. They play important role in induced systemic resistance due to physical and mechanical strength of cell wall and effects on biochemical and physiological reactions of the host plant through synthesis of chemical defense against fungal pathogens [23-26]. The genus Bacillus is distributed widely in environment and includes thermophilic, alkalophilic, and halophilic bacteria that utilize several sources of carbon for heterotrophic and autotrophic growth. Bacillus is one of the most common genera of gram-positive bacteria, isolated from several environmental habitats [27]. *B. formosus* and *B. brevis* are important species according to gene sequence study by Shida et al. [28] where some strains of them were studied for their activities as biocontrol agents against several plant pathogens due to their antibacterial and antifungal effects [29,30]. According to genome sequencing studies for taxonomy of genes and phylogenomics of Bacillus-like bacteria the *B. formosus* DSM 9885 was deposited in seven culture collections [31]. Some of these strains have biocontrol potential against different phytopathogenic fungi and can produce a hyperthermostable chitinase [32]. Also, several strains of *B. brevis* were studied as biocontrol agents for controlling a wide range of plant pathogens [33], different strains also evaluated and encouraged as potential plant growth for enhancing the growth and crop productivity [30,34,35]. Several studies recently reported that the use of *Brevibacillus* strains as biocontrol agents could reduce amounts of chemical fungicides applied for control of phytopathogenic fungi [20,31,33,34].

The aim of this work was to study the efficacy of two bacterial strains *B. formosus* "strain DSM 9885, and *B. brevis* "strain NBRC 15304" as control agents of *A. alternata* to reduce fungicide applications in brown leave spot diseases management. The changes of soluble protein in potato leaves due to Brevibacillus strains application were studied through protein profiling by sodium dodecyl sulfate polyacrylamide gel electrophoresis (SDS-PAGE).

Materials and Methods

Survey, collection and identification of the pathogens

Potato leaves showing typical brown spot symptoms were collected from different growing areas in four Egyptian Governorates viz, North Sinai (Baloza), Beheira (El-Nubaria; Wadi El Natrun); Ismailia (Abu Suweir, Fayed and Tell El Kebir); Sharqia (New Salheya; El Husseiniya) during 2015-2016 to identify the variability of the pathogen and their ability to control by biological agents. The most aggressive isolates were selected to other *in vitro* and greenhouse experiments. The pathogens were identified based on their cultural and morphological characters.

Fungal cultures and inoculation

A single spore isolate of *A. alternata* which caused symptoms in potato leaf tissue was used. Conidia were maintained on filter paper at 4°C. Pure cultures of the isolate were produced by placing a small section of filter paper containing conidia on Potato Dextrose Agar (PDA). For inoculums production, cultures of *A. alternata* were cultured on Potato Dextrose Agar (PDA) at 18°C for 14 days. Conidia were collected by flooding the surface of the Petri dish with 5 ml sterile distilled water, and gently scraping the surface of the media with an L-shaped glass rod to collect the conidia. Then the conidial suspension was stirred with a magnetic stirrer for 1 h and strained through cheesecloth to exclude the mycelial fragments. The concentration was then adjusted to 1×10^5 conidia/ml using a hemocytometer, the fungal purification and inoculum were prepared according to Soleimani and Kirk [9].

Pathogenicity test and virulence of *Alternaria alternata* isolates

The pathogenicity of purified *A. alternata* isolates was tested and proved by Koch's Postulates. Potatoes were planted in each pot with three replicates under greenhouse conditions. The conidial suspension from two weeks old culture of *A. alternata* was used. The conidial concentration adjusted to (5×10^5 spores ml^{-1}) using haemocytometer. Spore suspensions were sprayed on the plants of 30-day-old plants. The plants sprayed with sterile water served as control. Inoculated plants were covered tightly with plastic sheet, after 24 hours the cover was removed and the humidity was maintained by spraying tap water. The plants were grown in a greenhouse for the symptoms appeared and developed. The severe symptoms were observed on 12 to 15 days after inoculation and the disease intensity was recorded. The symptoms were observed and compared with the original symptoms. The fungal isolates were reisolated from artificially inoculated potato leaves and compared with original culture isolates and they were the same. The pathogenicity test was carried out according to Stammler [36]. The disease index was calculated using nine grade scale from 0-9 where, where 0=no spots and 9=brown spots visible more than 60% as leaf area spotted. The Per cent Disease Index (PDI) was calculated by using formula of McKinney [37]:

$$PDI = \frac{Over\ all\ of\ numerical\ rating}{Total\ number\ of\ leaves\ observed} \times \frac{100}{Maximum\ disease\ grade}$$

Morphological and molecular identification of biocontrol agents

The bacterial isolates used in this study were kindly provided by Soil Fertility and Microbiology Department, Desert Research Center, which evaluated against different fungal pathogens, and being obtained in previous investigation [20]. Selected isolates used in present study were identified to molecular level using partial 16S rRNA gene sequence technique based on Berg [38]. In Sigma Scientific Services Co., bacterial 16S rRNA gene sequences were amplified by PCR using the eubacterial primer pair 27f (5'-AGA GTT TGA TCC TGG CTC AG-3') and 1492r (5'-TAC GGY TAC CTT GTT ACG ACT T-3') [39]. The PCR product was sequenced with Genetic Analyzer sequencer, Data Collection v3.0, Sequencing Analysis v5.2 (Foster City, USA). Obtained sequences were aligned with reference RNA sequences from National Center for Biotechnology Information (NCBI) data base [20].

Bacterial inoculum preparation

Bacterial strains were maintained in 80% glycerol (v/v) at -80°C as stock cultures. In order to culture process, a loopful of inoculum was streaked on nutrient medium (NA) plates. Each strain was evaluated to control of *A. alternata* isolates. To harvest the metabolites, fresh cells were obtained from stock cultures and grown in nutrient broth medium at room temperature. 100 ml of nutrient broth was inoculated and incubated for 48 h at room temperature in a rotary shaker (80 round/min). The bacterial culture was centrifuged 10000 rpm for 10 min, and the supernatant was discarded. The cell pellets were suspended in sterile 0.85% NaCl then centrifuged again under the same conditions. The supernatant was discarded and washed bacterial cells were re-suspended in sterile distilled water. The concentration of cells in the suspension was spectrophotometrically adjusted to 108 CFU/ml and used for greenhouse pot experiments [40].

Antagonistic effect of bacterial isolates against *A. alternata*

The antagonistic activity of two bacterial strains was investigated against *A. alternata* by dual culture technique [41, 42] using PDA medium. For testing the antagonistic effect of bacterial strains, each of them was streaked in center of sterile petri dish on potato dextrose agar

(PDA). One disc (0.5 cm in diameter) of *A. alternata* was placed on the side of the same petri dish at 10 mm distance. Petri dishes with fungal cultures and free of bacteria were used as control. Each treatment was carried out with four plates per replicate. Periodical observations on the ability of bacterial strains to colonize the pathogen were calculated as percent inhibition of mycelial growth of pathogen by using the following formula [43].

$$\text{Percent Inhibition} \left(\text{PI} \right) = \frac{C-T}{C} \times 100$$

Where, T: Growth of pathogen in dual culture plates and C: control plates

Effects of biocontrol agents on spore germination of *A. alternata*

Inhibition of spore germination *in vitro*: The effect of bacterial strains on spore germination was studied *in vitro* according to the method of Nair and Ellingboe [44]. For the test, concentration of bacterial suspension (10^8 cells mL^{-1}) was prepared. A drop of bacterial cells was deposited on dried clean glass slides as a film. A drop of the spore suspension of the pathogen was spread over this film. Control treatment was prepared as a film of sterilized distilled water. Percentage of spore germination was determined microscopically using 400 folds magnification [45]. Percentage of germination was obtained using the following formula [46]

$$\text{Percentage of germination} = \frac{\text{Number of germinated spores}}{\text{Total number of spores}} \times 100$$

Inhibition of spore germination on detached leaflets: Effects of a separate and combined application of the tested bacterial strains were studied against spore germination of *A. alternata* on detached potato leaflets. Suspension of each bacterial strain and their mixture were sprayed on potato leaflets by using of an atomizer sprayer. Directly after spraying, drops (20 μl each) containing spore suspension of *A. alternata* were placed on the leaflets. Then the leaflets were put in plastic boxes with humid filter paper and covered to maintain high relative humidity. The boxes were placed in incubators. After one day of incubation, germination of *A. alternata* spores was determined where the leaflet bearing a drop of the interacting microorganisms were placed on glass slides and incubated for 2 h at room temperature (22°C), and examined microscopically, germination was determined in samples of 50 conidia of *A. alternata* for each treatment that were examined in each of 5 drops from three different leaflet replicates [47].

Greenhouse experiment

The experiment was conducted as a randomized complete block design with three replicates (ten plants/replicate) for each treatment. Potato tubers of Spunta cultivar were used. Potato seed tubers were planted in plastic pots (50 cm diameter) containing sandy loam soil. The treatments were added as single and mixture of two bacterial strains (*B. formosus* strain DSM 9885, and *B. brevis* strain NBRC 15304. Through this experiment the disease severity and disease index were calculated on plant under greenhouse conditions. Two days after foliar application of biocontrol agents as well as control treatment by water, the percentage of foliage protection against *A. alternata* was evaluated by using the detached leaf techniques where four leaves per each of the three different leaf positions (top, middle and lower) part were detached from ten plants per treatment and replication then transferred to the laboratory. The detached leaves were artificially inoculated with *A. Alternata* by placing a 50 μl droplet of conidial suspension (1×10^5 conidia/ml) on the center of the leaflet and incubated with humid filter

paper in a growth chamber in darkness at 18°C for one day, then the treated leaflets were incubated continually in a growth chamber at 21°C and fluorescent tube light for 16 h day. Disease symptoms development we observed daily from the third to seventh day after inoculation by visual assessment of the leaf area showing brown leaf spot. Also, Disease incidence and severity were calculated based on percentage of damaged potato leaf area and affected number of plants under greenhouse conditions. Disease severity was recorded by estimating the lesions on a scale from 1 to 7, where: 1=no lesions, 2=a few circles, 3=up to 30%, 4=31% to 40%, 5=41% to 50%, 6=51% to 60%, 7=61% to 100%, (most severe symptoms) of leaf area with brown leaf spot symptoms. Then the following formula was applied:

$$DS = \frac{\sum (n \times c)}{N}$$

Where, DS=disease severity, n=number of infected plants per category, c=category number and N=total number of examined plants.

Protein profiling and gel preparation

Protein of potato leaves extracted from treated plants by pathogen and biocontrol agents, the samples washed several times with distilled water and blotter dried before protein extraction. Amount of 1.0 g of each sample was grinded by mortar using 1:5 leaves: extraction buffer. The suspension was centrifuged at 10000 rpm for 30 min at 4°C. The supernatant was collected and used for profiling of protein [48]. SDS-PAGE was done to get banding pattern of soluble protein. Soluble protein was electrophoresed by 12% SDS polyacrylamide gel, based on the method of Laemmli [49]. Stacking, resolving gel and sample loading were prepared according to Rajik et al. [50].

Statistical analysis

The data obtained was subjected to analysis of variance technique using completely randomized design (CRD) following Gomez and Gomez [51].

Results and Discussion

In this study, laboratory and greenhouse experiments were performed to determine the effects of two bacterial strains *B. formosus* strain DSM 9885, and *B. brevis* strain NBRC 15304 which designated in this paper (*Brf1 and Brb2*) respectively, as potential biocontrol agents against screened eight *A. alternata* isolates designated (*Alt1–Alt8*).

Isolates of *A. alternata* and their virulence

Potato leaves showing typical brown leaf spot symptoms were collected from some potato growing areas in four Egyptian governorates viz., North Sinai, Beheira, Ismailia, Sharqia during 2015-2016. Forty-two isolates were obtained from infected potato plants. The isolates were grown on PDA and screened based on variations in culture morphology then preliminary test of pathogenicity (data not shown). Eight *A. alternata* isolates were selected for experiments of the present study. The tested isolates were isolated from eight different locations Baloza (North Sinai), El-Nubaria; Wadi El Natrun (Beheira); Abu Suweir, Fayed and Tell El Kebir (Ismailia); New Salheya; El Husseiniya (Sharqia) to assess their pathogenicity and their ability to controlled by tested bacterial strains as potential biocontrol agents. Isolates of *A. alternata* varied in pathogenicity on potato (Table 1). The most virulent isolate was *Alt5* (70.3% PDI), followed by *Alt2* (65.5% PDI). While *Alt1* was the least pathogenic. These results are agreed with some previous studies on pathogenicity of *A. alternata* proved

Sr. No.	Isolates	Isolation place	PDI %*
1	Alt1	North Sinai (Baloza)	28.33e
2	Alt2	Beheira (El-Nubaria)	63.66b
3	Alt3	Beheira (Wadi El Natrun)	38.5d
4	Alt4	Ismailia (Abu Suweir)	61.67b
5	Alt5	Ismailia (Fayed)	71.33a
6	Alt6	Ismailia (Tell El Kebir)	30.66de
7	Alt7	Sharqia (New Salheya)	42.67cd
8	Alt8	Sharqia (El Husseiniya)	46c
* Means with the same letter are not significantly different.			

Table 1: Sources and virulence of different *A. alternata* isolates.

Treatments	Fungal Isolate															
	Alt1		Alt2		Alt3		Alt4		Alt5		Alt6		Alt7		Alt8	
	Growth	Reduc-tion %	Growth	Reduc-tion %	Growth	Reduc-tion %	Growth	Reduc-tion %	Growth	Reduc-tion %	Growth	Reduc-tion %	Growth	Reduc-tion %	Growth	Reduc-tion %
Brf1	2.50	70.9	4.60	51.1	3.40	64.2	4.20	54.3	3.60	62.1	2.80	68.2	2.20	76.8	1.93	79.7
Brb2	5.90	31.4	6.20	34	4.57	51.9	8.20	10.9	7.40	22.1	5.50	37.5	4.80	49.5	4.57	51.9
Brf1+Brb2	2.23	74.1	2.50	73.4	2.23	76.5	3.30	64.1	2.93	69.2	2.60	70.5	1.90	80	2.17	77.2
Cont.	8.60	0	9.40	0	9.50	0	9.20	0	9.50	0	8.80	0	9.50	0	9.50	0
$LSD_{0.05}$	0.25	-	0.46	-	0.42	-	0.44	-	0.76	-	0.46	-	0.39	-	0.41	-

Table 2: Effect of *Brevibacillus formosus* strain DSM 9885 (*Brf1*), and *Brevibacillus brevis* strain NBRC 15304 (Brb2) and their mixture (*Brf1+Brb2*) on mycelial growth of *A. alternata* isolates (Alt1-Alt8).

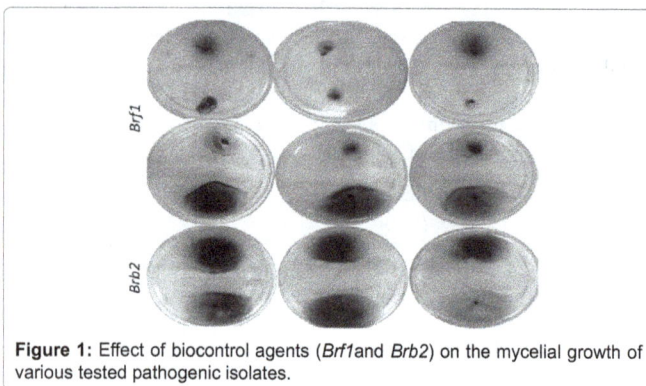

Figure 1: Effect of biocontrol agents (*Brf1* and *Brb2*) on the mycelial growth of various tested pathogenic isolates.

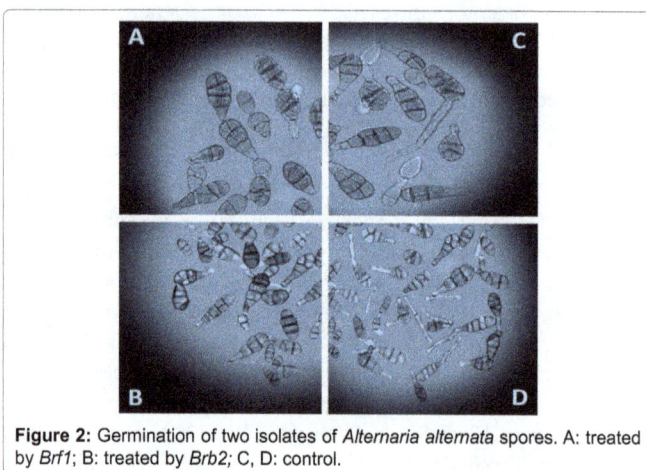

Figure 2: Germination of two isolates of *Alternaria alternata* spores. A: treated by *Brf1*; B: treated by *Brb2*; C, D: control.

under semi controlled condition where the variation of pathogenicity was reported with typical brown spot symptoms when observed on all the inoculated plants of several crops such as apple [52]. In present study, the maximum pathogenicity caused by *Alt2* and *Alt5* may be due to more toxin production as well as their adaptability to favorable environment conditions [53,54].

Effect of biocontrol agents on linear growth of *Alternaria alternata*

The antagonistic effect of tested bacteria against *A. alternata* was studied. *B. formosus* "strain DSM 9885, and *B. brevis* "strain NBRC 15304" were used *in vitro* to evaluated their effects on mycelial growth of *A. alternata* isolates using dual culture technique. PDA medium without adding the bacteria served as control. The mycelial growth reduction of fungi was calculated according to of the inhibition zones as a distance between fungal growth and bacterial colony [42]. The interactions on solid medium revealed the antagonistic effect of the bacterial strains used throughout this study. In dual cultures with evaluated bacterial strains, a more evident inhibitory action (clear zone of mycelial inhibition) was observed. The highest inhibitory effect was recorded in the six or seven days of cultivation with *A. alternata* isolates. The highest inhibition effect was detected with mixture of *Brf1* and *Brb2*, while the *Brb2* as a single treatment presented the least effect against all tested *A. alternata* isolates. The growth reduction rates were ranged from 51% to 79.7% in *Alternaria* isolates treated by *Brf1*, while the fungal growth was reduced by 10% to 51.9% with Brb2 compared to the mixture treatments where the fungal growth was reduced from 64.1% in *Alt4* to 80% in *Alt7* (Table 2). In the case of *Brf1*, the inhibition zone was clearer than the other that appears in the presence of *Brb2*, this aspect being indicated to more stable and a higher inhibitory activity. However, in early stage of the mycelium development, both of tested strains inhibited of mycelium extension and restricted the growth of fungal (Figure 1). It seems that *Brevibacillus* strains excreted metabolites that act as a barrier between the fungi and bacteria, the mycelium development being restricted due to the synthesis of compounds with antifungal activity surrounding colonies, at the same time, the mixture of two isolates can inhibit the fungal growth with higher effect than single treatment by each other, that may be due to the excretion of lytic enzymes or other compounds with fungicidal activity [27,29,55].

Effects of biocontrol agents on spore germination of *A. alternata*

Spore germination is one of most principal factors of survival, dispersal and virulence of pathogenic fungi. Effects of the *Brevibacillus* strains on spore germination of *A. alternata* ware tested through slide test *in vitro* and detached leave test. Most of *A. alternata* spores germinated in control samples were ranged from 64% to 86% germination in slide test, and 52% to 78% on detached leaves (Tables 3 and 4). Regarding to the effect of treatments in slide test, the germination spore rates were 25%, 30%, 33% in *Alt2, Alt4, Alt5* in case of *Brf1* treatment, and 28%, 51%, 40% by *Brb2*, compared to control (84%, 64%, 86%) germination spores of *Alt2, Alt4 and Alt5* (Table 3 and Figures 2A-2D). In detached leave test, the obtained results were in line with slide test. The means of spore germination of *A. alternata* were ranged from 25% to 54.6% in case of treatments by *Brf1*, while the germination rates were found from 34% to 66% resulted by *Brb2*. In combined treatment, the germination spore rates were ranged from 20% to 47% compared to control (52% to 78%). Significant differences were observed between effects of each bacterial isolate individually and mixed on inhibition of spore germination compared to mixture treatment (Table 4). Previous experimental results indicated that *Bacillus sp.* produce antibiotics such as *bacilysin, iturin, mycosubtilin* and siderophores which are responsible for the inhibition of fungal spore germination [56-58].

Assessment of *Brevibacillus* strains on brown leaf spot disease of potato

Disease suppression in the detached leaves which treated in greenhouse: This experiment was designed to test the hypothesis that disease progress is affected by the leaf age or leaf position in the lower, middle and upper parts of the leaf position. In terms of disease suppression, all of the treatments were effective compared to the untreated, inoculated control (Table 5). The results obtained from the detached leaves experiment showed significant difference regarding disease index reduction among biological control agents of *Brevibacillus*. However, it was clear that in both treated potato plants, application of the mixture of *Brf1* and *Brb2* was more encouraging to enhancement of disease resistance compared to the separate treatment. These two *Brevibacillus* strains have indicated higher reduction of disease index in treated detached leaves where High reduction was obtained by *Brf1+Brb2* as treatments combined on upper leaves against pathogenic isolates *Alt2* followed by *Alt5* where the mean of disease index recorded (1.49 and 1.7) respectively. *Brf1* and *Brb2* or their mixture showed significant reduction in disease index on detached leaves. Also, the observed symptoms indicated that leaf position has

Fungal isolate	Treatments	Age	*Mean of disease index
Alt2	Brf1	Upper	2.23 nop
		Middle	3.70 jk
		Lower	6.22 de
	Brb2	Upper	1.82 qr
		Middle	4.20 hi
		Lower	6.33 de
	Brf1+Brb2	Upper	1.49 r
		Middle	2.80 lm
		Lower	4.35 h
	Cont.	Upper	3.50 k
		Middle	5.77 f
		Lower	9.73 a
Alt4	Brf1	Upper	2.53 mno
		Middle	4.30 h
		Lower	6.60 d
	Brb2	Upper	2.40 no
		Middle	4.93 g
		Lower	7.10 c
	Brf1+Brb2	Upper	1.90 pq
		Middle	3.10 l
		Lower	5.20 g
	Cont.	Upper	3.90 ij
		Middle	6.60 d
		Lower	9.87a
Alt5	Brf1	Upper	2.20 op
		Middle	4.00 hij
		Lower	6.00 ef
	Brb2	Upper	2.60 mn
		Middle	4.20 hi
		Lower	6.50 d
	Brf1+Brb2	Upper	1.70 qr
		Middle	3.00 l
		Lower	5.10 g
	Cont.	Upper	3.00 l
		Middle	5.00 g
		Lower	8.80 b

* Means with the same letter are not significantly different.

Table 5: Effect of biocontrol agents on potato brown spot disease index. Pre-treated, detached potato leaves from different positions on plants were artificially inoculated *in vitro* with 50 µl suspensions containing 5×10^5 spore/ml of *A. alternata*.

a significant effect on the lesion growth rate of *A. alternata* on leaves from upper part of the plant. However, disease index was significantly greater on untreated plots. Disease progress was showed on leaves from three parts of the plant. A disease symptom developing was low at the apex, moderate in the middle, and high in the lower part of the plants in both treatments. This is in agreement with assessments by Visker [59], Soleimani and Kirk [9]. They have found that older leaves in the lower part of the plant seemed to be more susceptible to brown leaf spot disease than younger leaves in the upper part. So, they reported that leaf position is a significant factor in potato resistance. This may be due to induce the systemic acquired resistance as a result of treatment with biocontrol agents, and their ability to stimulate the plant resistance in younger leaves faster than the older leaves [9,27,60]. There are several reports on the reliability of the detached leaflet method as a screening technique and its correlation with laboratory and field or greenhouse disease data [59,61,62]. The detached leaflet screening method indicated that this technique provides a reasonable assessment of brown leaf spot resistance, and could be a reliable screening system [9].

Treatments	Fungal isolate							
	Alt1	Alt2	Alt3	Alt4	Alt5	Alt6	Alt7	Alt8
Brf1	20.00	25.00	16.00	30.00	33.00	21.33	22.00	30.00
Brb2	35.00	28.00	20.00	51.00	40.00	33.00	30.00	38.00
Cont.	78.00	84.00	72.00	64.00	86.67	68.00	82.00	76.00
LSD₀.₀₅	6.21	4.76	4.00	7.74	4.80	6.86	4.76	8.63

Table 3: Effect of *Brevibacillus formosus* strain DSM 9885 (*Brf1*), and *Brevibacillus brevis* strain NBRC 15304 (*Brb2*) on spore germination of *A. alternata in vitro*.

Treatments	Fungal isolate							
	Alt1	Alt2	Alt3	Alt4	Alt5	Alt6	Alt7	Alt8
Brf1	43.00	25.33	31.00	53.00	54.67	40.00	38.00	42.00
Brb2	62.67	37.00	34.00	66.00	61.00	48.00	47.00	51.00
Brf1+Brb2	35.00	20.00	24.67	45.00	47.00	33.00	30.00	38.00
Cont.	52.00	55.33	64.00	75.00	78.00	64.00	78.00	61.00
LSD₀.₀₅	8.17	6.50	6.41	5.41	3.69	4.80	4.31	3.39

Table 4: Effect of *Brevibacillus formosus* strain DSM 9885 (*Brf1*), and *Brevibacillus brevis* strain NBRC 15304 (*Brb2*) on spore germination of *A. alternata* on detached leaves.

Assessment of *Brevibacillus* strains effects on brown leaf spot disease of potato under greenhouse conditions

Greenhouse experiment was carried out to evaluate the antifungal activity of two *Brevibacillus* strains against *A. alternata*. According to data obtained from *in vitro* experiments, the most pathogenic *Alternaria* isolates (*Alt2*, *Alt4* and *Alt5*) were selected and tested under greenhouse conditions. Three treatments were compared: *Brevibacillus*1 (*Brf1*), *Brevibacillus* (*Brb2*), mixture of both strains (*Brf1+Brb2*), and water spray as control treatment. The two *Brevibacillus* starins were effective to disease index and severity of brown leaf spot symptoms. This observed reduction was different due to treatments, Superior disease reduction effect was observed when the two bacterial strains were combined. The most effective strain was *Brf1* that reduced the disease incidence by 58.3%, 54.5% and 66%, respectively in case of *Alt2*, *Alt4* and *Alt5* respectively. Superior effect of treatments in disease reduction was observed when they were combined. The highest record of disease reduction in order of 62.5% and 71.7% were obtained for the applied combined treatment against *Alt2* and *Alt5* isolate (Table 6). Similar trend was recorded concerning the severity of brown leaf spot disease. Most of potato plants receiving *Brevibacillus* treatments have significant reduction in disease severity with combined of both two strains application. High reduction as 37% and 50% were obtained from treatments by *Brf1* against *Alt2* and *Alt4* and reached up to 58% when both bacterial strains combined. Individual application *Brf1* and *Brb2* or their mixture showed significant reduction in disease incidence as well as severity. The results also showed that there was more disease development in case of treatment by *Brb2*, than at the *Brf1*. However, it was clear that, application of tested *Brevibacillus* strains as mixture was the most effective of all of the treatments against tested pathogenic isolates for enhancing disease resistance. Similar results were obtained on two different potato cultivars against different fungal pathogens [9,21]. Application of *Brevibacillus* strains has indicated higher resistance of plants against tested *Alternaria* isolates on the potato. However, the greenhouse data has indicated that the potato plants which treated with combined treatment were much healthier than for the other treatments. Previously some studies reported similar effects with the application of Bacillus strains against different pathogenic fungi on sage plants and potato, respectively [20-22]. In present study, the results showed that these two *Brevibacillus* strains are able to decrease *A. alternata* disease infection when given as a foliage spray. The encouraging effect and performance of tested biocontrol agents on disease severity and incidence on potato plants and pathogen growth as

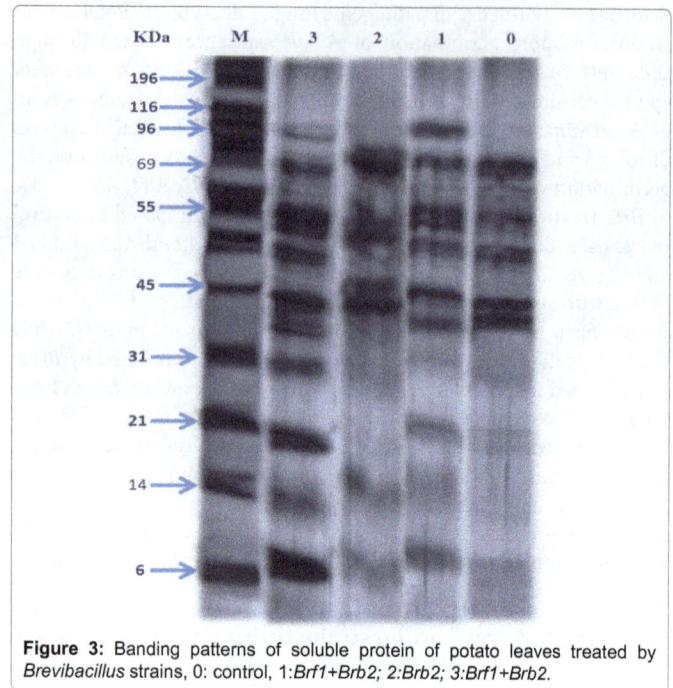

Figure 3: Banding patterns of soluble protein of potato leaves treated by *Brevibacillus* strains, 0: control, 1:*Brf1+Brb2*; 2:*Brb2*; 3:*Brf1+Brb2*.

well as spore germination might be due to the ability of *Brevibacillus* to suppress the fungal pathogen. Also, these effects may be associated with the activation of some novel defense pathways. Some previous studies on genomic sequences of different *Brevibacillus* strains mentioned to their responsibility for synthesis of antifungal compounds [28,29,57]. The systemic resistance also has been reported in other plant path systems, against leaf disease [63-66].

Protein profiling

Leaf protein contents of control, *Alternaria* infected, and *Brevibacillus* treated leaves are shown in Figure 3. In response to biocontrol agent's inoculation, the soluble protein contents increased significantly in comparison to control. SDS-PAGE is used for finding the banding pattern of proteins. Protein profiling was done to determine whether some new protein was associated with treatment and resistant to *A. alternata* in potato cultivar (Spunta) or not. The banding patterns of protein of different treatments were (14,16,19) bands in treatments with *Brf1*, *Brb2* and *Brf1+Brb2* respectively compared to control (11) bands. The highest number of bands was found in treated potato leaves by mixture of two tested *Brevibacillus* strains, and minimum number of bands was in control plant. The banding pattern of proteins from figure represented that some proteins of different molecular weight was found in treated plant which was not found in control. Similarly, some new bands were also found in mixture treatment not found in treatment by bacteria individually. The presence or absence of protein bands might be due to the inducing effects of *Brevibacillus* strains in plant which may also be responsible factors for enhancing of potato defense mechanism against *A. alternata*. Similar results were obtained by Biswas [67] reported that some new proteins were associated with resistance to *Bipolaris sorokiniana* induced by crude extracts of *Chaetomium globosum*. Also, Rajik et al. [50] and Romeiro [68] reported that protein profiling by SDS-PAGE revealed that some new protein is synthesized due to application by some induce resistance agents against *F. oxysporum f. sp. lycopersici* in tomato. Subsequently, given the results obtained by Silva et al. [69] where they found increased activity of some enzymes in a tomato rhizobacteria interaction against the pathogen *Pseudomonas syringae*, which was interpreted to mean

Fungal isolate	Treatment	DI	Reduction %	DS	Reduction %
Alt2	Brf1	*20.0 mno	58.3	36.0 fg	37.9
	Brb2	28.0 ijk	41.7	42.0 e	27.6
	Brf1+Brb2	19.3 no	62.5	28.0 ijk	51.7
	Cont.	48.0 d	0	58.0 abc	0
Alt4	Brf1	25.0 klm	54.5	30.0 hijk	50.8
	Brb2	31.0 ghij	43.6	34.0 fgh	31.7
	Brf1+Brb2	21.0 lmn	61.8	26.0 jkl	58.7
	Cont.	55.0 bc	0	63.0 a	0
Alt5	Brf1	18.0 no	66.0	33.0 fghi	45.0
	Brb2	21.0 lmn	60.4	37.0 f	38.3
	Brf1+Brb2	15.0 o	71.7	29.0 hijk	51.7
	Cont.	53.0 c	0	60.0 ab	0

* Means with the same letter are not significantly different.

Table 6: Effect of biocontrol agents on potato brown spot disease index (DI) and severity (DS) under greenhouse conditions.

that the rhizobacteria induced the systemic resistance in the tomato plants. However, the genus *Brevibacillus* includes a high diversity of thermophilic and halophilic strains which have ability to survive in harsh conditions and able to suppress wide range of plant pathogens. Therefore, some of *Brevibacillus* strains can be used as a source of many biotechnologically important enzymes such as a-amylase, xylanase and chitosanase [27,60], and can be play an effective role in control of some phytopathogenic fungi like *A. alternata* in potato plants.

Conclusion

In this study, the greenhouse and laboratory experiments have been performed to characterize the potential effect of two *Brevibacillus* strains against important fungal pathogen *A. alternata*. These experiments demonstrated that the use of the tested bacterial strains can enhance resistance to brown leaf spot in potato. The infection caused by *A. alternata* were observed in treated plants by biocontrol agents. Brown leaf spot severity was most significantly reduced by mixture of tested bacterial strains. The linear mycelial growth and spore germination of pathogenic fungi were inhibited by treatments which were confirmed by the results of *in vitro* and greenhouse experiments. Both of tested Brevibacillus strains reduced disease symptoms, and the effect was determined *in vitro* through detached leaves and under greenhouse conditions. Protein profiling by SDS-PAGE revealed that some bands of protein are produced due to application of biocontrol agents. The presence or absence of the bands in protein profiling might be responsible for induce resistance of potato plants against *A. alternata*. It may be concluded that, *B. formosus* strain DSM 9885, and *B. brevis* strain NBRC 15304 could be considered as part of management tools for reducing the impact of *A. alternata* causing brown leaf spot disease on potato.

References

1. Food and Agriculture Organization Corporate Statistical Database (2016), Rome, Italy.

2. Rotem J (1994) The Genus Alternaria: Biology and pathogenicity. American Phytopathological Society, St Paul, MN, USA.

3. Jayapradha C, Yesu RYI (2016). A Review of eco-friendly management of Alternaria species. Ind J Hill Farm. 29: 1-14.

4. Thomma BPH (2003) Alternaria spp.: From general saprophyte to specific parasite. Mol Plant Pathol 4: 225-236.

5. Reifschneider FJB, Lopez CA, Cobbe RV (1989) Integrated management of potato diseases. Circ Téc 7, Embrapa Hortaliças, Brasília.

6. Simmon EG (2000) Alternaria themes and variations (244–286): Species on Solanaceae. Mycotaxon 75: 1-115.

7. Möller K, Habermeyer J, Zinkernagel V, Reents HJ (2007) Impact and interaction of nitrogen and Phytophthora infestans as yield-limiting and yield-reducing factors in organic potato (Solanum tuberosum L.) crops. Potato Res 49: 281-301.

8. Bouws H, Finckh MR (2008) Effects of strip intercropping of potatoes with non-hosts on late blight severity and tuber field in organic production. Plant Pathology 57: 916-927.

9. Soleimani MJ, Kirk W (2012) Enhance resistance to Alternaria alternata causing potato brown leaf spot disease by using some plant defense inducers. J Pl Prot 52: 83-90.

10. Nowicki M, Nowakowska M, Niezgoda A (2012) Alternaria black spot of crucifers: Symptoms, importance of disease and perspectives of resistance breeding. Veg Crop Res Bull 76: 5-19.

11. Stevenson WR, Loria R, Franc GD, Weingartner DP (2001) Compendium of potato diseases. APS Press St. Paul, MN, USA.

12. Nash AF, Gardner RG (1988) Heritability of tomato early blight resistance derived from Lycopersicon hirsutum P.I. 126445. J Am Soc Hort Sci. 113: 264-268.

13. Douglas D, Groskopp M (1974) Control of early blight in eastern and south-central Idaho. Am Potato J 51: 361-367.

14. Harrison MD, Venette JR (1970) Chemical control of potato early blight and its effect on potato yield. Am Potato J 47: 81-86.

15. Pasche JS, Wharam CM, Gudmestad NC (2004) Shift in sensitivity of Alternaria solani in response to QoI-fungicides. Plant Dis 88: 181-187.

16. Campo ARO, Zambolim L, Costa LC (2007) Potato early blight epidemics and comparison of methods to determine its initial symptoms in a potato field. Rev Fac Nac Agron Medellin Colomb 60: 3877-3890.

17. Horsfield A, Wicks T, Davies K, Wilson D, Paton S (2010) Effect of fungicide use strategies on the control of early blight (Alternaria solani) and potato yield. Austral Plant Pathol 39: 368-375.

18. Heydari A, Pessarakli M (2010) A review on biological control of fungal plant pathogens using microbial antagonists. J Biol Sci 10: 273-290.

19. Bouizgarne B, El-Hadrami I, Ouhdouch Y (2006) Novel production of isochainin by a strain of Streptomyces sp. isolated from rhizosphere soil of the indigenous Moroccan plant Argania spinosa L. World J Microb Biotech 22: 423-429.

20. Omar, Amal M, Ahmed AIS (2014) Antagonistic and inhibitory effect of some plant rhizo-bacteria against different Fusarium isolates on Salvia officinalis. American-Eurasian J Agric and Environ Sci 14: 1437-1446.

21. Agha MKM, Gomaa SS, Ahmed AIS (2016) Effect of bacterial isolates and phosphite compounds on disease incidence of late blight (Phytophthora infestans) and improve productivity of some potato cultivars. Europ J Academic Essays 2: 21-36.

22. Agha MKM, Ahmed AIS, Gomaa SS (2016b) Effect of bacterial isolates and phosphite compounds on diseases incidence of early blight (Alternaria solani) and improve productivity of some potato cultivars. IOSR J Agric Veterin Sci 9: 48-58.

23. Bolwerk A, Lugtenberg BJJ (2005) Visualization of interactions of microbial biocontrol agents and phytopathogenic fungus Fusarium oxysporum f. sp. radicis lycopersici on tomato roots. In: Siddiqui ZA, (eds). PGPR: biocontrol and biofertilization. Springer, Berlin, pp. 217-231.

24. Ahmad F, Ahmad I, Khan MS (2008) Screening of free-living rhizospheric bacteria for their multiple plant growth promoting activities. Microbiol Res 163: 173-181

25. Aseri GK, Jain N, Panwar J, Rao AV, Meghwal PR (2008) Biofertilizers improve plant growth, fruit yield, nutrition, metabolism and rhizosphere enzyme activities of pomegranate (Punica granatum L.) in Indian Thar Desert. Sci Hortic 117: 130-135.

26. Amkraz N, Boudyach EH, Boubaker H, Bouizgarne B, Ait Ben Aoumar A (2010) Screening for fluorescent pseudomonades, isolated from the rhizosphere of tomato, for antagonistic activity toward Clavibacter michiganensis subsp. michiganensis. World J Microb Biotech 26: 1059-1065.

27. Panda AK, Bisht SS, Mondal SD, Kumar NS, Gurusubramanian G, et al. (2014) Brevibacillus sp. as biological tool: A short review. Antonie van Leeuwenhoek 105: 623-639.

28. Shida O, Takagi H, Kadowaki K, Komagata K (1996) Proposal for two new genera, Brevibacillus gen. nov. and Aneurinibacillus gen. nov. Int J Syst Bacteriol 46: 939-946.

29. Edwards SG, Seddon B (2001) Mode of antagonism of Brevibacillus brevis against Botrytis cinerea in vitro. J Appl Microbiol 91: 652-659.

30. Majeed A, Abbasi MK, Hameed S, Imran A, Rahim N (2015) Isolation and characterization of plant growth-promoting rhizobacteria from wheat rhizosphere and their effect on plant growth promotion. Front Microbiol 6: 198.

31. Wang JP, Liu B, Liu GH, Chen QQ, Zhu YJ, et al. (2015) Genome sequence of Brevibacillus formosus F12T for a genome-sequencing project for genomic taxonomy and phylogenomics of bacillus-like bacteria. Genome Announc 3: e00753-15.

32. Meena S, Gothwal RK, Krishna Mohan M, Ghosh P (2014) Production and purification of a hyperthermostable chitinase from Brevibacillus formosus BISR-1 isolated from the Great Indian Desert soils. Extremophiles 18: 451-462.

33. Nehra V, Choudhary M (2015) A review on plant growth promoting rhizobacteria acting as bioinoculants and their biological approach towards the production of sustainable agriculture. J Appl Nat Sci 7: 540-556.

34. Bhardwaj D, Ansari MW, Sahoo RK, Tuteja N (2014) Biofertilizers function as

key player in sustainable agriculture by improving soil fertility, plant tolerance and crop productivity. Microb Cell Fact 13:66.

35. Nehra V, Saharan BS, Choudhary M (2016) Evaluation of *Brevibacillus brevis* as a potential plant growth promoting rhizobacteria for cotton (*Gossypium hirsutum*) crop. Springer Plus 5: 1-10.

36. Stammler G, Böhme F, Philippi J, Miessner S, Tegge V (2013) Pathogenicity of Alternaria-species on potatoes and tomatoes, Fourteenth Euroblight Workshop, Germany.

37. Mckinney HH (1923) A new system of grading plant diseases. J Agric Res 26: 195-218.

38. Berg G, Roskot N, Steidle A, Eberl L, Zock A, et al. (2002) Plant-dependent genotypic and phenotypic diversity of antagonistic rhizobacteria isolated from different Verticillium host plants. Appl Environ Microbiol 68: 3328-3338.

39. Lane DJ (1991) 16S/23S rRNA sequencing. In: Stackebrandt E, Goodfellow M, (eds). Nucleic acid techniques in bacterial systematics. Chichester: John Wiley and Sons, USA. pp. 115-175.

40. Prashar P, Kapoor N, Sachdeva S, (2013) Isolation and characterization of *Bacillus* sp. within *vitro* antagonistic activity against *Fusarium oxysporum* from *Rhizoctonia* of tomato. J Agr Sci Tech 15: 1501-1512.

41. Alabouvette C, Lemanceau P, Stein-berg C (1993) Recent advances in the biological control of Fusarium wilts. Pestic Sci 37: 365-373.

42. Coskuntuna A, Özer N (2008) Biological control of onion basal rot disease using *Trichoderma harzianum* and induction of antifungal compounds in onion set following seed treatment. Crop Prot 27: 330-336.

43. Bekker TF, Kaiser C, Merwe PVD, Labuschagne N (2006) *In-vitro* inhibition of mycelial growth of several phytopathogenic fungi by soluble potassium silicate. S Afr J Plant Soil 23: 169-172.

44. Nair K, Ellingboe A (1962) A method of controlled inoculations with conidiospores of Erysiphe graminis var. tritici. Phytopathol 52: 714.

45. Palma-Guerrero J, Jansson HB, Salinas J, Lopez-Llorca LV (2008). Effect of chitosan on hyphal growth and spore germination of plant pathogenic and biocontrol fungi. J Appl Microbiol 104: 541-553.

46. Srivastava R (2009). Measuring germination percentage and rate. Int J Microbiol 7: 2.

47. Guetsky R, Shtienberg D, Elad Y, Dinoor A (2001) Combining biocontrol agents to reduce the variability of biological control. Phytopathol 91: 621-627.

48. Brymgelsson T, Gustavson M, Ramos Leal M, Bartonek E (1988) Induction of pathogenesis-related proteins in barley during the resistance reaction to mildew. J Phytopathol 123: 193-198.

49. Laemmli UK (1970) Cleavage of structural proteins during the assembly of the head of bacteriophage T4. Nature 227: 680-685.

50. Rajik M, Biswas SK, Shakti S (2012) Biochemical basis of defense response in plant against *Fusarium* wilt through bio-agents as inducers. Afr J Agric Res 7: 5849-5857.

51. Gomez KA, Gomez AA (1984) Statistical procedures for agricultural research. (2nd edn), John Wiley & Sons, Hoboken, New Jersey, USA.

52. Sharma JN, Gupta D, Bhardwaj LN, Kumar R (2005) Occurrence of Alternaria leaf spot (*Alternaria alternata*) on apple and its management. Int Pl Dis Manag 25-31.

53. Babu S (1994) Studies on leaf blight of tomato (*Lycopersicon esculentum* Mill.) caused by *Alternaria solani* (Ell. and Mart.) Jones and Grout. M.Sc. (Ag.) Thesis, Tamil Nadu Agricultural University, Coimbatore, India. p. 193.

54. Karthikeyan M (1999) Studies on onion (*Allium cepa varaggregatum* L.) leaf blight caused by *Alternaria palandui* Ayyangar. M.Sc., (Ag.) Thesis, Tamil Nadu Agricultural University, Coimbatore, India. p. 120.

55. Terpe K (2006) Overview of bacterial expression systems for heterologous protein production: From molecular and biochemical fundamentals to commercial systems. Appl Microbiol Biotechnol 72:211–222.

56. Shoda M (2000) Bacterial control of plant diseases. J Biosci Bioeng 89: 515-521.

57. Costa RG, Santos NM, De Medeiiros AN, Do Egypto Queiroga RCR, Madruga MS (2006) Microbiological evaluation of precooked goat "buchada". Braz J Microbiol 37:362-367.

58. Sivanantham T, Rasaiyah V, Satkunanathan N, Thavaranjit AC (2013) *In vitro* screening of antagonistic effect of soil borne bacteria on some selected phytopathogenic fungi. Arch Appl Sci Res 5:1-4.

59. Visker MHPW, Keizer LCCP, Budding DJ, Van Loon LC, Colon LT, Struik PC (2003) Leaf position prevails over plant age and leaf age in reflecting resistance to late blight in potato. Phytopathol 93: 666-674.

60. Mizukami M, Hanagata H, Miyauchi A (2010) *Brevibacillus* expression system: Host-vector system for efficient production of secretory proteins. Curr Pharm Biotechnol 11: 251-258.

61. Goth RW (1997) A detached-leaf method to evaluate late blight resistance in potato and tomato. Am Potato J 74: 347-352.

62. Vleeshouwers VGAA, Van Dooijewweert W, Keizer LCP, Sijpkes L, Govers F, et al. (1999) A laboratory assay for *Phytophthora infestans* resistance in various Solanum species reflects the field situation. Eur J Plant Pathol 105: 241-250.

63. Gorlach J, Volrath S, Knauf-Beiter G, Hengy G, Beckhove U, et al. (1996) Benzothiadiazole, a novel class of inducers of systemic acquired resistance, activates gene expression and disease resistance in wheat. The Plant Cell 8: 629–643.

64. Ishii H, Tomita Y, Horio T, Narusaka Y, Nakazawa Y, et al. (1999) Induced resistance of acibenzolar-S-methyl (CGA-245704) to cucumber and Japanese pear diseases. Eur J Plant Pathol 105: 77-85.

65. Bokshi AI, Morris SC, Deverall BJ (2003) Effects of benzothiadiazole and acetylsalicylic acid on β-1,3-glucanase activity and disease resistance in potato. Plant Pathol 52: 22–27.

66. Battu PR, Reddy MS (2009) Isolation of secondary metabolites from *Pseudomonas fluorescens* and its Characterization. Asian J Research Chem 2: 26-29.

67. Biswas SK, Srivastava, KD, Aggarwal R, Shelly P, Singh DV (2003) Biochemical changes in wheat induced by *Chaetomium globosum* against spot blotch pathogen. Indian Phytopath 54: 374-379.

68. Romeiro RS, Lanna Filho R, Vieira Junior JR, Silva HSA, Baracat-Pereira MC, et al. (2005) Macromolecules released by a plant growth-promoting rhizobacterium as elicitors of systemic resistance in tomato to bacterial and fungal pathogens. J Phytopathol 153:120-123.

69. Silva HSA, Romeiro RS, Macagnan D, Halfeld-Vieira BA, Baracat-Pereira MC, Mounteer A (2004) Rhizobacterial induction of systemic resistance in tomato plants: non-specific protection and increase in enzyme activities. Biological Control 29:288-295.

7

Identification, Mapping and Pyramiding of Genes/Quantitative Trait Loci (QTLs) for Durable Resistance of Crops to Biotic Stresses

Mekonnen T[1]*, Haileselassie T[1] and Tesfaye K[2]

[1]Addis Ababa University, Institute of Biotechnology, Addis Ababa University, Ethiopia
[2]Ethiopian Biotechnology Institute, Ethiopia

Abstract

Biotic stresses significantly limit global crop production. Identification and use of resistant cultivars is currently seen as the best strategy, cheapest, durable and environmentally friendly method to manage biotic stresses. However, resistance gained through single gene/quantitative trait loci (QTLs) transfer leads to resistance breakdown within a short period. Hence, current breeding programs targeted at developing durable and/ broad spectrum resistant cultivars by pyramiding multiple resistant genes/QTLs. Despite its significant contributions to crop improvement, gene pyramiding through conventional breeding suffers from being laborious, time consuming, costly and less efficient. Recently, the use of modern molecular tools like molecular markers and genetic engineering has dramatically enhanced the gene pyramiding strategy for biotic stress resistance. Molecular markers are very helpful for precise identification, mapping and introgression of multiple desirable genes/QTLs underlying trait of interest. Moreover, Genetic engineering has enabled scientists to transfer novel genes from any source into plants in a single generation to develop cultivars with the desired agronomic traits. Therefore, the current paper targeted to review the different types of biotic stress resistance in plants and the methodologies for identification, mapping and pyramiding of resistance genes/QTLs to develop durable and/or broad spectrum biotic stress resistant cultivars. So far, numerous crops with durable/broad spectrum resistance to pathogens, insect pests and herbicides have been developed by pyramiding multiple resistant genes/QTLs using marker assisted selection and genetic engineering techniques to contribute to increased crop production and productivity to maintain food security globally.

Keywords: Biotic stress; Durable resistance; Linkage mapping; Gene pyramiding; Marker assisted selection; Genetic engineering

Introduction

Biotic stresses remain the greatest constraint to crop production [1] accounting for 52% of the global yield loss [2]. Bacteria, viruses, fungi, nematodes, insect pests and weeds are considered to be biotic factors that limit crop production [2-4]. For years, chemicals have been used to control biotic damage of crop plants. Nowadays, interest in the use of chemicals against biotic stress is decreasing because of its various limitations such as the requirement for more than one chemical application, an investment that is not affordable by most small-scale farmers [5]. Besides, using chemical spray may have adverse effects on human health and the environment, including beneficial organisms and may lead to the development of chemical-resistant pathogen races, insects, and weeds [4,6]. On the other hand, the use of resistant cultivars is currently seen as the best strategy, durable, economical, and environmentally friendly means of biotic stress control [7-9].

Usually, breeding efforts made to incorporate single resistant gene leads to resistance breakdown within a short period [10]. Hence, recent breeding programs have targeted at developing cultivars that can withstand multiple stresses by assembling series of genes from different parents into a single genotype in a phenomenon called gene pyramiding or stacking [1,2]. Malav et al. [11] stated that gene pyramiding is a breeding method that aimed at assembling multiple desirable genes from multiple parents into a single genotype. The technique is very helpful for developing crops that confer broad spectrum resistance against different races of pathogens or pests or combination of stresses [12]. For several years, traditional breeding has been used to identify and incorporate multiple resistant genes/QTLs into cultivars of interest to develop durable resistance to biotic stresses [7]. However, conventional method of crop improvement has been complained to be slow, less precise, less flexible, labor-intensive and expensive [13,14]. With traditional breeding, breeder's capability to track the presence or absence of the target genes is very slow and limited. This limits the number of genes to be stacked into elite cultivars at any times [11].

Hence, a technological interventions that can reduce the time and costs necessary to develop and release new cultivars with durable resistance are always welcome. Recently, biotechnological tools like molecular markers and genetic engineering are widely used in crop improvement program for rapid and efficient accumulation of desirable genes from various sources into a single background to produce broad spectrum/durable resistance [2,7,11,15]. The advent and application of molecular marker technology made it easier to identify, map and efficiently pyramid resistant genes/QTLs into crop plants [16]. DNA markers tightly linked (<5 cM) to the desired gene serve as chromosomal landmark, 'signs' or 'flags' to track the introgression of the desired gene in progenies in a cross [17]. Hence, identification of resistant genes/ QTLs with closely linked DNA-markers is useful for successful transfer of the gene/QTLs into improved cultivars via marker-assisted selection (MAS) [18].

So far, various resistance genes/QTLs of crop plants have been identified and mapped using marker assisted selection. For instance, Yadav et al. [19] identified and mapped nine QTLs associated with sheath blight resistance in rice using MAS. Similarly, Perchepied et al. [20] identified and mapped two new pear resistance loci against the fungal pathogen *Venturia pirina* using MAS. Moreover, molecular markers are widely used for successful pyramiding of several resistance genes

*Corresponding author: Mekonnen T, Addis Ababa University, Institute of Biotechnology, Addis Ababa University, Ethiopia. King George VI street, P.O. Box 1176, Addis Ababa, Ethiopia, E-mail: mekonnentilahun27@yahoo.com

into crops including powdery mildew resistance genes (*Pm2+Pm4a, Pm2+Pm21, Pm4a+Pm21*) into wheat line [21], bacterial blight resistance genes (xa5, xa13, and Xa21) into rice [22], rust resistance genes (*Lr41, Lr42* and *Lr43*) into wheat [23], late blight resistance genes (*Rpi-mcd1* and *Rpi-ber*) into potato [24] etc. Therefore, it is important to deduce that molecular markers have remarkable applications in resistance gene/QTLs identification, mapping and pyramiding into crop plants to develop durable/ broad spectrum resistance to biotic stresses.

Moreover, the advent of genetic engineering (GE) has enabled scientists to transfer novel genes from any source to crop plants in a single generation [12]. Unlike conventional and MAS breeding methods which allow the transfer of desired genes between related species [25], Genetic engineering allows the specific transfer of gene of interest from any source (from animals, viruses, bacteria, or even from totally man-made sequences) into crop plants [26]. It has been reported that single gene transformation results in insufficient or narrow spectrum disease resistance [27], and hence a genetic transformation of crop plants with a combination of resistance genes would be more logical [1]. So far, a number of transgenic crops with durable resistance to bacterial diseases [28,29], viral diseases [30-32], fungal diseases [33], insect pests [34,35] and herbicides [36,37] have been developed. Thus, it is possible to deduce that genetic engineering is also another useful tool to pyramid novel resistance genes into crop plants to develop durable resistance to biotic stresses [2]. Therefore, the present review paper is aimed at reviewing the methodologies involved in identification, mapping and pyramiding of resistance genes/QTLs into to crop plants to develop durable and/or broad spectrum biotic stresses resistant cultivars.

Biotic stress resistance in crop plants

Being sessile organisms, plants are often exploited as a source of food and shelter by a wide range of parasites including viruses, bacteria, fungi, nematodes, insects, and even other plants [38]. Hence, biotic stress resistance in plants refers to the collective heritable characteristics of plant species to reduce the possibility of successful utilization of that plant as a host by these parasites [2]. As successful establishment of these biotic factors can cause severe damage on crop production, identification of the resistance genes and their utilization in breeding program makes the crop production system sustainable, economical, and environmentally friendly strategy [9]. Resistance in plants can be classified into two major categories and various terms have been used to describe the two categories of resistance, such as vertical versus horizontal resistance [39], qualitative versus quantitative resistance [40], and complete versus partial resistance [41].

Vertical resistance

Vertical resistance also called major-gene or single-gene resistance is a type of resistance where the plant possesses one or a few specific, well-defined genes that confer a high level of resistance to a specific pathogen. In this type of resistance, a particular gene gives the plant resistance to only one race of a pathogen and if other race comes, the plant needs different major genes for resistance to each race. It is sometimes called qualitative resistance because plants are either resistant or susceptible, without intermediate levels.

The simple model for how the host- pathogen recognition operates is that there is a dominant resistance (R) gene in the plant encoding a product that recognizes a pathogenicity factor (produced by a dominant *Avirulent or Avr* gene) in the pathogen to confer resistance [38,42,43]. There is mutual signaling between hosts and pathogens. Briefly, up on landing to the plant surface, the pathogen avirulence gene leads to

Virulence or avirulence genes in the pathogen	Resistance or susceptibility genes in the plant	
	R (resistant) dominant	r (susceptible) recessive
A (avirulent) dominant	AR (-)	Ar (+)
A (virulent) recessive	aR (+)	ar (+)

aMinus signs indicate incompatible (resistant) reactions and therefore no infection. Plus, signs indicate compatible (susceptible) reactions and therefore infection develops.

Table 1: Summary of host–pathogen reaction types based on the gene-for-gene concept.

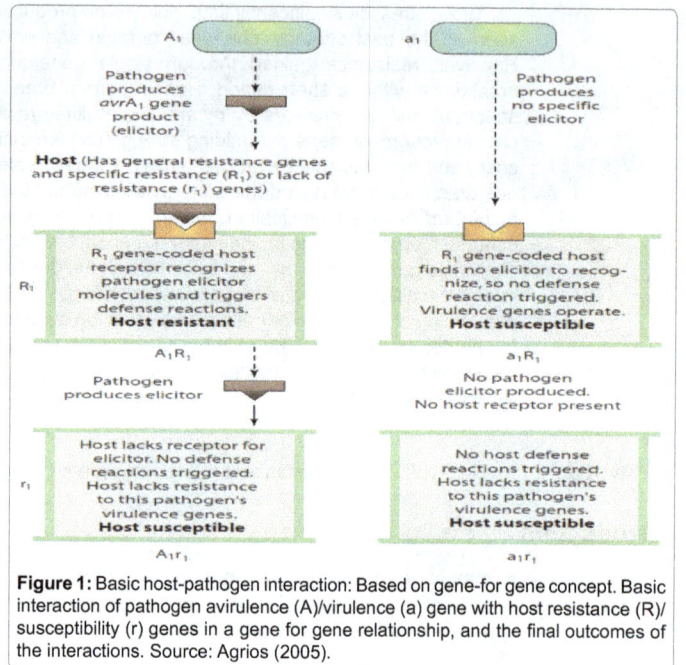

Figure 1: Basic host-pathogen interaction: Based on gene-for gene concept. Basic interaction of pathogen avirulence (A)/virulence (a) gene with host resistance (R)/susceptibility (r) genes in a gene for gene relationship, and the final outcomes of the interactions. Source: Agrios (2005).

the production of some "signal" molecule called elicitors (pathogen-associated molecular patterns (PAMPs)). The elicitors bind specifically to the plant Pattern Recognition Receptors (PRRs). This activates the PRRs and triggers a signal-transduction pathway leading to expression of the plant R gene to be expressed [44]. This ultimately results in recognition of the pathogen by the plant to be destroyed. Such type of resistance that depends on a precise match-up between a genetic allele in the plant and an allele in the pathogen is called gene- for- gene resistance [45]. On the other hand, absence of R gene in the plant and/or absence of the avirulence gene in the pathogen make the pathogen to be unrecognized by the plant. This results in the pathogen virulent gene to operate and makes the plant susceptible (diseased) [45]. Table 1 and Figure 1 summarize the gene-for gene concept when two cultivars, one with resistant gene (R) and the other with susceptible gene (r) are inoculated with two pathogen races; one carrying an avirulence (A) gene and the other with virulent (a) gene against the resistance gene R. According to the gene-for-gene concept, when the plant is resistant, the pathogen is called avirulent and the interaction is incomplete. While when the plant is susceptible, the pathogen is virulent and the interaction is complete [38].

As the effect of major gene is easy to recognize and select, most of the resistance exploited by plant breeders is of the major gene type [43]. However, major gene resistances are easier for a pathogen or an insect pest to break down in short period [10,46-51]. Because of this, recent plant breeding programs targeted at the identification and pyramiding of several major genes against a number of pathogenic races [11,15].

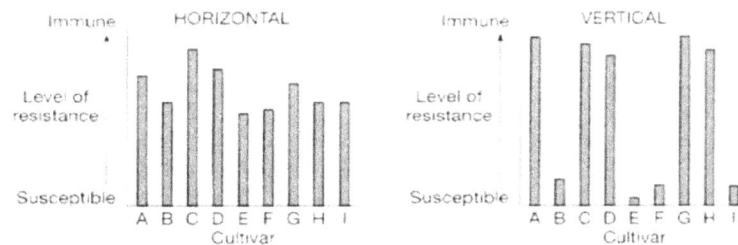

Figure 2: Horizontal versus vertical (gene-for-gene) resistance. In horizontal resistance, numerous genes have small additive effects so that the resistance varies by small amounts between cultivars. In vertical resistance, controlled by single genes, resistance is either close to complete immunity if the gene is present, or complete susceptibility if it is absent. Source: Dickinson (2005).

Horizontal resistance

Horizontal or quantitative resistance is defined as a race non- pacific or general resistance to a range of pathogens or pests [41] as a result of many genes expression with minor additive effects. As it is controlled by the collective effects of numerous genes known as quantitative trait loci (QTLs), horizontal resistance is important to control a broad range of pathogen races. Hence, horizontal resistance is durable and never breaks down. In a crop containing both major- and minor genes derived resistance, the minor gene resistance becomes visible after the "breaks down" of major gene resistance [43,52]. However, unlike vertical resistance that can protect the crop completely from the parasite, horizontal resistance does not protect plants from becoming infected. Rather it reduces the rate of disease development and spread. There is little difference in the level of horizontal resistance among crops (Figure 2).

Most reports indicate that horizontal resistance is polygenically inherited: does not obey the simple Mendelian inheritance. Mundt [53] reported that the resistance to the leaf rust pathogen *P. hordei* in barely is inherited polygenically and controlled by five or six minor genes. As a rule, a combination of major (R) genes and minor genes or QTLs for resistance against a pathogen is the most desirable makeup for any plant variety. Therefore, to successfully transfer desired resistance genes through modern breeding techniques, their precise location in the genome shall be known through genome mapping. The following section of this review presents the methodology involved in mapping genes/QTLs controlling important agronomic traits. Nowadays, molecular markers are becoming very helpful tools for precise detection and mapping of genes/QTLs controlling trait of interest.

Identification and Linkage Mapping of Resistance Genes/QTLs in Crop Plants

Gene mapping describes the methods used to identify the locus of a gene and the distances between genes. There are two distinctive types of "maps" used in the field of genome mapping: genetic maps and physical maps. They differ in techniques used to construct them and in the degree of resolution. Genetic map distances are constructed based on the genetic linkage information while physical maps use actual physical distances (has high resolution) usually measured in number of base pairs [54]. QTL map is a type of genetic map, which indicates the approximate location of a quantitative trait locus (QTL) within an interval delineated by two or more markers on a genetic map.

Genetic mapping/linkage mapping of genes/QTLs

Genetic mapping can be defined as the process of determining the linear order of molecular markers or genes (generally, loci) along a stretch of DNA or chromosome [55]. Linkage map indicate the relative position of markers on chromosome or linkage groups (LGs) based on the frequencies of recombination that occur between markers on

homologous chromosomes during meiosis. Recombination frequency between two markers is proportional to the distance separating the markers. The greater the frequency of recombination, the greater the distance between two genetic markers; conversely, the smaller the recombination frequency, the closer the markers are to one another. The distance between markers on a genetic map is given as Morgan (M) or centimorgan (cM), where one cM is the distance that separates two markers (or genes), between which a 1% chance of recombination exists (corresponding to one recombination event in 100 meioses). That means 99% of the times these two markers (genes) co-segregate, and hence MAS can be applied to select progenies with desired traits during crossing. The following steps are prerequisites for a successful linkage or genetic mapping of a target genome [55].

Selection of parent plants: The first step in linkage mapping is the selection of genetically divergent parents that exhibit sufficient polymorphisms for the trait of interest, but are not so distant as to cause sterility of the progeny [55]. Accordingly, in determining the chromosomal position of resistant genes/QTLs toward a particular pathogen, parental lines with sufficient polymorphism (pure resistant and pure susceptible parental liens) should be selected phenotypically in the field and/or using marker system [55].

Developing mapping population: Following the selection of polymorphic parental lines, the next key step is developing a mapping population [55]. Several types of mapping populations may be suitable for a particular project [56] including:

1. Double haploid lines (DHLs): Regenerated plants from pollen (which is haploid) of the F_1 plants and treated to restore diploid condition in which every locus is homozygous.

2. Backcross (BC) population: The F_1 plants are backcrossed to one of the parents.

3. F_2 population: F_1 plants are selfed.

4. Recombinant inbred lines (RILs): Inbred generation derived by selfing individual F_2 plants and further single seed descent. A population of RILs represents an 'immortal' or permanent mapping population.

Each of the above mapping populations has both advantages and disadvantages, and the choice of the type of mapping population depends on many factors such as the plant species, type of marker system used, and the trait to be mapped [56]. Accordingly, F_2 populations and BC populations are simple and can be developed in short period for self-pollinating species. While RIL population takes six to eight generations. Although development of a DH population takes much less time than RIL; it is only possible in species that are amenable to tissue culture. RIL and DH populations are good in that they produce homozygous or 'true-breeding' lines that can be multiplied and reproduced without genetic change occurring. This allows undertaking replicated trials across different locations and years. With regard to the marker choice, co-dominant markers are best informative in F_2 population, while information obtained by dominant marker systems can be maximized by using RILs or DHLs. Double haploids, F_2 families, or RILs are advantageous if the trait to be mapped cannot be accurately measured on a single-plant basis but must be assessed in replicated field experiments [56].

Determining mapping population size: In linkage mapping, the resolution of a map and the ability to determine marker order largely depend on population size [55]. A vague lower threshold that can localize quantitative trait loci (QTL) is a size of 100 individuals. However, high-resolution maps for map-based cloning of target genes ideally require population sizes of more than 500 or even 1000 individuals. Yadav et al. [19] used 210 F_2 and 150 BC_1F_2 mapping population to map QTLs governing the sheath blight resistance in rice. Similarly, Klarquist et al. [8] used 151 $F_{2:5}$ RIL populations to identify and map QTLs involved in stripe rust resistance in wheat. Moreover, Perchepied et al. [20] mapped two new pear resistance loci using three F_1 segregating populations (182, 144 and 81). Hence, it is important to decide the appropriate mapping population size required in locating chromosomal position of trait of interest, and generally the larger (>100) the mapping population, the better the map resolution would be [55].

Phenotype evaluation: Once a population segregating traits of interest is obtained, mapping the trait typically involves measuring the phenotype. Phenotypic evaluation can be undertaken in the field under natural condition (where high disease pressure can be expected) or in greenhouse/growth room in which the plants are inoculated with specific pathogen strains. Compared to the field evaluation, a greenhouse seedling inoculation can assess disease reactions quickly, reduce some sources of environmental variation by use of characterized pathogen strains and defined inoculum concentrations, and avoid confounding effects from other pests or diseases [9].

Genotype profiling: Generation of genotypic data for the mapping population involve two steps. First, DNA samples from the parental lines are screened for polymorphisms, using markers that span the chromosome(s) of interest. To scan the whole genome, polymorphic markers spaced approximately every 25 cM to 30 cM are needed. The second step is genotyping the mapping population with the selected polymorphic markers [55]. It is important to include many markers as much as possible [56].

Construction of linkage maps: The marker data collected through genotyping of the mapping population are used to construct the linkage map. Linkage analysis is based on the fact that two marker loci that are close to each other on the same chromosome tend to co-segregate; i.e., will be inherited together [55]. The frequency of recombinant (non-parental) genotypes is used to calculate recombination frequency, which is then used to infer the genetic distance between markers. By analyzing the segregation of markers, the relative order and distances between markers can be determined. The lower the frequency of recombination between two markers, the closer they are situated on a chromosome; conversely, the higher the frequency of recombination between two markers, the further away they are situated on a chromosome [56].

Simple statistical tests such as a χ^2 analysis will test the independent assortment of two loci and hence linkage. For two loci, a recombination frequency <50% indicates linkage. Usually, Kosambi's mapping function is used to derive genetic distances (cM) between linked loci from their recombination frequency. Linkage between two loci is usually calculated with an odds ratio (i.e., the ratio of linkage versus no linkage). This ratio is more conveniently expressed as the logarithm

of the ratio and is called a logarithm of odds (LOD) value or LOD score. A LOD score of 3 is normally accepted as a lower significance threshold to assert linkage [55], and the QTLs of interest are thought to exist at positions where an LOD score exceeded the corresponding significant threshold. Linked markers are grouped together into linkage groups (LG). In QTL analysis, the proportion of phenotypic variation explained by each QTL is calculated as R^2 value, and the degree of dominance of a QTL is estimated as the ratio of dominance effect to additive effect. A number of mapping computer programs are available for mapping traits controlled by single genes as well as quantitative traits like MAPMAKER/EXP [57] and JoinMap v.4.0 [58].

Applications

Identification and mapping of resistant genes/QTLs in two selected crops

A) Identification of QTLs and possible candidate genes conferring sheath blight resistance in rice (*Oryza sativa* L.)

Sheath blight is one of the most devastating diseases of rice caused by the fungus *Rhizoctonia solani Kühn*. Wang et al. [16] stated that pyramiding of diverse Sheath blight resistant (ShBR) QTLs could help to achieve higher levels of resistance to ShB. In line with this, Yadav et al. [19] aimed at identifying and mapping QTLs and candidate genes associated with sheath blight resistance in rice. As a procedure, two mapping populations namely 210 F_2 (derived from the cross between the susceptible BPT-5204 and moderately resistant ARC10531) and 151 BC_1F_2 populations (derived from the same cross) were developed. After greenhouse phenotypic evaluation in the presence of the pathogen *R. solani*, the F_2 population was genotyped using 70 polymorphic SSR markers. A linkage map was constructed using MAPMAKER 3.0 and significance threshold of >3 was considered for linkage grouping. Finally, 9 ShBR QTLs have been identified and mapped to five chromosomes (1, 6, 7, 8 and 9) with phenotypic variance ranging from 8.40% to 21.76% (Table 2). They identified new markers linked to the ShB resistances QTLs on chromosome 1, 6 and 8 (Figure 3). The study also identified two major ShBR-QTLs: *qshb7.3* (explained 21.76% of the total phenotypic variance) and *qshb9.2* (explained 19.81% of the phenotypic variance) that can be transferred using MAS into elite cultivars.

Validation of linked microsatellite markers associated with sheath blight resistance in rice

Another crucial step in linkage mapping is validation of the co-segregation of the identified marker and the trait. Usually, Bulk Segregant Analysis (BSA) has been employed to identify the DNA markers linked to the sheath blight resistance gene. Accordingly, in their validation analysis Yadav et al. [19] pooled the DNA from 10 extremes resistant and 10 extreme susceptible plants of the BC_1F_2 separately. And then, amplified along with both parents using the same SSR markers: RM336 and RM205 (Figures 4A and 4B). Finally, it was found that the resistance alleles show co-segregation among the parents ARC10531and BPT-5204 i.e., presence of the markers confirm presence of the resistant genes (Figure 4). Moreover, an *in-silico* analysis using rice data base RAP-DB for search of defense responsive gene identified 32 genes within QTL region near to the marker RM205 on chromosome 9. Functional annotation of predicted genes by blastp revealed one defense responsive gene ß 1-3 glucanase like protein present in a single copy within the cluster and it may be responsible for sheath blight resistance in the rice line ARC-10531. This shows that the identified markers are very efficient and helpful to select progenies carrying the desired genes/QTLs in crop breeding program. Hence, genetic mapping is helpful to

S.no.	QTLs	Chr.	Marker	Marker interval	LOD	%R²
1	qshb1.1	1	RM151	RM151-RM12253	10.7	10.99
2	qshb6.1	6	RM400	RM400-RM253	4.43	13.25
3	qshb7.1	7	RM81	RM81-RM6152	8.8	10.52
4	qshb7.2	7	RM10	RM10-RM21693	6.7	9.72
5	qshb7.3	7	RM336	RM336-RM427	4.12	21.76
6	qshb8.1	8	RM21792	RM21792-RM310	4.2	10.52
7	qshb9.1	9	RM257	RM257-RM242	5.9	8.4
8	qshb9.2	9	RM205	RM205-RM105	7	19.81
9	qshb9.3	9	RM24260	RM24260-RM3744	3.5	12.58

Source: Yadav et al. [20].

Table 2: QTLs identified for Sheath Blight resistance by Composite Interval Mapping (CIM).

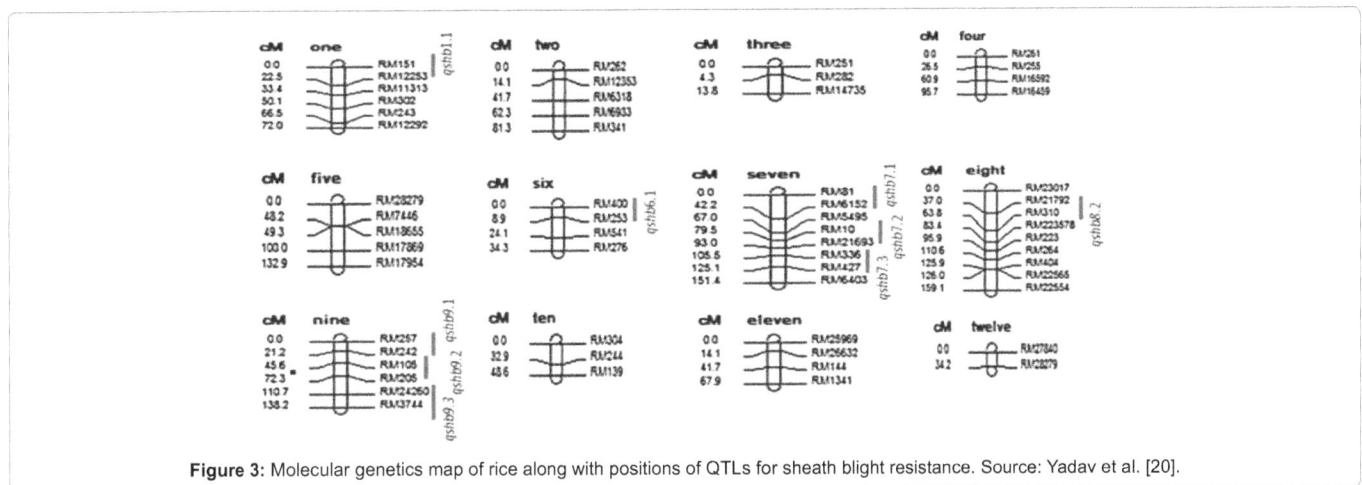

Figure 3: Molecular genetics map of rice along with positions of QTLs for sheath blight resistance. Source: Yadav et al. [20].

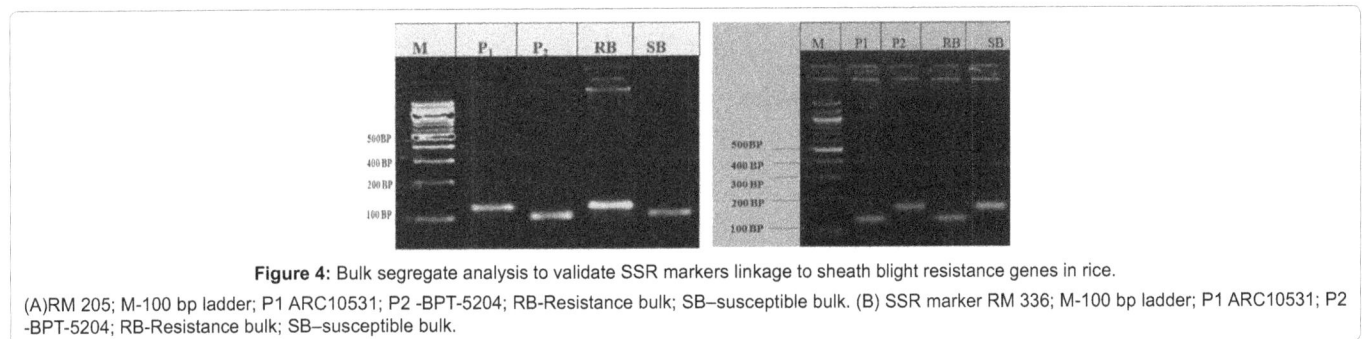

Figure 4: Bulk segregate analysis to validate SSR markers linkage to sheath blight resistance genes in rice.

(A)RM 205; M-100 bp ladder; P1 ARC10531; P2 -BPT-5204; RB-Resistance bulk; SB–susceptible bulk. (B) SSR marker RM 336; M-100 bp ladder; P1 ARC10531; P2 -BPT-5204; RB-Resistance bulk; SB–susceptible bulk.

identify and map markers linked to desired agronomic traits to be used in genome-assisted crop improvement.

B) Identification and mapping of new pear resistance loci against the fungal pathogen *Venturia pirina*

Scab is one of the major fungal diseases infecting pear trees. Perchepied et al. [20] targeted to identify and map new pear resistance loci against the fungal pathogen *Venturia*. As a procedure, they developed three F₁ segregating populations derived from the cross of: (1) Ange´lys (scab susceptible) × P3480 (scab-resistant), (2) Euras (resistant) × P2896 (susceptible) and (3) Euras × P3480. After phenotypic evaluation through artificial inoculation, the mapping populations including the parental lines were genotyped using 153 SSR markers. Linkage map was constricted using JoinMap v.4.0 software [58] at LOD significance threshold of 3. Kosambi's function was used to calculate genetic distances (cM). The position of marker-trait association was identified using CIM by the software MapQTL 5.0 at a significance threshold of LOD>3. A QTL with the largest LOD value

is the major QTL controlling the trait. A position on the LG where the LOD plot reaches its peak is the position of the major QTL (Figures 5A and 5B). The proportion of phenotypic variation explained (PVE) by each significant QTL indicated by the R^2. Hence, the study identified two new major QTLs namely, qrvp-LG01 on LG01 with a LOD score of 36.5 at the QTL peak located close to the SAmsCO865608 marker, and a second significant QTL qrvp-LG04 on LG04 with a LOD score of 19.2 (Table 3).

The percentage of phenotypic variation explained by qrvp-LG01 is 67.0 (Table 3). It was reported that the qrvp-LG04 QTL is located between the TsuGNH244 and TsuGNH076 markers and it is responsible for 52.8% of the phenotypic variation (Table 3).

Pyramiding of genes /QTLs for biotic stress resistance

Concept of gene pyramiding: Gene pyramiding is defined as a method of transferring multiple desirable genes/QTLs from multiple parents into a single genotype [2]. It is a breeding technique amid at assembling several genes with known effect on target trait [59]. The

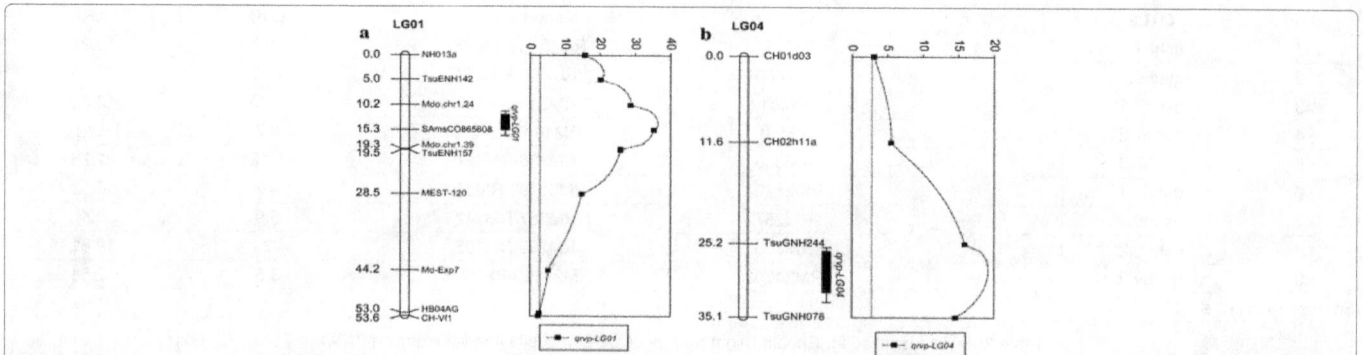

Figure 5: Genetic maps of the linkage groups 01 and 04, and LOD plot for the quantitative trait locus (QTL) detected for scab resistance (sporulation severity). The one- and two-LOD support intervals of the QTL are shown. a) Linkage group 01 and QTL, named *qrvp-LG01*, detected for the pear scab-resistant hybrid P3480 in the F1 segregating population deriving from the cross 'Angélys' × P3480. b) Linkage group 04 and QTL, named *qrvp-LG04*, detected for the resistant cultivar 'Euras' in the F1 segregating population deriving from the cross 'Euras' × P2.

The percentage of phenotypic variation explained by qrvp-LG01 is 67.0 (Table 3). It was reported that the qrvp-LG04 QTL is located between the TsuGNH244 and TsuGNH076 markers and it is responsible for 52.8% of the phenotypic variation (Table 3).

Progeny	Parental map	LG[a]	Position (cM)[b]	Marker closest to the QTL peak	LOD	R^2
A × P3480	P3480	1	13.33	SAmsCO865608	36.5	67.0
E × P2896	Euras	4	28.18	TsuGNH244	19.2	52.8
E × P3480	P3480	1	10.95	SAmsCO865608	4.3	22.6
	Euras	4	14	TsuGNH244	7.2	35.5

[a]Linkage group
[b]Position of the QTL peak on the LG

Table 3: Parameters associated with the quantitative trait loci (QTL) detected on the linkage groups 1 and 4 of the F1 segregating populations deriving from the crosses 'Angélys' × P3480 (A × P3480), 'Euras' × P2896 (E × P2896), and 'Euras' × P3480 (E × P3480).

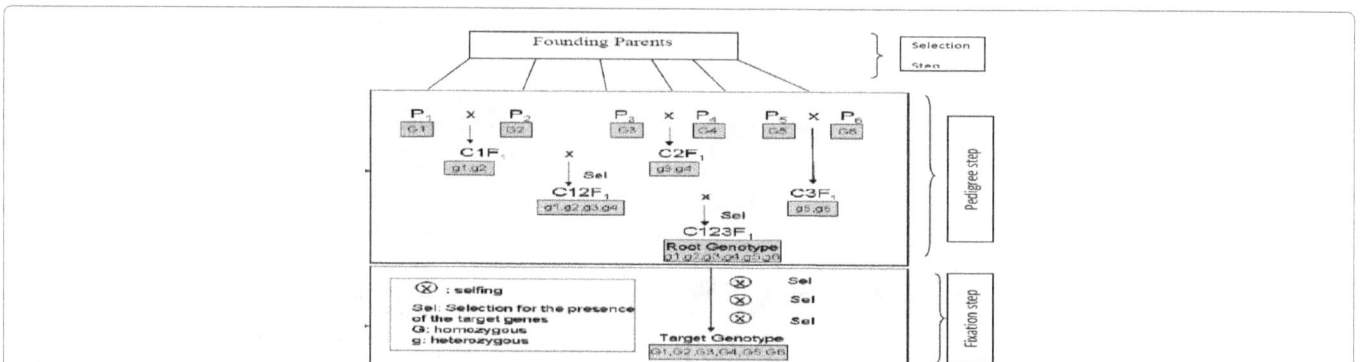

Figure 6: Diagrammatic representation of a gene-pyramiding scheme cumulating six target genes from six parental lines. Source: Suresh and Malathi [2].

technique is helpful in conferring broad spectrum resistance against different races of pathogens or pests or combination of stresses. Similarly, Ye and Smith [59] stated that genes are pyramided for one or combination of the following objectives: 1) enhancing trait performance by combining two or more complementary genes, 2) remedying deficits by introgression of genes from other sources, 3) increasing the durability of disease resistance, and 4) broadening the genetic basis of released cultivars.

Rationale behind gene/QTL pyramiding: The rationale behind gene pyramiding originates from the age-old philosophy of the use of insecticide mixtures to broaden the spectrum of insects controlled in one spray event [60]. In similar fashion, if two or more genes are stacked into a single variety, it is less probable for the plant to lose both resistant genes at the same time or a pathogen race with resistance to two genes to evolve. Nowadays, gene pyramiding is becoming an important breeding approach for developing durable or broad-spectrum resistance in crop plants against biotic stresses. It is a cost effective and environmentally friendly strategy to manage crop production loss due to biotic factors.

Recently, molecular techniques like molecular markers and genetic engineering are widely used for rapid and efficient accumulation of novel resistant genes from various sources into a single background to produce broad spectrum/ durable resistance in crop plants [1,8,10].

Designing a gene pyramiding strategy: The ultimate objective of a gene pyramiding program is to generate an ideal genotype having all desirable genes brought from various sources [12]. Successful gene pyramiding involves three steps [12,61] (Figure 6). The first step is identification/selection of parents containing the desirable genes (founding parents). This will be followed by a second step also called the pedigree step, which involves assembling single copy of (heterozygous) of the targeted genes (g1, g2, g3, g4, g5 and g6 in the example) through successive crossings to produce root genotype. The final step is called the fixation step which aims at fixing the target genes into a homozygous state to avoid their segregation in successive generations. Frequently, double haploid (DH) production and recombinant inbred line (RIL) techniques are used for homozygous line production. In this regard, the DH production technique that involves *in vitro* culturing of

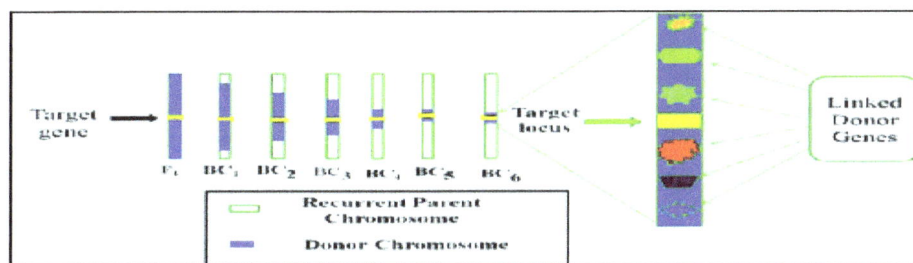

Figure 7: Schematic representation of transferring undesirable genes with target gene.

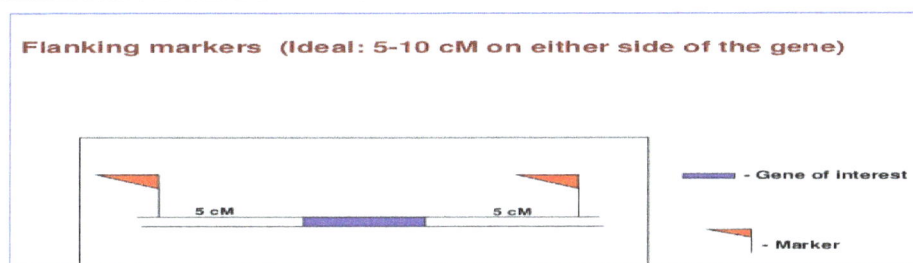

Figure 8: DNA markers tightly linked to the target gene.

gametes (anther, microspores or ovules) of the root genotype produces a population of fully homozygous individuals in single generation, among which the target genotype/ideotype can be found [62,63]. Application of co-dominant markers makes the ideotype selection process fast, efficient and cost effective [64]. However, producing large population of doubled haploid is difficult and cumbersome in certain plant species.

Alternatively, RIL development method could be used to fix the pyramided genes in root genotype. This involves selfing of the root genotype followed by intensive selection of progenies carrying the desired genes (Figure 6). It may take several generation selfing to develop fully homozygous lines of the pyramid genes [12]. Traditionally, breeders differentiate progenies carrying the desired genes based on phonotype which makes the screening process difficult. However, application of marker technology simplifies the pyramiding process by assisting the identification and maintenance of plants that carries the desired allele combination and discarding those that don't have [64].

Gene Pyramiding Methods

Gene pyramiding through traditional backcrossing

Recurrent backcrossing is a breeding method used to incorporate one or a few desirable traits into an elite variety containing large number of desirable traits but deficient in only a few traits [65,66]. Thus, the target of backcrossing is to transfer one or more genes of interest from donor parent into the genetic background of the improved variety and recover the recurrent parent genome (RPG). During backcrossing, together with the target gene, some unwanted genomic regions (gene drag) of the donor parent can transfer into the backcross progenies. Removing the linkage drag and recovering the recurrent parent genome requires six to eight backcrossing [67]. At each backcross generation, the proportion of recurrent parent genome recovered could be estimated using the formula $1-(1/2)^{n+1}$, where n is the number of backcross generations. At backcross six (BC6) up to 99.2% of the RPG would be recovered. Then, the resulting F_7 populations will be selected and selfed to generate three genotypes: homozygous resistant, heterozygous resistant and the

susceptible ones [7]. After one generation field screening, progenies homozygous for the resistant gene will be identified and maintained as improved line for resistance [64]. Surprisingly, the linkage drags (Figure 7) may remain even after six generations of backcrossing [1]. That is why conventional method of gene pyramiding for crop improvement is complained to be slow, tedious and inefficient. Hence a technology that can circumvent these limitations of conventional backcrossing and promotes the crop improvement program is always welcome [2].

Gene pyramiding using marker assisted selection

Molecular markers are identifiable DNA sequences found at specific location of the genome and inherited by the standard laws of inheritance [2,14]. DNA markers tightly linked (<5 cM) to the desired gene (Figure 8) serve as chromosomal landmark, 'signs' or 'flags' [17] to track the introgression of the desired gene in progenies in a cross i.e., identification of the marker indicates presence of the desired gene. The use of molecular marker technology in breeding to select progenies with the desired genes is called marker assisted selection (MAS), marker-assisted breeding or 'smart breeding' [20].

Application of MAS in plant breeding program has multiple advantages over the conventional phenotypic selection [64]. First, DNA based selection allows breeders to identify and select desirable plants at seedling stage, savings resources like greenhouse and/or field space, water, and fertilizer. Secondly, when assembling multiple genes for resistance to the same disease, using phenotypic selection alone, it can be difficult to distinguish those plants that carry all desired alleles from those that only have some of them. However, molecular markers are very powerful to precisely identify genotypes carrying the stacked desired genes. Thirdly, unlike phenotypic selection, genotypic selection is not affected by environmental factors. In addition to this, molecular markers are very important in backcross breeding to pyramid two or more genes associated with biotic stress resistance [1]. And marker assisted backcrossing (MABC) involves three levels of selection (Figure 9) [7]. The firs level of selection is called foreground selection where progenies carrying the desired gene would be selected using markers linked to the target gene. The second level of selection is called

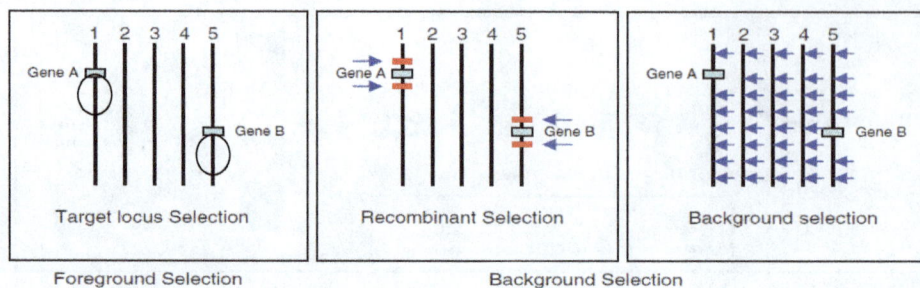

Figure 9: Schematic diagram showing whole genome selection process: Foreground selection, recombinant selection and background selection respectively. Source: Jain and Brar [1].

Features	Some DNA Markers				
	RFLPs	**RAPDs**	**AFLPs**	**SSRs**	**SNPs**
DNA quality	High	High	Moderate	Moderate	High
PCR-based	No	Yes	Yes	Yes	Yes
Ease of use	Not easy	Easy	Easy	Easy	Easy
Amenable to automation	Low	Moderate	Moderate	High	High
Reproducibility	High	Unreliable	High	High	High
Development cost	Low	Low	Moderate	High	High
Cost per analysis	High	Low	Moderate	Low	Low

Table 4: Comparison of most commonly used marker systems. Source: Korzun [59].

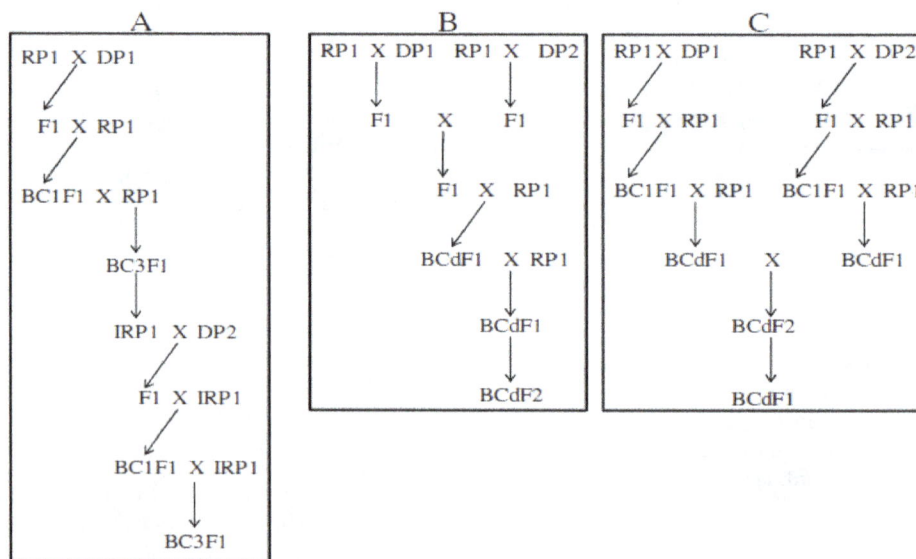

Figure 10: Different schemes of backcrossing for gene pyramiding. RP- Recurrent parent; DP- Donor parent; BC- Backcross; IRP- Improved recurrent parent. A. Stepwise transfer; B. Simultaneous transfer; C. Simultaneous and stepwise transfer. Source: Joshi and Nayak [1].

recombination selection where homozygous alleles of the recurrent parent will be selected using tightly liked markers flanking the target alleles. This step is important to reduce linkage drag [67]. The third level of selection is called the background selection which involves the selection of individuals carrying homozygous alleles of the recurrent parent at a number of unlinked marker loci covering the entire genome (chromosome 2, 3 and 4 in Figure 9 in background selection) [59]. Background selection accelerates the recovery of the recurrent parent genome and hence the use of MAS in backcross breeding reduce the time required to recover the recurrent genome by two to four backcrosses.

Commonly used marker systems in crops: So far, several marker systems have been developed and are applied to a range of crop species.

These include Restriction Fragment Length Polymorphisms (RFLPs), Random Amplification of Polymorphic DNAs (RAPDs), Amplified Fragment Length Polymorphisms (AFLPs), Simple Sequence Repeats (SSRs) or microsatellites, inter- SSRs (ISSRs), and most recently single nucleotide polymorphism (SNP) [14,68]. It is beyond the scope of this review to discuss the technical methods of how each of these DNA markers is functioning. However, each marker system has its own pros and cons, and the various factors to be considered in selecting one or more of these marker systems have been described in Table 4.

MAS gene pyramiding strategies: Generally, there are three possible MAS gene pyramiding strategies: Stepwise transfer, simultaneous transfer, and simultaneous and stepwise transfer (Figure 10) [2,12]. In stepwise transfer method, the recurrent parent (RP_1) is crossed with donor parent (DP$_1$) to produce the F$_1$ hybrid which then backcrossed up to third backcross generation (BC$_3$) to produce the improved recurrent parent (IRP$_1$). This improved recurrent parent is then crossed with other second donor parent (DP$_2$) to pyramid multiple genes (Figure 10A). Although this pyramiding strategy is very precise, the method is less accepted as it is time taking. In simultaneous gene pyramiding strategy, the recurrent parent (RP$_1$) is crossed with series of donor parents (DP1, DP2, etc.) to get the F1 hybrids which are then intercrossed to produce improved F1 (IF1). This improved F1 is then backcrossed with the recurrent parent to get the improved recurrent parent (IRP) (Figure 10B). The third strategy is a combination of the first two strategies which involve simultaneous crossing of recurrent parent (RP1) with many donor parents and then backcrossing them up to the BC$_3$ generation. The backcross populations with the individual gene are then intercrossed with each other to get the pyramided lines (Figure 10C) [12]. This is the most acceptable way as in this method not only time is reduced but fixation of genes is fully assured.

Efficiency of MAS gene pyramiding: Compared to conventional method, marker assisted backcrossing have been found to be rapid and efficient method of transferring multiple desirable genes into an elite cultivar. In line with this, Tanksley et al. [69] stated that using traditional backcross breeding, it takes six backcross generation to recover 99.2% of the recurrent parent genome and the same proportion of recurrent genome reconstruction can be achieved in less than two to four backcrossing generation using marker assisted backcross breeding [66]. Similarly, Jain and Brar [1] stated that MAS enable to monitor the introgression of many traits (up to six traits) at a time which often needs to conduct separate trials to screen for individual traits with conventional method. Moreover, MAS make the gene pyramiding process cost effective by allowing breeders to identify and select desirable plants very early (at seedling stage), saving resources like greenhouse and/or field space, water, and fertilizer. Furthermore, when multiple genes for resistance to the same disease are assembled, molecular markers are so efficient and powerful to discriminate those plants carrying all desired alleles from those that only have some of them. Therefore, it is possible to deduce that MAS gene pyramiding is a rapid, efficient, cost effective and a straight forward strategy in plant breeding for pyramiding genes/QTLs to crop plants to increase their durable resistance to biotic stresses.

Major achievements of marker aided gene pyramiding: Marker-assisted gene pyramiding has been used extensively for pyramiding major or qualitative disease resistance genes in plants [70]. So far, a number of promising achievements have been reported in developing durable disease resistance in crop plants through marker assisted gene pyramiding. Table 5 summarize some of the achievements made in controlling biotic stresses by pyramiding multiple resistant genes into crop plants through marker assisted breeding. Therefore, it is possible to deduce that molecular marker technology is a very helpful tool in the identification and introgression of multiple desirable genes into cultivars of interest to develop durable and broad spectrum biotic stress resistant cultivars to boost crop reduction and productivity.

Gene pyramiding through genetic engineering

Concepts of genetic engineering and its applications in crop improvement program: Genetic engineering or recombinant DNA technology (rDNA) is defined to be the technology that is used to cut a known DNA sequence from one organism and introduce it into another organism thereby altering the genotype (hence the phenotype) of the recipient [26,71,72]. The organism whose genes have been artificially altered for a desired trait is often called genetically modified organism

Crop	Trait	Pyramided genes	References
Rice	Bacterial blight resistance	xa5, xa13, Xa21	Pradhan et al. [13]
	Bacterial blight resistance	Xa4, xa5, xa13 and Xa21	Shanti et al. [98]
	Blast resistance	Pi1, Pi2 and Pi33	Usatov et al. [99]
Wheat	Leaf rust resistance	Lr41, Lr42 and Lr43	Cox et al. [24]
	Powdery mildew resistance	Pm2+Pm4a, Pm2+Pm21, Pm4a+Pm21)	Liu et al. [22]
	Stripe rust resistance	Yr5 and Yr15	Santra et al. [100]
Barley	Barley Yellow Mosaic Virus resistance	rym4, rym5, rym9 and rym11	Werner et al. [101]
Potato	Late blight resistance	Rpi-mcd1 and Rpi-ber	Tan et al. [25]
Soybean	Soybean mosaic virus (SMV)-resistance	Rsv1, Rsv3, and Rsv4	Zhu et al. [102]

Source: Joshi and Nayak [12] and Suresh and Malathi [2].

Table 5: Summary of the success history of pyramiding genes using MAS to develop biotic stress tolerant cultivars.

(GMO). The techniques use highly sophisticated laboratory tools and specific enzymes to cut out, insert, and alter pieces of DNA that contain one or more genes of interest [72].

Unlike conventional and MAS breeding methods which allow the transfer of desired genes between related species [25], genetic engineering allows the specific transfer of gene of interest from any source (from animals, viruses, bacteria, or even from totally man-made sequences) into crop plants [26], to generate crops with the desired agronomic trait/s. Jain and Brar [1] stated that genetic engineering is the only option to transfer genes of interest originates from cross barrier species, distant relatives, or from non-plant sources in a very fast way than through conventional or molecular breeding. Although, genetic engineering is a universal, precise, and fast method to transfer desired gene/s into crop plants [72], it will not replace conventional breeding but it will add to the efficiency of crop improvement.

Major steps in plant genetic engineering: The process of genetic engineering requires the successful completion of the following series of steps [73].

1) Identifying the target gene and isolating the DNA from the desired organism: To identify a desirable new trait or gene it is important to look to nature. It means in searching for a trait that would allow a crop to survive in a specific environment, it is important to look for organisms that naturally are able to survive in that specific environment. For instance, Monsanto created "Roundup Ready" plants after finding bacteria growing near a Roundup factory that contained a gene that allowed them to survive in the presence of the herbicide [74]. The other desired gene identification techniques are comparative genome analysis of organisms showing the trait and lacking the trait as well as mutational analysis i.e., purposeful deletion, or "knock out," of parts of the genome of interest until the desired trait is lost [75].

2) Gene Cloning: After the target gene has been identified and isolated, it will be multiplied by inserting it into bacterial plasmid (cloning vector). Plasmids are small circular DNA capable of replicating independently [73]. During gene cloning, both the DNA with target gene and the vector are cut open with the same restriction endonuclease. After the insertion of the target gene, the cut ends will be sealed by molecular glue called DNA ligase. Then, the construct (vector plus target gene) will be reintroduced into bacterial cells to allow it to replicate together with the replication of the host cell. There are also other types of coning vectors used to transport target gene into

resistance breeding and also start the use of modern molecular tools to speed up the crop improvement program.

Acknowledgment

All the literatures used in this review paper are greatly acknowledged.

Conflicts of Interest

We the authors have not declared any conflict of interests.

References

1. Jain SM, Brar DS (2010) Molecular techniques in crop improvement. Springer Science+Business Media, Berlin, Germany.

2. Suresh S, Malathi D (2013) Gene pyramiding for biotic stress tolerance in crop plants. Weekly Sci Res J 1: 2321-7871.

3. Goodwin SB (2007) Back to basics and beyond: Increasing the level of resistance to Septoria tritici blotch in wheat. Australas Plant Pathol 36: 532-538.

4. Vincelli P (2016) Genetic engineering and sustainable crop disease management: Opportunities for case-by-case decision-making. Sustain 8: 495.

5. Brading PA, Verstappen ECP, Kema GHJ, Brown JKM (2002) A gene for-gene relationship between wheat and Mycosphaerella graminicola, the Septoria tritici blotch pathogen. Phytopathol 92: 439-445.

6. Miedaner T, Zhao Y, Gowda M, Longin CFH, KorzunV, et al. (2013) Genetic architecture of resistance to Septoria tritici blotch in European wheat. BMC Genomics 14: 858.

7. Ragimekula N, Varadarajula NN, Mallapuram SP, Gangimeni G, Reddy RK, et al. (2013) Marker assisted selection in disease resistance breeding. J Plant Breed Genet 1: 90-109.

8. Klarquist E, Chen XM, Carter AH (2016) Novel QTL for stripe rust resistance on chromosomes 4A and 6B in soft white winter wheat cultivars. Agronomy 6: 1-14.

9. Hansona P, Lua SF, Wanga JF, Chena W, Kenyona L, et al. (2016) Conventional and molecular marker-assisted selection and pyramiding of genes for multiple disease resistance in tomato. Sci Hort 201: 346-354.

10. Kottapalli KR, Narasu LM, Jena KK (2010) Effective strategy for pyramiding three bacterial blight resistance genes into fine grain rice cultivar, Samba Mahsuri, using sequence tagged site markers. Biotechnol Lett 32: 989-96.

11. Malav AK, Kuldeep I, Chandrawat S (2016) Gene pyramiding: An overview. Int J Curr Res Biosci Plant Biol 3: 22-28.

12. Joshi RK, Nayak S (2010) Gene pyramiding-A broad spectrum technique for developing durable stress resistance in crops. Biotechnol Genet Eng Rev 5: 51-60.

13. Wieczorek A (2003) Use of biotechnology in agriculture- Benefits and risks. College of Tropical Agriculture and Human Resources (CTAHR), University of Hawaii, USA

14. Choudhary K, Choudhary OP, Shekhawat NS (2008) Marker assisted selection: A novel approach for crop improvement. American-Eurasian J Agronomy 1: 26-30.

15. Campbell MM, Brunner AM, Jones HM, Strauss SH (2003) Forestry's fertile crescent: The application of biotechnology to forest trees. Plant Biotechnol J 1: 141-154.

16. SH (2003) Forestry's fertile crescent: The application of biotechnology to forest trees. Plant Biotechnol J 1: 141-154.

17. Asad MA, Xia XC, Wang CS, He ZH (2012) Molecular mapping of stripe rust resistance gene YrSN104 in Chinese wheat line shaannong104. Hereditas 149: 146-152.

18. Collard BCY, Jahufer MZZ, Brouwer JB, Pang ECK (2005) An introduction to markers, quantitative trait loci (QTL) mapping and marker-assisted selection for crop improvement: The basic concepts. Euphytica 142: 169-196.

19. Asins M (2002) Present and future of quantitative trait locus analysis in plant breeding. Plant Breed 121: 281-291.

20. Yadav S, Anuradha G, Kumar RR, Vemireddy LR, SudhakarR, et al. (2015) Identification of QTLs and possible candidate genes conferring sheath blight resistance in rice (Oryza sativa L.). SpringerPlus 4:175.

21. Perchepied L, Leforestier D, Ravon E, Gue´rif P, Denance C, et al. (2015)

22. Liu J, Liu D, Tao W, Li W, Wang S, et al. (2000) Molecular marker-facilitated pyramiding of different genes for powdery mildew resistance in wheat. Plant Breeding 119: 21-24.

23. Pradhan SK, Nayak DK, Mohanty S, Behera L, Barik SR, et al. (2015) Pyramiding of three bacterial blight resistance genes for broad-spectrum resistance in deep water rice variety. Jalmagna. Rice. 8: 19.

24. Cox TS, Raupp WJ, Gill BS (1993) Leaf rust-resistance genes, Lr41, Lr42 and Lr43 transferred from Triticum tauschii to common wheat. Crop Sci 34: 339-343.

25. Tan MYA, Hutten RCB, VIsser RCF, Van Eck HJ (2010) The effect of pyramiding Phytophtora infestans resistance genes Ppi-mcd1 and Ppi-ber in potato. Theor Appl Genet 121: 117-125.

26. Lemaux PG (2008) Genetically engineered plants and foods: A scientist's analysis of the issues (Part I). Annu Rev Plant Biol 59: 771-812.

27. FAO (2011) Biotechnologies for agricultural development. Proceedings of the FAO international technical conference on "Agricultural Biotechnologies in Developing Countries: Options and Opportunities in Crops, forestry, livestock, fisheries and agro-industry to face the challenges of food insecurity and climate change" (ABDC -10), Rome.

28. Anand A, Zhou T, Trick HN, Gill BS, Bockus WW et al. (2003) Greenhouse and field testing of transgenic wheat plants stably expressing genes for thaumatin-like protein, chitinase and glucanase against Fusarium graminearum. J Exp Botany 54: 1101-1111.

29. Horvath DM, Stall RE, Jones JB, Pauly MH, Vallad GE, et al. (2012) Transgenic resistance confers effective field level control of bacterial spot disease in tomato. PLoS ONE 7: e42036.

30. Hummel AW, Doyle EL, Bogdanove AJ (2012) Addition of transcription activator-like effector binding sites to a pathogen strain-specific rice bacterial blight resistance gene makes it effective against additional strains and against bacterial leaf streak. New Phytol 4: 883-93.

31. Ferreira SA, Pitz KY, Manshardt R, Zee F, Fitch MM, et al. (2002) Virus coat protein transgenic papaya provides practical control of Papaya ringspot virus in Hawaii. Plant Dis 86: 101-105.

32. Malinowski T, Cambra M, Capote N, Zawadzka B, Gorris MT, et al. (2006) Field trials of plum clones transformed with the Plum pox virus coat protein (PPV-CP) gene. Plant Dis 90: 1012-1018.

33. Bravo-Almonacid F, Rudoy V, Welin, B (2012) Field testing, gene flow assessment and pre-commercial studies on transgenic Solanum tuberosum spp. tuberosum (cv. Spunta) selected for PVY resistance in Argentina. Transgenic Res 21: 967.

34. Krens FA, Schaart JG, Groenwold R, Walraven AEJ, Hesselink T, et al. (2011) Performance and long-term stability of the barley hordothionin gene in multiple transgenic apple lines. Transgenic Res 20: 1113-1123.

35. Maqbool SB, Christou P, Riazuddin S, Loc NT, Gatehouse AMR, et al. (2001) Expression of multiple insecticidal genes confers broad resistance against a range of different rice pests. Mol Breeding 7: 85-93.

36. Horvath H, Rostoks N, Brueggeman R, Steffenson B, Von Wettstein D, et al. (2003) Genetically engineered stem rust resistance in barley using the Rpg1 gene. Proc Natl Acad Sci USA 100: 364-369.

37. Green JM, Micheal DK (2011) Herbicide-resistant crops: Utilities and limitations for herbicide-resistant weed management. Journal of Agric Food Chem 59: 5819-5829.

38. Heap I (2016) International survey of herbicide resistant weeds. Weeds resistant to EPSP synthase inhibitors.

39. Gachomo EW, Shonukan OO, Kotchoni SO (2003) The molecular initiation and subsequent acquisition of disease resistance in plants. Afr J Biotechnol 2: 26-32.

40. Van der Plank JE (1968) Disease Resistance in Plants. Academic, New York, London.

41. Ou SH, Nuque FL, Bandong JM (1975) Relationship between qualitative and quantitative resistance in rice blast. Phytopathology 65: 1315–1316.

42. Parlevliet JE (1979) Components of resistance that reduce the rate of disease epidemic development. Annu Rev Phytopathol 17: 203-232.

43. Keen NT (1990) Gene-for-gene complementarity in plant-pathogen interactions. Annu Rev Genet 24: 447-463.

44. Do Vale FXR, Parlevliet JE, Zambolim L (2001) Concepts in plant disease resistance. Fitopatol Bras 26.

45. Bogdanove AJ (2002) Protein-protein interactions in pathogen recognition by plants. Plant Mol Biol 50: 981-989.

46. Dickinson M (2005) Molecular plant pathology. Bios, Scientific publishers Taylor and Francis Group, 29 West 35th Street, New York, USA; pp. 175-177.

47. Marshall DR (1977) The advantages and hazards of genetic homogeneity. In: PR Day (eds). The genetic basis of epidemics in agriculture. New York Academy of Sciences, New York, USA.

48. Wolfe MS, Barrett JA (1977) Population genetics of powdery mildew epidemics. In: PR Day (eds). The genetic basis of epidemics in agriculture. New York Academy of Sciences, New York, USA.

49. Simmonds NW (1979) Principles of crop improvement. Longman, London and New York. p. 408.

50. De Boef WS, Berg T, Haverkort B (1996) Crop genetic resources. In J. Bunders, B. Haverkort and W. Hiemstra (eds). Biotechnology; building on farmers' knowledge. Macmillan, London and Basingstoke. pp. 103-128.

51. Rubenstein DK, Heisey P, Shoemaker R, Sullivan J Frisvold G (2005) Crop genetic resources: An economic appraisal. United States Department of Agriculture (USDA). Economic Information Bulletin.

52. Smolders H (2006) Enhancing farmers' role in crop development: Framework information for participatory plant breeding in farmer field schools. Center for Genetic Resources, Wageningen University and Research Center, Netherlands.

53. Mundt, CC (2014) Durable resistance: A key to sustainable management of pathogens and pests. Infect Genet Evol 27: 446–455.

54. Parlevliet JE (1978) Further evidence of polygenic inheritance of partial resistance in barley to leaf rust, Puccinia hordei. Euphytica 27: 369-379.

55. Dixit R, Rai DV, Agarwal R, Pundhir A (2014) Physical Mapping of genome and genes. J biol eng res rev 1: 06-11.

56. Weising K, Nybom H, Wolff K, Kahl G (2005) DNA Fingerprinting in plants principles, methods, and applications second edition. CRC Press Taylor and Francis Group. pp. 277-291.

57. Sehgal D, Singh R, Rajpal VR (2016) Quantitative trait loci mapping in plants: Concepts and approaches. Mol Breeding for Sustainable Crop Improvement, Sustainable Development and Biodiversity, Springer International Publishing Switzerland.

58. Lander ES, Green P, Abrahamson J, Barlow A, Daly MJ, et al. (1987). MAPMAKER: An interactive computer package for constructing primary genetic linkage maps of experimental and natural populations. Genomics 1: 174–181.

59. Van Ooijen JW (2006) JoinMap 4, Software for the calculation of genetic linkage maps in experimental populations. Wageningen, Netherlands.

60. Ye G, Smith KF (2008) Marker assisted gene pyramiding for inbreed line development: Practical applications. Int J Plant Breed 2: 11-22.

61. Manyangarirwa W, Turnbull M, McCutcheon GS, Smith JP (2006) Gene pyramiding as a Bt resistance management strategy: How sustainable is this strategy? Afr J Biotechnol 5: 781-785.

62. Serajazari M, Schaafsma AW, Falk DE (2013) Pyramiding disease resistance genes in Wheat (Triticum aestivum L.) A PhD Thesis dissertation in Plant Agriculture, The University of Guelph, Guelph, Ontario, Canada.

63. Dunwell JM (2010) Haploids in flowering plants: Origins and exploitation. Plant Biotechnol J 8: 377-424.

64. Khan MA, Ahmad J (2011) In Vitro wheat haploid embryo production by wheat x maize cross system under different environmental conditions. Pak J Agri Sci 48: 39-43.

65. Francis DM, Merk HL, Namuth D (2012) Gene Pyramiding using molecular markers.

66. Brumlop S, Finckh MR (2010) Applications and potentials of marker assisted selection (MAS) in plant breeding.

67. Hasan MM, Rafii MY, Ismail MR, Mahmood M, Rahim HA, et al. (2015) Marker-assisted backcrossing: A useful method for rice improvement. Biotechnology Equipment 29: 237-254.

68. Hasan MM, Rafii MY, Ismail MR, Mahmood M, Rahim HA, et al. (2015) Marker-assisted backcrossing: A useful method for rice improvement. Biotechnology Equipment 29: 237-254.

69. Korzun V (2003) Molecular markers and their applications in cereals breeding. A paper presented during the FAO international workshop on marker assisted selection: A fast track to increase genetic gain in plant and animal breeding? Turin, Italy.

70. Tanksley SD, Young ND, Paterson AH, Bonierbale MW (1989) RFLP mapping in plant breeding: A new tools for an old science biotechnology. Nat Biotechnol 7: 257-264.

71. Thakur RP (2007) Host plant resistance to diseases: Potential and limitations. Indian Journal of Plant Protection 35: 17-21.

72. Nicholl DST (2008) An introduction to genetic engineering. (3rd edn), Cambridge University Press, Cambridge, UK.

73. (ISAAA) (International Service for the Acquisi tion of Agri-biotech Applications) (2010) Agricultural Biotechnology (A Lot More than Just GM Crops).

74. Powell C (2015) How to Make a Gmo. Blog, Special Edition on Gmos.

75. Rebecca B (2011) How to genetically modify a seed, step by step. Popular Science. Popular Science, 24 January.

76. Jacqueline AP, Shipton CA, Chaggar S, Howells RM, Kennedy MJ, et al. (2005) Improving the nutritional value of golden rice through increased pro-vitamin a content. Nat Biotechnol 23: 482-87.

77. (ISAAA) (2013) Bt brinjal In India (Pocket K No. 35).

78. Halpin C (2005) Gene stacking in transgenic plants- The challenge for 21st century plant biotechnology. Plant Biotechnol J 3: 141-155.

79. Que Q, Chilton M, De Fontes C, He C, Nuccio M, et al. (2010) Trait stacking in transgenic plants- Challenges and opportunities. GM Crops 1: 220-229.

80. Shelton AM, Zhao J, Roush RT (2002) Economic, ecological, food safety and social consequences of the deployment of Bt transgenic plants. Annu Rev Entomol 47: 845-881.

81. Ferry N, Edwards MG, Mulligan EA, Emami K, Petrova AS, et al. (2004) Engineering resistance to insect pests. In: Christou P, Klee H (eds). Handbook of Plant Biotechnology. John Wiley and Sons, Chichester 1: 373-394.

82. Tabashnik BE, Brevault T, Carriere Y (2013) Insect resistance to Bt crops: Lessons from the first billion acres. Nat Biotechnol 31: 510-521.

83. Carrière Y, Crickmore N, Tabashnik BE (2015) Optimizing pyramided transgenic Bt crops for sustainable pest management. Nat Biotechnol 33: 161-168.

84. Jiang F, Zhang T, Bai S, Wang Z, He K (2016) Evaluation of Bt Corn with pyramided genes on efficacy and insect resistance management for the Asian Corn Borer in China. PLoS one 11: e0168442.

85. Schnepf E, Crickmore N, Van Rie J, Lereclus D, Baum J, et al. (1998) Bacillus thuringiensis and its pesticidal crystal proteins. Microb Mol Biol Reviews 62: 775-806.

86. Puja ZK (2006) Recent developments toward achieving fun gal disease resistance in transgenic plants. Can J Plant Pathology 28: 298-308.

87. Collinge DB, Lund OS, Thordal-Christensen H (2008) What are the prospects for genetically engineered, disease resistant plants? Eur J Plant Pathol 121: 217-231.

88. Wally O, Punja ZK (2010) Genetic engineering for increasing fungal and bacterial disease resistance in crop plants. GM Crops 1: 199-206.

89. Herrmann KM, Weaver LM (1999) The shikimate pathway. Annual Review of Plant Physiology Plant Molecular Biology 50: 473-503.

90. Rodemeyer M (2001) Harvest on the horizon: Future uses of agricultural biotechnology. The pew Initiative on food and biotechnology. (www.animalbiotechnology.org/future uses.pdf). Accessed on: 30 April 2014.

91. Young BG (2006) Changes in herbicide use patterns and production practices resulting from glyphosate-resistant crops. Weed Technology 20: 301-307.

92. Gianessi LP (2008) Economic impacts of glyphosate-resistant crops. Pest Management Society 64: 346-352.

93. James C (2014) Global status of commercialized biotech/GM crops: 2014. ISAAA Brief No. 49.

94. (USDA) U.S. Department of Agriculture. (2015). Crop Acerage.

95. Brookes G, Barfoot P (2015) Environmental impacts of genetically modified (GM) crop use 1996-2013: Impacts on pesticide use and carbon emissions. Genetically Modified Crops Food: Biotech Ag Food Chain 6: 103-133.

96. Powles SB (2008) Evolved glyphosate-resistant weeds around the world: Lessons to be learnt. Pest Manage Sci 64: 360-365.

97. Heap I (2014) Global perspective of herbicide-resistant weeds. Pest Manage Sci 70: 1306-1315.

98. Shanti ML, Shenoy VV, Devi GL, Kumar VM, Premalatha P, et al. (2010) Marker-assisted breeding for resistance to bacterial leaf blight in popular cultivar nad parental lines. J Plant Pathol 92: 495-501.

99. Usatov AV, Kostylev PI, Azarin KV, Markin NV, Makarenko MS, et al. (2016) Introgression of the rice blast resistance genes Pi1, Pi2 and Pi33 into Russian rice varieties by marker-assisted selection. Indian J Genet Plant Breed 76: 18-23.

100. Santra D, DeMacon VK, Garland-Campbell K, Kidwell K (2006) Marker assisted backcross breeding for simultaneous introgression of stripe rust resistance genes yr5 and yr15 into spring wheat *(triticum aestivum* L.). In 2006 International meeting of ASA-CSSA-SSSA. pp. 74-75.

101. Werner K, Friedt W, Ordon F (2005) Strategies for pyramiding resistance genes against the barley yellow mosaic virus complex (BaMMV, BaYMV, BaYMV-2). Mol Breed 16: 45-55.

102. Zhu Y, Yu DY, Chen SY, Gai JY (2006) Inheritance and gene mapping of resistance to SMV strain SC-7 in soybean. Acta Agron Sinica 32: 936-938.

103. Ahmad A, Maqbool SB, Riazuddin S, Sticklen MB (2002) Expression of synthetic cry1Ab and cry1Ac genes in basmati rice (*Oryza sativa* L.) variety 370 via Agrobacterium-mediated transformation for the control of the European corn borer (*Ostrinia nubilalis*). *In Vitro* Cell Dev Biol 38: 213-220.

104. Cheng X, Sardana R, Kaplan H, Altosaar I (1998) Agrobacterium-transformed rice plants expressing synthetic cryIA(b) and cryIA(c) genes are highly toxic to striped stem borer and yellow stem borer. Proc Natl Acad Sci USA 95: 2767-2772.

105. Datta K, Vasquez A, Tu J, Torrizo L, Alam MF, et al. (1998) Constitutive and tissue–specific differential expression of CryIA(b) gene in transgenic rice plants conferring enhanced resistance to insect pests. Theor Appl Genet 97: 20-30.

106. Li L, ZhuY, Jin S, Zhang X (2014) Pyramiding Bt genes for increasing resistance of cotton to two major lepidopteran pests: Spodoptera lituraand *Heliothis armigera*. Acta Physiol Plant 36: 2717-2727.

107. Jackson RE (2004) Performance of feral and Cry1Ac-selected Helicoverpa zea (Lepidoptera: Noctuidae) strains on transgenic cottons expressing one or two Bacillus thuringiensis ssp kurstaki proteins under greenhouse conditions. Journal of Entomology Society 39: 46-55.

108. Gahan LJ, Ma YT, Cobble MLM, Gould F, Moar WJ, et al. (2005) Genetic basis of resistance to Cry 1Ac and Cry 2Aa in *Heliothis virescens* (Lepidoptera: Noctuidae). J Econ Entomol 98: 1357- 1368.

109. Cao J, Shelton AM, Earle ED (2008) Sequential transformation to pyramid two Bt genes in vegetable Indian mustard (*Brassica juncea* L.) and its potential for control of diamondback moth larvae. Plant Cell Rep 27: 479-487.

110. Meenakshi M, Aditya KS, Indraneel S, Illimar A, Amla DV (2011) Pyramiding of modified cry1Ab and cry1Ac genes of *Bacillus thuringiensis* in transgenic chickpea (*Cicer arietinum* L.) for improved resistance to pod borer insect H*elicoverpa armigera*. Euphytica 182: 87-102.

111. Borejsza-Wysocka E, Norelli JL, Aldwinckle HS, Malnoy M (2010) Stable expression and phenotypic impact of attacin E transgene in orchard grown apple trees over a 12-year period. BMC Biotechnol 10: 41.

112. Tripathi L, Mwaka H, Tripathi JN, Tushemereirwe WK (2010) Expression of sweet pepper Hrap gene in banana enhances resistance to *Xanthomonas campestris* pv. musacearum. Mol Plant Pathol 11: 721-31.

113. Foster SJ, Park TH, Pel M, Brigneti G, Sliwka J, et al. (2009) Rpi-vnt1.1, a Tm-2(2) homolog from *Solanum venturii*, confers resistance to potato late blight. Mol Plant Microbe Interact 22: 589-600.

114. Bradeen JM, Iorizzo M, Mollov DS, Raasch J, Kramer LC, et al. (2009) Higher copy numbers of the potato RB transgene correspond to enhanced transcript and late blight resistance levels. Mol Plant Microbe Interact 22: 437-446.

115. Halterman D, Kramer L, Wielgus S, Jiang J (2008) Performance of transgenic potato containing the late blight resistance gene RB. Plant Dis 92: 339-343.

116. Truve E, Aaspollu A, Honkanen J, Puska R, Mehto M, et al. (1993) Transgenic potato plants expressing mammalian 2'–5' oligoadenylate synthetase are protected from potato virus X infection under field conditions. Biotechnol 11: 1048-1052.

117. Zhao BY, Lin XH, Poland J, Trick H, Leach J, et al. (2005) A maize resistance gene functions against bacterial streak disease in rice. Proc Natl Acad Sci USA 102: 15383-15388.

118. Lius S, Manshardt RM, Fitch MM, Slightom JL, Sanford JC, et al. (1997) Pathogen-derived resistance provides papaya with effective protection against papaya ringspot virus. Mol Breed 3: 161-168.

119. Lacombe S, Rougon-Cardoso A, Sherwood E, Peeters N, Dahlbeck D, et al. (2010) Interfamily transfer of a plant pattern-recognition receptor confers broad-spectrum bacterial resistance. Nat Biotechnol 28: 365-369.

120. Brunner S, Stirnweis D, Quijano DC, Buesing G, Herren G, et al. (2012) Transgenic Pm3 multiline of wheat show increased powdery mildew resistance in the field. Plant Biotechnol J 10: 398-409.

8

Genetic and Pathogenic Variability of *Ascochyta rabiei* Isolates from Pakistan and Syria as Detected by Universal Rice Primers

Hina Ali*, Syed Sarwar Alam and Nayyer Iqbal

Nuclear Institute for Agriculture and Biology, Faisalabad, Pakistan

Abstract

Ascochyta rabiei is the casual agent of blight disease of chickpea (*Cicer arietinum* L.). The study was aimed to assess the genetic diversity of highly aggressive *Ascochyta rabiei* (AR) isolates (pathotypes III and IV) from Syria and its comparison with highly aggressive isolates from Pakistan. AR isolates were characterized for pathogenicity assay and genetic variability. Previously genetic variability of AR isolates have been checked with RAPD and SSR markers, here we are reporting for the first time diversity using ten Universal rice primers (URP) derived from the repeat sequence of rice genome. URP proved very useful for the characterization of isolates and clearly differentiated Syrian pathotypes from Pakistani ones. URP can be helpful in studying the population variability from AR pathotypes worldwide.

Keywords: *Ascochyta rabiei*; Chickpea; URP; Pathotypes

Introduction

Blight disease in chickpea is caused by a fungus that exists both in sexual (*Didymella rabiei*) and asexual stages (*Ascochyta rabiei*) [1]. Pathotype variability is necessary to select the appropriate pathotype for screening genotypes in resistance breeding programme [2]. Differentiation of *Ascochyta rabiei* (AR) into 3 classes (pathotype I, II and III) was reported in Syria and has been widely accepted [3] and recently highly aggressive pathotype IV has been reported by Imtiaz et al. [4]. In Pakistan three pathotypes were also identified by Jamil et al. [5] and Ali et al. [6] using chickpea differential genotypes (ILC1929, ILC482 and ILC3279) and (Spanish white, Dwelley and ICC12004) respectively. In this study we have compared pathogenic behavior of Pakistani AR isolates (PAR) with pathotypes III and IV from Syria on chickpea differential genotype ICC12004 (resistant to pathotype III as mentioned by Taleei et al. [7]. The isolates were also characterized using 3 SSR markers [8] and 10 Universal rice primers (URP) that are repeat motifs obtained from Korean weedy rice and have been utilized in diverse genome like animals, plants and microbes [9]. Previous studies were based on RAPD and SSR markers for the assessment of genetic diversity of AR isolates [3,5,6,10-13]. Only SSR markers were reported by Imtiaz et al. (2011) to differentiate pathotype III and IV from Syria.

Materials and Methods

For this study, 24 isolates including 18 from different geographical locations of Pakistan alongwith 6 Syrian AR (SAR) isolates (three each of pathotypes III and IV) were used. DNA was extracted from isolates using CTAB method [14] from the pycnidial growth of fungus on V8 medium and was diluted to 10 ng/µl for PCR. URP-PCR conditions were followed as described by Aggarwal et al. [15] with slight modifications and DNA amplification was performed in a gradient thermal cycler Infinigen (Germany) programmed for 40 cycles with initial denaturation at 94°C for 4 min then 1 min at 94°C, 1 min at 55°C and 2 min at 72°C. Cycling was concluded with final extension at 72°C for 10 min. Amplified products were separated on 1.5% ethidium stained agarose gels and photographed under UV light. Un-weighted pair group method with arithmetic means (UPGMA) cluster analysis was performed on binary data from URP banding profile based on

Jaccard's coefficient using Genstat Programme, 10th ed.; Rothamsted Experimental Station, Harpenden, United Kingdom

Pathogenicity assay was performed using 18 PAR isolates on chickpea differential genotypes viz. Spanish white (susceptible to pathotype I), Dwelley (susceptible to pathotype II) and ICC12004 (susceptible to pathotype III). One pathotype IV from Syria was also used for pathogenicity assay in this study using a mini dome assay described by Chen and Muehlbauer [16]. For pathogenicity assay, AR isolates were grown on autoclaved and boiled chickpea seeds [17] for two weeks. Seeds were then crushed with glass rod and autocalved distilled water was added to get pycnidial colonies. Spores concentration was adjusted with the help of haemocytometer and set to 2×10^5 spores mL^{-1} and sprayed with the help of spray pump (approx. 2ml per plant) on two weeks old seedlings of chickpea differential genotypes. Three replicates of each genotype were used and placed in growth room at 21°C for 12 h and 16°C for 12 h night at 100% relative humidity. Each replicate comprised of two plants per pot and these were immediately covered with plastic dome to create high humidity for 48 hrs to spread disease. Control plants were sprayed with water and treated in the same way as inoculated plants. Data was recorded after 14 days of inoculation based on 1-9 rating scale defined by Reddy & Singh [18] and modified for seedling assay by Chen et al. [19]. Disease severity was analyzed by Analysis of variance (ANOVA) using the program Statistix version 8.1 [20] and the means were compared with least significance difference (LSD) at 1% level of significance for comparison of treatment mean.

---

*Corresponding author: Hina Ali, Nuclear Institute for Agriculture and Biology, Faisalabad, Pakistan, E-mail: hinali991@hotmail.com

Results and Discussions

Out of 18 isolates tested, PAR 4 was found to be pathotype I based on its susceptible reaction on Spanish white only (susceptible to pathotype I) but did not cause any disease symptoms on Dwelley and ICC12004. 17 isolates showed virulence on Dwelley (susceptible to pathotype II and III) and ICC12004 (susceptible to pathotype III). Disease incidence on the differential lines was highly significantly (P<0.01) caused by treatment with different PAR isolates. PAR Isolates including SAR (pathotype IV) infected ICC12004 with a disease rating ranging from 5.3 to 8.0 as shown in Figure 1.

SSR marker ArH05T was recently reported to amplify Syrian pathotypes III and IV specific bands [4] while in our study two other SSR markers (ArA06T and ArH02T) also differentiated Syrian pathotypes III from IV but unable to distinguish PAR from SAR (Figure 2a). On the other hand the URP primers were highly polymorphic and not only differentiated least aggressive pathotype I (PAR 4) from rest of highly aggressive pathotypes III and IV but also differentiated PAR from SAR (Figure 2b). The cluster analysis based on UPGMA utilizing URP banding profile differentiated 24 isolates into three major groups (Figure 3). Group 1 comprised of least aggressive pathotype I (PAR 4) that out grouped singly. The second major group consisted of 2 sub-groups, 11 PAR isolates in one and 6 SAR isolates in another sub-group. The third group comprised of remaining 6 PAR isolates. The clustering of PAR isolates was not consistent with the geographical distribution of the isolates which is in agreement with our previous results [6]. URPs discriminated Syrian AR isolates from Pakistani AR that may be owing to their geographical boundaries whereas SSR markers showed same banding pattern for isolates from both countries used in this study (Figure 2a). One set of highly pathogenic 11 PAR isolates formed a cluster with Syrian isolates while other set comprising of 6 highly aggressive formed a separate group. The distinction of highly aggressive isolates into two groups might be due to complex nature of genetics of AR pathogenicity. Variation in disease reaction among different isolates was observed on ICC12004 which might be due to natural distribution of aggressiveness in the

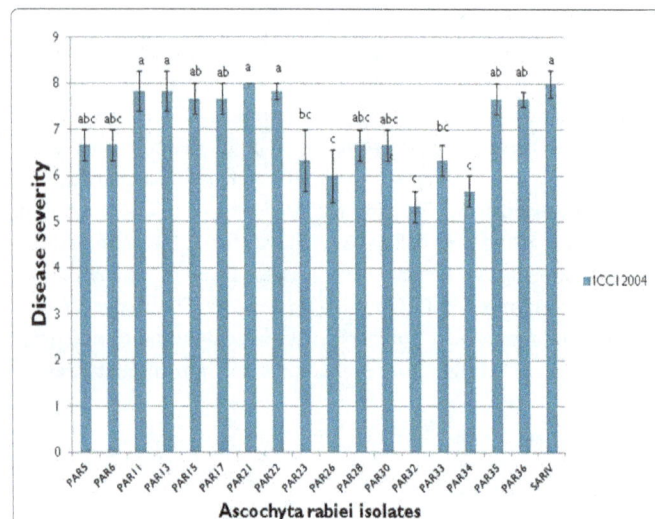

Figure 2a: Banding profile of SSR marker (ArH05T) of AR isolates on 1.5% agarose. Lane 1-23 (PAR4, PAR6, PAR11, PAR15, PAR17, PAR21, PAR22, PAR23, PAR26, PAR28, PAR30, PAR32, PAR33, PAR34, PAR35, PAR36, PAR5, PAR13, SARIII, SARIII, SARIII, SARIV, SARIV), M- 100 bp Mol wt. marker.

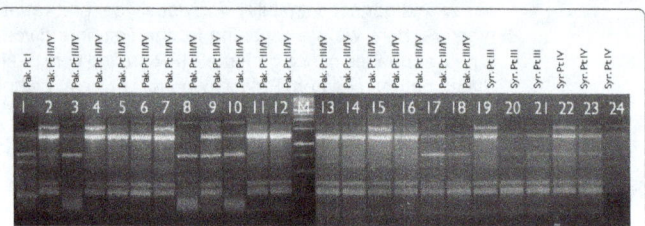

Figure 2b: Banding profile of URP marker (URP-9F) of AR isolates on 1.5% agarose. Lane 1-24 (PAR4, PAR6, PAR11, PAR15, PAR17, PAR21, PAR22, PAR23, PAR26, PAR28, PAR30, PAR32, PAR33, PAR34, PAR35, PAR36, PAR5, PAR13, SARIII, SARIII, SARIII, SARIV, SARIV, SARIV) M- 100 bp Mol wt. marker.

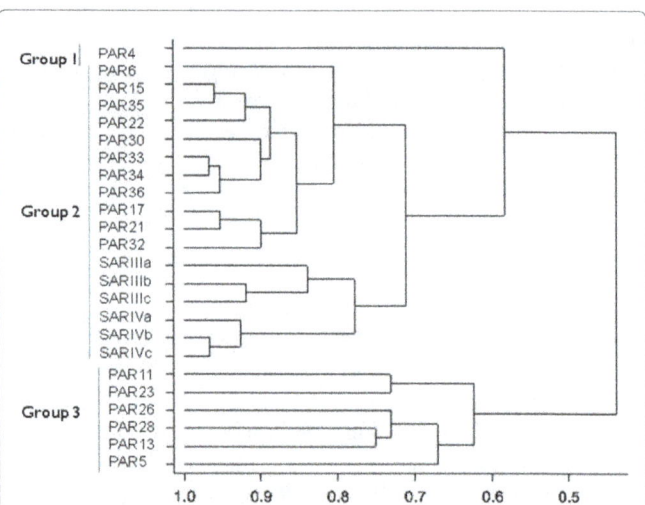

Figure 3: Dendrogram showing three main groups for 24 *Ascochyta rabiei* isolates using UPGMA based on ten universal rice primers (URP) banding profile.
PAR - Pakistani *Ascochyta rabiei*
SAR – Syrian *Ascochyta rabiei*

Figure 1: Pathogenic effect of different Pakistani AR isolates (PAR) on chickpea differential genotype ICC12004. Bars represent mean disease severity (± S.E.) at rating scale (1-9). Bars sharing the same letters are not significantly different at 1% level of significance (Least significant difference test [LSD], Statistix 8.1) Pairwise comparison of data. There are 3 groups (a, b, etc.) in which the means are not significantly different from one another.

pathogen population. Differences in aggressiveness among isolates on a particular differential genotype (ICC12004) seems complicated and needs further investigations as previously many races and pathotypes have been reported for AR [21]. Thus differential genotype/s other than ICC12004 need to be identified that can clearly differentiate highly aggressive pathotypes of AR, moreover DNA markers are required for differentiating pathotype III and IV that would help also in planning strategies for effective and continuous resistance breeding.

Acknowledgement

Pathotype III and IV of *Ascochyta rabiei* was kindly provided by Dr. Muhammad Imtiaz and Dr. Aladdin Hamwieh, ICARDA, Syria. Authors are also thankful to

Alexander Von Humboldt. (AVH) Foundation, Germany, for subsidizing equipment used to conduct molecular work in this study.

References

1. Wilson AD, Kaiser WJ (1995) Cytology and genetics of sexual incompatibility in *Didymella rabiei*. Mycologia 87: 795-804.

2. Sarwar N, Ashfaq S, Akhtar KP, Jamil FF (2013) Biological pathotyping and RAPD analysis of *Ascochyta rabiei*, from various chickpea growing areas of Pakistan. J Animal and Plant Sciences 23: 882-887.

3. Udupa SM, Weigand F, Saxena M, Kahl G (1998) Genotyping with RAPD and microsatellite markers resolves pathotype diversity in the Ascochyta blight pathogen of chickpea. Theor Appl Genet 97: 299-307.

4. Imtiaz M, Abang MM, Malhotra RS, Ahmed S, Bayaa B, et al. (2011) Pathotype IV, a new and highly virulent pathotype of *Didymella rabiei*, causing Ascochyta blight in chickpea in Syria. Plant disease 95: 1192-1192.

5. Jamil FF, Sarwar N, Sarwar M, Khan JA, Geistlinger J, et al. (2000) Genetic and pathogenic diversity within *Ascochyta rabiei* (Pass.) Lab. populations in Pakistan causing blight of chickpea (*Cicer arietinum* L.). Physiol Mol Plant Pathol 57: 243-254.

6. Ali H, Alam SS, Attanayake R, Rahman M, Chen W (2012) Population structure and mating type distribution of the chickpea blight pathogen *Ascochyta rabiei* from Pakistan and United States. J Plant Pathol 94: 99-108.

7. Taleei A, Kanouni H, Baum M (2009) Genetical analysis of Ascochyta blight resistance in chickpea. In: Bioscience and Biotechnology, Communications in Computer and Information ScienceSpringer-Verlag, Berlin, Heidelberg.

8. Geistlinger J, Weising K, Winter P, Kahl G (2000) Locus-specific microsatellite markers for the fungal chickpea pathogen *Didymella rabiei* (anamorph) *Ascochyta rabiei*. Mol Ecol 9: 1939-1941.

9. Kang HW, Park DS, Park YJ, You CH, Lee BM, et al. (2002) Fingerprinting of diverse genomes using PCR with universal rice primers generated from repetitive sequence of Korean weedy rice. Mol Cells 13: 281–297.

10. Navas-Cortés JA, Perez-Artes E, Jiménez-Díaz RM, Lobell A, Bainbridge BW, et al. (1998) Mating Type, pathotype and RAPDs analysis in *Didymella rabiei*, the agent of Ascochyta blight of chickpea. Phytoparasitica 26: 199-212.

11. Chongo G, Gossen BD, Buchwaldt L, Adhikari T, Rimmer SR (2004) Genetic diversity of *Ascochyta rabiei* in Canada. Plant Disease 88: 4-10.

12. Jamil FF, Sarwar M, Sarwar N, Khan JA, Zahid MH, et al. (2010) Genotyping with RAPD markers resolves pathotype diversity in the Ascochyta blight and Fusarium wilt pathogens of chickpea in Pakistan. Pak J Bot 42: 1369-1378.

13. Atik O, Ahmed S, Abang MM, Imtiaz M, Hamwieh A, et al. (2013) Pathogenic and genetic diversity of *Didymella rabiei* affecting chickpea in Syria. Crop Protection 46: 70-79.

14. Khan IA, Awan FS, Ahmad A, Khan AA (2004) A Modified mini-prep method for economical and rapid extraction of genomic DNA in plants. Plant Mol Biol Rep 22: 89a–89e

15. Aggarwal R, Sharma V, Kharbikar LL, Renu (2008) Molecular characterization of Chaetomium species using URP-PCR. Genet Mol Bio 31: 943-946.

16. Chen W, Muehlbauer FJ (2003) An Improved Technique for virulence assay of *Ascochyta rabiei* on chickpea. ICN 10: 31–33.

17. Alam SS, Strange RN, Qureshi SH (1987) Isolation of *Ascochyta rabiei* and a convenient method for inoculum production. Mycologist 21: 20.

18. Reddy MV, Singh KB (1984) Evaluation of a world collection of chickpea germplasm accessions for resistance to Ascochyta blight. Plant Disease 68: 900–901.

19. Chen W, Coyne CJ, Peever TL, Muehlbauer FJ (2004) Characterization of chickpea differentials for pathogenicity assay of Ascochyta blight and identification of chickpea accessions resistant to *Didymella rabiei*. Plant Pathol 53: 759-769.

20. Analytical Software (2005) Statistix 8.1 for Windows. Analytical Software, Tallahassee, Florida.

21. Taylor PWJ, Ford R (2007) Diagnostics, genetic diversity and pathogenic variation of Ascochyta blight of cool season food and feed legumes. Eur J Plant Pathol 119: 127-133.

9

Identification of a Universal Marker for Detecting Possible Mutation in *Botrytis cinerea* Isolates Associated with Virulence

Moytri Roy Chowdhury[1,2]*, Jake R. Erickson[1] , Peter Rafael Ferrer[2], Brian Foley[3], Shannon Piele[2], James Titius[2], Kshitij Shrestha[1] and Caleb Fiedor[2]

[1]*Department of Biological Sciences, Idaho State University, 650 Memorial Dr. Gale Life Sciences Building, Pocatello, ID 83209, USA*
[2]*Department of Biology, Fresno Pacific University, 1717 S Chestnut Ave, Fresno, CA 93702, USA*
[3]*Los Alamos National Laboratory, Santa Fe, NM 87545, USA*

Abstract

Botrytis cinerea sporadically infects plants in Mediterranean climates and contributes to a significant crop loss every year. Diseases caused by *B. cinerea* can affect many crops and are of particular concern to strawberry growers in California, which is the major state for fresh strawberry production in the United States. This study looks at genetic mutations, and the subsequent phenotypical changes, of several strains of *B. cinerea* obtained from plant tissues. It was found that strains with four nucleotide insertions were more virulent than their wild type counterparts. Strains with single nucleotide polymorphisms had conidia smaller or similar in size to the wild type strains and exhibited similar virulent properties as the wild type. We also observed identical mutations of fungal samples obtained from different plant tissues from Asia.

Keywords: *B. cinerea*; Isolates; Virulence; Strawberries; Mutation

Introduction

B. cinerea Pers.: Fr and related *Botrytis* species are common and economically important pathogens affecting a wide variety of vegetables and ornamental crops [1]. While there are 25-30 necrotropic plant pathogens closely related to *B. cinerea* Pers.: Fr (teleomorph *Botryotinia fuckelian* (de Bary) Whetzel, *B. cinerea* is the most comprehensively studied [2,3]. The added attention is due to the vast economic effects caused by the fungus's widespread host range, infecting over 200 plant species, and by its involvement in pre- and post-harvest crop losses worldwide, which cause vast economic effects [2,4]. The ability to infect diverse plant species and tissues under a wide-range of environmental conditions is aided by *B. cinerea's* ability to survive in topsoil, which cause it to be very persistent in nature. It is also aided by dispersed airborne conidia or ascospores, which allow *B. cineria* to infect host tissue and sexually reproduce [5].

The conidia, which are commonly found in large quantities in necrotophic host tissue, are the main sources of infection by *B. cinerea* [2]. In the presence of high levels of humidity (>93%), germination is activated by physical and chemical signals. This activation is a cAMP-dependent pathway and MAP kinase cascade, which are involved in signal transductions that lead to germination [2]. Because of their low degree of differentiation, the ability of *B. cinerea* appressoria to physically penetrate intact host tissue seems to be limited and probably needs to be supported by secreted lytic enzymes [6]. *B. cinerea* is found to secrete a number of endopolygalacturonases, which are used to soften cell walls [7]. Although, the ability to penetrate the cell wall of host tissues might need the action of specific chitin synthase isoenzymes, which is shown to be necessary for pathogenicity [8,9]. Once penetration has occurred, it is typically followed by the prompt death of the host cell.

The fungus is phenotypically characterized by grey mold on the tissue that it infects [10]. The pattern of growth can vary in color and density, but in general the pathogen has gray mycelium that differentiates it from other fungi [11]. It occurs sporadically in Mediterranean climates and contributes to a significant crop loss every year, making *B. cinerea* an extreme nuisance in California [12].

Out of all the Californian crops that *B. cinerea* is known to infect, it has one of the most undesirable effects on its sixth most valuable crop, the field strawberry (*Fragaria ananassa*). California is the major state for strawberry production in the U.S., accounting for over 80% of the fresh market product grown [13]. Second to grapes, strawberries are the most important cultivated fruit of the berry family worldwide with 3.82 million tons produced per year [14]. *B. cinerea* causes one of the most prominent post-harvest molds that cause quality deterioration of strawberries and other fruits in the field and during refrigeration storage [15,16]. Gray mold can cause up to a 30% to 40% loss of the harvest where chemical methods of pest management are not practiced. This loss can reach 50% to 60% in areas of acute infestations, in which the economic loss will be up to 100% [15].

Botryticides and fungicides are currently used to combat these losses, but both come with their own sets of problems. The use of botryticides, despite their latest availability, results in continued crop loss, while the use of fungicides is becoming increasingly restricted or forbidden because of their unfavorable effects on ecosystem and human health [12]. Fungicides are still used for disease control, but the fungi are adapting quickly to these drugs *via* mutations, including the insertion of retroposon sequences into gene promoters [17,18]. As a result of recurring fungicide treatments, resistant strains are quickly developing [19]. Three types of increasing multidrug resistance populations of *B. cinerea* have appeared since the mid 1990's in Europe [17]. After studying the immerging resistant strains, a few distinctive fungicide resistance mechanisms have been identified in both field and greenhouse environments [20,21]. It has been anticipated that by continually selecting multi-resistant strains, chemical control of *B. cinerea* in the field will become increasingly difficult [17].

With the increased resistance and concern for human health in the last decade, there is a need for a greater understanding of the development of this resistance epidemic in order to create resilient and

*Corresponding author: Moytri RoyChowdhury, Department of Biological Sciences, Idaho State University, 650 Memorial Dr. Gale Life Sciences Building, Pocatello, ID 83209, USA, E-mail: roycmoyt@isu.edu

Isolate number	Isolated from	Type of infection	Area	Ranch	Date	Note
BOT01	Strawberry	postharvest	Oxnard	04.2156	Feb. 2008	Fruit
BOT02	Strawberry	postharvest	Oxnard	04.2156	Feb. 2008	Fruit
BOT03	Strawberry	postharvest	Oxnard	04.2156	Feb. 2008	Fruit
BOT04	Strawberry	postharvest	Oxnard	04.2156	Feb. 2008	Fruit
BOT05	Strawberry	postharvest	Oxnard	04.2156	Feb. 2008	Fruit
BOT06	Strawberry	postharvest	Oxnard	04.2156	Feb. 2008	Fruit
BOT07	Strawberry	postharvest	Oxnard	04.2156	Feb. 2008	Fruit
BOT08	Strawberry	postharvest	Oxnard	04.2156	Feb. 2008	Fruit
BOT09	Strawberry, El Dorado	field	Oxnard	Sammis	Mar.2008	Fruit
BOT10	Strawberry, El Dorado	field	Oxnard	Lennox	Mar.2008	Fruit
BOT11	Strawberry, El Dorado	field	Oxnard	Sammis	Mar.2008	Fruit
BOT12	Strawberry, El Dorado	field	Oxnard	Beardsley	Mar.2008	Fruit
BOT13	Strawberry, El Dorado	field	Oxnard	Sammis	Mar.2008	Fruit
BOT14	Strawberry, El Dorado	field	Oxnard	Lennox	Mar.2008	Fruit
BOT15	Strawberry, El Dorado	field	Oxnard	Sammis	Mar.2008	Fruit
BOT16	Strawberry, El Dorado	field	Oxnard	Sammis	Mar.2008	Fruit
BOT17	Strawberry, El Dorado	field	Oxnard	Sammis	Mar.2008	Fruit
BOT18	Strawberry, El Dorado	field	Oxnard	Sammis	Mar.2008	Fruit
BOT19	Strawberry, El Dorado	field	Oxnard	Beardsley	Mar.2008	Fruit
BOT20	Strawberry, El Dorado	field	Oxnard	Sammis	Mar.2008	Fruit
BOT21	Strawberry, El Dorado	field	Oxnard	Lennox	Mar.2008	Fruit
BOT22	Strawberry, El Dorado	field	Oxnard	Lennox	Mar.2008	Fruit
BOT23	Strawberry, El Dorado	field	Oxnard	Beardsley	Mar.2008	Fruit
BOT24	Strawberry, El Dorado	field	Oxnard	Beardsley	Mar.2008	Fruit
BOT25	Strawberry, El Dorado	field	Oxnard	Beardsley	Mar.2008	Fruit
BOT26	Strawberry, El Dorado	field	Oxnard	Beardsley	Mar.2008	Fruit
BOT27	Strawberry, El Dorado	field	Oxnard	Beardsley	Mar.2008	Fruit
BOT28	Strawberry, El Dorado	field	Oxnard	Beardsley	Mar.2008	Fruit
BOT29	Strawberry, El Dorado	field	Oxnard	Beardsley	Mar.2008	Fruit
BOT30	Strawberry, El Dorado	field	Oxnard	Beardsley	Mar.2008	Fruit
BOT31	Strawberry, El Dorado	field	Oxnard	Sammis	Mar.2008	Fruit
BOT32	Strawberry, El Dorado	field	Oxnard	Sammis	Mar.2008	Fruit
BOT33	Strawberry, El Dorado	field	Oxnard	Lennox	Mar.2008	Fruit
BOT34	Raspberry, Pacifica	field	Oxnard	Lennox	Mar.2008	Fruit
BOT35	Raspberry, Pacifica	field	Oxnard	Lennox	Mar.2008	Fruit
BOT36	Raspberry, Pacifica	field	Oxnard	Lennox	Mar.2008	Fruit
BOT37	Raspberry, Pacifica	field	Oxnard	Sammis	Mar.2008	Fruit
BOT38	Raspberry, Pacifica	field	Oxnard	Beardsley	Mar.2008	Fruit
BOT39	Raspberry, Pacifica	field	Watsonville		May.2008	Cane infection
BOT40	Raspberry, Pacifica	field	Watsonville		May.2008	Cane infection
BOT41	Raspberry, Pacifica	field	Watsonville		May. 2008	Cane infection
BOT42	Raspberry, Pacifica	field	Watsonville		May. 2008	Cane infection
BOT43	Raspberry, Pacifica	field	Watsonville		May. 2008	Cane infection
BOT44	Raspberry, Pacifica	field	Watsonville		May.2008	Cane infection
BOT45	Raspberry, Pacifica	field	Watsonville		May.2008	Cane infection
BOT46	Strawberry	field	Watsonville		May. 2008	Cane infection
BOT47	Strawberry	field	Watsonville		May. 2008	Cane infection
BOT48	Strawberry	field	Watsonville		May. 2008	Cane infection
BOT49	Strawberry	field	Watsonville		May. 2008	Cane infection
BOT50	Strawberry	field	Salinas	Davis	Jun. 2008	Fruit
BOT51	Strawberry	field	Watsonville	TCR	Jun. 2008	Fruit
BOT52	Strawberry	field	Salinas	Davis	Jun. 2008	Fruit
BOT53	Strawberry	field	Salinas	Davis	Jun. 2008	Fruit
BOT54	Strawberry	field	Salinas	Davis	Jun. 2008	Fruit
BOT55	Strawberry	field	Salinas	Davis	Jun. 2008	Fruit
BOT56	Strawberry	field	Salinas	Davis	Jun. 2008	Fruit
BOT57	Strawberry	field	Watsonville	TCR	Jun. 2008	Fruit
BOT58	Strawberry	field	Salinas	Davis	Jun. 2008	Fruit
BOT59	Strawberry	field	Salinas	Davis	Jun. 2008	Fruit
BOT60	Strawberry	field	Salinas	Davis	Jun. 2008	Fruit

BOT61	Strawberry	field	Salinas	Davis	Jun. 2008	Fruit
BOT62	Strawberry	field	Salinas	Davis	Jun. 2008	Fruit
BOT63	Strawberry	field	Salinas	Davis	Jun. 2008	Fruit
BOT64	Strawberry	field	Watsonville	TCR	Jun. 2008	Fruit
BOT65	Strawberry	field	Watsonville	TCR	Jun. 2008	Fruit
BOT66	Strawberry	field	Watsonville	TCR	Jun. 2008	Fruit
BOT67	Strawberry	field	Watsonville	TCR	Jun. 2008	Fruit
BOT68	Strawberry	field	Salinas	Davis	Jun. 2008	Fruit
BOT69	Strawberry	field	Salinas	Davis	Jun. 2008	Fruit
BOT70	Strawberry	field	Salinas	Davis	Jun. 2008	Fruit
BOT71	Strawberry	field	Salinas	Davis	Jun. 2008	Fruit
BOT72	Strawberry	field	Salinas	Davis	Jun. 2008	Fruit
BOT73	Strawberry	field	Salinas	Davis	Jun. 2008	Fruit
BOT74	Strawberry	field	Watsonville	TCR	Jun. 2008	Fruit
BOT75	Strawberry	field	Watsonville	TCR	Jun. 2008	Fruit
BOT76	Strawberry	field	Watsonville	TCR	Jun. 2008	Fruit
BOT77	Strawberry	field	Watsonville	TCR	Jun. 2008	Fruit
BOT78	Blackberry	field	Watsonville	TCR	7/11/2014	Leaf Petiole
BOT79	Strawberry	field	Nursery	Tule Lake	6/19/2014	Fruit

Table 1: List of isolates collected and used for morphological, phylogenetic, and pathogenic analysis.

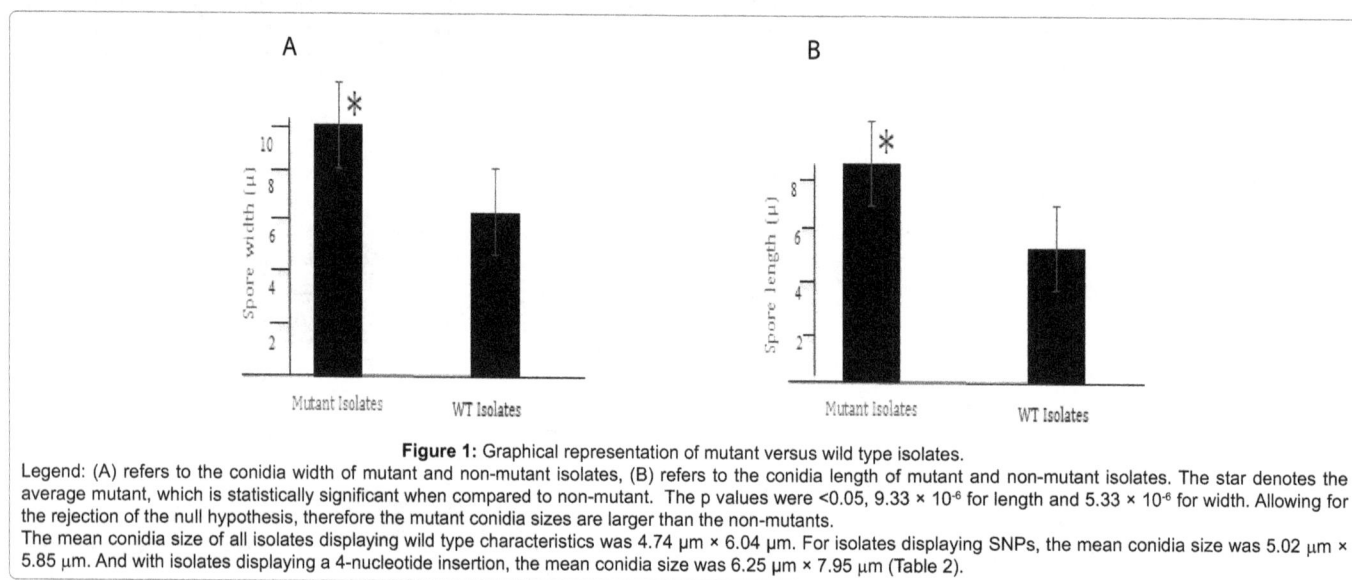

Figure 1: Graphical representation of mutant versus wild type isolates.
Legend: (A) refers to the conidia width of mutant and non-mutant isolates, (B) refers to the conidia length of mutant and non-mutant isolates. The star denotes the average mutant, which is statistically significant when compared to non-mutant. The p values were <0.05, 9.33×10^{-6} for length and 5.33×10^{-6} for width. Allowing for the rejection of the null hypothesis, therefore the mutant conidia sizes are larger than the non-mutants.
The mean conidia size of all isolates displaying wild type characteristics was 4.74 μm × 6.04 μm. For isolates displaying SNPs, the mean conidia size was 5.02 μm × 5.85 μm. And with isolates displaying a 4-nucleotide insertion, the mean conidia size was 6.25 μm × 7.95 μm (Table 2).

maintainable disease control methods [12]. A rapid and convenient method to detect resistant mutations is vital for understanding and managing fungicide resistance [22]. In using genomics to continue the study of mutations in *B. cinerea*, it is likely that the identification of resistance causing mutations can lead to effective disease control methods. Therefore, the objectives of this research were to: i) identify genetic diversity of *Botrytis* isolates obtained from strawberries in California using sequencing analysis, ii) identify morphological and phylogenetic diversity of California's population of *Botrytis* in strawberries, iii) use pathogenicity assay to detect if the diversity has an impact on virulence. This study was undertaken to characterize and identify the different species of *Botrytis* using sequencing analysis and morphology. In addition, pathogenicity studies were undertaken to understand the role of these species in fruit decay [23].

Materials and Methods

Fungal isolate collection

We obtained a total of 79 pure culture isolates from Driscoll's Strawberry Associates, Watsonville, California (Table 1). The isolates were grown on Potato Dextrose Agar (PDA) with 0.01% tetracycline at 27°C for a period of 21 days. Plugs (0.6 cm diameter) of developed hyphae were placed in the center of a PDA plate and the cultures were routinely transferred every 3 week.

Microscopy

Microscopy was done using a phase contrast microscope (M4000-D Swift Instruments Inc, Japan). Measurements were taken at 1000 x magnification using oil immersion. Mycelia used in this study were 21 days old. Mycelia were placed on a slide and stained with methylene blue prior to observation under the microscope. We measured the conidia length and width for all 79 isolates, each with three replicates. Measurements were done using the ocular micrometer after calibrating the microscope with the stage micrometer. Photographs were taken with 8-megapixel iSight camera with 1.5 u pixels.

DNA extraction

21-day-old Mycelium was used to extract DNA from each of the isolates. DNA was extracted from the fungal isolates using PrepMan

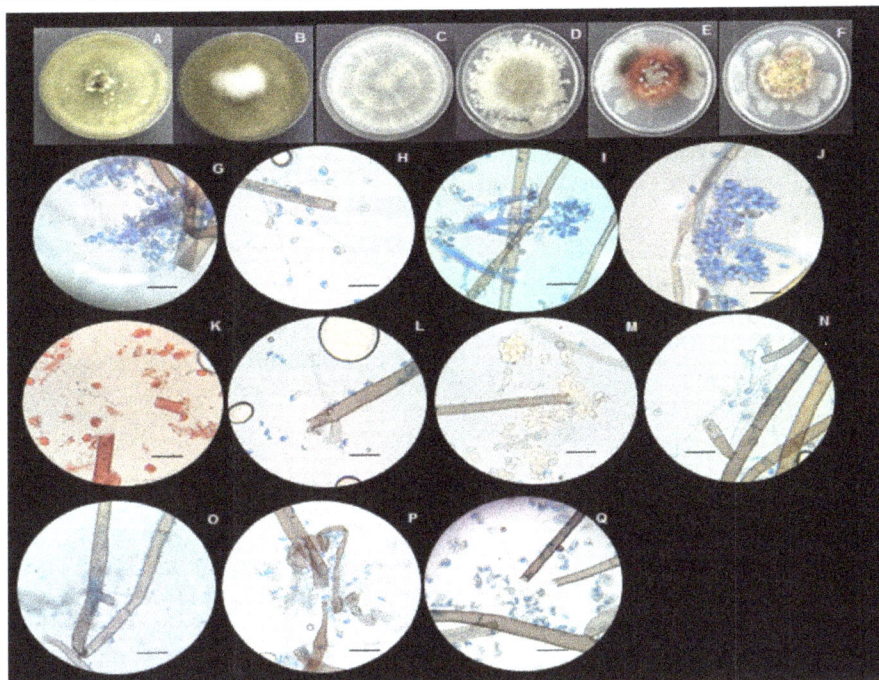

Figure 2: Morphological diversity of *Botrytis*.

Legend: A: Isolate 49 (Possible mutant) hyaline pattern in middle with spores dispersing only in center. Black center with dark yellow mycelium; B: Isolate 68 (mutant) powdery center with light brown mycelium in a circular form; C: Isolate 15 hyaline patter, radiating in circular growth from the center outward, very light yellow and white in color; D: Isolate 61 (mutant) visible growth of branching hyphae, brown in the middle and white/light yellow; E: Isolate 70 same branching patter as isolate 61 with very dark brown and bright red center; F: Isolate 58 similar to 70, but with less growth of mycelium outward, hints of red, green, and yellow; G isolate 30 (4 nucleotide insertion mutant) measurements 7.87 μm × 8.59 μm; H isolate 33 (4 nucleotide insertion mutant) measurements 6.35 μm × 10.96 μm; I isolate 35 (4 nucleotide insertion mutant) measurements 8.71 μm × 9.29 μm; J isolate 39 (4 nucleotide insertion mutant) measurements 7.92 μm × 10.03 μm; K isolate 49 (4 nucleotide insertion mutant) measurements 8.63 μm × 9.71 μm; L isolate 61 (3 SNP's) measurements 5.07 μm × 6.36 μm; M isolate 68 (4 nucleotide insertion mutant) measurements 8.42 μm × 10.75 μm; N isolate 77 (4 nucleotide insertion mutant) measurements 8.62 μm × 9.26 μm; O isolate 71 (wild type) measurements 4.94 μm × 6.2 μm; P isolate 62 (wild type) measurements 5.03 μm × 6.91 μm; Q isolate 44 (4 nucleotide insertion mutant) measurements 5.33 μm × 7.11 μm. All spore images are at 100 x magnification. Scale at 1 μm.

[TM] Ultra from Applied Biosystems by Life Technologies, Carlsbad, California [24,25]. The DNA was quantified using a nanodrop 2000c UV-Vis Spectrophotometer.

PCR and sequencing

PCR was done using a MJ Mini [TM] Personal Thermal Cycler from Bio-Rad. Amplification was done as follows: (1 cycle for 1 min at 95°C, 35 cycles of 45 sec at 94°C, 45 sec at 50°C, 45 sec for extension at 72°C; 1 cycle of final extension for 5 min at 72°C). Primers 729+(5'AGCTCGAGAGAGATCTCTGA3', bases 788,925–788,944) and 729-(5'CTGCAATGTTCTGCGTGGAA3', complement of bases 789,634–789,653), that are specific for *B. cinerea,* were used for amplification [26]. These primers amplify a region on chromosome 13 (bases 788,945–789,633), which is a part of the complete genome sequence of *B. cinerea* B05.10 (http://www.ncbi.nlm.nih.gov/bioproject/PRJNA264284). A single band of 0.7 kb that is specific to *B. cinerea* was amplified in our isolates. We also used a second set of primers: BC108+(5'-ACCCGCACCTAATTCGTCAAC-3', bases 789,015–789,035) and BC563-(5' GGGTCTTCGATACGGGAGAA-3', complement of bases 789,470–789,489). These new primers were used to amplify a DNA fragment of 0.48 kb to 0.36 kb in case there is a possible deletion of 0.12 kb that could not be detected with the primers C729+/– [27]. Upon verification of desired band sizes through gel electrophoresis, the PCR products were purified using the USB'EXOSAP-IT' reagent (USB Corporation, Cleveland Ohio) [28]. The purified PCR products were sequenced at UC Davis sequencing facility. Nucleotide sequences (bases 789,012 to 789,488 of chromosome

13) were analyzed and aligned using Vector NTi software version 8 (Life Technologies, Pleasanton, California).

Sequence analysis

The 476 bases (bases 789,012 to 789,488 of chromosome 13) and 688 bases (bases 788,945 to 789,633 of chromosome 13), which were sequenced and aligned to each other for analysis, are too small for accurate phylogenetic analysis. These gene regions contain a few polymorphic sites that are useful for distinguishing several genotypes. Thus, they are also useful for estimating population diversity on chromosome 13 among the isolates. We used the Highlighter tool to compress the sequence information into a figure that illustrates the diversity (Figure 1). (http://www.hiv.lanl.gov/content/sequence/HIGHLIGHT/HIGHLIGHT_XYPLOT/highlighter.html).

Pathogenicity assay

Isolates were grown on PDA plates at 27°C. Plugs (0.6 cm diameter) of developed hyphae were placed in the center of a PDA plate. The plates were grown for 21 days.

To produce inoculum, the spores were washed from the plates with 4 ml KH_2PO_4 glucose solution, filtered with 50 l m-mesh sieves, and adjusted to a concentration of 10^7 conidia mL^{-1} using Bright-Line haemocytometer (Cambridge Instruments Inc. USA) [3,29]. Before inoculation, the spore suspensions were vortexed thoroughly to ensure the homogeneous dispersal of conidia in the solution.

Whole *Fragaria ananassa*, the common field strawberry, were obtained commercially and used for inoculation. Uniform and undamaged strawberries were selected for experimentation. The fruits were surface-sterilized with 2% sodium hypochlorite (NaClO) for a period of 2 min, rinsed with sterile water, and then air dried on a clean bench for approximately 1 h [3]. Once dried, the fruits were injected with the inoculum (approximately 1-2 mm in depth) at the strawberry's equator with a sterile syringe [30]. The fruits were also inoculated with double distilled H_2O as a negative control.

Inoculations were done in egg cartons. The cartons were placed in a cardboard box and incubated at 20°C for 6 days. The egg cartons were randomized and their positions changed within the cardboard box. The fruits were monitored for lesion size every 12 h for a period of 5 days.

Results

Surveys

The host range and geographic distribution of *B.cinerea* in California are summarized in Table 1. There were 79 isolates collected from various locations in California and 63 of those isolates were sequenced. Agricultural crop hosts in this study on which *B. cinerea* were identified included: strawberry (*F. ananassa*) collected from Oxnard, Watsonville, Salinas, and Tule Lake; blackberry (*Rubus fruticosus*) from Watsonville; and raspberry (*Rubus idaeus*) from Oxnard and Watsonville. There were a total of 66 isolates collected from strawberries, 12 isolates collected from raspberries, and 1 from blackberry. *B. cinerea* samples were collected from infected areas of raspberry and strawberry crops. *B. cinerea* was collected from leaf petiole in the blackberry sample.

Morphology-pattern of growth

Various kinds of growth patterns were observed on the PDA agar plates at 27°C under light. Mycelia were produced in numerous patterns that displayed powdery, cottony, and radial characteristics (Figure 2). The colonies were varied, from white, dirty white or grayish in color to reddish, brown and black. Many displayed a hyaline pattern in the middle but as it dispersed it became light gray to dark brown. The pattern of growth was similar to the observations recorded by others [31].

Microscopy

Studies were conducted to measure the length and width of the conidia in the 79 studied isolates. We observed larger conidia sizes in the mutant isolates, identified through genomic studies, using sequencing analysis (Table 2) [32,33]. The larger conidia size was observed only in those mutant lines that had 4 nucleotide insertions (Table 3). However, isolates 04, 16, 59, 61, 63, and 74, which only had single nucleotide polymorphisms (SNPs), showed smaller/similar conidia size compared

Isolate number	Spore Width (um)			Mean	Standard Error	Spore Length (um)			Mean	Standard Error
	Replicate 1	Replicate 2	Replicate 3			Replicate 1	Replicate 2	Replicate 3		
BOT01	5.00	5.10	5.12	5.07	0.037	6.00	5.80	5.99	5.93	0.065
BOT02	5.00	5.20	5.11	5.10	0.058	6.14	6.10	6.20	6.15	0.029
BOT03	5.00	5.00	5.00	5.00	0.00	6.14	6.14	6.26	6.18	0.040
BOT04	4.71	4.70	4.74	4.72	0.012	6.86	6.82	6.65	6.78	0.064
BOT05	5.00	5.16	5.20	5.12	0.061	6.43	6.30	6.50	6.41	0.059
BOT06	4.71	4.69	4.68	4.69	0.0088	6.43	6.21	6.51	6.38	0.090
BOT07	5.00	5.10	5.00	5.03	0.033	5.00	4.97	5.20	5.06	0.072
BOT08	4.71	4.72	4.64	4.69	0.025	6.43	6.22	6.25	6.30	0.066
BOT09	5.00	5.30	5.20	5.17	0.088	6.43	6.43	6.49	6.45	0.020
BOT10	5.00	5.16	5.23	5.13	0.068	7.14	7.15	6.92	7.07	0.075
BOT11	4.71	4.66	4.69	4.69	0.015	5.71	5.66	5.55	5.64	0.047
BOT12	4.43	4.39	4.41	4.41	0.012	6.43	6.22	6.35	6.33	0.061
BOT13	5.00	5.11	5.16	5.09	0.047	6.43	6.70	6.55	6.56	0.078
BOT14	4.28	4.30	4.24	4.27	0.018	6.43	6.20	6.34	6.32	0.067
BOT15	4.43	4.41	4.42	4.42	0.0058	5.86	6.00	5.70	5.85	0.087
BOT16	5.00	4.90	5.20	5.03	0.088	5.00	5.23	5.13	5.12	0.067
BOT17	5.00	5.10	5.00	5.03	0.033	6.43	6.24	6.51	6.39	0.080
BOT18	4.71	4.71	4.83	4.75	0.040	6.14	6.00	6.20	6.11	0.059
BOT19	4.71	4.72	4.63	4.69	0.028	6.43	6.31	6.23	6.32	0.058
BOT20	5.00	5.18	5.20	5.13	0.064	6.43	6.59	6.55	6.52	0.048
BOT21	5.00	5.20	5.26	5.15	0.079	5.00	5.20	5.10	5.10	0.058
BOT22	5.00	5.01	5.14	5.05	0.045	5.00	5.00	5.00	5.00	0.00
BOT23	4.28	4.30	4.28	4.29	0.007	6.43	6.21	6.20	6.28	0.075
BOT24	5.00	5.10	4.90	5.00	0.058	6.13	6.31	6.13	6.19	0.060
BOT25	5.00	5.20	5.12	5.10	0.058	6.14	6.00	6.26	6.13	0.075
BOT26	5.00	5.10	5.10	5.07	0.033	6.00	6.19	6.12	6.10	0.055
BOT27	5.86	5.89	5.83	5.86	0.017	7.43	7.38	7.25	6.35	0.054
BOT28	5.00	5.00	5.00	5.00	0.00	5.90	6.00	5.99	5.96	0.032
BOT29	5.67	5.60	5.40	5.56	0.081	7.86	7.73	7.70	7.76	0.049
BOT30	7.86	7.90	7.86	7.87	0.013	8.57	8.65	8.56	8.59	0.028
BOT31	4.71	4.70	4.90	4.77	0.065	5.00	5.30	5.22	5.17	0.090
BOT32	4.43	4.40	4.23	4.35	0.062	5.28	5.25	5.20	5.24	0.023
BOT33	6.43	6.39	6.25	6.35	0.055	11.0	10.8	11.1	11.0	0.086
BOT34	4.71	4.69	4.88	4.76	0.060	6.14	6.80	6.50	6.48	0.19
BOT35	8.57	8.68	8.88	8.71	0.091	9.28	9.25	9.36	9.30	0.033

BOT36	5.00	5.10	5.00	5.03	0.033	5.71	5.63	5.80	5.71	0.049
BOT37	5.20	5.12	5.00	5.11	0.058	6.14	6.00	6.30	6.15	0.087
BOT38	5.20	5.14	5.16	5.17	0.018	6.43	6.40	6.58	6.47	0.056
BOT39	7.86	7.90	8.00	7.92	0.042	10.0	9.89	10.2	10.0	0.091
BOT40	5.20	5.18	5.00	5.13	0.064	6.57	6.50	6.60	6.56	0.030
BOT41	5.10	5.00	5.20	5.10	0.056	5.71	5.41	5.60	5.57	0.088
BOT42	5.89	5.82	5.86	5.86	0.020	7.14	7.26	7.30	7.23	0.048
BOT43	4.00	4.90	4.89	4.60	0.30	6.43	6.23	6.35	6.34	0.058
BOT44	5.00	5.89	5.10	5.33	0.28	7.14	7.00	7.21	7.12	0.062
BOT45	5.00	4.87	5.00	4.96	0.043	5.96	6.00	5.90	5.95	0.029
BOT46	5.00	4.90	5.10	5.00	0.058	6.43	6.16	6.16	6.25	0.090
BOT47	5.00	5.10	5.30	5.13	0.088	6.23	6.20	6.23	6.22	0.010
BOT48	5.00	5.20	5.20	5.13	0.067	6.00	6.00	6.11	6.04	0.037
BOT49	8.57	8.68	8.64	8.63	0.032	9.28	9.23	9.00	9.17	0.086
BOT50	4.98	5.19	5.19	5.12	0.07	6.43	6.40	6.50	6.44	0.030
BOT51	5.43	5.39	5.58	5.47	0.058	7.00	7.00	7.20	7.07	0.067
BOT52	0.00	0.00	0.00	0.00	0.00	6.76	6.70	6.72	6.73	0.018
BOT53	5.43	5.49	5.50	5.47	0.022	7.86	7.73	7.59	7.73	0.078
BOT54	5.00	5.20	5.00	5.07	0.067	5.42	5.86	5.80	5.69	0.14
BOT55	5.00	5.10	5.30	5.13	0.088	5.00	5.02	5.00	5.01	0.0067
BOT56	5.00	4.90	5.16	5.02	0.076	6.29	6.30	6.30	6.30	0.0033
BOT57	5.00	4.96	5.21	5.06	0.078	6.43	6.40	6.21	6.35	0.069
BOT58	5.00	4.89	5.10	5.00	0.061	6.43	6.12	6.25	6.27	0.090
BOT59	5.00	5.10	5.30	5.13	0.088	5.00	5.00	5.30	5.10	0.10
BOT60	0.00	0.00	0.00	0.00	0.00	0.00	0.00	0.00	0.00	0.00
BOT61	5.00	5.23	5.00	5.08	0.077	6.43	6.23	6.42	6.36	0.065
BOT62	5.00	5.11	5.00	5.04	0.037	6.86	6.89	7.00	6.92	0.043
BOT63	5.00	5.29	5.14	5.14	0.083	6.43	6.42	6.59	6.48	0.055
BOT64	4.43	5.39	4.58	4.80	0.30	6.00	6.02	6.23	6.08	0.074
BOT65	5.00	4.20	4.90	4.70	0.25	6.43	6.20	6.28	6.30	0.067
BOT66	5.00	4.80	5.00	4.93	0.067	6.23	6.20	6.23	6.22	0.010
BOT67	5.89	5.90	5.90	5.90	0.003	6.89	6.86	6.86	6.87	0.010
BOT68	8.57	8.20	8.50	8.42	0.11	10.7	10.8	10.8	10.8	0.030
BOT69	4.71	4.65	4.60	4.65	0.032	6.43	6.59	6.52	6.51	0.046
BOT70	5.00	4.89	5.10	5.00	0.061	6.14	6.10	6.11	6.11	0.012
BOT71	5.00	4.99	4.85	4.95	0.048	6.14	6.30	6.16	6.20	0.050
BOT72	5.00	5.29	5.39	5.23	0.12	7.86	7.88	7.85	7.86	0.0088
BOT73	4.43	4.49	4.50	4.47	0.022	6.30	6.28	6.24	6.27	0.018
BOT74	5.00	4.96	5.12	5.03	0.048	5.00	5.00	5.00	5.00	0.00
BOT75	4.90	4.97	5.09	4.99	0.055	7.00	6.42	6.42	6.61	0.19
BOT76	5.71	5.65	5.60	5.65	0.032	7.14	7.14	7.19	7.16	0.017
BOT77	8.57	8.60	8.70	8.62	0.039	9.28	9.20	9.30	9.26	0.031
BOT78	5.00	5.00	4.80	4.93	0.067	6.43	6.20	6.35	6.33	0.067
BOT79	5.20	5.20	5.10	5.17	0.033	6.14	6.00	6.10	6.08	0.042

Table 2: Measurements of width and length of conidial spores found in each isolate.

to the non-mutant lines. A t-test of unequal variance on the length and width of the conidia was done to show the significance of the size of the conidia for possible mutant isolates compared to the non-mutant isolates; p-values were <0.05 for length and width of the conidia. Results showed a p-value of (5.33×10^{-6}) for length and (9.33×10^{-6}) for width (Figures 1A and 1B). Thus, the size of mutant conidia with a 4-nucleotide addition was significantly larger than the non-mutant lines; this can be seen in the graph (Figures 3A-3C). Conidia with SNPs did not show significant difference when compared with wild type isolates (Figures 3D and 3E).

PCR amplification

PCR primers amplified a 0.7 kb intergenic sequence unique to *B. cinerea* [26]. A secondary set of primers amplified an internal region of the 720 bp, identified by C729+/- primer sets in order to identify all strains as *B. cinerea*. Because of the specificity of primer C729 +/- [26], we were able to get PCR bands for all our 79 samples. We were also able to obtain sequencing results for 65 samples. However, since it was reported that C729+/- primers were not able to identify some of the *Botrytis* species due to a deletion [27], we also used another set of primers: BC108/563 [26]. This was done in an effort to make sure that all our desired products would be identified correctly. Both sets of primers were also used to for each of the 79 samples to assure accuracy. Using the two sets of primers successfully amplified all of the samples, but it was impossible to sequence all of the samples. A possible explanation for 14 samples not having sequencing information is, a PCR reaction is exponential and therefore amplification can go beyond the linear part of the reaction, meaning if we had inefficient primers it will produce products similar to those of very efficient primers. However, sequencing is linear. Therefore, inefficient primers can produce weak

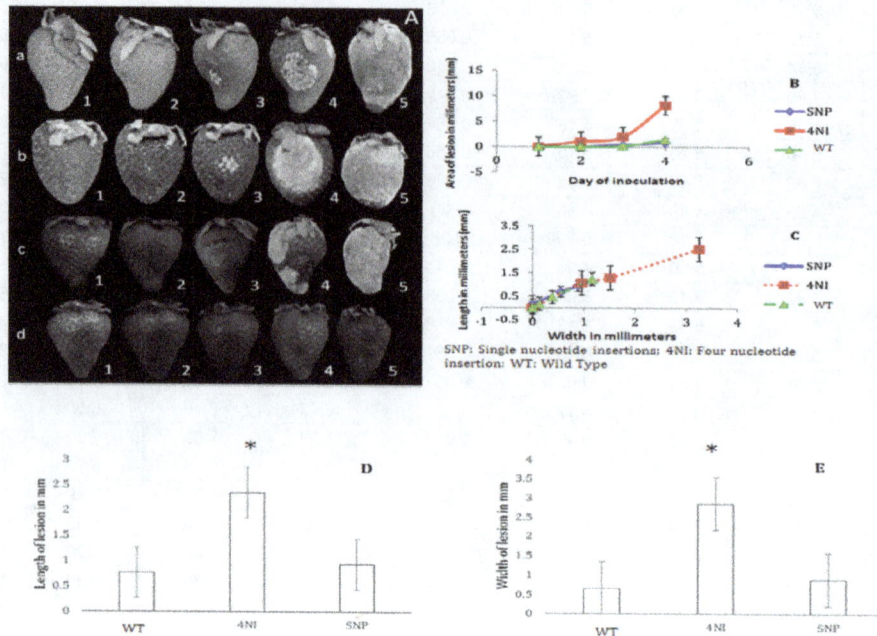

Figure 3: Virulence pattern of wild type and mutant isolates of *B. cinerea*.
Legend: A. Isolate 27 (4 nucleotide insertion) a1. Day 0, a2. Day 01, a3. Day 02, a4. Day 03, a5. Day 04; Isolate 49 (4-nucleotide insertion) b1. Day 0, b2 Day 01, b3. Day 02, b4. Day 03, b5. Day 04; Isolate 61 SNP (3- G → A) c1. Day 0, c2. Day 01, c3. Day 02, c4. Day 03, c5. Day 04; Control (H$_2$O) d1. Day 0, d2. Day 01, d3. Day 02, d4. Day 03, d5. Day 04; B Lesion growths in millimeters taken over a period of 4 days; C Length/Width growth over time (4 days); D T-test for unequal variance for the length and the width of lesion size. P-value<0.05, so the null hypothesis can be rejected. Rate of growth of mutants is greater than that of non-mutants. P-value for length 0.018, p-value for width 0.0025.
There were statistically significant differences in the virulence between our 15 isolates with the 4-nucleotide insertion than the other isolates without this insertion, as seen through lesion development (Figure 3A). The p-values were <0.05 for length and width of the lesion size, and the growth can be viewed in Figures 3D and 3E. Differences were not statistically significant between the other isolates (Figures 3D and 3E) when compared to the wild type sequence without the 4-base insertion. It was also noticed that the pattern of lesion development was similar to the wild type isolates (Figures 3D and 3E). These lesions were smaller in size than the lesions obtained from the 4-base nucleotide mutants.

or undetectable bands, which might have been the reason for the 14 samples with no sequencing information. Thus, inefficient primers can lead to failed sequencing although the PCR reaction was a success.

Sequence analysis

A total of sixteen (Table 3) of our isolate sequences shared the same 4-base insertion (GGGA at columns 282-285), which was also shared by the *Botrytis* identified from *Gladiolus* sp. of India. The 4-base insertion was seen in twelve of the isolates (27, 29, 30, 33, 35, 42, 44, 49, 51, 53, 68, and 72 when using both the BC108/563 and the C729+/- primers. Three isolates 67, 76, and 77 showed the 4-base insertion only when sequenced using the BC108/563 primers. Isolate 39 showed the 4-base insertion only when sequenced using the C729+/- primer (Table 3) (Figures 4A and 4B). Microscopic studies showed that each of the sixteen mutants shared identical phenotypic changes. The highlighter plot (Figures 4A and 4B) shows that there is as much variability within the population of *Botrytis* sampled in Californian strawberry fields, as there is between isolates collected in other areas of the world. For example, the 4-base insertion GGGA columns 282-285 in our alignment is shared with an isolate found on Gladiolus flowers from India with Gene Bank accession number KP141796 (as yet unpublished). The authors who submitted this isolate have yet to publish their phenotypic findings, thus we have yet to see if the 4 bp insertion corresponds with increased virulence in the Indian isolate of Gladiolus. The majority of our isolates had the same 4 single-base differences from the complete genome sequence (http://www.ncbi.nlm.nih.gov/bioproject/PRJNA264284 chromosome 13) used as the "MASTER" or reference sequence for this figure.

Pathogenicity

To evaluate the virulence of different *B. cinerea*, artificial inoculation experiments with the strawberry (*F. ananassa*) was done. Disease symptoms were always greater after inoculation of mutant isolates with 4 nucleotide insertion compared to the wild type and mutant isolates with SNPs in all our three replicates (Figures 3A, 3B, and 3C).

Surprisingly, individual fruits of the negative controls also showed slight disease symptoms 4 days after inoculation. It is to be noted here that *B. cinerea* (grey mold) appears on the surface of a strawberry fruit, only when the fruiting body of the pathogen is visible. However, it is to be noted that the fungus spreads quickly and can exist in a non-fruiting stage in the fruit (non-visible stage). It is quite possible that the strawberries selected for analysis may have the pathogen in the non-fruiting stage. Therefore, we made sure our control fruits, and the other fruits, went through the same sterilization technique to maintain uniformity. The graph indicated that the infection rate was three times higher than the wild type isolates, and five times higher than the negative control (Figures 3B and 3C).

Discussion

Considering the importance of *B. cinerea* and its significant damage to a wide variety of crops, its control management is necessary. The first step to manage this pathogen is to identify it [27]. Identification has traditionally been dependent on morphological and cultural characteristics coupled with host specificity [34]. Morphological characteristics are often influenced by environmental conditions, which are variable and therefore do not always provide accurate

Isolate number	Cultivar	Type of infection	Isolate origin	Primers C729+/-	Primers BC108+/563-	Type of mutation
BOT01	strawberry	postharvest	Oxnard	(-)	na	
BOT02	strawberry	postharvest	Oxnard	na	na	
BOT03	strawberry	postharvest	Oxnard	na	na	
BOT04	strawberry	postharvest	Oxnard	(+)	na	Single bp change (T→C)
BOT05	strawberry	postharvest	Oxnard	(-)	na	
BOT06	strawberry	postharvest	Oxnard	na	na	
BOT07	strawberry	postharvest	Oxnard	(-)	(-)	
BOT08	strawberry	postharvest	Oxnard	(-)	(-)	
BOT09	strawberry, El Dorado	field	Oxnard	(-)	(-)	
BOT10	strawberry, El Dorado	field	Oxnard	na	na	
BOT11	strawberry, El Dorado	field	Oxnard	(-)	na	
BOT12	strawberry, El Dorado	field	Oxnard	(-)	(-)	
BOT13	strawberry, El Dorado	field	Oxnard	(-)	(-)	
BOT14	strawberry, El Dorado	field	Oxnard	(-)	(-)	
BOT15	strawberry, El Dorado	field	Oxnard	(-)	(-)	
BOT16	strawberry, El Dorado	field	Oxnard	(+)	(+)	Single bp change (G-->A)
BOT17	strawberry, El Dorado	field	Oxnard	(-)	(-)	
BOT18	strawberry, El Dorado	field	Oxnard	(-)	(-)	
BOT19	strawberry, El Dorado	field	Oxnard	(-)	(-)	
BOT20	strawberry, El Dorado	field	Oxnard	(-)	na	
BOT21	strawberry, El Dorado	field	Oxnard	na	na	
BOT22	strawberry, El Dorado	field	Oxnard	(-)	(-)	
BOT23	strawberry, El Dorado	field	Oxnard	(-)	(-)	
BOT24	strawberry, El Dorado	field	Oxnard	(-)	(-)	
BOT25	strawberry, El Dorado	field	Oxnard	na	na	
BOT26	strawberry, El Dorado	field	Oxnard	(-)	(-)	
BOT27	strawberry, El Dorado	field	Oxnard	(+)	(+)	4 nucleotide insertion (GGGA)
BOT28	strawberry, El Dorado	field	Oxnard	(-)	na	
BOT29	strawberry, El Dorado	field	Oxnard	(+)	(+)	4 nucleotide insertion (GGGA)
BOT30	strawberry, El Dorado	field	Oxnard	(+)	(+)	4 nucleotide insertion (GGGA)
BOT31	strawberry, El Dorado	field	Oxnard	(-)	(-)	
BOT32	strawberry, El Dorado	field	Oxnard	(-)	(-)	
BOT33	strawberry, El Dorado	field	Oxnard	(+)	(+)	4 nucleotide insertion (GGGA)
BOT34	strawberry, El Dorado	field	Oxnard	(-)	na	
BOT35	strawberry, El Dorado	field	Oxnard	(+)	(+)	4 nucleotide insertion (GGGA)
BPT36	strawberry, El Dorado	field	Oxnard	(-)	(-)	
BOT37	strawberry, El Dorado	field	Oxnard	(-)	(-)	
BOT38	strawberry, El Dorado	field	Oxnard	(-)	(-)	
BOT39	raspberry, Pacifica	field	Watsonville	(+)	na	4 nucleotide insertion (GGGA)
BOT40	raspberry, Pacifica	field	Watsonville	(-)	(-)	
BOT41	raspberry, Pacifica	field	Watsonville	(-)	(-)	
BOT42	raspberry, Pacifica	field	Watsonville	(+)	(+)	4 nucleotide insertion (GGGA)
BOT43	raspberry, Pacifica	field	Watsonville	na	na	
BOT44	raspberry, Pacifica	field	Watsonville	(+)	(+)	4 nucleotide insertion (GGGA)
BOT45	raspberry, Pacifica	field	Watsonville	(-)	(-)	
BOT46	raspberry, Pacifica	field	Watsonville	(-)	na	
BOT47	raspberry, Pacifica	field	Watsonville	(-)	(-)	
BOT48	raspberry, Pacifica	field	Watsonville	(-)	na	
BOT49	raspberry, Pacifica	field	Watsonville	(+)	(+)	4 nucleotide insertion(GGGA)
BOT50	strawberry	field	Salinas	(-)	(-)	
BOT51	strawberry	field	Watsonville	(+)	(+)	4 nucleotide insertion (GGGA)
BOT52	strawberry	field	Salinas	(-)	(-)	
BOT53	strawberry	field	Salinas	(+)	(+)	4 nucleotide insertion (GGGA)
BOT54	strawberry	field	Salinas	(-)	(-)	
BOT55	strawberry	field	Salinas	(-)	(-)	
BOT56	strawberry	field	Salinas	(-)	(-)	
BOT57	strawberry	field	Watsonville	na	na	
BOT58	strawberry	field	Salinas	na	na	
BOT59	strawberry	field	Salinas	na	(+)	SNP (1) A→G
BOT60	strawberry	field	Salinas	na	na	

BOT61	strawberry	field	Salinas	(-)	(+)	SNP (3) G→A
BOT62	strawberry	field	Salinas	(-)	(-)	
BOT63	strawberry	field	Salinas	(+)	(-)	SNP (2) T→C; A→G
BOT64	strawberry	field	Watsonville	(-)	(-)	
BOT65	strawberry	field	Watsonville	(-)	na	
BOT66	strawberry	field	Watsonville	na	na	
BOT67	strawberry	field	Watsonville	na	(+)	4 nucleotide insertion (GGGA)
BOT68	strawberry	field	Salinas	(+)	(+)	4 nucleotide insertion (GGGA)
BOT69	strawberry	field	Watsonville	na	na	
BOT70	strawberry	field	Watsonville	na	na	
BOT71	strawberry	field	Salinas	(-)	na	
BOT72	strawberry	field	Salinas	(+)	na	4 nucleotide insertion (GGGA)
BOT73	strawberry	field	Watsonville	na	na	
BOT74	strawberry	field	Watsonville	(+)	(-)	SNP (3) C→T; C→T; A→G
BOT75	strawberry	field	Watsonville	na	(-)	
BOT76	strawberry	field	Watsonville	na	(+)	4 nucleotide insertion (GGGA)
BOT77	strawberry	field	Watsonville	na	(+)	4 nucleotide insertion (GGGA)
BOT78	Blackberry		Watsonville	(-)	(-)	
BOT79	Strawberry	Nursery	Tule Lake	na	na	

Table 3: List of all isolates and primers used to determine mutation in sequencing.
Legend: (+) denotes mutation using the specific primer. (-) denotes no mutation or 100% match with NCBI original sequence of *Botrytis cinerea*. The far-right column displays the type of mutation referring to the (+) samples showed according to sequencing. (T → C) means that a base "t" was replaced with "c." (A → G) means that base "a" was replaced with "g." The 4-nucleotide insertion mutation was present with bases (GGGA). SNP refers to the number of single nucleotide polymorphisms present for that isolate.

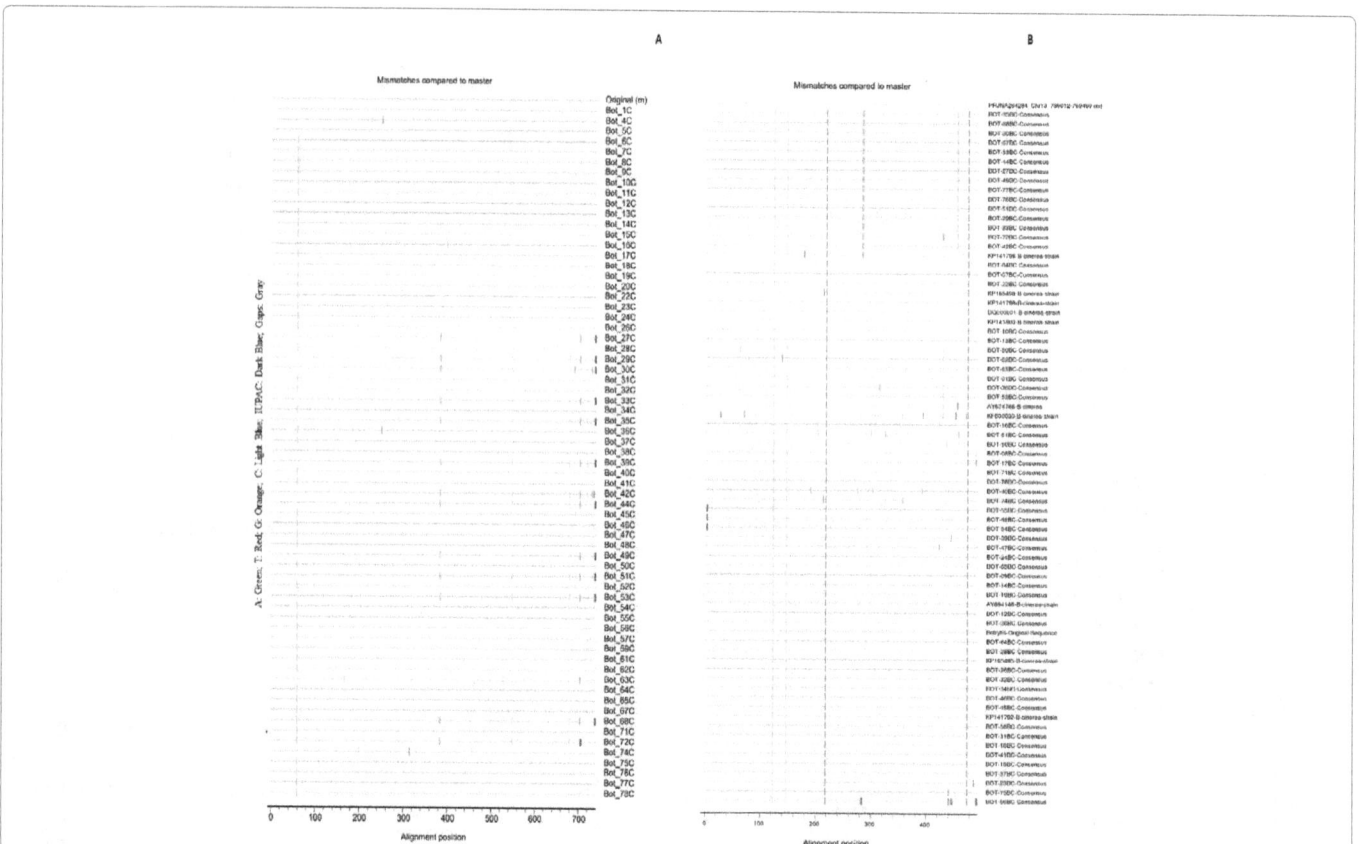

Figure 4: Nucleotide sequence comparison using highlighter plot.
Legend: Highlighter plot of nucleotide differences between the *Botritis cinerea* B05.10 complete genome record PRJNA264284 (http://www.ncbi.nlm.nih.gov/nuccore/CP009817.1) and sequences of the same genomic regions sequenced from isolates in Table 2 and other isolate sequences in GenBank (with accession numbers). Blank space equals identity to the master or reference sequence, while different colored ticks indicate single base changes or polymorphisms (SNPs) in each sequence. A: alignment sequence using primers C729+/-. B: alignment sequence using primers BC108/563.

information [35]. The taxonomy of *Botrytis* species has been used to show how cultural conditions could considerably modify taxonomic

characters such as dimension and shape of conidia [35]. Conidia size, form, and colony characters are temperature and media dependent and are often reversible [27,36]. Therefore, to eliminate variability, the use of molecular markers is a good alternative for identification [32]. Species-specific primers that had been used in *B. cinerea* detection [26] revealed a 4-nucleotide insertion and a single base pair substitution in sixteen isolates. Of the sixteen isolates, the isolates 67, 76, and 77 revealed the 4-nucleotide insertion only when using the BC108/563 primers. This is possibly because of a deletion that could not be detected by the C729+/- primers, as described previously [26]. However, the 4-nucleotide insertion for isolate 39 was only detected using C729+/- primers. This could be because of a failed sequencing reaction. From our experimental findings, we observed that these possible mutated isolates have a larger conidia size when compared to the wild type isolates in our study (Table 2). Virulence, as observed by lesion size on strawberries, was also significantly higher when these isolates were used to inoculate strawberries in the laboratory (Figures 3B and 3C). The question is, does mutation cause a change in conidia size and does this affect virulence? If so, is the mutation responsible for genetic diversity? Does this genetic diversity make epidemiology of the fungus difficult? Mutation is a change in the DNA at a particular locus in an organism. It is the ultimate source of new alleles in plant pathogen populations. It also is the source of new alleles that create new genotypes (such as new pathotypes) within clonal lineages. Small populations have fewer alleles. This is due to genetic drift, and due to fewer mutations generated in small populations. Old populations have more neutral alleles than new populations [16]. Therefore, the center of gene diversity for a species is most often the center of origin for a species. Plants and pathogens have coevolved for the longest time, leading to selection for a diversity of resistance alleles in the plant population [36]. This is why plant breeders seek resistant germplasm at centers of diversity. If the pathogen coevolved with its plant host at the center of origin, we predict that the pathogen population will also exhibit maximum diversity at the center of origin.

Mutation plays an important role in evolution. The ultimate source of all genetic variation is mutation. Mutation is an important first step of evolution because it creates a new DNA sequence for a particular gene, creating a new allele [37]. Recombination can also create a new DNA sequence (a new allele) for a specific gene through intragenic recombination [38]. Mutation acting as an evolutionary force by itself has the potential to cause significant changes in allele frequencies over very long periods of time. But, if mutation were the only force acting on pathogen populations, then evolution would occur at a rate that we could not observe [39,40].

In plant pathology, we are most often concerned with mutations that affect pathogen virulence, sensitivity to fungicides, or sensitivity to antibiotics. In pathogens that show a gene-for-gene interaction with plants, we are especially interested in the mutation from non-virulence to virulence, because this is the mutation that leads to pathogenicity [18,41]. Having that said, mutations from fungicide sensitivity to fungicide resistance are also important in agroecosystems, as are any mutations that affect fitness [42,43].

The marker we sequenced on chromosome 13 is not noted to be closely linked to any pathogenicity genes, so we have no explanation for why strains with one genotype at this locus would be extra pathogenic. The NEP2 pathogenicity gene is on chromosome 2, and the Gluconurodase pathogenicity gene is on chromosome 14.

Why are larger conidia more virulent? In a similar study on a zygomycetes fungus, it was reported that the short or absent isotropic growth period for larger spores, compared to the long phase observed prior to germ tube emergence for smaller spores, could be involved in the differences in virulence [23]. The larger spores are likely poised to

undergo rapid invasive hyphal growth compared to the smaller spores. The extended isotropic growth phase of the smaller spores [27,44] results in slow germ tube formation, a block, or a delay in germ tube emergence, and could reduce virulence in the host. The response of macrophages to spores further supported their hypothesis, where larger spores engulfed by macrophages were still able to send germ tubes [45]. In a recent study, it was observed that the mutation of a *Botrytis* gene has altered the size of conidia and altered virulence [44]. Since we were not able to see similar results in our wild type isolates, we can conclude that the mutation can be considered as a possible reason for larger conidia size and virulence. It has also been reported that the larger conidia are usually fungicide resistant, have a greater life span, and are more virulent in *Botrytis* [46]. Since there is currently no efficient alternative to control *Botrytis*, the fungal pathogen is recognized by the Fungicide Resistance Action Committee (FRAC) as a pathogen at high risk of fungicide resistance development. The risk of resistance is also due to numerous characteristics including: large population size, long distance dissemination of conidia by air currents, high genetic variability, the ability to reproduce sexually and asexually, abundant sporulation, polycyclic disease cycle, and wide host range [29]. The development of resistance threatens high yields and crop quality. Thus, it is a serious issue for growers, scientists, and manufacturers. It not only reduces growers' and manufacturers' income, but it also has consequences for the environment and human health [46]. Therefore, using a genomic approach to identify diversity [47] within the fungal species, then connecting that diversity to morphological variation and virulence, will provide the initial knowledge needed to better understand the pathogen. This initial knowledge on the pathogenicity factors is essential for fungal infections and is very important because it gives researchers targets in the fight against this pathogen [48].

The isolates with larger conidia had an identical 4-nucleotide insertion, and were observed both in strawberries and raspberries (Tables 2 and 3). Based on the available information of fungicide resistance in *B. cinerea*, we can possibly assume that the mutation was a result of fungicide resistance [17]. There is also evidence for long-distance migration of mutant strains of pathogens [17].

Conclusion

Our findings have provided valuable genotypic and phenotypic information on the *B. cinerea* field isolates. We were able to provide information on the pathogenicity of these field isolates subjected to several fungicides not known in a laboratory setting. It is well known that application of fungicides can result in mutations in a pathogen and therefore contribute to resistance. The genotypic results obtained in our study can therefore be used for obtaining basic knowledge on fungicide resistance of this pathogen, and can also indicate appropriate resistance management strategies to ensure continued effectiveness for *Botrytis* control. In addition, we have identified a genetic marker for detection of *Botrytis* across a wide variety of plant species.

Sequence Submission

Sequence data from this article have been deposited with the NCBI gene bank under accession numbers: KU145342-KU145392, KX772771-KX772776, KX781161-KX781167.

Conflict of Interest

The authors declare that there is no conflict of interest regarding the publication of this paper.

Author Contribution

J.RE. was responsible for sequence analysis, submission of sequences to NCBI gene bank, PCR, and manuscript preparation and submission. B.F. created the highlighter plot in Figure 1 and with bioinformatics software identified homology of the pathogen sequence in other plant species. S.P. performed microscopy analysis on all isolates used in this study. J.T. created Figure 3A and assisted with sequence analysis and inoculation. P.R.F. created Figures 3B, 3C and 3D, designed and conducted inoculation experiments. C.F. assisted with DNA extraction and PCR. K.S. edited and organized the manuscript and figures for manuscript submission. M.R.C. supervised the research project, wrote the first draft of the manuscript, designed experiments, and assisted with sequence analysis and submission.

Acknowledgments

We express our appreciation to Janet Broome and Mansun Kong at Driscoll's Strawberry Associates, Watsonville, CA for providing us with the fungal isolates. Table 1 in the manuscript was provided by Janet Broome, Driscoll's Strawberry Associates, Watsonville, CA. Thanks to Alison Frye and Jackie Hayes for assisting with PCR and microscopy. Thanks to Tanner Harding, Evan Harrison, Branden Robinson and Benjamin Cundick for uploading some of the sequences to NCBI and editing the first draft of the manuscript. Funding for this research was provided in part by Dr. Doug Gubler at UC Davis for funding our sequencing analysis at UC Davis Sequencing facility. The publication was made possible in part by the INBRE program, NIH grant No.P20 GM103408 (National Institute of General Medical Sciences) and SBOE/ISU funds for STEM undergraduate research project No. AHRC38.

References

1. Chilvers MI, du Toit LJ, Akamatsu H, Peever TL (2007) A real-time, quantitative PCR seed assay for Botrytis spp. that cause neck rot of onion. Plant Dis 91: 599-608.

2. Hahn M (2014) The rising threat of fungicide resistance in plant pathogenic fungi: Botrytis as a case study. J Chem Biol 7: 133-141.

3. Zhang Z, Qin G, Li B, Tian S (2014) Infection Assays of Tomato and Apple Fruit by the fungal Pathogen Botrytis cinerea. Bio Protoc 4: e131.

4. Valiuskaite A, Surviliene E, Baniulis D (2010) Genetic diversity and pathogenicity traits of Botrytis spp. isolated from horticultural hosts. Žemdirbystė-Agric 97: 85-90.

5. Richard F, Glass NL, Pringle A (2012) Cooperation among germinating spores facilitates the growth of the fungus, Neurospora crassa. Biol Lett 8: 419-422.

6. Bestfleisch M, Luderer-Pflimpfl M, Hofer M, Schulte E, Wunsche JN, et al. (2015) Evaluation of strawberry (Fragaria L.) genetic resources for resistance to Botrytis cinerea. Plant Pathol. 64: 396-405.

7. Have AT, Mulder W, Visser J, van Kan JA (1998) The endopolygalacturonase gene Bcpg1 is required for full virulence of Botrytis cinerea. Mol Plant Microbe Interact 11: 1009-1016.

8. Arbelet D, Malfatti P, Simond-Côte E, Fontaine T, Desquilbet L, et al. (2010) Disruption of the Bcchs3a Chitin Synthase Gene in Botrytis cinerea is Responsible for Altered Adhesion and Overstimulation of Host Plant Immunity. Mol Plant Microbe Interact 23: 1324-1334.

9. Morcx S, Kunz C, Choquer M, Assie S, Blondet E, et al. (2013) Disruption of Bcchs4, Bcchs6 or Bcchs7 chitin synthase genes in Botrytis cinerea and the essential role of class VI chitin synthase (Bcchs6). Fungal Genet Biol 52: 1-8.

10. Daughtrey ML, Wick RL, Peterson JL (1995) Compendium of flowering potted plant diseases. Am Phytopathol Soc (APS Press), USA.

11. Martinez F, Blancard D, Lecomte P, Levis C, Dubos B, et al. (2003) Phenotypic differences between vacuma and transposa subpopulations of Botrytis cinerea. Eur J Plant Pathol 109: 479-488.

12. Elmer PAG, Michailides TJ (2007) Epidemiology of Botrytis cinerea in orchard and vine crops. In Botrytis: Biology, Pathology, and Control. Springer, Netherlands, Germany.

13. Welch NC, Beutel JA, Bringhurst R, Gubler D, Otto H, et al. (1989) Strawberry production in California. Leaflet-University of California, Cooperative Extension Service, USA.

14. [FAO] Food and Agriculture Organization. FAO Statistics Division (2009) Available from: http://faostat.fao.org2009.

15. [CSC] California Strawberry Commission and [CMCC] (2003) The California Minor Crops Council. A Pest Management Strategic Plan for Strawberry Production in California.

16. Villa-Rojas R, Sosa-Morales M, Lopez-Malo A, Tang J (2012) Thermal inactivation of Botrytis cinerea conidia in synthetic medium and strawberry puree. Int J Food Microbiol 155: 269-272.

17. Kretschmer M, Leroch M, Mosbach A, Kretschmer MAS, Fillinger S, et al. (2009) Fungicide-driven evolution and molecular basis of multidrug resistance in field populations of the grey mould fungus Botrytis cinerea. PLoS Pathog 5: e1000696.

18. Fernández Acero FJ, Carbú M, El-Akhal MR, Garrido C, González-Rodríguez VE, et al. (2011) Development of proteomics-based fungicides: new strategies for environmentally friendly control of fungal plant diseases. Int J Mol Sci 12: 795-816.

19. Hsiang T, Chastagner GA (1991) Growth and virulence of fungicide-resistant isolates of three species of Botrytis. Can J Plant Pathol 13: 226-231.

20. Fungicide Resistance Action Committee (2011) FRAC Code List: Fungicides sorted by mode of action (including FRAC Code numbering).

21. Leroux P, Fritz R, Debieu D, Albertini C, Lanen C, et al. (2002) Mechanisms of resistance to fungicides in field strains of Botrytis cinerea. Pest Manag Sci 58: 876-888.

22. Banno S, Yamashita K, Fukumori F, Okada K, Uekusa H, et al. (2009) Characterization of QoI resistance in Botrytis cinerea and identification of two types of mitochondrial cytochrome b gene. Plant Pathol 58: 120-129.

23. Li X, Kerrigan J, Chai W, Schnabel G (2012) Botrytis caroliniana, a new species isolates from blackberry in South Carolina. Mycologia 104: 650-658.

24. Dittrich-Schroder G, Wingfield MJ, Klein H, Slippers B (2012) DNA extraction techniques for DNA barcoding of minute gall-inhabiting wasps. Mol Ecol Resour 12: 109-115.

25. Wasilenko JL, Fratamico PM, Narang N, Tillman GE, Ladely S, et al. (2012) Influence of Primer Sequences and DNA Extraction Method on Detection of Non-O157 Shiga Toxin–Producing Escherichia coli in Ground Beef by Real-Time PCR Targeting the eae, stx, and Serogroup-Specific Genes. J Food Prot 75: 1939-1950.

26. Rigotti S, Gindro K, Richter H, Viret O (2002) Characterization of molecular markers for specific and sensitive detection of Botrytis cinerea Pers.: Fr. in strawberry (Fragaria x ananassa Duch.) using PCR. FEMS Microbiol Lett 209: 169-174.

27. Rigotti S, Viret O, Gindro K (2006) Two New Primers Highly Specific for the Detection of Botrytis cinerea Pers.: Fr. Phytopathol Mediterr 45: 253-260.

28. Gachon C, Saindrenan P (2004) Real-time PCR monitoring of fungal development in Arabidopsis thaliana infected by Alternaria brassicicola and Botrytis cinerea. Plant Physiol Biochem 42: 367-371.

29. Holz G, Coertze S, Williamson B (2007) The ecology of Botrytis on plant surfaces. Botrytis: Biology, Pathology, and Control. Springer, Netherlands, Germany.

30. Khazaeli P, Zamanizadeh H, Morid B, Bayat H (2010) Morphological and molecular identification of Botrytis cinerea causal agent of gray mold in rose greenhouses in central regions of Iran. Int J Agric Sci Res 1: 19-24.

31. Rigotti S, Viret O, Gindro K (2006) Two New Primers Highly Specific for the Detection of Botrytis cinerea Pers. Fr. Phytopathol Mediterr 45: 253-260.

32. Bell J (2008) A simple way to treat PCR products prior to sequencing using ExoSAP-IT. Biotechniques 44: 834.

33. Skou JP, Jørgensen JH, Lilholt U (1984) Comparative studies on callose formation in powdery mildew compatible and incompatible barley. J Phytopathol 109: 147-168.

34. Jarvis WR (1977) Botryotinia and Botrytis species: taxonomy, physiology, and pathogenicity. Monograph, Research Branch Canada Department of Agriculture, Ottawa, Canada.

35. Menzinger W (1966) On the variability and taxonomy of species and forms of the genus Botrytis Mich. Investigations on the culture-dependent variability of morphological properties of the genus Botrytis. From the Institute for Plant Diseases and Plant Protection at the Hanover Technical University. Central leaf for bacteriology, parasites, infectious diseases, and hygiene 120: 141-178.

36. Fu YX, Li WH (1992) Statistical tests of neutrality of mutations. Genetics 133: 693-709.

37. Ellis J, Dodds P, Pryor T (2000) Structure, function, and evolution of plant disease resistance genes. Curr Opin Plant Biol 3: 278-284.

38. Sniegowski PD, Gerrish PJ, Johnson T, Shaver A (2001) The evolution of mutation rates: separating causes from consequences. Bioessays 22: 1057-66.

39. Hudson RR (1983) Properties of a neutral allele model with intragenic recombination. Theor Popul Biol 23: 183-201.

40. Lucht JM, Mauch-Mani B, Steiner HY, Metraux JP, Ryals J, et al. (2002) Pathogen stress increases somatic recombination frequency in Arabidopsis. Nat Genet 30: 311-314.

41. Sarkar SF, Guttman DS (2004) Evolution of the core genome of Pseudomonas syringae, a highly clonal, endemic plant pathogen. Appl Environ Microbiol 70: 1999-2012.

42. De Waard MA, Andrade AC, Hayashi K, Schoonbeek HJ, Stergiopoulos I, et al. (2006) Impact of fungal drug transporters on fungicide sensitivity, multidrug resistance and virulence. Pest Manag Sci 62: 195-207

43. Watson IA (1970) Changes in virulence and population shifts in plant pathogens. Annu Rev Phytopathol 8: 209-230.

44. Li X, Fernandez-Ortuno D, Chai W, Wang F, Schnabel G (2012) Identifcation and prevalence if Botrytis spp. From blackberry and strawberry fields of the Carolinas. Plant Dis 96: 1634-1637.

45. McDonald BA, Linde C (2002) The population genetics of plant pathogens and breeding strategies for durable resistance. Euphytica 124: 163-180.

46. Harren K, Brandhoff B, Knödler M, Tudzynski B (2013) The High-Affinity Phosphodiesterase BcPde2 Has Impact on Growth, Differentiation and Virulence of the Phytopathogenic Ascomycete Botrytis cinerea. PLoS One 8: e78525.

47. Walker AS, Micoud A, Rémuson F, Grosman J, Gredt M, et al. (2013) French vineyards provide information that opens ways for effective resistance management of Botrytis cinerea (grey mould). Pest Manag Sci 69: 667-678.

48. Rehner SA, Buckley E (2005) A Beauveria phylogeny inferred from nuclear ITS and EF1-α sequences: evidence for cryptic diversification and links to Cordyceps teleomorphs. Mycologia 97: 84-98.

10

Evaluation of Antifungal Activity of Plant Extracts against Papaya Anthracnose (*Colletotrichum gloeosporioides*)

Anteneh Ademe[1]*, Amare Ayalew[2] and Kebede Woldetsadik[2]

[1]Sekota Dryland Agricultural Research Center, Sekota, Ethiopia
[2]Department of Plant Sciences, Haramaya University, Haramaya, Ethiopia

Abstract

Antifungal activities of nineteen plant extracts were tested in 2010 with the objectives of screening potential plant extracts against *Colletotrichum gloeosporioides* under *in vitro* and anthracnose caused by *Colletotrichum gloeosporioides*, on papaya (*Carica papaya* L.) during storage. Ethyl acetate extracts of *Lantana camara* resulted in the highest inhibition (with inhibition zone of 35.3 mm) and showed strong activity against *C. gloeosporioides*. Inhibition levels of spore germination that reached 88.7, 85.8, 85.1 and 84.6% were recorded over the control by extracts of *Lantana camara*, *Lantana viburnoides*, *Echinops sp.* and *Ruta chalepensis*. Four aqueous extracts were evaluated for control of anthracnose under *in vivo* for 14 days, and *Echinops sp.* (25%) was found to be most effective in the reduction of disease development and maintaining the overall quality of papaya fruit. Further studies on isolation and characterization of the active (antifungal) compounds are needed.

Keywords: Anthracnose; *Colletotrichum gloeosporioides*; Ethyl acetate; Papaya

Introduction

Papaya (*Carica papaya* L.) is a popular and economically important fruit tree of tropical and subtropical countries [1]. The leading global producers of papaya are Brazil, Colombia, Democratic Republic of Congo, Ethiopia, Guatemala, India, Indonesia, Mexico, Nigeria and Philippines [2]. Papaya is known as "common man's fruits". It is rich sources of vitamin A, C and calcium. The ripe fruit is prone to many diseases, among which anthracnose caused by *Colletotrichum gloeosporioides* (Penz.) Penz & Sacc. is an economically important disease during transit, storage and market [3-5]. In general, the fungus initiates infection as soon as flowering starts and stays latent until the postharvest environment conditions favor colonization of fruit tissue [3,6]. According to Coursey [7], postharvest losses of approximately 40-100% have been generally reported in papaya in developing countries.

Synthetic fungicides are currently used as the primary means for the control of plant diseases. However, the alternative control methods are needed because of the negative public perceptions about the use of synthetic chemicals, resistance to fungicides among fungal pathogens and high development cost of new chemicals [8,9]. Application of higher concentrations of chemicals in an attempt to overcome anthracnose disease increases the risk of high levels of toxic residues, which is, particularly serious, since papaya fruit is consumed in relatively short time after harvest [10].

Bioactive products of plants are less persistent in environment and are safe for mammals, other non target organisms [11-13], and for the control of postharvest disease than synthetics [14]. A number of plant species have been reported to possess natural substances that are toxic to many fungi causing plant diseases [15,16]. Ranaware et al. [17] indicated the efficacy of aqueous plant extracts as potential inhibitors of *Alternaria carthami*. Similarly, Dwivedi and Shukla [18] reported the effectiveness of aqueous extracts of different species of plants against *Fusarium oxysporum*.

Papaya anthracnose is one of the major diseases of the crop in Ethiopia [19]. Hence, this study was conducted with objective of determining the *in vitro* effect of plant extracts on conidial germination,

mycelial growth of *Colletotrichum gloeosporioides* and their efficacy against the development of postharvest papaya anthracnose.

Materials and Methods

Isolation of target pathogen

Colletotrichum gloeosporioides was isolated from papaya fruits showing anthracnose lesions. An isolate of the pathogen grown in pure culture was maintained in PDA culture tubes at 4°C, and used as stock culture throughout the study [20,21].

In vitro evaluation of botanicals

Sample collection and extraction: The potential extracts were selected from a screening of nineteen plant species. The plants were collected from Haramaya and Ambo areas of Ethiopia, in 2010. The experiment was conducted at the Plant Pathology Laboratory of the School of Plant Sciences at Haramaya University. The plant specimens (leaves) were shade dried at a room temperature and milled into a fine powder. Following the procedures employed by Amare [22], 50 gram of the pulverized plant specimens were extracted with 250 ml ethyl acetate by stirring for 2 hrs on magnetic stirrer. The extract was filtered through folded filter paper into a 500 ml round bottom flask and reduced to dryness on a rotary evaporator at 40°C water bath temperature. About 50 mg of the ethyl acetate extracts of each plant was weighed, redissolved in 1 ml of the extraction solvent and then tested for antifungal activities.

*Corresponding author: Anteneh Ademe, Sekota Dryland Agricultural Research Center, P.O. Box 62, Sekota, Ethiopia
E-mail: ad.antish@gmail.com

Paper disc assay: Filter paper discs, 6 mm diameter, were sterilized by dry heat for 1 h at 160°C oven temperature and impregnated with each of the test extracts by applying 10 μL of the extract solution using a capillary pipette. The culture media containing spore suspension of *Colletotrichum gloeosporioides* was poured into 14.5 cm in diameter petri plate and allowed to solidify. After the carrier solvent evaporated from the paper discs, they were placed on the surface of the medium; the plates were incubated for 4 days. The diameter of inhibition zone was measured in mm and the degree of inhibition of the fungal growth expressed on a 0-4 scale was recorded, where 0=no inhibition zone visible, 1=inhibition zone barely distinct, fungal growth and sporulation only slightly inhibited, 2=inhibition zone well distinct, fungal growth ca. 50% of the control, slight sporulation, 3=inhibition zone with sparse (ca. 25% of the control) fungal growth, and 4=inhibition zone free of visible fungal growth [22] (Figure 1a and 1b).

Conidial germination test: Conidia of *C. gloeosporioides* were adjusted using hemacytometer to a concentration of 10^5 condia/ml. Ten μL of plant extracts and 90 μL of the conidial suspension were mixed and the mixtures were added to the surface of dried depression slides. The slides were then placed on a glass rod in petri dish under moistened conditions and incubated at 25°C for 24 h. Control conidia received an equivalent amount of the solvent. After incubation, slides were fixed in lactophenol cotton blue and observed microscopically for spore germination. The experiment was laid out in CRD with three replications. The number of conidia germinated was scored to calculate the percentage inhibition of conidial germination.

In vivo antifungal assay of plant extracts

Aqueous extracts were tested for their effect on the papaya anthracnose development on harvested fruit. "Solo" papaya was obtained from Yilma State Farm in Dire Dawa, Ethiopia. For this purpose, undamaged, matured fruits of comparable size, color class and free from any pesticide were used. Aqueous solutions of selected plant species were evaluated at a concentration of 10 and 25%. Conidial suspension of *C. gloeosporioides* was prepared and adjusted to 10^5 conidia/ml Papaya fruits were surface-sterilized by dipping in 1% sodium hypochlorite solution for 10 min, rinsed in sterile distilled water and inoculated by dipping into spore suspension of *C. gloeosporioides*. After incubation for 15 h in plastic bag, fruits were dipped into extracts, while the control fruits were dipped into sterile distilled water. Five fruits (i.e. replications) for each concentration of extracts were used and arranged in CRD [21]. As of first symptom appearance, data on incidence and severity of anthracnose was recorded. Disease incidence was expressed as the percentage of fruits showing symptom. Disease severity was rated on 1 to 5 scale, where 1=0% of surface fruit rotten, 2=1-25%, 3=26-50%, 4=51-75%, and 5=76-100% [23]. Overall quality was assessed according to the following score: 1-2= fruit not marketable; 3=poor quality, limited marketability; 4-5=fair quality, marketable; 6-7=good quality, marketable; 8-9=excellent quality [24]. Percentage of marketability was assessed as a ratio of the number of fruits with scores 6, 7 for overall quality against the total number of fruits [25].

Statistical analysis

Analysis of variance (ANOVA) was carried out with the statistical software SAS v. 9.0 and Least Significant Difference (LSD) at 5% probability level were used for mean comparison. Severity was square root transformed, while spore germination was arcsine transformed before statistical analysis.

Results and Discussion

In vitro effect of botanicals on mycelial growth and spore germination of *Colletotrichum gloeosporioides*

There was a highly significant difference (P<0.0001) among the antifungal effects of ethyl acetate extracts on inhibition zone and degrees of inhibition against the test fungus. Extracts of *Echinops* sp. and *Lantana camara* more strongly inhibited growth of the pathogen than the remaining extracts. On the other hand, the inhibition resulting from other extracts of both solvents ranged from weak to moderately active. Ethyl acetate extracts of *Lantana camara* had the highest inhibition zone among plants. This was followed by *Artemisia afara*, *Echinops* sp., *Lantana viburnoides*, *Ruta chalepensis* and *Vernonia amygdalina* (Table 1). *Lantana camara* were found to be superior in mycelial growth reduction among different botanical tested against anthracnose of papaya [5]. The preservative nature of some plant extracts has been known for centuries, and there has been renewed interest in the antimicrobial properties of extracts from aromatic plants [12,26]. Numerous investigations on the genus *Echinops* have resulted in the isolation of thiophenes. Thiophenes from *Echinops* have been reported to possess many biological activities, including insecticidal and fungicidal [13]. Bautista-Banos et al. [23] tested leaf and stem

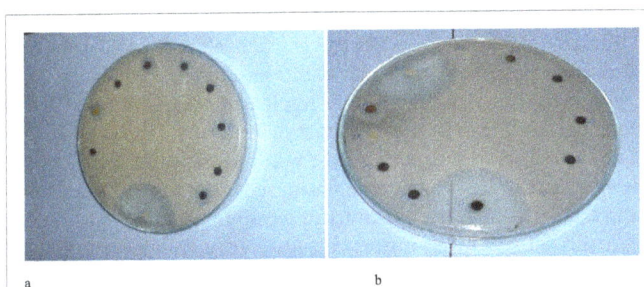

Figure 1: a and 1b: *In vitro* inhibition effect of botanicals on *Colletotrichum gloeosporioides* on PDA (Paper disc labeled as 2, 4, 5, 6, 7, 9, 10, 12, 13, 15, 16, 19, 20 and 21 was *Echinops* sp., *Artemisia afara*, *Ruta chalepensis*, *Thymus serrulatus*, *Lantana viburnoides*, *Vernonia amygdalina*, *Ocimum* sp. *Citrus limon*, *Nicotinia tabacum*, *Lantana camara*, *Ocimum lamifolium*, *Zingiber officinale*, carbendazim and ethyl acetate, respectively).

Species	Family	DI (mm)[a]	IE[b]	Spore germination (%)[c]
Artemisia afara	Asteraceae	4.5	3	18.8
Citrus limon	Rutaceae	1.3	1	46.1
Echinops sp.	Asteraceae	5.7	4	13.3
Lantana camara	Verbenaceae	35.3	4	10.1
Lantana viburnoides	Verbenaceae	5.0	3	12.6
Nicotinia tabacum	Solanaceae	1.0	3	42.6
Ocimum lamifolium	Lamiaceae	2.5	3	24.7
Ocimum sp.	Lamiaceae	2.2	3	30.9
Ruta chalepensis	Rutaceae	4.5	3	13.7
Thymus serrulatus	Lamiaceae	1.8	2	18.1
Vernonia amygdalina	Asteraceae	5.3	3	16.3
Zingiber officinale	Zingiberaceae	4.0	3	16.8
Control	-	0.0	0	89.1
LSD (0.05)		1.41		4.32

[a]diameter of inhibition zone in mm measured after 4 days of incubation
[b]inhibition effect on a 0-4 scale, where 0=none and 4=strong inhibition
[c]spore germination 24 h after treatment
Values are means of three replications

Table 1: Antifungal activity of some plant species from Ethiopia against *C. gloeosporioides*.

extracts of various plant species against *Colletotrichum gloeosporioides*. In *in vitro* experiment, leaf extracts of *Citrus limon* were found to be inhibitive the *in vitro* radial growth of *C. gloeosporioides*. In general, the presence of antimicrobial substances in the different extracts which caused the inhibition of radial growth *in vitro* agrees with reports of other studies [27-29].

The result of the *in vitro* screening tested against *C. gloeosporioides* revealed that there was a highly significant difference (P<0.0001) in effects among ethyl acetate extracts of plants on spore germination (Table 1). From tested extracts, *Lantana camara* gave the lowest spore germination (10.1%), followed by *Lantana viburnoides* (12.6%), *Echinops* sp. (13.3%) and *Ruta chalepensis* (13.7%), with no significant difference among them. The remaining ethyl acetate extracts showed also varying degrees of inhibition of spore germination ranging from 84.6% in *Ruta chalepensis* to 48.3% in *Citrus limon* (Table 1). Commercial essential oils of *Ruta chalepensis* and *Thymus vulgaris* and extracts of *Ocimum basilicum* and *Vernonia amygdalina* were found to be effective in reducing conidial germination of *C. gloeosporioides* [14,30]. Antifungal activities of 13 plant extracts were tested against conidial germination of *C. gloeosporioides* and *Zingiber officinales* were reported to be effective in minimizing conidial germination [31].

Effect of extracts on anthracnose development and quality of papaya

Extracts of botanicals evaluated for their efficacy against papaya anthracnose on papaya fruit that had been artificially inoculated by *C. gloeosporioides* showed a highly significant difference (P<0.0001) among the treatments in the incidence and severity of the disease. The incidence and severity of anthracnose was lowest in fruits treated with *Echinops* sp. extract, which were statistically at par with the positive control. Within the aqueous extract concentrations, fruits treated with 25% had relatively lower incidence and severity of anthracnose than those treated with 10% aqueous extracts, with the exception of *Ruta chalepensis* and *Thymus serrultus*. Overall, fruits treated with aqueous plant extracts had lower severity than the untreated control (Figure 2a and 2b). Water is a universal solvent, used to extract plant products with antimicrobial activity. Nearly all of the identified antimicrobial compounds from plants are aromatic or saturated organic compounds, thus water is among other solvents that are most commonly used for preliminary investigation of antimicrobial activity in plants [32]. The presences of Furoquinolines and coumarins are reported in Rutaceae family to exhibit antifungal activity [33,34]. The antimicrobial properties of extracts from various species have been proven to affect fungal development *in vivo* [24,30,35].

In this study, there was a highly (P<0.0001) significant difference in marketability of fruits treated with extracts. The results showed that extracts of different plant species substantially varied in their antifungal potentials and the difference might accrue from the variability in chemical constituents of the plants. The highest marketability was achieved in fruits treated with *Echinops* sp. extracts at a concentration of 25% (Figure 2c). Bhaskara et al. [36] reported the antifungal activity of thyme oil against *B. cinerea* and *R. stolonifer*. Similarly, the inhibitory effects of thyme oil against *C. gloeosporioides* growth in *in vitro* evaluations and over development of storage rots in papaya fruit [37].

Postharvest diseases like *C. gloeosporioides* greatly reduce the storage life papayas. However, dipping fruit in plant extracts inhibited rot development during storage [35,38]. Navel orange fruits treated with aqueous extract kept on quality of navel orange under cold storage

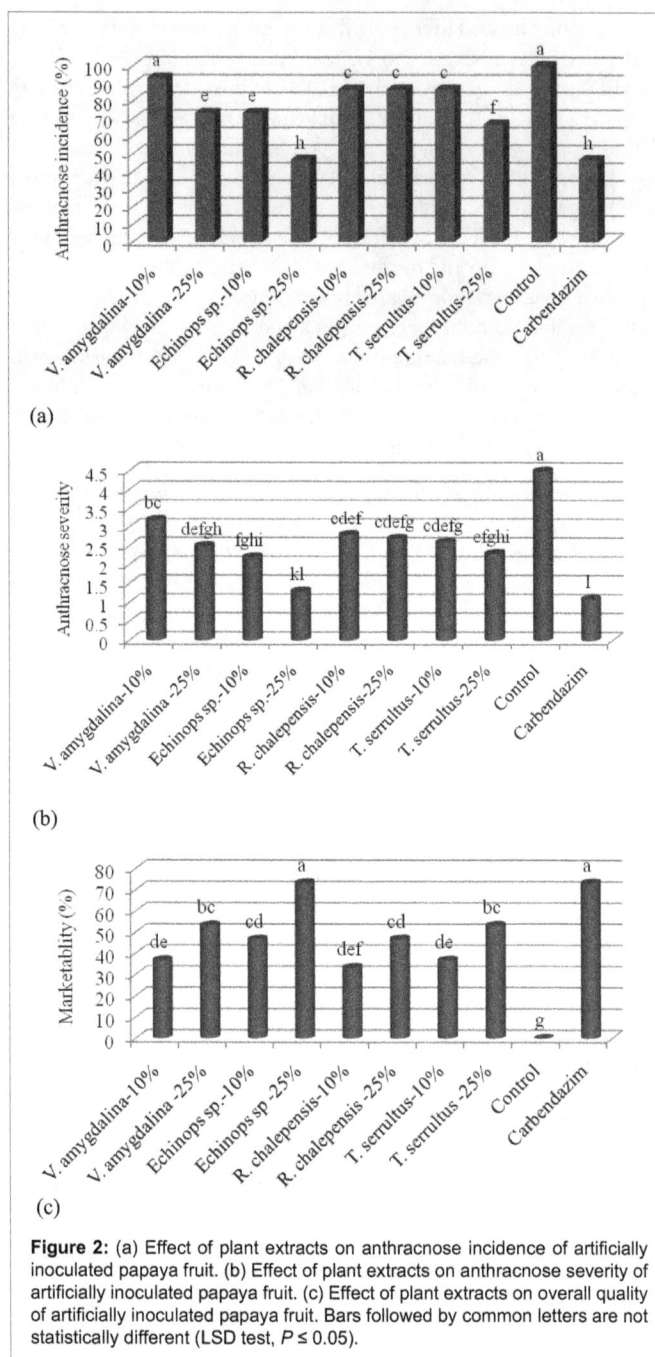

(a)

(b)

(c)

Figure 2: (a) Effect of plant extracts on anthracnose incidence of artificially inoculated papaya fruit. (b) Effect of plant extracts on anthracnose severity of artificially inoculated papaya fruit. (c) Effect of plant extracts on overall quality of artificially inoculated papaya fruit. Bars followed by common letters are not statistically different (LSD test, *P* ≤ 0.05).

condition, and reduced the incidence and severity of green rot disease comparing with the control treatment [27]. The report by Anthony et al. [39] also showed that extracts was effective in controlling postharvest diseases, while maintaining the fruit quality. Plants are known to contain a number of secondary substances like phenols, flavonoids, quinines, essential oils, alkaloids, saponins and steroids. Some of these plant-based metabolites have antimicrobial properties, and are toxic to phytopathogens [40].

Conclusions

Examination of plant extracts on *C. gloeosporioides* in this study showed promising prospects for the utilization of plant extracts in postharvest disease control. *In vivo* experiments showed that *Echinops* sp. (25%) extract reduced postharvest diseases on papaya caused by *C. gloeosporioides*, while maintaining overall quality of the fruit. Further studies on isolation and characterization of the active (antifungal) compound are needed.

References

1. Teixeira da Silva JA, Rashid Z, Nhut DT, Sivakumar D, et al. (2007) Papaya (*Carica papaya* L.) biology and biotechnology. Tree For Sci Biotech 1: 47-73.

2. Lustria JUJ, Nacional A, Morillo AE (2009) Commodity situation report: Papaya. Working Paper 34.

3. Alvarez AM, Nishijima WT (1987) Postharvest diseases of papaya. Plant Dis 71: 681-686.

4. Paull RE, Nishijima W, Reyes M, Cavaletto C (1997) A review of postharvest handling and losses during marketing of papaya (*Carica papaya* L.). Postharvest Biol Tec 11: 165-179.

5. Tasiwal V, Benagi VI, Hegde YR, Kamanna BC, Naik KR (2009) *In vitro* evaluation of botanicals, bioagents and fungicides against anthracnose of papaya caused by *Colletotrichum gloeosporioides* (Penz.) Penz. and Sacc. Karnataka J Agric Sci 22: 803-806.

6. Capdeville G de, Souza Jr MT, Santos JRP, Miranda SP, Caetano AR, et al. (2007) Scanning electron microscopy of the interaction between *Cryptococcus magnus* and *Colletotrichum gloeosporioides* on papaya fruit. Pesq Agropec Bras 42: 1537-1544.

7. Coursey DG (1983) Post-harvest losses in perishable foods of the developing world. NATO Advance Study Institute Series A46 485.

8. Lee SO, Choi GJ, Jang KS, Lim HK, Cho KY, et al. (2007) Antifungal activity of five plant essential oils as fumigant against postharvest and soilborne plant pathogenic fungi. Plant Pathol J 23: 97-102.

9. Anand T, Bhaskaran R (2009) Exploitation of plant products and bioagents for ecofriendly management of chilli fruit rot disease. J Plant Prot Res 49: 195-203.

10. Hernandez-Albiter RC, Barrera-Necha LL, Bautista-Banos S, Bravo-Luna L (2007) Antifungal potential of crude plant extracts on conidial germination of two isolates of *Colletotrichum gloeosporioides* (Penz.) Penz. And Sacc. Mex J Phytopathol 25: 180-185.

11. Meepagala KM, Sturtz G, Wedge DE (2002) Antifungal constituents of the essential oil fraction of *Artemisia dracunculus* L. var. *dracunculus*. J Agric Food Chem 50: 6989-6992.

12. Sharma N, Trivedi PC (2002) Screening of leaf extracts of some plants for their nematicidal and fungicidal properties against *Meloidogyne incognita* and *Fusarium oxysporum*. Asian J Exp Sci 16: 21-28.

13. Fokialakis N, Cantrell CL, Duke SO, Skaltsounis AL, Wedge DE (2006) Antifungal activity of thiophenes from *Echinops ritro*. J Agri Food Chem 54: 1651-1655.

14. Barrera-Necha LL, Bautista-Banos S, Flores-Moctezuma HE, Estudillo AR (2008) Efficacy of essential oils on the conidial germination, growth of *Colletotrichum gloeosporioides* (Penz.) Penz. and Sacc. and control of postharvest diseases in papaya (*Carica papaya* L.). Plant Pathol J 7: 174-178.

15. Amadioha AC (2000) Controlling rice blast *in vitro* and *in vivo* with extracts of *Azadirachta indica*. Crop Prot 19: 287-290.

16. Sateesh K, Marimuthu T, Thayumanavan B, Nandakumar R, Samiyappan R (2004) Antimicrobial activity and induction of systemic resistance in rice by leaf extract of *Datura metel* against *Rhizoctonia solani* and *Xanthomonas oryzae* pv.oryzae. Physiol Mol Plant Pathol 65: 91-100.

17. Ranaware A, Singh V, Nimbkar N (2010) *In vitro* antifungal study of the efficacy of some plant extracts for inhibition of *Alternaria carthami* fungus. Indian J Nat Prod Resour 1: 384-386.

18. Dwivedi BP, Shukla DN (2000) Effect of leaf extracts of some medicinal plants on spore germination of some *Fusarium* species. Karnataka J Agric Sci 13: 153-154.

19. Yesuf M, Mandefro W, Ahmed E, Adugna G, Tadesse D, et al. (2009) Review of Research on fruit crop diseases in Ethiopia. In: Increasing crop production through improved plant protection-Volume II, Abraham Tadesse (Ed.), Plant protection society of Ethiopia (PPSE), PPSE and EIAR, Addis Ababa, Ethiopia.

20. Gamagae SU, Sivakumar D, Wijeratnam RSW, Wijesundera RLC (2003) Use of Sodium bicarbonate and *Candida oleophila* to control anthracnose in papaya during storage. Crop Prot 22: 775-779.

21. Yonas K, Amare A (2008) Postharvest biological control of anthracnose (*Colletotrichum gloeosporioides*) on mango (*Mangifera indica* L.). Postharvest Biol Tec 50: 8-11.

22. Amare AM (2002) Mycoflora and mycotoxins of major cereal grains and antifungal effects of selected medicinal plants from Ethiopia. Doctoral Dissertation. Georg-August University of Gottingen. Cuvillier Verlag Gottingen.

23. Bautista-Banos S, Barrera-Necha LL, Bravo-Luna L, Bermudez-Torres K (2002) Antifungal activity of leaf and stem extracts from various plant species on the incidence of *Colletotrichum gloeosporioides* of papaya and mango fruit after storage. Mex J Phytopathol 20: 8-12.

24. Sivakumar D, Hewarathgamagae NK, Wijeratnam RSW, Wijesundera RLC (2002) Effect of ammonium carbonate and sodium bicarbonate on anthracnose of papaya. Phytoparasitica 30: 486-492.

25. Gamagae SU, Sivakumar D, Wijesundera RLC (2004) Evaluation of post-harvest application of sodium bicarbonate incorporated wax formulation and *Candida oleophila* for the control of anthracnose of papaya. Crop Prot 23: 575-579.

26. Tripathi P, Shukla AK (2007) Emerging non-conventional technologies for control of postharvest diseases of perishables. Fresh Prod 1: 111-120.

27. Abd-El-Khair H, Omima, MH (2006) Effect of aqueous extracts of some medicinal plants in controlling the green mould disease and improvement of stored "Washington" navel orange quality. J Appl Sci Res 2: 664-674.

28. Nashwa SMA, Abo-Elyousr KAM (2012) Evaluation of various plant extracts against the early blight disease of tomato plants under greenhouse and field conditions. Plant Prot Sci 48: 74-79.

29. Al-Samarrai GF, Harbant S, Mohamed S (2013) Extacts some plants on controlling green mold of orange and on postharvest quality parameters. World Appl Sci J 22: 564-570.

30. Ogbebor ON, Adekunle AT, Enobakhare DA (2007) Inhibition of *Colletotrichum gloeosporioides* (Penz) Penz. and Sacc. causal organism of rubber (*Hevea brasiliensis* Muell. Arg.) leaf spot using plant extracts. Afr J Biotechol 6: 213-218.

31. Imtiaj A, Rahman SA, Alam S, Parvin R, Farhana KM, et al. (2005) Effect of fungicides and plant extracts on the conidial germination of *Colletotrichum gloeosporioides* causing mango anthracnose. Mycobioloby 33: 200-205.

32. Gurjar MS, Shahid A, Masood A, Kangabam SS (2012) Efficacy of plant extracts in plant disease management. Agric Sci 3: 425-433.

33. Gray AI, Watermann PG (1978) Coumarins in the Rutaceae. Phytochemistry 17: 845-864.

34. Michael JP (2003) Quinoline, quinazoline and acridone alkaloids. Nat Prod Rep 20: 476-493.

35. Bautista-Banos S, Hernandez-Lopez M, Bosquez-Molina E, Wilson CL (2003) Effects of chitosan and plant extracts on growth of *Colletotrichum gloeosporioides*, anthracnose levels and quality of papaya fruit. Crop Prot 22: 1087-1092.

36. Bhaskara MV, Angers P, Gosselin A, Arul J (1998) Characterization and use of essential oil from *Thymus vulgaris* against *Botrytis cinerea* and *Rhizopus stolonifer* in strawberry fruits. Phytochemistry 47: 1515-1520.

37. Bosquez-Molina E, Ronquillo-de Jesus E, Bautista-Banos S, Verde-Calvo JR, Morales-Lopez J (2010) Inhibitory effect of essential oils against *Colletotrichum gloeosporioides* and *Rhizopus stolonifer* in stored papaya fruit and their possible application in coatings. Postharvest Biol Tec 57: 132-137.

38. Bautista-Banos S, Hernandez-Lopez M, Diaz-Perez JC, Cano-Ochoa CF (2000) Evaluation of the fungicidal properties of plant extracts to reduce *Rizopus stolonifer* of 'ciruela' (*Spondias purpurea* L.) during storage. Postharvest Biol Tec 18: 67-73.

39. Anthony S, Abeywickrama K, Wijeratnam SW (2003) The effect of spraying essential oils of *Cymbopogon nardus*, *Cymbopogan flexuosus* and *Ocimum basilicum* on postharvest diseases and storage life of Embul banana. J Hortic Sci Biotech 78: 780-785.

40. Tripathi P, Shukla AK (2010) Exploitation of botanicals in the management of phytopathogenic and storage fungi. In: Management of fungal plant pathogens, Arya A, Perello AE (Ed.), CAB International, USA 36-50.

11

Bacterial Structure of Agricultural Soils with High and Low Yields

Michelli de Souza dos Santos[1]*, Kavamura VN[1], Reynaldo ÉF[1], Souza DT[1], da Silva EHFM[2] and May A[3]

[1]*Laboratory of Environmental Microbiology, Brazilian Agricultural Research Corporation, EMBRAPA Environment, SP 340, Km 127.5, 13820-000, Jaguariúna, SP, Brazil.*
[2]*College of Agriculture "Luiz de Queiroz", University of São Paulo, Av. Pádua Dias, 11, 13418-900, Piracicaba, SP, Brazil.*
[3]*Laboratory of Plant Physiology, Brazilian Agricultural Research Corporation, EMBRAPA Maize and Sorghum, MG 424, Km 45, 35701-970, Sete Lagoas, MG, Brazil.*

Abstract

The purpose of this study was to evaluate the structure of bacterial communities at two agricultural fields in Brazil (Paraná (PR) and Bahia (BA) states) with a history of high and low productivity of soybean. 16S rRNA gene amplicons revealed that plots with low yield of grains showed greater bacterial richness than plots with high yield. The phylum Acidobacteria was more abundant in soil samples from PR site. The rhizosphere of plants presented a similar bacterial community for both high and low yield plots. Soil samples from BA showed differences in the diversity between the plots with high and low productivity. The use of 16S rRNA amplicon sequencing allowed the assessment of differences between plots with different soybean yields. This might be useful in the future to harness plant microbiomes for increased crop productivity.

Keywords: 16S rRNA amplicons; Bacterial community; Productivity; Soybean

Introduction

There is growing evidence that plants recruit microorganisms to protect themselves from biotic and abiotic factors [1]. Since rhizosphere of plants contain a plethora of microorganisms, this makes them excellent model systems for studying the assembly and regulation of a beneficial microbiome throughout the productivity process of crops.

Although soil microorganisms play important roles in ecosystems multifunctionality [2] it is reported that changes in land use, management practices and fertilisation regime affect soil diversity [3-4]. Modifications in microbial diversity can be assessed with the use of next-generation sequencing technologies, such as the analysis of 16S rRNA gene amplicons [5]. Decreased soil microbial diversity may be an important indicator of the loss of soil quality, revealing a balance among organisms and the functional domains in soils [6].

Conceivably, much of the ecosystem services provided by microorganisms has evolved as a result of their interactions with other microorganisms in highly diverse environments and this often indicates the type of activity that occurred in the studied area, as portrayed by [7]. Thus, according to the literature, there is an increase in the diversity of soil bacteria when plant diversity is high, probably due to the different composition of the exudates coming from the different plant species present in the system [8].

In this way, the knowledge about soil microbes can help identifying potential phytosanitary and yield problems. Probably, soils with high and low productivity present different structure and bacterial composition [9]. Thus, based on the results of 16S rRNA gene amplicons, the present study had the objective of evaluating the structure and composition of bacterial communities at two agricultural fields in Brazil with a history of high and low productivity of soybean.

Methodology

Study area and soil sampling

The sampling was performed in two Brazilian states (Paraná (PR) and Bahia (BA)) with all features of each site described in Table 1. Bulk soil samples consisted of soil without plant interference. Three replicates of each plot were obtained, with each one corresponding to ten subsamples collected in zig-zag. Rhizosphere samples consisted of soil closely attached to roots of soybean plants at flowering stage.

For the area of Paraná (PR), six bulk soil (BS) and six soybean rhizosphere (RZ) samples were collected for plots with low (Lp) and high (Hp) productivity (3 replicates each). For the area of Bahia (BA), six bulk soil (BS) samples were collected for plots with low (Lp) and high (Hp) productivity. Additionally, bulk soil samples were collected from a native forest adjacent to both areas (FBS).

All samples were placed in plastic bags and stored in a styrofoam box and immediately sent to the laboratory. The soil and climatic characteristics of the places where the samples were collected are shown in Table 2. Chemical analysis of soil is shown in Table 3.

Metagenomic DNA extraction and sequencing of 16S rRNA gene

Metagenomic DNA extraction was performed for soil and rhizosphere samples using Power Soil™ DNA Isolation Kit (MoBio Laboratories, Inc., Carlsbad, CA, USA), according to the protocol provided by the manufacturer. In total, twenty-four samples were processed (Table 3). For sequencing, the samples were PCR-amplified using the primer set 967F [10] and 1193R [11] to generate amplicons included in the V6-V7 region of the 16S rRNA gene. The PCR reactions and purifications were performed according to [12]. The amplicon libraries were sequenced on an Ion Torrent PGM system of Life Technologies using the Ion 316™ Chip according to manufacturer's instruction.

Sequence processing and data analysis

Raw sequences were manipulated using Galaxy software (https://usegalaxy.org/). After processing, 2,387,087 sequences were analyzed using the QIIME (Quantitative Insights into Microbial Ecology) software version 1.8.1 [13]. To identify Operational Taxonomic Units (OTUs) with 97% similarity, UCLUST tool [14] was used. A representative sequence of each OTU was aligned against Greengenes database using the NAST algorithm [14]. Chimeric sequences were

---

*Corresponding author: Michelli de Souza dos Santos, Pós-doctor Embrapa Environment Street SP 340, KM 127,5, S/N - Tanquinho Velho, Jaguariúna-SP, Brazil, E-mail: michellisantos30@hotmail.com

Sampling sites		Candói, Paraná (PR)	São Desidério, Bahia (BA)
Coordinates		S-25° 31' 15,6', W-51° 47' 19.8"	S-13°15′01′′, W-46°13′18′′
Climate features	Type	Cfb Rainy during winter and summer	Aw Dry winter and rainy summer
	Average annual temperature	16.9°C	24.7°C
	Dry season	June to August	May to September
	Rainy season	September to February	October to March
	Monthly rainfall	150 mm -190 mm	100 mm -220 mm
Soil features	Type	cambic aluminum Bruno Latosol	dystrophic Red-Yellow Latosol
	Texture	Clay	Medium
Soil management		Crop rotation: soybean, oat, maize, wheat, barley	Monoculture: soybean
Sampling	Soil type	Bulk soil and rhizosphere	Bulk soil

Table 1. Characteristics of sampling sites.

Samples	Productivity	Soil type	Area	Barcodes
Hp.BS.PR1	High	Bulk soil	Paraná	GATCT
Hp.BS.PR2	High	Bulk soil	Paraná	ATCAG
Hp.BS.PR3	High	Bulk soil	Paraná	ACACT
Hp.RZ.PR1	High	Rhizosphere	Paraná	AGATG
Hp.RZ.PR2	High	Rhizosphere	Paraná	CACTG
Hp.RZ.PR3	High	Rhizosphere	Paraná	CAGAG
Lp.BS.PR1	Low	Bulk soil	Paraná	AGCTA
Lp.BS.PR2	Low	Bulk soil	Paraná	CACAC
Lp.BS.PR3	Low	Bulk soil	Paraná	ACAGA
Lp.RZ.PR1	Low	Rhizosphere	Paraná	CGCAG
Lp.RZ.PR2	Low	Rhizosphere	Paraná	CTGTG
Lp.RZ.PR3	Low	Rhizosphere	Paraná	GTGAG
FBS.PR1	Forest	Bulk soil	Paraná	TCATG
FBS.PR2	Forest	Bulk soil	Paraná	AGCAT
FBS.PR3	Forest	Bulk soil	Paraná	CAGCT
Hp.BS.BA1	High	Bulk soil	Bahia	CATGT
Hp.BS.BA2	High	Bulk soil	Bahia	CTGAT
Hp.BS.BA3	High	Bulk soil	Bahia	CTGCA
Lp.BS.BA1	Low	Bulk soil	Bahia	GATGA
Lp.BS.BA2	Low	Bulk soil	Bahia	TACGC
Lp.BS.BA3	Low	Bulk soil	Bahia	ACTGC
FBS.BA.1	Forest	Bulk soil	Bahia	GTCAC
FBS.BA.2	Forest	Bulk soil	Bahia	CGTAC
FBS.BA.3	Forest	Bulk soil	Bahia	TGCGT

Table 2. Description of soil samples used for metagenomic DNA extraction.

Samples	pH	Al (cmolc/dm³)	V (%)	OM (dag/kg)	Ca (cmolc/dm³)
Hp.BS-PR	6.23	0.02	47.75	7.07	5.42
Hp.RZ-PR	6.57	0.01	67.68	8.87	7.15
Lp.BS-PR	6.00	0.02	43.11	4.27	3.67
Lp.RZ-PR	6.53	0.00	62.49	5.27	3.20
FBS-PR	5.40	1.56	10.13	13.5	1.7
Hp.BS-BA	5.87	0.00	42.96	1.51	1.33
LP.BS-BA	5.63	0.02	34.11	1.12	0.77
FBS-BA	5.27	0.19	3.91	1.74	0.01

Samples	Mg(cmolc/dm³)	K (mg/dm³)	P(mg/dm³)	C(%)
Hp.BS-PR	1.53	181.70	13.62	4.11
Hp.RZ-PR	2.5	267.47	16.93	5.16
Lp.BS-PR	0.95	131.30	7.92	2.48
Lp.RZ-PR	1.68	203.97	11.69	3.07
FBS-PR	0.45	137.4	10.48	7.85
Hp.BS-BA	0.43	43.89	12.87	4.85
LP.BS-BA	0.29	31.44	8.62	3.14
FBS-BA	0.07	9.36	0.0	9.1

High productivity, bulk soil, Paraná state (Hp.BS-PR); Low productivity, bulk soil, Paraná state (Lp.BS-PR); High productivity, soybean rhizosphere, Paraná state (Hp.RZ-PR); Low productivity, soybean rhizosphere, Paraná state (Lp.RZ-PR); Forest, bulk soil, Paraná state (FBS-PR); High productivity, bulk soil, Bahia state (Hp.BS-BA); Low productivity, bulk soil, Bahia state (Lp.BS-BA); Forest, bulk soil, Bahia state (FBS-BA).

Table 3. Mean (n=3) of the chemical analyzes of bulk soil and soybean rhizosphere samples collected in Paraná (PR) and Bahia (BA) with high (Hp) and low productivity (Lp) plots. As the control, a native forest bulk soil (FBS) was used for both areas (PR and BA).

removed by the UCHIME method [15]. The taxonomic classification was performed using the UCLUST taxonomy assigned method with the Greengenes reference sequence database [16]. Non-target sequences (i.e. chloroplast, singleton and sequences that failures for alignment) were removed from the dataset. After processing, 358,563 sequences were assigned to 20,061 different OTUs and a sample vs. OTU table was created and used as input data for downstream analysis. Sequences are available in the MG-RAST server under accession numbers 317970 to 317993. Diversity indexes based on the OTU table were calculated and PCoA plots were generated using PAST software [17]. In addition, SIMPER (Similarity Percentage) test was performed to weigh the contribution of each phylum in the similarity/dissimilarity among the samples [18].

Results

Variation in OTU richness

Bulk soil samples collected from low productivity (Lp) plots displayed a 20% higher richness than both bulk soil samples collected from high productivity (Hp) plots and bulk soil (FBS) samples from the forest. For soybean rhizosphere samples, the difference in richness between high and low productivity plots was less pronounced (Figure 1).

PCoA plot clearly shows that bacterial communities from bulk soil samples are different, with the first two axes corresponding to more than 69% of the variation (Figure 2). The first axis explains 54.57% of the variation, thus forest soil samples are very different from bulk soil samples collected from agricultural field. Besides the difference between bacterial communities obtained from high and low productivity plots is explained by more than 14%. For soybean rhizosphere samples, the first axis itself explains the difference of bacterial communities between high and low productivity plots (Figure 2).

These differences can be better observed through SIMPER test, which compares the relative frequencies of the phyla found in the samples. There is a dissimilarity of 64.11% for bulk soil samples collected from high and low productivity plots, with Acidobacteria corresponding to 9.35% of the total difference and Proteobacteria to 1.02%. Soybean rhizosphere samples showed a lower dissimilarity than bulk soil samples (49.41%), with Acidobacteria being responsible for 1.41% of the total difference and Proteobacteria to 1.12%.

Bulk soil samples from Bahia state collected in the field with a high productivity history presented a 50% lower richness than the microbial communities found in samples from the low productivity plot. The richness of the OTUs of the native forest soil sample was similar to that of samples from the low yield plot (Figure 1). PCoA analysis showed that a separation of the samples took place, due to the history of productivity, with samples being separated by approximately 54% (Figure 2). These differences can be better analyzed by performing the SIMPER test, which compares the relative frequencies of the obtained phyla for the samples. Thus, in general, when comparing the samples by productivity history (high and low), there was a dissimilarity among samples of 86.46%. The main phyla that contributed to this differentiation were: Acidobacteria (9.03%), Proteobacteria (6.12%), Actinobacteria (6%) and Chloroflexi (1.80%) (Figure 3).

Bacterial structure and composition

Sequences from the domain Bacteria found in bulk soil and soybean rhizosphere samples collected from Paraná state were classified into forty-one phyla, whereas thirty-eight phyla were assigned to samples from Bahia state. Of these total, nine phyla (Acidobacteria, Actinobacteria, Chloroflexi, Firmicutes, Gemmatimonadetes, Planctomycetes, Proteobacteria, Chlorobi and Verrucomicrobia) and two candidate phyla (AD3 e GOUTA4) had a frequency greater than

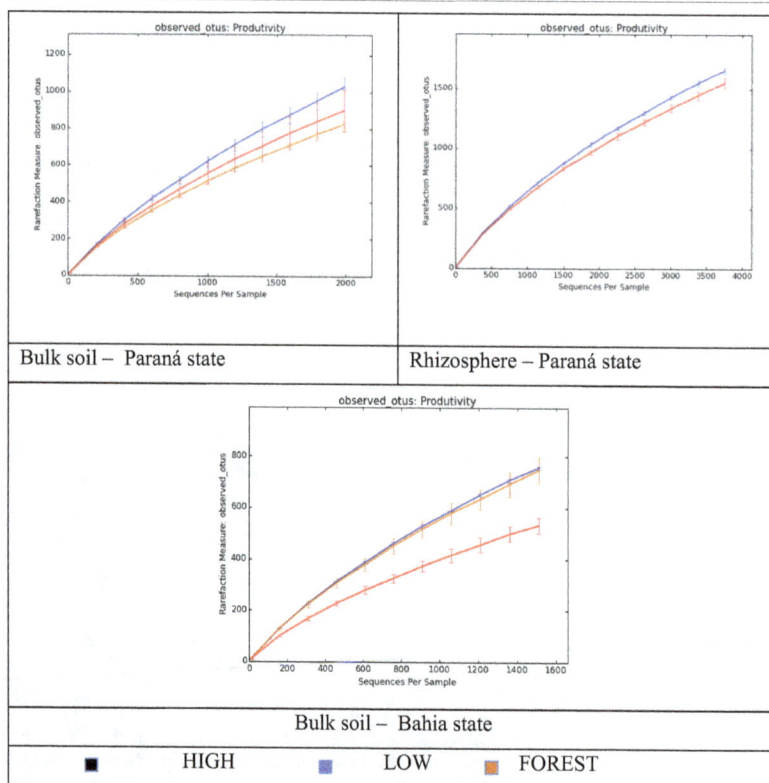

Figure 1. Number of OTUs obtained for bulk soil and soybean rhizosphere samples for high and low productivity plots for the states of Paraná and Bahia.

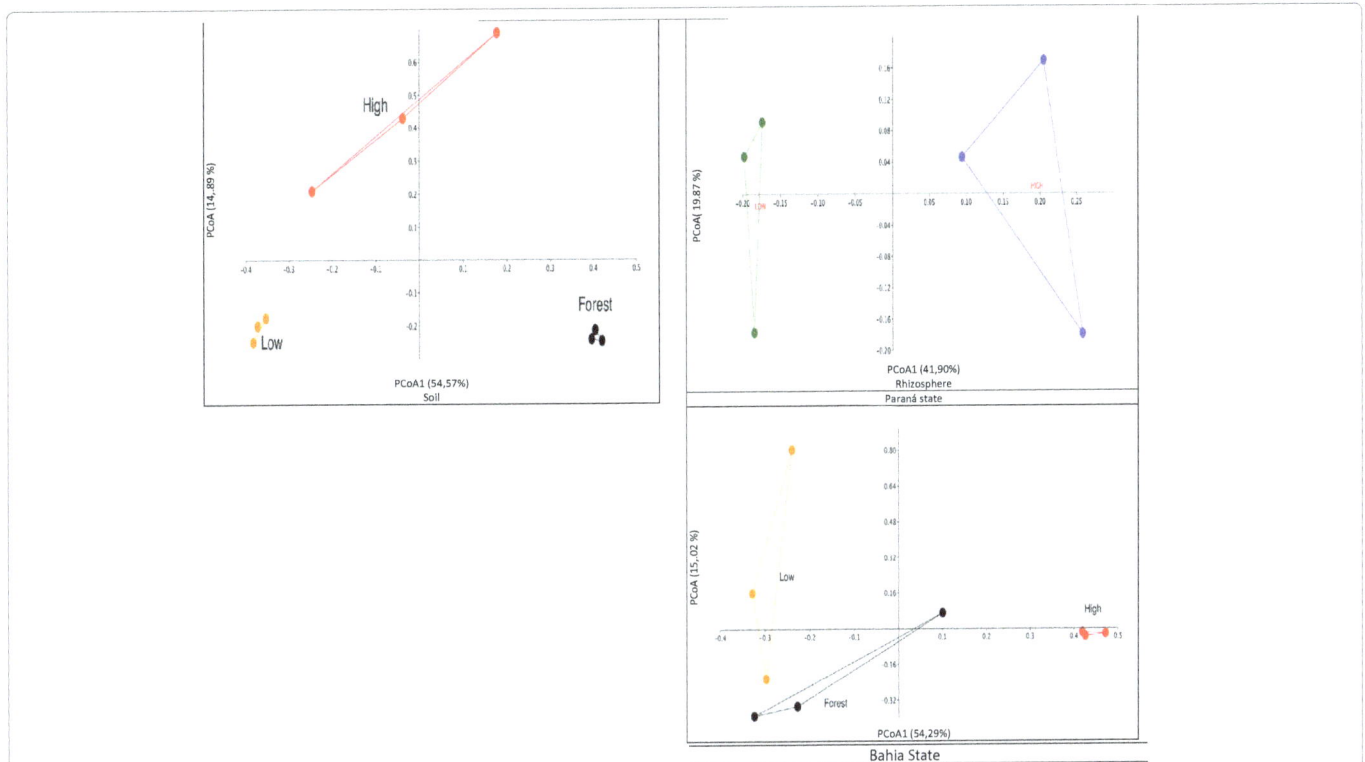

Figure 2. Principal Coordinates Analysis (PCoA) plot showing the dissimilarity of OTUs found in bulk soil and soybean rhizosphere samples collected in the soybean farms in the States of Paraná and Bahia.

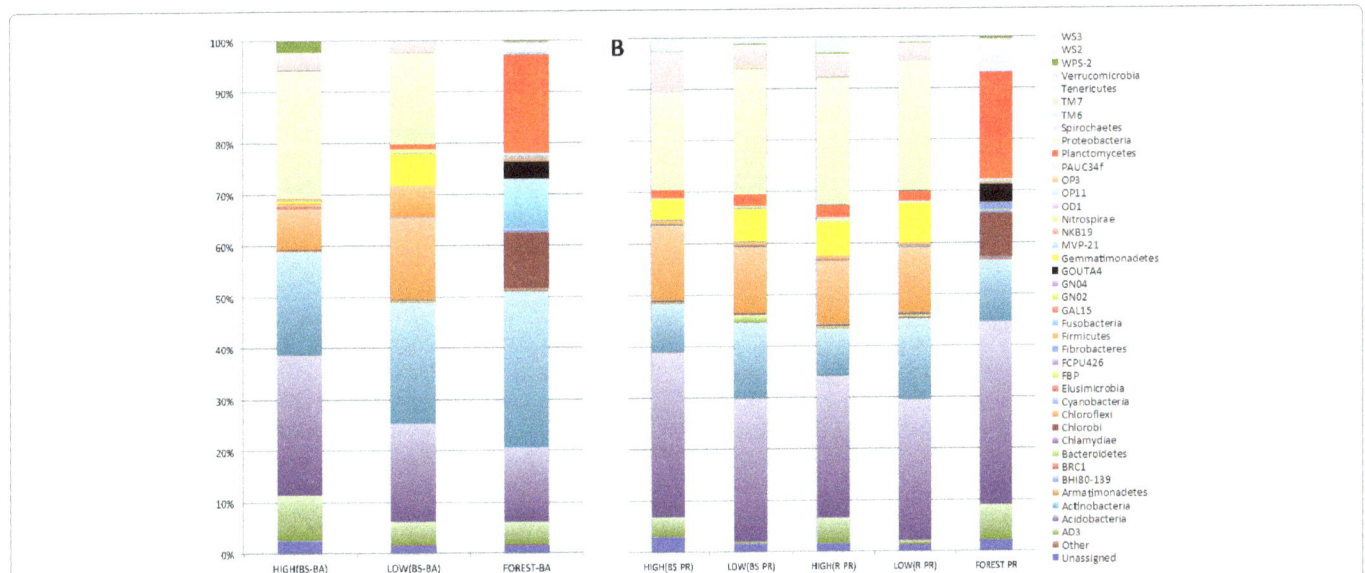

Figure 3. Taxonomic assignments at the phylum level showing the relative frequency of OTUs from bulk soil and soybean rhizosphere samples collected from high and low-productivity areas as well as a forest area in Bahia **(A)** and Paraná **(B)** states, in Brazil. High productivity, bulk soil, Paraná state (Hp.BS-PR); Low productivity, bulk soil, Paraná state (Lp.BS-PR); High productivity, soybean rhizosphere, Paraná state (Hp.RZ-PR); Low productivity, soybean rhizosphere, Paraná state (Lp.RZ-PR); Forest, bulk soil, Paraná state (FBS-PR); High productivity, bulk soil, Bahia state (Hp.BS-BA); Low productivity, bulk soil, Bahia state (Lp.BS-BA); Forest, bulk soil, Bahia state (FBS-BA).

1%. The candidate phylum WS3 (samples from Paraná) and WPS-2 (samples from Bahia) presented frequency greater than 1% (Table 4). For bulk soil samples, the phylum Acidobacteria had higher abundance (32% and 27%) in the high productivity plots than in the low productivity plots (28% and 19%) for PR and BA states, respectively. The opposite trend was observed for the phyla Actinobacteria and Gemmatimonadetes. Members from Actinobacteria phylum were more abundant (15% and 24%) in samples from low productivity plots than in plots with high productivity (10% and 20%) for PR and BA states, respectively. Members from the phyla Chloroflexi and Firmicutes were more abundant in plots with low productivity (16% and 6%) from Bahia state when compared to high productivity plots from the same site. This trend is not seen for samples from PR state. For Proteobacteria, the abundance of members from this phylum is higher (24%) in low

Phylum/ Candidates phylum	Hp.BS PR	Lp.BS PR	Hp.RZ PR	Lp.RZ PR	FBS PR	Hp.BS BA	Lp.BS BA	FBS BA
Unassigned	3%	2%	2%	1%	2%	3%	1%	1%
AD3	4%	0%	5%	1%	7%	9%	4%	4%
Acidobacteria	32%	28%	28%	27%	35%	27%	19%	15%
Actinobacteria	10%	15%	9%	16%	12%	20%	24%	30%
Chlorobi	0%	0%	0%	0%	8%	0%	0%	11%
Chloroflexi	15%	13%	12%	13%	0%	8%	16%	0%
Firmicutes	1%	1%	1%	1%	0%	1%	6%	0%
GOUTA4	0%	0%	0%	0%	3%	0%	0%	3%
Gemmatimonadetes	4%	7%	6%	7%	0%	1%	6%	0%
Planctomycetes	1%	2%	2%	2%	21%	0%	1%	19%
Proteobacteria	19%	24%	25%	25%	0%	25%	18%	0%
Verrucomicrobia	8%	4%	5%	3%	0%	4%	2%	0%
WPS-2	0%	0%	0%	0%	0%	2%	0%	0%
WS3	3%	1%	3%	1%	0%	0%	0%	0%

High productivity, bulk soil, Paraná state (Hp.BS-PR); Low productivity, bulk soil, Paraná state (Lp.BS-PR); High productivity, soybean rhizosphere, Paraná state (Hp.RZ-PR); Low productivity, soybean rhizosphere, Paraná state (Lp.RZ-PR); Forest, bulk soil, Paraná state (FBS-PR); High productivity, bulk soil, Bahia state (Hp.BS-BA); Low productivity, bulk soil, Bahia state (Lp.BS-BA); Forest, bulk soil, Bahia state (FBS-BA).

Table 4. Values of the relative frequencies of phyla and candidate's phylum with frequency greater than 1% in the soil and rhizosphere samples of the Paraná and Bahia states.

productivity plots than in high productivity plots (19%) for PR state. On the other hand, bulk soil samples from high productivity plots from BA state showed more abundance (25%) of Proteobacteria than bulk soil samples from low productivity plots (18%). The frequency of members from the phylum Verrucomicrobia was two times larger in the high productivity samples from PR state when compared to both low productivity plots from the same site and high productivity plots from BA state.

The results also show that agricultural soils present a higher frequency of members belonging to Gemmatimonadetes, Chloroflexi, Proteobacteria and Verrucomicrobia when compared to the forest soils. Yet, soils collected in the native forests present a higher frequency of the Tenericutes, Planctomycetes, GOUTA4 and Chlorobi when compared to agricultural soils.

Discussion

In this study, the differences among the structure of bacterial communities from bulk soil and soybean rhizosphere samples were evaluated, as well as bulk soil from native forests close to the agricultural fields. It was observed that bulk soil and rhizosphere of soybean plants are different niches, hosting distinct microbial communities. The results show that there are significant differences between bacterial communities from bulk soil and soybean rhizosphere based on high and low productivity history for both areas. Bulk soil samples showed a greater differentiation between the plots with a history of high and low productivity, whereas for rhizosphere samples this difference is less pronounced. This behavior can be as expected due to the close association of plant roots and microbes via the production of molecules known as exudates, which are beneficial to microbial life [19]. Soil microorganisms are attracted to the roots of plants through a well-known mechanism, which involves cross signaling between roots and microbes [20]. However, a certain selection might occur. Thus, plants or improved genotypes of cultivated plants have the ability to act on the microbial community in their rhizosphere, due to the distinction in signaling, especially under stressful situations, such as the physical changes of the soil, which are capable of harming the development of the plant [21].

Bacterial structure of samples collected in Paraná

Soil samples from the high productivity plots showed a higher relative frequency of members from the phylum Acidobacteria, compared to the soil samples collected in the low productivity plots. It is known that species of this phylum are capable of reducing nitrates and nitrites and may also form a biofilm, which can improve soil structure. Besides, they can produce compounds that catalyze several proteins and use of soil carbon [22]. However, the phylum Acidobacteria is still little understood, although its abundance in the studied samples may suggest their importance in nutrient cycling, since the nutrients available in the soil for the plants and /or other organisms are one of the attributes that most interfere with soil quality [23]. Thus, the decrease of Acidobacteria in the low productivity plots may have some relation to the productivity of the crop in these areas. This phylum also displayed a higher frequency in forest samples, resembling to the soil of the field of high productivity. In general, the frequencies of native forest phyla resembled that of the high productivity field, rather than the low productivity, as well as edaphic factors. The native forest bulk soil collected in Paraná state is characterized as Atlantic Forest soil, known to have one of the largest biodiversity on the planet, able to maintain its vegetation in full equilibrium, being considered, therefore, a hotspot [24].

Proteobacteria and Actinobacteria phyla appeared more frequently in samples of low productivity plots. In this way, soils with high nitrogen and carbon content usually present a higher occurrence of Proteobacteria, whereas, in soils with lower levels of nutrients, Acidobacteria appear more frequently [25]. It is observed that soils with higher nutrient contents are those resulting from the high productivity fields or the native forest. Thus, the bacterial composition of soil samples collected within plots with different yields could act as soil quality bioindicators, through the evaluation of the frequency of existing phyla.

Bacterial structure of samples collected in Bahia

Soil samples from high yielding plots showed a greater relative frequency of Proteobacteria and Acidobacteria. The phylum Proteobacteria is the largest and most distinguished group of bacteria known, being very diverse morphologically and metabolically. Their representatives are easily found in cultivated soils, being highly important in the nitrogen and sulfur cycles [26]. These two phyla are the most abundant in soil samples, with the phylum Proteobacteria being more commonly found in nutrient-rich soils. This might explain their higher frequency in soil samples with a high productivity history. The class β-Proteobacteria congregates copiotrophic microorganisms, being more frequently observed in soils with greater carbon content, i.e., greater amount of organic matter [27].

In soil samples from low productivity plots, a higher frequency of the Actinobacteria phylum occurs. This phylum is related to Gram-positive bacteria, generally known as decomposers of organic material (cellulose, lignin and chitin), producing a mass of proteins that serves to nourish other organisms [28]. The phylum Actinobacteria is composed of microbes able to produce antimicrobial compounds. However, production of these substances in excess may eventually impair the development of plants or microorganisms beneficial to the development of the crops [29].

Variation in OTU richness in soil samples collected in Paraná and Bahia states

All the soil samples presented a greater richness of OTUs in the samples collected in the areas with low productivity history, showing a possible imbalance in these environments. This might help explain the productivity differences in these plots. Soil samples from fields with low yield history have a higher number of species; however, changes in soils may impair sustainability, causing anomalies in plant groups, and also changes in bacterial communities [30].

Soil richness of the native forest in Paraná state resembled the soil of the field of high productivity plot for Paraná state. The opposite was observed for Bahia, where the native forest soil resembled the soil samples from low productivity plots. The high productivity and native forest soils of the Paraná state are chemically similar, whereas in Bahia State, the native forest soil is poor in nutrients, more similar to the plots of low productivity history than to the high productivity ones. Thus, chemical changes in soils can interfere in bacterial communities, such as, for example, pH and soil phosphorus content [31].

Conclusions

There are fluctuations of bacterial communities in soils with different history of productivity. The diversity and richness of bacterial communities can be used as bioindicators of soil quality.

References

1. Agler MT, Ruhe J, Kroll S, Morhenn C, Kim S-T, Weigel D, et al. (2016) Microbial Hub Taxa Link Host and Abiotic Factors to Plant Microbiome Variation. PLoS Biol 14(1): e1002352.

2. Delgado-Baquerizo MF, Maestre P, Reich T, Jeffries J, Gaitan D, et al. (2016) Microbial diversity drives multifunctionality in terrestrial ecosystems. Nat Commun 7: 10541.

3. Wall DH, Bardgett RD, Kelly E (2010) Biodiversity in the dark. Nat Geosci 3: 297-298.

4. Wang S, Chen HYH, Tan Y, Fan H, Ruan H (2016) Fertilizer regime impacts on abundance and diversity of soil fauna across a poplar plantation chronosequence in coastal Eastern China. Sci Rep 6: 20816.

5. Roesch LFW, Fulthorpe RR, Riva A, Casella G, Hadwin AKM, et al. (2007) Pyrosequencing enumerates and contrasts soil microbial diversity. ISME J 1: 283-290.

6. Benizri E, Amiaud B (2005) Relationship between plants and soil microbial communities in fertilized grasslands. Soil Biol Biochem 37: 2055-2064.

7. Quince C, Curtis TP & Sloan WT (2008) The rational exploration of microbial diversity. ISME J 2: 997–1006.

8. El Moujahid L, Le Roux X, Michalet S, Bellvert F, Weigelt A, Poly F (2017) Effect of plant diversity on the diversity of soil organic compounds. PLoS ONE 12(2): e0170494.

9. Ehrenfeld JG, Ravit B, Elgersma K (2005) Feedback in the plant-soil system. Annu Rev Environ Resour 30: 75-115.

10. Sogin ML, Morrison HG, Huber JA, Welch DV, Huse SM, et al. (2006) Microbial diversity in the deep sea and theunderexplored "rare biosphere". Proc Natl Acad Sci U S A 103:12115-12120.

11. Wang Y, Qian PY (2009) Conservative fragments in bacterial 16S rRNA genes and primer design for 16S ribosomal DNA amplicons in metagenomic studies. PloS One 4:e7401.

12. Souza DT, Genuário DB, Silva FSP, Pansa CC, Kavamura VN, Moraes FC, Taketani RG, Melo IS (2017) Analysis of bacterial composition in marine sponges reveals the influence of host phylogeny and environment. FEMS Microbiol Ecol; 93 (1): fiw204.

13. Caporaso JG, Kuczynski J, Stombaug J, Bittinger K, Bushman FD (2010) QIIME allows analysis of high-throughput community sequencing data. Nat Methods 7: 335-336.

14. Edgar RC (2010) Search and clustering orders of magnitude faster than BLAST. Bioinformatics 26: 2460-2461.

15. Edgar RC, Haas BJ, Clemente JC, Quince C, Knight R (2011) UCHIME improves sensitivity and speed of chimera detection. Bioinformatics 27: 194-200.

16. McDonald D, Price MN, Goodrich J, Nawrocki EP, DeSantis TZ, et al. (2012). An improved Green genes taxonomy with explicit ranks for ecological and evolutionary analyses of bacteria and archaea. ISME J 6: 610-618.

17. Hammer Ø, Harper DAT, Ryan PD (2001) Past: Paleontological Statistics Software package for education and data analysis. Palaeont Elec 4: 1-9.

18. Mesel I, Derycke S, Moens T, Van der Gucht K, Vincx M, et al. (2004) Top-down impact of bacterivorous nematodes on the bacterial community structure: a microcosm study. Environ Microbiol 6: 733-744.

19. Lundberg DS, Lebeis SL, Paredes SH, Yourstone S, Gehring J, et al. (2012) Defining the core Arabidopsis thaliana root microbiome. Nature 488: 86-90.

20. Bais HP, Park SW, Weir TL, Callaway RM, Vivanco JM (2004) How plants communicate using the underground information superhighway. Trends Plant Sci 9: 26-32.

21. Barea JM, Pozo MJ, Azoón R, Azoón-Aguilar C (2005) Microbial cooperation in the rhizosphere. J Exp Bot 56: 1761-1778.

22. Ward NL, Challacombe JF, Janssen PH, Hentrissat B, Coutinho B, et al. (2009) Three genomes from the phylum Acidobacteria provide insight into the lifestyles of these microorganisms in soils. Appl Environ Microbiol 75: 2046-2056.

23. Vezzani FM, Mielniczuk J (2009) An insight into soil quality. Rev Bras Cienc Solo 33: 743-755.

24. Myers N, Mittermeier RA, Mittermeier CG, Da Fonseca GAB, Kent J (2000) Biodiversity hotspots for conservation priorities. Nature 403: 853-858.

25. Smit E, Leeflang P, Gommans S, Broek JVD, Mil SV, et al. (2001) Diversity and seasonal fluctuations of the dominant members of the bacterial soil community in a wheat field as determined by cultivation and molecular methods. Appl Environ Microbiol 67: 2284-91.

26. Nüsslein K, Tiedje JM (1999) Soil bacterial community shift correlated with change from forest to pasture vegetation in a tropical soil. Appl Environ Microbiol 65: 3622-26.

27. Fierer N, Breitbart M, Nulton J, Salamon P, Lozupone C, et al. (2007) Metagenomic and small-subunit rRNA analyses reveal the genetic diversity of bacteria, archaea, fungi, and viruses in soil. Appl Environ Microbiol 73: 7059-7066.

28. Gava CAT, Pereira JC, Fernandes MC, Neves MCP (2002) Selection of streptomycetes isolates for control of Ralstonia solanacearum in tomato. Pesqui Agropecu Bras 37: 1373-1380.

29. Kennedy AC (1999) Bacterial diversity in agroecosystems. Agric Ecosyst Environ 74: 65-76.

30. Requena N, Perez-Solis E, Azcón-Aguilar C, Jeffries P, José-Barea M (2001) Management of indigenous plant-microbe symbioses aids restoration of desertified ecosystems. Appl Environ Microbiol 67: 495-498.

31. Lindström ES, Vrede K, Leskinen E (2004) Response of a member of the Verrucomicrobia, among the dominating bacteria in a hypolimnion, to increased phosphorus availability. J Plankton Res 26: 241-246.

12

Identification, Validation of a SSR Marker and Marker Assisted Selection for the Goat Grass Derived Seedling Resistance Gene *Lr28* in Wheat

Pallavi JK[1], Anupam Singh[1], Usha Rao I[2] and Prabhu KV[3]*

[1]National Phytotron Facility, Indian Agricultural Research Institute, New Delhi-110012, India
[2]Department of Botany, University of Delhi-110007, New Delhi, India
[3]Joint Director (Research), Directorate, Indian Agricultural Research Institute, India

Abstract

The goat grass (*Aegilops speltoides*) derived seedling leaf rust resistance gene *Lr28* is effective in providing resistance against infection to leaf rust including its most virulent strain, 77-5 (121R63-1) of the pathogen. A polymorphic SSR marker specific to *Lr28* was identified by employing bulk segregant analysis on an F_2 population derived from the cross between PBW343-*Lr28*, a leaf rust resistant near isogenic line of the most cultivated variety PBW343 and CSP44-*Lr48*, the Australian cultivar Condor derived CSP44 line carrying the APR gene *Lr48*. The marker amplified a polymorphic fragment which was particular to the presence of the seedling resistance gene and it was mapped at a distance of 2.9 cM from the *Lr28* resistance locus on chromosome 4AL. It was also validated on a set of 42 NILs which carried other potent leaf rust resistance genes of diverse origin. Such a polymorphic codominant SSR marker will be useful in wheat breeding programmes to differentiate plants homozygous at the *Lr28* locus from those that are heterozygous.

Keywords: Microsatellite markers; Seedling leaf rust resistance; Bread wheat

Introduction

Leaf rust disease caused by the fungal pathogen *Puccinia triticina* syn. *P. recondita* Rob. Ex. Desm. f.sp. *tritici* Eriks. & E. Henn is a significant threat to the yield of wheat crop in all major wheat growing parts of the world. Reports of yield loss in wheat due to damage by leaf rust range from 30-50% [1]. Plant breeders utilize the model of transferring leaf rust resistance genes (*Lr* genes) into the host in order to confer it with genetic resistance. However, the pathogen has been able to throw up physiological races to cause virulence against the deployed *Lr* genes and convert the resistant variety into a susceptible one. Since it is expected that *Lr* genes sourced from wild relatives are likely to be more durable, several have been transferred into wheat from its wild relatives and many of these have been documented as located on different chromosomes [2,3]. The gene *Lr28* is one such gene transferred from *Aegilops speltoides*, which is assigned into bread wheat through a chromosomal translocation T4AS.4AL-7S #2S located on chromosome 4AL [2]. *Lr28* is an effective gene for resistance from seedling stage through the entire lifespan of wheat crop in most parts of the world including the South Asian wheat regions [4]. There are more than 60 *Lr* genes available with varying degrees of resistance of which many are indistinguishable from each other in their phenotypic expression. Molecular markers serve the purpose by detecting only those plants that carry the distinct genes. In breeding populations, the phenotypic expression of resistance would be identical in plants which are either heterozygous or homozygous at the resistance locus but distinction between these categories is essential since the latter only are desirable to be carried forward. Dominant molecular markers such as RAPD, SCAR or AFLP markers also do not serve that purpose. The currently available *Lr28* linked markers are only dominant type markers [5]. Though reported a null allelic SSR marker; it cannot be useful for direct selection. Such a marker could only be used for confirmation or zygosity determination in those plants which are already identified as *Lr28* positive through phenotyping or marker assisted selection utilizing other dominant markers. It has been already proved by that the codominant STS marker reported by was actually not associated with *Lr2* [6-8]. Pyramiding resistance genes in combination is an effective way of thwarting the breakdown of resistance and in providing diversity that limits race evolution. The current investigation to identify a codominant SSR marker polymorphic for *Lr28* gene locus employs one F_2 breeding population targeted at combining APR gene *Lr48* with the seedling resistance gene *Lr28*. It is anticipated that combinations of effective seedling resistance genes with race non-specific APR genes may provide a longer lasting resistance [9].

The codominant SSR marker, *Xwmc497* which is being reported in this paper as linked to *Lr28* locus was used to select plants which carried homozygous *Lr28* resistance alleles. The two dominant flanking RAPD markers, $S3_{450}$ linked to the recessive resistance allele and $S336_{775}$ linked to the dominant susceptibility allele at the *Lr48* locus, which span a distance of 11.3 cM were employed to identify the plants carrying *Lr48* recessive resistant allele alone [10]. Wheat genotypes from diverse genetic backgrounds which have been testified to carry various other alien and native genes were included in the study for validating the marker for *Lr28*.

Materials and Methods

Plant material

An F_2 population developed from the cross between the most widely cultivated and successful Indian wheat cultivar PBW343 carrying the gene *Lr28* (PBW343-*Lr28*) developed at IARI, India and the Australian cultivar Condor derived CSP44 line (with WW80/2*WW1511Kalyansona parentage) carrying the gene *Lr48*

*Corresponding author: Prabhu KV, Joint Director (Research), Directorate, Indian Agricultural Research Institute, New Delhi, India
E-mail: jd_research@iari.res.in

(CSP44-*Lr48*) was used for the study. *Lr28* is a seedling resistance gene thus conferring resistance in all stages of the plant and *Lr48* is an adult plant resistance gene, effective only from the time the plant reaches booting stage. The zygosity of each of the F_2 individual plants was established both by F_3 progeny testing and co-dominant molecular marker analysis. A set of 30 plants per each F_2 family were sown to erect the F_3 population. The experiments were conducted in the controlled conditions of National Phytotron Facility, IARI and New Delhi.

Pathotype of the fungal pathogen

The inoculum of the most virulent *Puccinia recondita* pathotype, 77-5 (121R63-1) was obtained from the Directorate of Wheat Research, Regional Station, Flowerdale, Shimla. Inoculation of the spores of the pathotype was done by spraying inoculum suspended in water fortified with Tween-20˚ (0.75 µl/ml) at an average concentration of 20 urediospores/microscopic field (10x × 10x).

DNA extraction

Young leaves from parents and individuals of the segregating population were collected, lyophilized and ground in liquid nitrogen using a pestle and mortar. DNA extraction was performed by the micro-extraction method described by Prabhu et al. [11]. Final concentration of DNA samples was maintained at 10 µg/µl for PCR reactions.

Seedling test

After sampling for DNA extraction, seedlings 8-10 days old at decimal code DC 11 stage were inoculated during the evening hours [12]. Prior to inoculation, the plants were sprayed with water to provide a uniform layer of moisture on the leaf surface. After inoculation, the seedlings were incubated for 36 h in humid glass chambers at a temperature of 23 ± 2˚C and more than 85% relative humidity after which, the pots were shifted to muslin cloth chambers in the same green house. The disease reaction was recorded 12-14 days after inoculation, using the scoring method described by Stakman et al. [13].

PCR Amplification using molecular markers

Ten SSR markers specific to the 4A chromosome were selected from published data [14,15]. The SSR markers (custom synthesized at Biobasic Inc, Canada) were used to screen the parents (PBW343-*Lr28* and CSP44-*Lr48*), F_2 population (comprising homozygous resistant, homozygous susceptible and heterozygous plants) and bulks (resistant and susceptible).

PCR amplification was done following the protocol developed by Williams et al. [16]. The PCR reactions with SSR markers were performed in a 20 µl volume which consisted of 10 mM Tris HCl (pH 8.3), 50 mM KCl, 2 mM MgCl$_2$, 200 µM of each dNTP (MBI Fermentas, Germany), 40 ng of each of the forward and reverse primers, 0.75 U Taq DNA polymerase (Banglore Genei Pvt. Ltd., India) and 50 ng template DNA. PCR amplifications for RAPD markers were performed in 20 µl reaction volume containing 10 mM Tris-HCl (pH 8.3), 50 mM KCl, 2 mM MgCl$_2$, 200 µM of each dNTP (MBI Fermentas, Germany), 0.2 µM of primer, 0.75 U Taq DNA Polymerase (Bangalore Genei Pvt. Ltd., India) and 10-15 ng of genomic DNA. The amplification reactions were carried in a PTC-200 thermal cycler (MJ Research, Las Vegas, NV, USA) with the following thermal profile – initial denaturation of 94˚C for 10 min followed by 44 cycles of 94˚C for 1 min (denaturation), 61˚C and 36˚C (for SSR markers and RAPD markers respectively) for 1 min (annealing), 72˚C (extension) and a final extension step of 72˚C for 10 min. This was followed by 4˚C for 10 min.

The amplified products from SSR markers and RAPD markers were separated on a 3% Metaphor˚ agarose gel and 2% Agarose gel respectively, in 1X TAE buffer at 80 V for 3 hrs to separate the fragments. The gels were later stained with 10 mg/ml ethidium bromide and viewed in a digital gel documentation system (Alpha Innotech, San Leandro, CA, USA).

Bulked segregant analyses were done to identify the markers' linkage to the dominant resistance gene [17]. Ten randomly selected plants from the homozygous resistant and homozygous susceptible F_2 plants were used to prepare bulks. The bulks differentiated for the presence and absence of the leaf rust resistance gene *Lr28* (Figure 1).

Statistical Analysis

Segregation ratios were analyzed using a chi-square test. The individuals from the crosses that were scored as resistant and susceptible in the progeny populations were subjected to chi-square test for goodness of fit to test the deviation from the theoretically expected Mendelian segregation ratios. Mean and standard error of the grain yield of the F_2 plants was calculated on the basis of standard formulae. The linkage analysis was carried out using Mapmaker version 3.0 [18].

Results

The parent PBW343-*Lr28* showed resistance to the 77-5 (121R63-1) race of *Puccinia triticina* with a resistant infection type of 0; while the APR parent, CSP44 showed a typical seedling susceptibility with a reaction type of 33+ (Growth stage 11 of Zadoks growth scale). 61 seedlings of the F_2 population showed susceptibility to the leaf rust infection while the remaining 193 plants remained resistant by expressing the seedling resistance conferred by the dominant resistance allele of the *Lr28* locus and the population followed a monogenic segregation ratio (P = 0.6645). All the susceptible F_2 derived F_3 families remained susceptible whereas only 67 out of the 193 resistant F_2 derived F_3 families were homozygous for resistance. The remaining 126 families were heterozygous thus distributing the F_2 genotypes into 1R:2R:1S monogenic segregation ratio (P = 0.6467). The phenotypic expression of adult plant resistance could not be examined due to the interference of the dominant seedling resistance gene *Lr28* in the same genetic background.

Out of ten SSR markers specific to the 4AL chromosome, only *Xwmc497* (Forward: 5'CCCGTGGTTTTCTTTCCTTCT3', Reverse: 5'AACGACAGGGATGAAAAGCAA3') with annealing temperature of 61˚C was identified to be polymorphic between the parents. 10 randomly selected samples were taken from the resistant and susceptible plants to prepare bulks for bulk segregant analysis (Figure

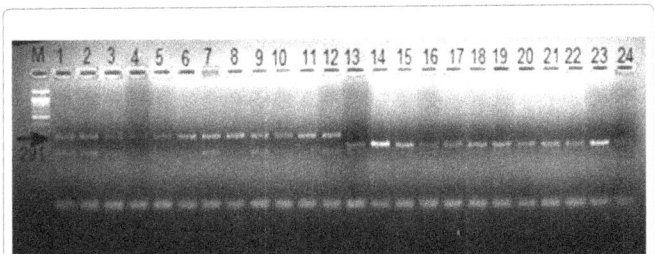

Figure 1: Screening of the SSR marker Xwmc497$_{291}$ on the bulked DNA constituent F_2 plants of the cross PBW343 X CSP44 for genetic linkage analysis. M: 100bp DNA ladder, Lanes1-10: F_2 seedling resistant individual plants, 11: Resistant Bulk, 12: Resistant parent, PBW343+ *Lr28*, 13-22: F_2 seedling susceptible individual plants, 23: Susceptible Bulk 24: Seedling susceptible parent, CSP44+ *Lr48*.

1). The marker was found putatively linked to the *Lr28* locus. This polymorphic SSR marker was analysed on the 254 F$_2$ plants for linkage analysis with the *Lr28* locus. The marker *Xwmc497* was associated with the *Lr28* locus and was located at a distance of 2.9 cM from it. The PBW343-*Lr28* resistance allele linked SSR marker allele amplified a 291 bp fragment and the CSP44 susceptibility allele linked marker allele amplified a 226 bp fragment.

The 291 bp fragment was specific to the *Lr28* resistance allele and did not amplify in other *Lr* genes carrying lines from other native and alien sources.

By employing the flanking RAPD markers $S3_{450}$ (5'CATCCCCTG3') and $S336_{775}$ (5'TCCCCATCAC3') linked respectively to the recessive resistance allele and dominant susceptible allele of the *Lr48* locus; plants which were homozygous for recessive APR gene *Lr48* were identified, as these two markers served as one co-dominant marker system capable of identifying both dominant and recessive alleles of heterozygous plants. 70 F$_2$ plants were found to possess the homozygous recessive resistance allele of *Lr48* out of the 254 plants (Table 1). Of these, only 14 plants were homozygous for the gene *Lr28* also and were identified to be carried forward as breeding lines.

The grain yield of each plant was recorded in order to advance only those which were comparable to PBW343 in mean yield/plant and displayed rust resistance imparted by both *Lr28* and *Lr48* (Table 1). PBW343 is a high yielding Indian cultivar and had a mean single plant yield of 9.50 gm while the APR parent CSP44 recorded a lower yield of 8.78 gm. The mean yield of the 14 plants homozygous for *Lr28*+ *Lr48* was 9.49 gm. These would be advanced as pyramided lines and followed for ear-to-row progeny analysis without elimination to select for high yielding recombinants through pedigree selection approach as the two genes are fixed in these progenies.

Discussion

Gene pyramiding holds its base on the concept that the probability of mutation at more than one avirulence gene locus in the pathogen is low for it to turn virulent for all the pyramided resistance genes. This enables a host variety which possesses more than one gene to remain durably resistant to the disease relatively for a long period compared to the single gene based resistance. In addition, when the added gene is from wild species the resistance is expected to last long as matching virulence is less likely to be present in the pathogen population. Further, if the resistance is race non-specific such as APR, there would be still less chance for virulence development for all the prevailing races. Thus a pyramided combination of alien seedling resistance and APR would be an ideal means to ensure durable resistance. In the past three decades, combinations of alien and APR genes such as *Lr16* and *Lr13*, *Lr13* and *Lr34*, *Lr13* and *Lr37*, *Lr34* and *Lr37* have been achieved through conventional means as there were available pathogen virulence differentials or phenotypic differences in reaction types to distinguish each gene [19,20]. However, in a case where the presence of both genes cannot be detected due to lack of such differences as in the case of *Lr28*, *Lr24*, etc, a selection process which employs molecular markers tagged to the genes is a reliable methodology as has been demonstrated by in pyramiding *Lr24* and *Lr48* in wheat by marker assisted selection utilizing dominant SCAR and RAPD markers in consecutive generations till homozygosity was achieved at both loci [10]. We were able to identify plants fixed for both genes *Lr28* and *Lr48* in F$_2$ generation itself owing to the codominant SSR marker in combination with the flanking RAPD marker set linked to both recessive resistance and dominant susceptibility alleles at the *Lr48* locus. Gene pyramiding

is well utilized in rice breeding programmes also to develop plants carrying *Xa21* and *xa13* resistant to bacterial blight which has also led to commercial release of the pyramided variety in India. Marker assisted pyramiding is also reported against fungal blast (*Pi1* and *Pi2*) and brown plant hopper (*Qbph1* and *Qbph2*) [21]. This strategy is being followed in many other breeding programmes with various crops for a range of beneficial phenotypes.

Seedling resistance genes such as *Lr28* are important to control the pathogen infection during the entire crop duration. There are previous reports of identified markers tagged to *Lr28*. The SCAR marker $SCS421_{570}$ is being successfully employed in various wheat breeding programmes in India. A recent publication by has suggested the utility of two SSR markers, *Xbarc327* and *Xbarc343* to identify the presence of *Lr28* [5,22]. However, these two markers were found to be monomorphic amplifying the critical marker fragment in both the parents. A null allelic microsatellite marker, *Xgwm160* has also been reported to be specific to the *Lr28* gene. $Xgwm160_{196}$ and $Xwmc497_{291}$ are positioned at a distance of 144.9 cM and 149.9 cM respectively, from the centromere on the long arm of the 4A chromosome [6,14].

The microsatellite marker reported in this paper will be helpful for breeding purposes since it differentiates the presence of the gene in homozygous resistant and heterozygous resistant plants (Figure 2). It has been suggested by that the markers should be within 10 cM of the gene of interest for effective marker-assisted selection breeding [23,24]. The marker *Xwmc497* mapped at a distance of 2.9 cM will therefore be especially useful for those breeding programmes in wheat where pyramiding is performed to stack more than one resistant gene into a single background. In the current study, molecular markers were effectively used to identify pyramided single plants in the F$_2$ generation itself which otherwise would have needed a laborious and time consuming selection process consisting a combination of phenotype based selection and a dominant marker based selection till the F$_5$/F$_6$ generations.

Gene(s)	Generation	Marker(s) employed	Marker alleles	No. of plants	Mean yield
Lr28	F$_2$	Xwmc497§	R	62	9.32 ± 0.1842
		Xwmc497§	H	132	9.23 ± 0.1933
		Xwmc497§	S	60	9.40 ± 0.2028
Lr48	F$_2$	S3#	+	70	9.22 ± 0.2143
		S336¶	-		
		S3#	+	117	9.32 ± 0.1352
		S336¶	+		
		S3#	-	67	9.01 ± 0.2205
		S336¶	+		
Lr28 + *Lr48*	F$_2$	Xwmc497§	R	14	9.49 ± 0.1827
		S3#	+		
		S336¶	-		
PBW343-*Lr28*	Parent	Xwmc497§	R	25	9.50 ± 0.1314
CSP44-*Lr48*	Parent	S3¶	+	25	8.78 ± 0.0980
		S336¶	-		

Table 1: Mean grain yield of the F$_2$ plants pooled with reference to the segregation of the resistant alleles of the marker loci. §Codominant microsatellite marker; R: Homozygous resistant; ¶Dominant RAPD marker; H: Heterozygous resistant; S: Homozygous susceptible; +: Presence of RAPD marker fragment; -: Absence of RAPD marker fragment.

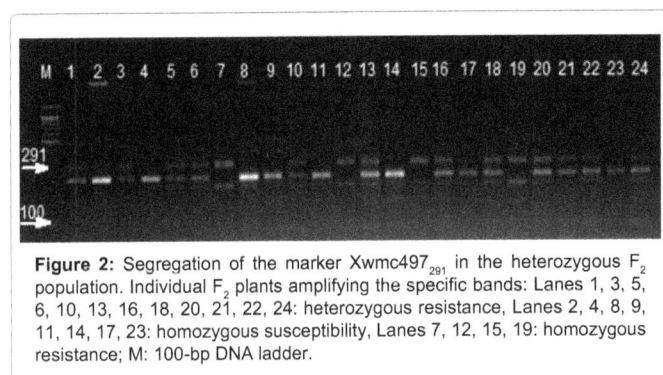

Figure 2: Segregation of the marker Xwmc497$_{291}$ in the heterozygous F$_2$ population. Individual F$_2$ plants amplifying the specific bands: Lanes 1, 3, 5, 6, 10, 13, 16, 18, 20, 21, 22, 24: heterozygous resistance, Lanes 2, 4, 8, 9, 11, 14, 17, 23: homozygous susceptibility, Lanes 7, 12, 15, 19: homozygous resistance; M: 100-bp DNA ladder.

The RAPD marker pair S3$_{450}$ and S336$_{775}$ which we used in the study had an advantage enabling us to successfully identify the plants which carried only the recessive adult plant resistance allele pair of the *Lr48* locus. From among 254 F$_2$ plants, we could select 14 plants carrying both the genes.

The grain yield of a plant follows a quantitative inheritance pattern and the expression of resistance is a qualitative character and there is no available information suggesting the influence of the leaf rust resistance loci on the grain yield of the plant. In this experiment we have also scrutinized the plants on the basis of their yield and only those plants with adequate grain number and with the presence of both the resistant genes were chosen. The 14 plants were comparable with PBW343 for mean yield/plant. The progeny of these plants will be carried forward through marker assisted pedigree breeding procedure.

Acknowledgements

The authors are grateful to the Indian Council of Agricultural Research for sponsoring the project and funding the fellowship to JK and AS under Molecular Breeding Network Project. We acknowledge Dr R. G. Saini for supplying the parental material of the *Lr* gene donors. The authors are grateful to Head, Regional Station, Indian Agricultural Research Institute, Wellington for providing pure seed of the near-isogenic lines of wheat and Directorate of Wheat Research, Flower dale, Shimla for providing pure inoculums of leaf rust pathogen.

References

1. McIntosh RA, Wellings CR, Park F (1995) In Wheat rusts: An Atlas Resistance Genes CSIRO Publishers, Australia pp. 1-20.

2. McIntosh RA, Yamazaki Y, Devos KM, Dubcovsky J, Rogers WJ, et al. (2003) Catalog of gene symbols for wheat. Proceedings of the 10th International Wheat Genetics Symposium.

3. McIntosh RA, Yamazaki Y, Dubcovsky J, Rogers J, Morris C (2008) Catalog of gene symbols for wheat. 11th International Wheat Genetics Symposium.

4. Tomar SMS, Menon MK (1998) Adult plant response of near isogenic lines and stocks of wheat carrying specific Lr genes against leaf rust. Indian Phytopathol 51: 61-67.

5. Cherukuri DP, Gupta SK, Ashwini C, Sunita K, Prabhu KV, et al. (2005) Molecular mapping of Aegilops speltoides derived leaf rust resistance gene Lr28 in wheat. Euphytica 143: 19-26.

6. Vikal Y, Chhuneja P, Singh R, Dhaliwal HS (2004) Tagging of an Aegilops speltoides derived leaf rust resistance gene Lr28 with a microsatellite marker in wheat. J Plant Biochem. Biotechnol 13: 47-49.

7. Prabhu KV, Gupta SK, Charpe A, Koul S, Cherukuri DP, et al. (2003) Molecular markers detect redundancy and miss-identity in genetic stocks with alien leaf rust resistance genes Lr32 and Lr28 in bread wheat. J Plant Biochem and Biotech 12: 123-129.

8. Naik S, Gill KS, Prakasa Rao VS, Gupta VS, Tamhankar SA, et al. (1998) Identification of a STS marker linked to the Aegilops speltoides derived leaf rust resistance gene Lr28 in wheat. Theor Appl Genet 97: 535-540.

9. Nazari K, Wellings CR (2008) Genetic analysis of seedling stripe rust resistance in the Australian wheat cultivar 'Batavia'. The 11th International Wheat Genetics Symposium proceedings. Sydney University Press.

10. Samsampour D, Maleki Zanjani B, Singh A, Pallavi JK, Prabhu KV (2009) Marker assisted selection to pyramid seedling resistance gene Lr24 and adult plant resistance gene Lr48 for leaf rust resistance in wheat. Indian journal of genetics and plant breeding 69: 1-9.

11. Prabhu KV, Somers DJ, Rakow G, Gugel RK (1998) Molecular markers linked to white rust resistance in mustard Brassica juncea. Theoretical and Applied Genetics 97: 865-870.

12. Zadoks JC, Chang TT, Konzak CF (1974) A decimal code for the growth stages of cereals. Weed Res 14: 415-421.

13. Stakman EC, Stewart DM, Loegering WQ (1962) Identification of physiological races of Puccinia graminis var. tritici, USDA-ARS-Bulletin E617

14. Torada A, Koike M, Mochida K, Ogihara Y (2006) SSR-based linkage map with new markers using an intra specific population of common wheat. Thoer Appl Genet 112: 1042-1051.

15. Roder MS, Victor K, Wendehake K, Plaschke J, Tixier MH, et al. (1998) A microsatellite map of wheat. Genetics 149: 2007-2023.

16. Williams JGK, Kubelik AR, Livak KJ, Rafalski JA,Tingey SV (1990) DNA polymorphisms amplified by arbitrary primers are useful as genetic markers. Nucleic Acids Research 18: 6531-6535.

17. Michelmore RW, Paran I, Kesseli RV (1991) Identification of markers linked to disease-resistance genes by bulked segregant analysis: A rapid method to detect markers in specific genomic regions by using segregating populations. Proc Natl Acad Sci 88: 9828-9832.

18. Lander ES, Green P, Abrahamson J, Barlow A, Daley MJ, et al. (1987) MAPMAKER: an interactive computer package for constructing primary genetic maps of experimental and natural populations. Genomics 174-181.

19. Samborski DJ, Dyck PL (1982) Enhancement of resistance to Puccinia recondite by interaction of resistance gene in wheat. Canadian Journal of Plant Pathology 4: 152-156.

20. Kloppers FJ, Pretorius ZA (1997) Effects of combinations amongst Lr13, Lr34 and Lr37 on components of resistance in wheat to leaf rust. Plant Pathology 46: 737-750.

21. He Y, Li X, Zhang J, Jiang G, Liu S, et al. (2004) Proceedings of the 4th International Crop Science Congress.

22. Cakir M, Drake Brockman F, Shankar M, Golzar H, McLean R, et al. (2008) Molecular mapping and improvement of rust resistance in the Australian wheat germplasm. 11th International Wheat Genetics Symposium.

23. Timmerman GM, Frew TJ, Weeden NF, Miller AL ,Goulden DS (1994) Linkage analysis of er-1, a recessive Pisumsativum gene for resistance to powdery mildew fungus (Erysiphe pisi D.C.). Theor Appl Genet 85: 1050-1055.

24. Cheng FS, Weeden NF, Brown SK, Aldwinckle HS, Gardiner SE, et al. (1998) Development of a DNA marker for Vm, a gene conferring resistance to apple scab. Genome 41: 208-214.

13

Expressional Regulation of the Virulence Gene eglXoA Encoding Endoglucanase, Dependent on HrpXo and Cyclic AMP Receptor-Like Protein (Clp) in *Xanthomonas oryzae* pv. *oryza*

Temuujin U[1] and Kang HW[2,3]*

[1]*Department of Biotechnology and Breeding, Mongolian University of Life science, Ulaanbaatar 17024, Mongolia*
[2]*Department of Horticulture, Hankyong National University, Ansung 17579, Korea*
[3]*Institute of Genetic Engineering, Hankyong National University, Ansung 17579, Korea*

Abstract

An *eglXoA* encoding endoglucanase that clustered with *eglXoB* and *eglXoC* in *Xanthomonas oryzae* pv. *oryza* genome (accession No. AE013598) is a pathogenicity related gene. RT-PCR showed that the *in trans eglXoA* was transcriptionally regulated by *HrpX*, a type III secretion regulator, and cyclic AMP receptor-like protein in *X. oryzae* pv. *oryzae* (*ClpXo*), which has been known as a global regulator. Western blot analysis showed that *EglXoA* is secreted *via* a type II secretion system and was detected in wild-type strain KACC10859, but not in the mutant strains *hrpX*::Tn5 and *clpXo*::Tn5. In an electrophoretic mobility shift assay, the promoter region of *eglXoA* directly bound to *ClpXo*. The two consensus *eglXoA* upstream regions were found to include putative Clp-binding sites with a perfect TCACA-N block in the left arm and a 2/5 matched block, TGT, in the right arm. *eglXoA*, which encodes endoglucanase, appears to be the first gene of *Xoo* known to be activated by *ClpXo via* direct binding to the promoter region. Molecular interaction between *HrpX* and *ClpXo* shows that *ClpXo* acts as transcription regulator of *hrpX* and binds to promoter region of *hrpX*.

Keywords: *ClpXo*; *eglXoA* gene; Expression; *HrpX*; *Xanthomonas oryzae* pv. *oryzae*

Introduction

Xanthomonas oryzae pv. *oryzae* (*Xoo*), which causes bacterial leaf blight, is the most economically important bacterial disease in rice. Bacterial blight is prevalent in many rice-growing countries [1]. During pathogenesis, plant cell walls act as the first barrier of defense against bacterial invasion and multiplication. Nevertheless, the enzymatic activities of Cell Wall-Degrading Enzymes (CWDEs) may facilitate pathogen invasion into the cells of host plants by digesting the plant cell walls [2-4]. Research regarding CWDEs in phytopathogenic bacteria has mainly focused on enzymatic activities, the identification of genes encoding them, and their roles in virulence [5,6]. The type II secretion system (T2SS) allows most gram-negative bacteria to secrete extracellular hydrolytic enzymes and toxins [7,8], many of which are responsible for pathogenesis in plants, into their surroundings and hosts. CWDEs such as cellulases, pectinases, xylanases, and proteases are secreted by plant pathogens to degrade the components of host cell walls and may play a crucial role in virulence and bacterial nutrition [9,10]. Currently, genes encoding CWDEs such as cellulase and xylanase are thought to play a role in the virulence of *Xoo* [11-14]. The T2SS-related gene cluster consists of 11 genes, *xpsEFGHIJKLMND*, in the *Xoo* genome. Mutations in the *xpsD* and *xpsF* structural genes of the *Xoo* T2SS reduce virulence and cause xylanase accumulation in the periplasmic space [5,12].

The *hrp* genes encode proteins involved in the type III secretion system (T3SS), which is involved in the secretion of effector proteins from bacteria to plants [15]. The *hrp* gene cluster in *Xoo* is composed of 27 genes, from *hrpA2* to *hrpF* [16], and the expression of these genes is regulated by two regulators, *hrpG* and *hrpX*, which are separate from the *hrp* gene cluster [17]. HrpG belongs to the OmpR family and activates the expression of *hrpX*, an AraC-like transcription activator that controls *hrp* genes along with some effector proteins [17,18]. Moreover, *HrpXo* regulates the transcriptional expression of genes associated with T2SS proteins such as cysteine proteases [19]. Recently, it was reported that polygalacturonase and extracellular proteases in *X. campestris* pv. *campestris* (*Xcc*) are regulated by *HrpX* [20,21]. These reports suggest

that *HrpX* can potentially regulate expression of other genes in addition to the *hrp* genes. It has been reported that purified cellulase and lipase induce defense responses in rice that are suppressible by *Xoo* in a T3SS-dependent manner [22]. Therefore, it is plausible that genes encoding CWDEs can participate in diverse virulence functions associated directly or indirectly with the expression of key pathogenicity-related genes, such as *hrp* genes.

Catabolite-Activator Protein (CAP), also called the cAMP receptor protein (CRP)-like protein (Clp), belongs to the CRP/FNR superfamily of transcriptional factors, which is one of the largest groups of bacterial environmental sensors [23-25]. Also, it has been known to be a global transcriptional regulator for the expression of virulence factors in *Xcc*. Clp transcriptionally activates more than 150 genes, including those that encode for extracellular enzymes and the production of exopolysaccharides (EPS), and other macromolecules such as flagellin and Hrp proteins [24]. Clp contains nucleotide- and DNA-binding domains and binds to the promoters of an endoglucanase (*engA*) from Xcc [26].

Recently, 12 genes that encode cellulases, including endoglucanases and exoglucanases, were isolated from the *Xoo* genome and mutated, and novel pathogenicity-related cellulase genes were identified and characterized [14]. Interestingly, the *eglXoABC* genes arranged as a cluster in genome of *X. oryzae* pv. *oryzae*. Of them, it was revealed that transposon insertion mutant of *eglXoA* and *eglXoA* displayed virulence-

---

***Corresponding author:** Kang HW, PhD, Professor, Department of Horticulture, Hankyong National University, Ansung 17579, Korea
E-mail: kanghw2@hknu.ac.kr*

deficient phenotype, but not in the *eglXoC*. However, little is known about expressional regulation of *eglXo* genes in the pathogenesis of *Xoo*. The goal of this study was to elucidate expressional regulation of *eglXoA*. We demonstrate that *HrpX* and *ClpXo* act as regulators of expression of *eglXoA*. Furthermore, the electrophoretic mobility shift assay (EMSA) showed that *ClpXo* binds to the promoter region of *eglXoA*.

Materials and Methods

Bacterial strains, plasmids, and culture conditions

The bacterial strains and plasmids used in this study are listed in Table 1. Wild-type strain *Xoo* KACC10859 was obtained from the Korean Agricultural Culture Collection (KACC) at the National Institute of Agricultural Biotechnology, Suwon, Korea. *Xoo* strains were cultured at 28°C on peptone sucrose agar (PSA: peptone, 10 g/L; sucrose, 10 g/l; and agar, 15 g/L) or XOM2 medium [27]. *E. coli* was grown in Luria-Bertani (LB) broth (Difco, Detroit, MI, USA) at 37°C for 18 h. Antibiotics were added at the following final concentrations for *E. coli* and *Xoo*, respectively: ampicillin, 80 µg/mL and 50 µg/mL; gentamycin, 50 µg/mL and 20 µg/mL; and kanamycin, 50 µg/mL and 20 µg/mL.

Pathogenicity assays

Inoculums (approximately 1×10^6 cells/ml) prepared from wild-type and mutant strains of *Xoo* were grown on PSA for 3 days. Pathogenicity assays were performed on 60-days-old leaves of a susceptible rice cultivar (Milyang 23) by the leaf-punching method by Temujin et al. [14]. Pathogenicity was observed at 14 days post inoculation.

Northern blot analysis

Xoo strains were first incubated in NB to OD600=1.0 and then collected by centrifugation. The pelleted cells were washed twice with XOM2 media and resuspended in XOM2 to OD_{600}=0.5 and cultured in XOM2 to OD_{600}=1.0. The total RNAs were extracted with TRIzol Reagent (Invitrogen, Carlsbad, CA, USA) and electrophoresed. Probes were labeled with the DIG Northern Starter Kit (Roche) according to the manufacturer's protocol. Hybridization signals were detected on exposure of the samples to X-ray film (Fujifilm, Tokyo, Japan).

Reverse transcription (RT)-PCR analysis

The transposon mutants and wild-type *Xoo* KACC10859 were cultured in nutrient broth (NB) until OD_{600}=0.5, pelleted by

Bacterial strains, plasmids and PCR primers	Characteristics	References
Bacterial strains		
E. coli BL21 (DE3)	*fhuA2 [lon] ompT gal (λ DE3) [dcm] ΔhsdSλ DE3=λ sBamHIo ΔEcoRI-B int::(lacI::PlacUV5::T7 gene1) i21 Δnin5*	Lab collection
Xanthomonas oryzae pv. *oryzae* strains		
KACC10859	Wild type strain, Korean race 1	KACC
eglxoA::Tn5	Transposon insertion in *eglXoA* of KACC10859, Km^r	This study
clpXo::Tn5	Transposon insertion in *clpXo* of KACC10859, Km^r	This study
hrp X::Tn5	Transposon insertion in *hrpX* of KACC10859, Km^r	This study
xps F::Tn5	Transposon insertion in *xpsF* of KACC10859, Km^r	This study
CeglXoA	*eglXoA::Tn5* harboring pMLeglXoA, Gm^r Km^r	
Plasmids		
pGEM-TEasy vector	T- cloning vector, Amp^r	Promega
pQE-80L	Overexpression vector, Ampr	Lab collection
pQE-clpXo	pQE-80L harboring clpXo, Ampr	This study
pET15b	Overexpression vector, Amp^r	Novagen
pET-eglXoA	pET15b harboring *eglxoA*, Amp^r	This study
pML122	Broad-host-range vector, Gm^r	Lab collection
pMLeglxoA	pML122 harboring *eglxoA*	This study
Primers for RT-PCR		
eglXoA	F: 5'-GCA TCC ATC GAG AGA AAC CAC-3' R: 5'-CAA TAG CGT GAA CTG CCT TC-3'	This study
hrpX	F: 5'-AGG AGC AGT TTC GCG AAC TC-3' R: 5'-TCT GCG TCC TGC TCA TCC AA-3'	This study
xpsF	F: 5'-GTT GCG CAA GAA GCC GTT CG-3' R: 5'-GTG CCACAT CCA GGC TTT CG-3'	This study
clpXo	F: 5'-GGT TGT GAC TAC GAC GGT AC-3' R: 5'-GCT TCC GGC TCT TTG GAA AG-3'	This study
16S rDNA	F: 5'-TCG TGA TCG CGACCG TAA CC-3' R: 5'-GTT GAG CTC CTC CAC CTT CT-3'	This study
Primer for probes used in EMSA		
hrpX	Probe 1: F: 5'-CTT ACA TAA CGG GCA TGT GGG-3' Probe 2: F: 5'-CTG CCG CTC ATC ATT AAG CCA-3' Probe 3: F: 5'-GAC GTG CTC GTT TGA GAA CAG-3' R: 5'- CAA CGC AGA GAT CGC TGC AAA-3'	This study
eglXoA	Probe 1: F: 5'- GTG CTC ATC TGA AAA CTC CGG -3' Probe 2: F: 5'- CGC AGA GAA AGG ATC GAT AGC -3' Probe 3: F: 5'- ACG CAG CAG CCG ATC ACC CTG -3' R: 5'- CAG GCC AGC GGT TTC CTT CTT -3'	This study

Km^r: Kanamycin Resistant; Amp^r: Ampicillin Resistant; KACC: Korea Agricultural Culture Collection

Table 1: Bacterial strains, plasmids and PCR primers used in the study.

centrifugation at 3,000 g for 10 min, and washed with distilled water. The bacterial cells were suspended in 5 mL of XOM2 medium and additionally cultured in a shaking incubator (180 rpm) at 28°C for 36 h. Total RNA was extracted by Trizol Reagent according to manufacturer's instructions and treated with RNase-free water. Then, DNase I (Promega, Madison, WI, USA) was used to remove potential traces of DNA according to the manufacturer's instructions. The cDNA synthesis and PCR were conducted using a SuperScript First-Strand RT-PCR kit (Invitrogen) with the RT-PCR primers listed in Table 1 under the following conditions: 1 cycle of 1 min at 94°C; 30 cycles of 30 sec at 94°C, 30 secs at 60°C, 1 min at 72°C; and a final extension cycle of 10 min at 72°C. PCR products were visualized in agarose gels by staining with ethidium bromide.

Overexpression and purification of *EglXoA* and *ClpXo*

eglXoA was amplified from genomic DNA of *Xoo* using a forward primer (*eglXoA*-F: 5′-CAGAATCTCATATGTCCAACCGCACCAC-3′) containing an *Nde*I restriction site (underlined) at the start codon of the ORF and a reverse primer (*eglXoA*-R: 5′-CTGCTCGAGTCAATTTTGATTCACCAAC-3′) containing an *Xho*I restriction site after the stop codon. The PCR amplicons were double digested with *Nde*I and *Xho*I, ligated into the pET-15b expression vector (Novagen) containing a 6× His tag upstream of a thrombin cleavage site and the multiple cloning site, and transformed into *E. coli* BL21 (DE3) pLysS, yielding the recombinant clone pET-eglA. To construct the *clpXo* overexpression vector, *clpXo* was amplified with the forward primer containing a *Bam*HI restriction site (underlined) at the start codon of the ORF *clpXo*-F: (5′-GGATCCATGAGCTCAGCAAAC-3′) and the reverse primer *clpXo*-R: (5′-AAGCTTTTAGCGCGTGCCGTA-3′) containing a *Hind* III restriction site spanning the stop codon. The PCR product was digested with *Bam*HI and *Hind* III, ligated in the pQE-80L vector, and then transformed into *E. coli* BL21 (DE3) pLysS, yielding recombinant clone pQE-*ClpXo*. For protein overexpression of *EglXoA* and *ClpXo* in *E. coli*, clones pET-*eglXoA* and pQE-*ClpXo* were grown in 1 L of LB liquid medium containing ampicillin at 37°C until OD$_{600}$=0.5 and overexpression was induced by adding 0.5 mM IPTG for 3 h. To purify *EglXoA*, the bacterial cells were pelleted and suspended in 200 mL buffer (20 mM Tris, 5 mM imidazole, pH 8.0), sonicated for 2 min at 20 kHz and the acoustic power ranged between 35 W and 95 W, and centrifuged at 6,500 g for 15 min. The supernatant was loaded on a column packed with nickel-nitrilotriacetic acid (Ni-NTA) equilibrated with buffer solution (20 mM Tris-HCl, 0.5 M NaCl, 5 mM imidazole, 8.0 M urea, pH 8.0). The column was first washed with the same buffer containing 50 mM imidazole and the fusion protein was then eluted using a 250-mM imidazole in the same buffer. To purify *ClpXo*, the bacterial cells were pelleted and suspended in 200 mL buffer (300 mM NaCl, 50 mM NaH$_2$PO$_4$, 10 mM imidazole, pH 8.0), sonicated 6 times, and centrifuged at 6,500 g for 15 min. The pelleted cells were washed with 0.5% Triton X-100 in 1 × PBS and solubilized by the buffer. The mixture was centrifuged at 6,500 g for 15 min, and the supernatant purified using an Ni-NTA column as for *EglXoA*.

Western blot analysis

The 30 N-terminal amino acids (NH2-LILYQKNAKAAELSKKILGLQAQDLPGNLA) corresponding to residues 98 to 127 in *EglXoA* were deduced from nucleotide sequence data and synthesized, and antiserum against the peptide was commercially produced (Peptron, Daejeon, Korea). To isolate, extraction of extracellular secreted proteins from bacterial cells, *Xoo* strains were precultured in 5 mL NB media for 3 days and pelleted and then suspended in 30 mL XOM2 liquid media and cultured additionally

to OD$_{600}$=0.8. The bacterial cells were harvested by centrifugation at 15,000 rpm for 20 min and further filtrated by membrane filter (0.45 µm pore size) to remove remnant cells. The supernatant was precipitated by 30% ammonium sulfate addition on ice for 30 min. After centrifugation at 15,000 rpm for 30 min, protein precipitates were washed with ester and suspended in 1/30 original volume of 50 mM Tris-HCl (pH 8.0) and resuspended in 2x Laemmli buffer. Protein samples were boiled for 5 min and separated by 10% of sodium dodecyl sulfate SDS-PAGE by CBB and Silver staining methods. For western blot analysis, protein samples from different *Xoo* strains were electrophoresed in 12% SDS-PAGE gels and transferred onto a polyvinylidene fluoride (PVDF) membrane (GE Healthcare, Little Chalfont, UK). The blotted membrane was incubated at room temperature for 2 h in a blocking solution and then hybridized with anti-EglXo antibody. After 3 washes with the washing solution, the PVDF membrane was incubated with anti-rabbit secondary antibody HRF (GE Healthcare) for 1 h according to the manufacturer's protocol. The signals on the membrane were detected by exposure to X-ray film (Fujifilm).

Electrophoretic mobility shift assay

The DNA probes used for the EMSA were prepared by PCR amplification of the desired regions of the *eglXoA* promoter under the following conditions: 1 cycle of 4 min at 94°C; 30 cycles of 30 secs at 94°C, 30 secs at 58°C, 1 min at 72°C; and a final extension cycle of 10 min at 72°C, using 5′-end biotin-labeled synthetic oligonucleotides as the primers in Table 1. The amplicons were purified from agarose gels and used for gel-shift experiments. The EMSA reaction mixture (10 µL) contained ca. 1.0 pmol of biotin-labeled probe and various amounts of *ClpXo* in a 5 × binding buffer containing 1 µg of nonspecific competitor DNA poly d(I-C) (Panomics, Inc). Following incubation at room temperature for 5 min, the DNA-protein complexes were resolved by electrophoresis in a 6% non-denaturing polyacrylamide gel in 0.5 × TBE buffer (40 mM Tris-Cl, pH 8.3, 45 mM boric acid, 1 mM EDTA). The gel was transferred onto a positively charged Hybond N$^+$ nylon membrane (GE Healthcare) for 30 min at 300 mA. After UV cross-linking, biotinylated probes in the membrane were detected corresponding to protocol provided by the EMSA Gel Shift Kit (Panomics, Inc).

Results

eglXoA is essential for the virulence and its expression is non-polar

The virulence of the *eglXoA* mutant strain assayed on leaves of a susceptible rice variety, Milyang 23. The degree of pathogenicity was checked in 14 days after inoculation. Figure 1, shows the disease symptom of brown stripes was observed on rice leaf inoculated with the wild-type strain. In contrast, the mutant strains *eglXoA*::*Tn5* was virulence deficient. Mutation was confirmed by complementation with the entire sole *eglXoA* gene. The complemented *X. oryzae* pv. *oryzae* strain *CeglXoA*::*Tn5* was constructed by introducing the recombinant plasmid pML*eglxoA* containing the entire wild-type *eglXoA* gene into the mutant *eglXoA*::*Tn5* (Table 1). Consequently, *CeglXoA*::*Tn5* recovered virulence and produced disease lesions similar to those of the wild-type strain KACC10859 (Figure 1), suggesting that *eglXoA* is essential for the virulence.

The *eglXoA* was located upstream *eglXoB* and it was observed long intergenic regions of 682 bp between *eglXoA* and *eglXoB* and 649 bp between *eglXoB* and *eglXoC* (Figure 1). The transcriptional linkage in the gene cluster was analyzed by RT-PCR. Two cDNA products synthesized from total RNA samples of *eglXoA*::*Tn5* and wild-type strain KACC10859 were used as template. The primer pair targeting *eglXoABC* genes amplified RT-PCR products of the predicted length in the wild-type RNA sample, while mutation in *eglXoA* did not affect transcriptional expression of *eglXoB* and *eglXoC*, as they continued to generate RT-PCR products (Figure 2). These results reasonably assumed that each *eglXoA, eglXoB,* and *eglXoC* is monocistronically transcribed without polar expressional fashion dependent on a promoter.

eglXoA expression is regulated by *HrpX* and *ClpXo*

In this experiment, the *in trans* transcriptional regulation of *eglXoA* was investigated with regard to *HrpX* and *ClpXo*. The total RNA samples were extracted from the mutants and wild-type strain KACC10859 after cultivation in *hrp*-inducing XOM2 medium. RT-PCR analysis was performed by using primer pairs targeting *eglXoA*. The primer set targeting 16S rRNA, which was used as a positive control, amplified relatively intense RT-PCR bands from the wild-type strain as well as the *hrpX::Tn5* and *clpXo::Tn5* mutants. On the other hand, the primer pair targeting *eglXoA* amplified the RT-PCR band only in the wild-type strain KACC10859 RNA sample and not in those of the *hrpX::Tn5* and *clpXo::Tn5* mutants (Figure 3A). These results indicate that transcriptional expression of *eglXoA* is regulated by *hrpX* and *clpXo*.

An N-terminal peptide consisting of 30 amino acids corresponding to residues 98 to 127 of the *EglXoA* protein (357 amino acids) was inferred from the nucleotide sequence of *eglXoA* (1,074 bp), which was artificially synthesized, and polyclonal antibodies were raised against

Figure 2: Reverse transcription (RT)-PCR analysis of *eglXo* genes of *Xanthomonas oryzae* pv. *oryzae* on transcriptome of eglXoA::*Tn5* and wild type strains.

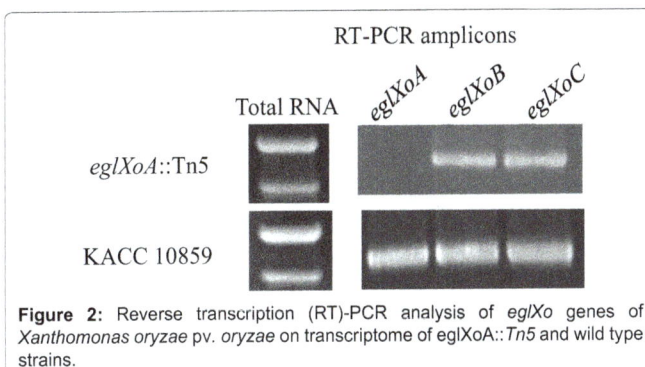

Figure 3: Expression of *eglXoA* dependent on HrpXo and ClpXo. Total RNA samples were extracted from wild-type strain KACC10859, *hrpX*::Tn5, and *clpXo*::Tn5. RT-PCR was conducted using a primer pair targeting *eglXoA*. (A) For western blot analysis, (B) the extracellular proteins were obtained from different *Xanthomonas oryzae* pv. *oryzae* strains cultured in XOM2 medium that were blotted onto PVDF membranes and then probed with an anti-EglXoA antibody.

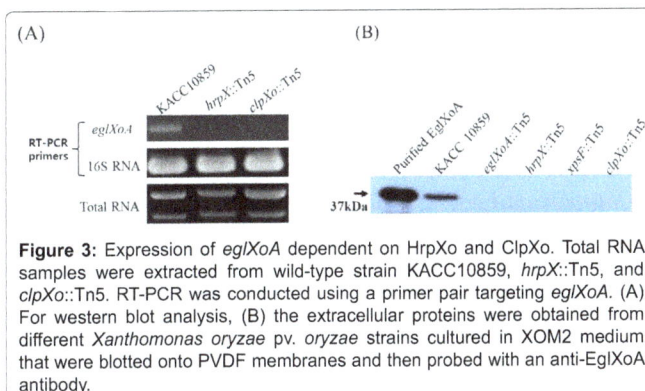

Figure 1: Pathogenicity assay of *eglXoA* in a gene cluster encoding endoglucanases of *Xanthomonas oryzae* pv. *oryzae*. The mutant strain *eglXoA::Tn5,* wild type strain KACC10859 and complement strain pMLeglA were inoculated on leaves of rice variety, Milyang 23 using leaf-punching method and pathogenicity was checked after 14 days of inoculation. Distilled water (DW) was used as negative control.

the peptide. The wild-type strain KACC10859 as well as *xpsF::Tn5, clpXo::Tn5,* and *hrpX::Tn5* mutant strains were cultured in XOM2 medium. Their extracellular proteins were isolated, resolved using SDS-PAGE, and then transferred to PVDF membranes. In western blot analysis using anti-*EglXoA* antibodies, the predicted 37-kDa band was detected in the positive controls, the extracellular protein of the wild-type strain, and the purified protein, but not in extracellular proteins of the *xpsF::Tn5, hrpX::Tn5, clpXo::Tn5,* and *eglXoA::Tn5* mutants (Figure 3B). The putative signal peptide consisting of 38 amino acids was observed on deduced sequence of *eglXoA* (Figure 3). Consequently, these findings suggest that *EglXoA* is a secreted protein dependent on T2SS and regulated by *HrpX* and *ClpXo*.

ClpXo binds to the promoter region of *eglXoA*

RT-PCR and western blot analyses showed that *ClpXo* regulates transcriptional expression of *eglXoA* and correspondingly that the gene encoding the protein was not expressed. It has been reported that the Clp protein directly binds to the promoter regions of the endoglucanase (*engA*) and polygalacturonase (*pheA*) genes of *X. campestris,* presumably enhancing transcriptional expression [26,28]. Therefore, it is reasonable to assume that *ClpXo* can also bind to the promoter region of *eglXoA*. EMSA was employed to evaluate the binding between soluble *ClpXo* protein and the promoter region of *eglXoA*. To obtain this protein, *clpXo* (693 bp) was subcloned into plasmid pQE-80L, yielding the recombinant clone pQE-Clp, which was then overexpressed in *E. coli*. The soluble *ClpXo* protein (27.7 kDa) was obtained by purification with a His-binding affinity column.

Promoter regions (probe A, −316/+127; probe B, −263/+127; and probe C, −85/+127) upstream of *eglXoA* were generated by nested PCR (Figure 4A). The EMSA was performed with 50 ng of *ClpXo* protein per reaction. The protein-DNA complex was electrophoresed in a 6% non-denaturing polyacrylamide gel, blotted onto a nylon membrane, and hybridized with biotin-labeled promoter regions. A probe B-*ClpXo* protein complex was formed, which migrated more slowly than the unbound probe B (Figure 4B). *ClpXo* do not bind to reside in the region −85/+127 and thus the region −256/−85 appears to possess the complete sequence for the binding of *ClpXo*. The *E. coli* CRP-binding site (5′-AAATGTGA-N6-TCACATTT-3′) is 22-bp long, exhibiting perfect 2-fold sequence symmetry, with the bold-faced bases representing the left and right arms for the binding of one subunit of the active CRP dimer. Two putative Clp-binding sites, including a perfect TCACA-N block in the right arm and a 2/5 matched block, TGT, in the left arm, were found from −133 to−153 sequences (Figure 4C).

ClpXo regulates transcriptional expression of *hrpX*

In our results, it was revealed genes, *hrpX* and *clpXo* are closely related to *eglXoA* expression. However, it was wonder how *hrpX* and *clpXo* is interracially associated on each gene expression. To investigate expressional control on both genes, Northern blot hybridization was done by using probes *hrpX* and *clpXo* against total RNA samples that were extracted from the *hrpX*::Tn5 and *clpXo*::Tn5 mutants and wild-type strain. Probe *clpXo* hybridized to RNA samples of wild-type and *hrpX*::Tn5 strains, but not to *clpXo*::Tn5 (Figure 5). However, probe *hrpX* detected hybridized signal on wild-type strain KACC10858, but not on the *clpXo*::Tn5 and *hrpX*::Tn5 mutant strains, suggesting *clpXo* regulates transcriptional expression of *hrpX*. Furthermore, the EMSA was conducted to determine if the promoter region of the *hrpX* gene binds to *ClpXo*, three types promoter regions (-539/+9, -429/+9 and -238/+9) generating probes A, B and C reacted with *ClpXo* protein of 50 ng. The protein-DNA complex showing retarded gel migration observed on probe A and B, but not on probe C (Figures 6A and 6B), suggesting *ClpXo* directly binds to promoter region of *hrpXo*. The putative *ClpXo* binding sequence sites with TGTCG-N-TCACA, including a perfect TCACA block in the right arm and a 4/5 matched block, TGTCG, in the left arm, were found in the −388/−409 sequences (Figure 6C).

Discussion

Novel pathogenicity-related genes including *eglXoA*, which encodes endoglucanase, have been isolated and characterized [14]. Whole genome sequence information of *Xoo* KACC10331 is available on NCBI's GenBank. *eglXoA* (1074 bp), *eglXoB* (1133 bp), and *eglXoC*

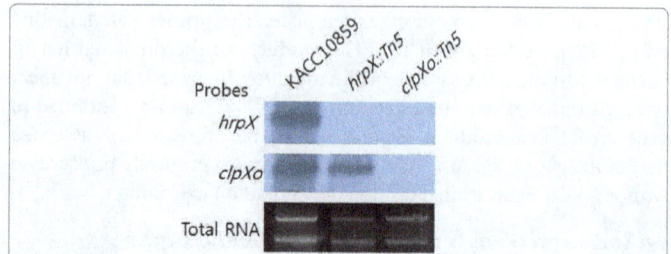

Figure 5: Northern blot profile of *X. oryzae* pv. *oryzae* strains using probes *hrpX* and *clpXo*. Total RNA was extracted from wild-type KACC10859 and mutants *hrpX*::Tn5 and *clpXo*::Tn5. For northern blot analysis, total RNA was loaded in each lane of 1.5% formaldehyde agarose gels, blotted onto nylon membranes, and hybridized with *hrpX* and *clpXo* genes.

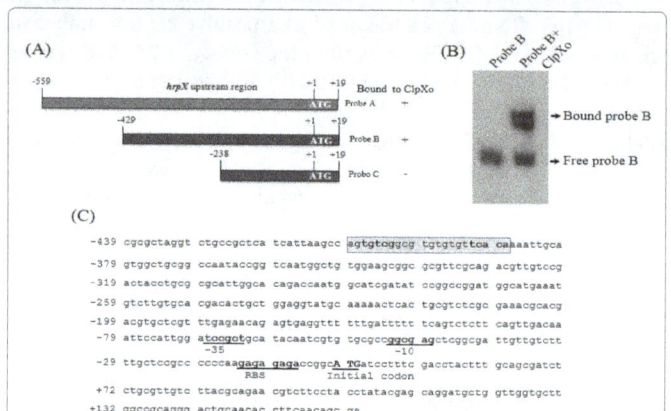

Figure 6: Electrophoretic Mobility Shift Assay (EMSA) for evaluating the binding of *hrpX* promoter region using probe ClpXo. The promoter regions (−539/+19) upstream *hrpX*, (A) was reacted with ClpXo protein (50 ng), the ClpXo protein-DNA complex was electrophoresed in a polyacrylamide gel, and the DNA-protein complex formation was observed as a retardation of migration in the gel, (B) The sequence region upstream hrpX, (C) includes ClpXo-binding sites (boxed sequence), predicted consensus sequence, −10/−35 and putative ribosome-binding site (RBS); capitalized sequence: initial codon (ATG).

(1131 bp) are organized as a cluster in the same region of the *Xoo* KACC10331 genome [16]. *EglXoA* was classified into cellulase family 5, which exhibits endo-1,4-glucanase activities, and had identity over 88.8% to endoglucanase genes, *egl1* from *X. campestris* pv. *vesicatoria* [14].

Currently, genes encoding CWDEs such as cellulase and xylanase are thought to play a role in the virulence of *Xoo* [12,14,29]. It has been reported that purified cellulase and lipase proteins induce defense responses in rice that are suppressible by *Xoo* in a T3SS-dependent manner [22]. Therefore, cellulase genes may play a role in diverse virulence traits that are directly or indirectly associated with the expression of key pathogenicity-related genes, such as *hrp* genes.

HrpX is the key regulator in the T3SS that controls the expression of *hrp* and some effector genes. Previous studies have revealed that expression of the virulence genes *pghAxc* and *pghBxc*, which encode extracellular polygalacturonase in *Xcc*, are regulated by the T3SS regulator *HrpX* [20]. It was also reported that *hrpX* negatively regulates the α-amylase isozymes in *X. axonopodis* pv. *citri* [29] and extracellular proteases in *Xcc* [21]. In *Xoo*, only a gene encoding the extracellular T2SS enzyme Cysp2, which is related to pathogenicity, has so far been demonstrated to be regulated by *HrpX* [19]. These reports support that *HrpX* is also involved in regulating the expression of some T2SS-related extracellular enzymes. Using RT-PCR, we discovered that transcriptional products of *eglXoA* from *Xoo* did not amplify in the *hrpX*

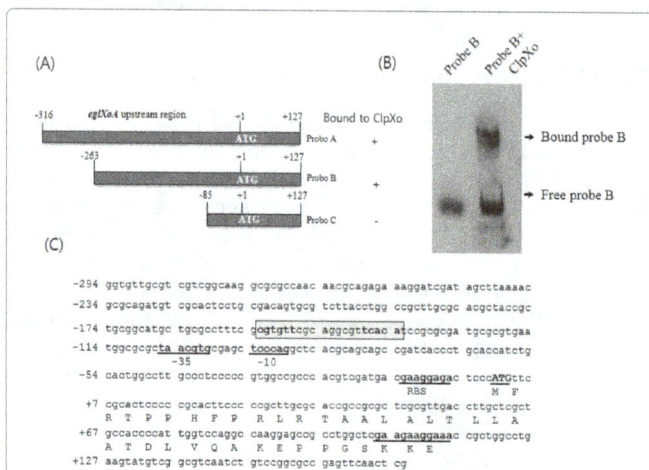

Figure 4: Electrophoretic mobility shift assay (EMSA) evaluating the binding of the *eglXoA* promoter region to the ClpXo protein. Different promoter regions upstream of *eglXoA*, probes A, B, and C, (A) bound to *ClpXo*. The probe B-*ClpXo* complex was retarded in SDS-PAGE, (B) The promoter sequence upstream of *eglXoA*, (C) includes putative ClpXo-binding sites (Boxed sequence), predicted consensus sequence, −10/−35, putative Ribosome-Binding Site (RBS). Amino acids of N-teminal; indicate putative signal peptide.

mutant, showing *HrpX*-dependent expression. This finding strongly indicates that *HrpX* is the key regulator of *eglXoA* expression. The *HrpX* regulons of *Xanthomonas* species include a consensus sequence motif called the PIP box (TTCGC-N15-TTCGC) around the promoter regions [30,31]. However, the PIP box was not observed in *eglXoA*. The PIP box can be an effective marker for screening *HrpX* regulons from the entire genomic sequence database, and several of these regulons are predicted to be involved in the pathogenicity of xanthomonads and *R. solanacearum* [31]. Twelve and 20 candidate genes for *HrpX* regulons, which did not include the genes in *hrp* clusters, were found in *Xanthomonas campestris* pv. *campestris* and *Xanthomonas axonopodis* pv. *citri* [32], respectively. However, genes with an imperfect PIP box and genes without a PIP box have been found to be expressed in an *HrpX*-dependent manner [33]. This regulation by *HrpX* indicates that the signal transduction networks of pathogens are cross-linked and that the T3SS and T2SS may cooperate *via* various regulators to promote virulence of the pathogen in the host. *Xanthomonas* protein secretion (*xps*) genes encode structural proteins that form the T2SS, which is essential for the secretion of T2SS extracellular enzymes [11,12]. Immunoblot analysis using anti-*EglXoA* antibodies in this study provided direct evidence of T2SS/Xps-dependent secretion in the culture media of wild-type *Xoo* and *xps* mutants.

The transcription factor Clp is a member of a conserved global-regulator family that regulates the expression of approximately 300 genes involved in pathogenesis of *Xanthomonas* spp. [23]. Clp is a homologue (45% amino acid sequence identity) of the model transcription factor CRP of *E. coli*. Clp in *Xcc* also influences the expression of a number of genes, especially the genes in the T2SS [23,24]. The *clp* gene in *Xoo* was isolated and characterized [34], and a mutation in *clpXo* resulted in a significant decrease in the production of cellulase, xylanase, and EPS. Moreover, a previous study demonstrated the direct binding of Clp to promoter regions [26]. The data from the present study indicate that *eglXoA* is regulated by *ClpXo*, which is consistent with previous *Xcc* studies that show that Clp is involved in the expression of extracellular enzymes of the T2SS [24,26]. Experimental evidence was provided by EMSA, wherein the *ClpXo-eglXoA* promoter region complex showed gel retardation, indicating that *ClpXo* directly binds to the *eglXoA* promoter region. In sequence analysis, a potential Clp-binding site (GTGTT-N9-TCACA) was identified in the *eglXoA* promoter region. The Clp in *X. campestris* is homologous to the CRP of *E. coli*. It was reported that Clp upregulates the transcription of endoglucanase-encoding *engA* in *X. campestris* by direct binding to the upstream region of Clp. Two consensus Clp-binding sites were determined on the *engA* promoter region by site-directed mutagenesis. In this study, two putative Clp-binding sites, including a perfect TCACA-N block in the left arm and a 2/5 matched block in the right arm, TGT, were found in sequence upstream of *eglXoA*. *ClpXo* bound to the promoter region that possesses the Clp-binding sites, whereas the promoter region that does not contain Clp-binding sites does not result in a DNA-protein complex, assuming that these sites are responsible for *ClpXo* binding to DNA. The transcriptional regulator Clp contains nucleotide- and DNA-binding domains that bind to promoters of target genes; DNA binding of Clp from *X. axonopodis* pv. *citri* is inhibited *in vitro* by cyclic di-GMP [35]. In *X. campestris* pv. *campestris*, Clp induces the expression of genes belonging to the diffusible signal factor (DSF) regulon, which encodes extracellular enzymes, components of T2SS and T3SS, and genes involved in EPS synthesis. Previous findings regarding DSF-dependent quorum sensing, including the transcriptional self-regulation of Clp, depict a detailed DSF signaling model of the regulation of bacterial virulence. Consequently, the cellular level of free Clp increases, then the regulator acts as a positive transcription factor to induce its own gene transcription and virulence gene expression.

Conclusion

In present study, we concluded that two regulatory genes, *hrpX* and *clpXo*, are interracially associated with expression of the endoglucanase gene *eglXoA* of *X. oryzae* pv. *oryzae*. Furthermore, we investigated the interplay between *hrpX* and *clpXo*. In northern blot analysis, *clpXo* mutant inhibited transcriptional expression of *hrpX*, suggesting *ClpXo* acts as transcription regulator of *hrpX* expression. In EMSA, we found that the *hrpX* promoter region directly binds to the *ClpXo* protein (Figure 6B). In conclusion, our results suggest that *ClpXo* leads to dual regulation by binding to the promoter regions of *hrpX* and *eglXoA*. In the future, it would be interesting to evaluate how the binding target sequences of *ClpXo* play a role in regulating expression of *eglXoA*, either as an activator or an enhancer.

References

1. Ezuka A, Kaku H (2000) A historical review of bacterial blight of rice. Bull Natl Inst Agrobiol Resour Japan 15:1-207.

2. Mäe A, Heikinheimo R, Palva ET (1995) Structure and regulation of the *Erwinia carotovora* subspecies carotovora SCC3193 cellulase gene *celV1* and the role of cellulase in phytopathogenicity. Mol Gen Genet 247: 17-26.

3. Saile E, McGarvey JA, Schell MA, Denny TP (1997) Role of extracellular polysaccharide and endoglucanase in root invasion and colonization of tomato plants by *Ralstonia solanacearum*. Phytopathol 87: 1264-1271.

4. Vian B, Reis D, Gea L, Grimault V (1996) The plant cell wall, first barrier or interface for microorganism: In site approaches to understanding interactions. In: Nicole M, Gianinazzi-Pearson V (eds). Histology, ultrastructure and molecular cytology of plant-microorganisms interactions. Kluwer Academic Publisher, Dordrecht, Netherlands. pp. 99-115.

5. Kang HW, Park YJ, Lee BM (2009) Functional description and identification of virulence related genes from whole genome sequence of *Xanthomonas oyzae* pv. oryzae KACC10331. Kor J Microbiol 44: 1-9.

6. Liu H, Zhang S, Schell MA, Denny TP (2005) Pyramiding unmarked deletions in *Ralstonia solanacearum* shows that secreted proteins in addition to plant cell-wall-degrading enzymes contribute to virulence. Mol Plant Microbe Interact 18: 1296-1305.

7. Pugsley AP (1993) The complete general secretory pathway in gram-negative bacteria. Microbiol Rev 57: 50-108.

8. Pugsley AP, Francetic O, Possot OM, Sauvonnet N, Hardie KR (1997) Recent progress and future directions in studies of the main terminal branch of the general secretory pathway in gram- negative bacteria- A review. Gene 192: 13-19.

9. Barras F (1994) Extracellular enzymes and pathogenesis of soft-rot *Erwinia*. Annu Rev Phytopathol 32: 201-234.

10. Jahr H, Dreier J, Meletzus D, Bahro R, Eichenlaub R (2000) The endo-beta-1,4-glucanase CelA of *Clavibacter michiganensis* sub sp. michiganensis is a pathogenicity determinant required for induction of bacterial wilt of tomato. Mol Plant Microbe Interact 13: 703-714.

11. Hu NT, Hung MN, Chiou SJ, Tang F, Chiang DC, et al. (1992) Cloning and characterization of a gene required for the secretion of extracellular enzymes across the outer membrane by *Xanthomonas campestris pv. campestris*. J Bacteriol 174: 2679-2687.

12. Rajeshwari R, Jha G, Sonti RV (2005) Role of an in planta-expressed xylanase of *Xanthomonas oryzae pv. oryzae* in promoting virulence on rice. Mol Plant Microbe Interact 18: 830-837.

13. Sun QH, Hu J, Huang GX, Ge C, Fang RX, et al. (2005) Type-II secretion pathway structural gene xpsE, xylanase- and celllulase secretion and virulence in *Xanthomonas oryzae pv. oryzae*. Plant Pathol 54: 15-21.

14. Temuujin U, Kim JW, Kim JK, Lee BM, Kang HW (2011) Identification of novel pathogenicity-related cellulase genes in *Xanthomonas oryzae pv. oryzae*. Physiol Mol Plant Pathol 76: 152-157.

15. Buttner D, Bonas U (2010) Regulation and secretion of Xanthomonas virulence factors. FEMS Microbiol Rev 34:107-133.

16. Lee BM, Park YJ, Park DS, Kang HW, Kim JG, et al. (2005) The genome sequence of *Xanthomonas oryzae* pathovar oryzae KACC10331, the bacterial blight pathogen of rice. Nuc Acids Res 33: 577-586.

17. Tsuge S, Furutani A, Kaku H (2006) Gene involved in transcriptional activation of the hrp regulatory gene hrpG in *Xanthomonas oryzae pv. oryzae*. J Bacteriol 188: 4158-4162.

18. Wengelnik K, Bonas B (1996) HrpXv, an AraC-type regulator, activates expression of five of the six loci in the hrp cluster of *Xanthomonas campestris pv. vesicatoria*. J Bacteriol 178: 3462-3469.

19. Furutani A, Tsuge S, Ohnishi K, Hikichi Y, Oku T, et al. (2004) Evidence for HrpXo-dependent expression of type II secretory proteins in *Xanthomonas oryzae pv. oryzae*. J Bacteriol 186: 1374-1380.

20. Wang L, Wei R, Chaozu H (2008) Two Xanthomonas extracellular polygalacturonases, *PghAxc* and *PghBxc*, are regulated by type III secretion regulators *HrpX* and *HrpG* and are required for virulence. Mol Plant Microbe Interact 21: 555-563.

21. Wei K, Tang JL (2007) hpaR, a putative marR family transcriptional regulator, is positively controlled by *HrpG* and *HrpX* and involved in the pathogenesis, hypersensitive response, and extracellular protease production of *Xanthomonas campestris pv. campestris*. J Bacteriol 189: 2055-2062.

22. Jha G, Rajeshwari R, Sonti RV (2007) Functional interplay between two *Xanthomonas oryzae pv. oryazae* secretion systems in modulating virulence on rice. Mol Plant Microbe Interact 20: 31-40.

23. Chin KH, Lee YC, Tu ZL, Chen CH, Tseng YH, et al. (2010) The cAMP receptor-like protein CLP is a novel c-di-GMP receptor linking cell–cell signaling to virulence gene expression in *Xanthomonas campestris*. J Mol Biol 396: 646-662.

24. He YQ, Zhang L, Jiang BL, Zhang ZC, Xu RQ, et al. (2007) Comparative and functional genomics reveals genetic diversity and determinants of host specificity among reference strains and a large collection of Chinese isolates of the phytopathogen *Xanthomonas campestris pv. campestris*. Genome Biol 8: R218.

25. Körner H, Sophia J, Zumft WG (2003) Phylogeny of the bacterial superfamily of Crp-Fnr transcription regulators: Exploiting the metabolic spectrum by controlling alternative gene programs. FEMS Microbiol Rev 27: 559-592.

26. Hsiao YM, Liao HY, Lee MC, Yang TC, Seng YH (2005) Clp upregulates transcription of engA gene encoding a virulence factor in *Xanthomonas campestris* by direct binding to the upstream tandem Clp sites. FEBS Lett 579: 3525-3533.

27. Tsuge S, Furutani A, Fukunaka R, Oku T, Tsuno K, et al. (2002) Expession of *Xanthomonas oryzae pv. oryzae* hrp genes in a novel synthetic medium, XOM2. J Gen Plant Pathol 68: 363-371.

28. Hsiao YM, Zheng MH, Hu RM, Yang TC, Tseng YH (2008) Regulation of the *pehA* gene encoding the major polygalacturonase of *Xanthomonas campestris* by *Clp* and *RpfF*. Microbiology 154: 705-713.

29. Hu J, Qian W, He C (2007) The *Xanthomonas oryzae pv. oryzae* eglXoB endoglucanase gene is required for virulence to rice. FEMS Microbiol Lett 269: 273-279.

30. Furutani A, Nakayama T, Ochiai H, Kaku H, Kubo Y, et al. (2006) Identification of novel *HrpXo* regulons preceded by two cis-acting elements, a plant-inducible promoter box and a 10 box-like sequence, from the genome database of *Xanthomonas oryzae pv. oryzae*. FEMS Microbiol Lett 259: 133-141.

31. TsugeS, Terashima S, Furutani A, Ochiai H, Oku H, et al. (2005) Effects on promoter activity of base substitutions in the cis-acting regulatory element of *HrpXo* regulons in *Xanthomonas oryzae pv. oryzae*. J Bacteriol 187: 2308-2314.

32. Da Silva ACR, Ferro JA (2002) Comparison of genomes of two Xanthomonas pathogens with differing host specificities. Nature 417: 459-463.

33. Noeïl L, Thieme F, Ninnstiel D, Bonas U (2001) cDNA-AFLP analysis unravels a genome-wide hrpG-regulon in the plant pathogen Xanthomonas. Mol Microbiol 41: 1271-1281.

34. Cho JH, Jeong KS, Han JW, Kim WJ, Cha JS (2011) Mutation in *clpxoo4158* reduces virulence and resistance to oxidative stress in *Xanthomonas oryzae pv. oryzae* KACC10859. Plant Pathol J 27: 89-92.

35. Leduc JL, Roberts GP (2009) Cyclic di-GMP allosterically inhibits the CRP-like protein (Clp) of *Xanthomonas axonopodis* pv. citri. J Bacteriol 91: 7121-7122.

14

Characterization of *Xylella fastidiosa popP* Gene Required for Pathogenicity

Xiangyang Shi[1,2]* and Hong Lin[2]*

[1]*Department of Plant Science, University of California, Davis, CA 95616, USA*

[2]*Crop Diseases, Pests & Genetics Research Unit, San Joaquin Valley Agricultural Sciences Center, USDA-ARS, Parlier, CA 93648, USA*

Abstract

Xylella fastidiosa (*Xf*) possess a two component regulatory system (TCS) PopP-PopQ which differentially regulates genes in response to environmental stimuli. To elucidate the role of *popP* in the pathogenicity of *X. fastidiosa* causing Pierce's disease (PD) of grapevine, a site-directed deletion method and chromosome-based genetic complementation strategy were employed to create *popP* deletion mutant *XfΔpopP* and its complementary strain *XfΔpopP*-C. *In vitro* studies showed that while all strains had similar growth curves, *XfΔpopP* showed significant reduction in cell-cell aggregation and cell-matrix adherence. Biofilm production of *XfΔpopP* was about 42% less than that of wild type *X. fastidiosa* and *XfΔpopP*-C. No symptoms were observed in grapevines inoculated with *XfΔpopP*, whereas grapevines inoculated with wild type *X. fastidiosa* and *XfΔpopP*-C showed typical PD symptoms. Several biofilm-related genes and genes involving protein secretary systems were down regulated in *XfΔpopP* in compared with wild type *X. fastidiosa* and *XfΔpopP*-C. These *in vitro* and *in planta* assay results provide strong evidence that the role of PopP is required for pathogenicity of *X. fastidiosa* on grapevine.

Keywords: *Xylella fastidiosa;* Pierce's disease; Pathogenicity; Two component regulatory system (TCS); *PopP*

Introduction

Xylella fastidiosa (*Xf*) is a xylem-limited gram-negative plant pathogenic bacterium, which causes diseases in many plants, including Pierce's disease (PD) in grapevine [1]. In diseased plants, *X. fastidiosa* cells are embedded in the plant vessel matrix in clumps (biofilm) and result in the blockage of the water flow within the xylem vessels. The formation of biofilms allows the pathogenic bacteria to adapt in low nutrition and sub-optimal osmolarity conditions, potentially protecting them from a hostile environment [2]. The bacteria to sense changes in the environment and to differentially regulate genes in response to environmental stimuli employ the regulatory systems. These regulatory circuits are generally involved in the two-component systems (TCS) in pathogenic bacteria [3]. One of the TCS used by bacteria to sense the environment is TCS PhoP-PhoQ, which senses specific nutrients and regulates responses in the bacteria [3].

The PhoP-PhoQ is a well-studied and highly conserved TCS responsible for the regulation of genes involved in the adaptation to new environments. PhoQ is a transmembrane histidine kinase protein with a long C-terminal tail residing in the cytoplasm, involved in sensing for extracellular signal [3]. Upon activation via environmental stimuli, PhoQ phosphorylates the corresponding response regulator PhoP, which regulates gene expression in response to environment stress [3]. In phytopathogenic *Xanthomonas oryzae pv. oryzae*, TCS PhoP-PhoQ was recently shown to be a requirement for the induction of density-dependent gene expression, including genes related with biofilm formation [4-7]. PhoP-PhoQ controls virulence functions in the plant pathogen *Erwinia carotovora* and *E. chrysanthemi* [8-11]. A cell wall degrading enzyme endopolygalacturonase (designated PehA) secreted by *Erwinia* is transcriptionally regulated by the PhoP-PhoQ homologue PehR-PehS, and the *pehR* or *pehS* deficient of *Erwinia* are attenuated for virulence in tobacco seedlings [8].

Evidently, the PhoP-PhoQ system is required for virulence in several animal pathogens. *Salmonella* species are facultative intracellular pathogens that can infect a wide variety of animals causing different diseases [12]. A *phoP* mutant of the etiologic agent of bubonic plagues *Yersinia pestis* [13,14], reduced the abilities to survival within macrophages, and increased sensitivity to low pH, oxidative stress, and high osmolarity [15]. A *phoP* deficient of *Shigella* is hypersensitive to killing by neutrophils [16]. The *phoP* or *phoQ* deletion mutant of *Salmonella* is highly attenuated for virulence [17-19]. More than forty genes regulated by phosphorylated PhoP have been shown to play important roles in the survival and the pathogenesis of *Salmonella* within macrophages [19,20]. While *Salmonella*, *Shigella*, and *Yersinia* cause different diseases in animal, inactivation of the *phoP* gene results in the defective for survival and virulence in their hosts.

The homologue of PopP-PopQ was identified in the *X. fastidiosa* genome [21]. To elucidate the role of TCS regulator *popP* in *X. fastidiosa*, here, we characterize *X. fastidiosa popP* gene by comparing the phenotypes of a deletion mutant *XfΔpopP*, complementary strain *XfΔpopP*-C and its wild-type *X. fastidiosa in vitro* and *in planta* studies. These results confirm the role of *popP* pathogenicity in *X. fastidiosa* on grapevine.

*Corresponding author: Xiangyang Shi, Department of Plant Science, University of California, Davis, CA 95616, USA, E-mail: shixy2100@yahoo.com

Hong Lin, Crop Diseases, Pests & Genetics Research Unit, San Joaquin Valley Agricultural Sciences Center, USDA-ARS, Parlier, CA 93648, USA E-mail: hong.lin@ars.usda.gov

Materials and Methods

Construction of *XfΔpopP* mutant and complementation strain *XfΔpopP-C*

A crossover PCR-based strategy for a site-directed deletion was used to construct an *XfΔpopP* mutant of *X. fastidiosa* [22]. The replacement of the *popP* ORF with a Gentamicin cassette in the genome of *XfΔpopP* was confirmed by sequencing. The complementation strain *XfΔpopP-C* was generated by the chromosome-based genetic complementation strategy [23]. The location of *popP* gene incorporating into the chromosome of *XfΔpopP-C* was confirmed by PCR. All amplified DNA fragments were confirmed by resequencing. The bacterial strains, plasmids, and primers used in this work are listed in Tables 1 and 2.

The detection of expression of *popP*

A modified hot-phenol RNA preparation procedure was used to extract the total RNA from wild type *X. fastidiosa*, *XfΔpopP*, and *XfΔpopP-C* [24]. Bacterial cultures were incubated in 50 ml of PD2 broth at 28°C for 5 days under constant agitation at 200 rpm. After the hot-phenol extraction, RNA was dissolved in RNase-free distilled H_2O and treated by Turbo DNA-free DNase (2U/μl) (Ambion, TX). The quality of isolated RNAs was determined by denaturing RNA formaldehyde gel electrophoresis [25]. The expression of *popP* was analyzed by reverse transcription polymerase chain reaction (RT-PCR) with primers popPmRNAXf-F/R (Table 2), using the AccessQuick RT-PCR System according to the manufacturer's instructions (Promega, WI). The expression of DNA polymerase III related gene *dnaQ* served as a positive control (Table 1). RT-PCR was conducted in three independent experiments.

Pathogenicity assays

Wild type *X. fastidiosa*, *XfΔpopP*, and *XfΔpopP-C* were grown on PD2 agar medium for 5 days at 28°C, suspended in sterile deionized water, and adjusted to an OD_{600} of 0.10. A 20-μl of cell suspension was used to inoculate five plants of Cabernet Sauvignon by a needle inoculation procedure described previously [26]. A water inoculation served as a negative control. The inoculated grapevines were kept on benches in a greenhouse at 24 to 32°C with 18 hr of exposure supplemented with High-Pressure Sodium lamp (20 watts per sq.

ft.) [27]. Pierce's disease symptoms were observed two months post inoculation. The symptoms were rated on a scale from 0 to 5 as described previously [28] with 0 representing healthy grapevines without any scorched leaves (water control) and 5 representing plants with severely scorching symptoms. The final disease index was an average from 5 independent replications in each *X. fastidiosa* strain. The pathogenicity assays on grapevine were conducted in three independent inoculation experiments in greenhouse from May 2012 to July 2014.

Bacterial titer assessment

To confirm and detect the bacterial population in inoculated grapevines, 10-week post inoculation petioles (2 to 3 cm) from each grapevine inoculated with wild type *X. fastidiosa*, *XfΔpopP*, and *XfΔpopP-C* were harvested at 50cm above the inoculation points. Total DNAs (plant and bacteria) were prepared from the petioles according to standard DNA extraction procedure [29]. The DNA samples were amplified by PCR using specific Rst31/33 primers [30] to confirm *X. fastidiosa* in the samples. The concentrations of DNA extracted from the leaf petiole of healthy and inoculated grapevine were quantified by PicoGreen Dye using DNA Auant Kit (Invitrogen, CA), and diluted to 5 ng/μl. The copy numbers of *X. fastidiosa* genomic DNA in the samples were estimated using quantitative PCR (qPCR) according to the method described earlier [29]. The assessment of bacterial titer was evaluated from three independent experiments.

Phenotypic analyses

In vitro growth curves of wild type *X. fastidiosa*, *XfΔpopP*, and *XfΔpopP-C* were determined after 3 to 21 days of growth at 28°C as previously described [22]. Cell concentration was determined by measuring turbidity at OD_{600}. Cell aggregations were analyzed as described previously [31,32]. Due to the aggregation of the cells in broth, cells of all tested strains cultured in PD2 broth were dispersed by repeated pipetting or vortexing, and processed to measure the turbidity of bacterial cells at the OD_{540nm} according to the previously described methods [31,32]. Biofilm formation analyses were done by culturing all tested stains in 96-well plates as described previously [32]. The data were averaged from three replications.

Strain or plasmid	Characteristic(s)	Source or reference
Strains		
Escherichia coli DH5	DH1 F⁻ Φ80ΔlacZΔM15Δ(lacZYA-argF)U169	Promega
X. fastidiosa (Xf)		
Temecula1	*X. fastidiosa* wild type	
XfΔpopP	Gentamicin (Gm) cassette replacing entire *popP* ORF (ΔpopP::Gm) of *X. fastidiosa* wild type	This work
XfΔpopP-C	GmʳCmʳ; a fragment including chloramphenicol (Cm) cassette and the *popP* promoter and ORF of *X. fastidiosa* insert the chromosome of *XfΔpopP*	This work
Plasmids		
pGEM-T Easy	Apʳ; cloning vector	Promega
pBBR1MCS-5	Gmʳ; broad-range plasmid	Kovach et al., [62]
pGEM-T-GM	Apʳ Gmʳ; Gm cassette from pBBR1MCS-5 cloned into pGEM-T	This work
pUC129	Apʳ; cloning vector	New England Biolabs
pUC1679	Apʳ; mutagenized PCR fragment of the flanking regions of *popP* ORF of *X. fastidiosa* cloned into pUC129	This work
pUC16791	Apʳ Gmʳ; Gm cassette from pGEM-T-GM cloned into the *AscI* site of pUC1679	This work
pBBR1MCS	Cmʳ; broad-range plasmid	Kovach et al., [62]
pUC129Cm	Cmʳ; Cm cassette from pBBR1MCS-3 cloned into pUC129	This work
pUC*popP*Xf-Exp	Apʳ Cmʳ; a fragment including the *popP* promoter and ORF of *X. fastidiosa* cloned into pUC129Cm	This work

Table 1: Bacterial strains and plasmids used in this study.

popPA	5'- AGTAATAGTACGATGCCAGCA-3'	This work
popPB	5'- *CGGCGCG*CCGGATCTGCTGTGCACCATGTT-3'	This work
popPC	5'- *CGGCGCG*CCGTATCTAAAGGTTATCGGCAC-3'	This work
popPD	5'- ATTAGAGCTTCTCCTCCAAT-3'	This work
popPORF For	5'- GCGTCAAGCGCATAACCAGC-3'	This work
popPORF Rev	5'- CTCTTCACGCATGGACGTTG-3'	This work
Gm-F	5'-GAATTGACATAAGCCTGTTC-3'	This work
Gm-R	5'-CGTTGTGACAATTTACCGAA-3'	This work
popPXFExpFor	5'- ACATGGTGCACAGCAGATCT-3'	This work
popPXFExpRev	5'- ATACGATAGATTTTGTGGCT-3'	This work
popPmRNAXf-F	5'-TTAATTTTCGTTGCGGGCAA-3'	This work
popPmRNAXf-R	5'-TGCCGTTCGATGTTGGTATT-3'	This work
CmF	5'-GGATGCATATGATCAGATCTT-3'	This work
CmR	5'-TCACTTATTCAGGCGTAGCAC-3'	This work
PD0702For	5'-CACGCCCGTTATTAATCGAA-3'	This work
PD0703Rev	5'-TAACCTTGTCAGCGTAGATG-3'	This work
Rst31	5'-GCGTTAATTTTCGAAGTGATTCGATTGC-3'	Minsavage et al., [30]
Rst33	5'-CACCATTCGTATCCCGGTG-3'	Minsavage et al., [30]
pUCFor	5'-GTTTTCCCAGTCACGAC-3'	Promega
pUCRev	5'-CAGGAAACAGCTATGAC-3'	Promega
M13For	5'-CGCCAGGGTTTTCCCAGTCACGAC-3'	Promega
M13Rev	5'-TCACACAGGAAACAGCTATGAC-3'	Promega
GacA-F	5'-TGAGTGCCTTCTAAGTACCT-3'	This work
GacA-R:	5'-TGCGTAGCGCAGTATCTACT-3'	This work
dnaQFor	5'-TTACGCAACTTGGCCAAACG-3'	This work
dnaQRev	5'-TGGAATGGAGCAAGGGGAAC-3'	This work

Table 2: Primers used in this study.

The differentially expressed genes between wild type, *XfΔpopP*, and *XfΔpopP*-C *in vitro*

Previously, studies showed that PhoP differentially regulated genes involving pathogenesis on the host [20]. Based on the genome sequences of *X. fastidiosa* 9a5c (a CVC strain) [21] and *X. fastidiosa* Temecula1 (a PD strain) [33], genes associated with putative roles in *X. fastidiosa* virulence, the metabolism of nucleic acids and proteins, and cellular transport and stress tolerance were selected for differential expression analysis (Table S1). Total RNA was extracted from wild type *X. fastidiosa*, *XfΔpopP*, and *XfΔpopP*-C strains grown in PD2 at an initial OD_{600nm} of 0.05 of in final volumes of 2 ml in glass tubes, and were agitated at 200 rpm at 28°C for 7 days [34,35]. Total RNA was extracted, DNase-treated, and purified to assure no contamination of DNA as described previously [26,34,35]. RT-PCR was conducted with primers specifically designed to amplify internal regions of the ORFs of the selected genes (S1). Ten microliters of amplified product was run on agarose gels and visualized under UV light. RT-PCR was conducted for three biological replicates from three independent experiments.

Results

Physiological properties of XfΔpopP in vitro

Sequence analysis confirmed that gentamicin cassette physically replaced the entire *popP* ORF from start codon ATG to terminal codon TAA in the wild type *X. fastidiosa* genome. The expression of *popP* in *XfΔpopP* was not detectable with RT-PCR. Successfully complemented cells from chloramphenicol-resistance clones were confirmed by PCR. Sequence analysis further confirmed that *popP* gene (promoter and ORF) was incorporated into the site between PD0702 and PD0703 of chromosome of *XfΔpopP* as expected. Stable *XfΔpopP* and *XfΔpopP*-C colonies were obtained after five to eight streaks on PD2 agar medium supplemented with 10μg/ml Gm or 10μg/ml Cm, respectively.

In vitro growth curves showed that *XfΔpopP* reached the exponential and stationary phase in a manner similar to wild type *X. fastidiosa* after twelve days in culture (data not shown). Whereas there was no obvious difference in colony morphology between wild type *X. fastidiosa* and *XfΔpopP*-C, *XfΔpopP* had less sticky colonies when touched with a bacteriological loop (data not shown). The quantitative assessment of cell aggregation showed that *XfΔpopP* had about a 36% reduction in cell-to-cell aggregation comparing to that of wild type *X. fastidiosa* (Figure 1A). However, *XfΔpopP*-C had about a 19% reduction in cell-to-cell aggregation in compared with the wild type *X. fastidiosa*. Biofilm formation by *XfΔpopP* was found to have about 42% less than that of wild type *X. fastidiosa* after ten days of static incubation determined by the crystal violet assay while complement strain *XfΔpopP*-C had about only 16.6% of reduction in biofilm formation in comparison with that of wild type *X. fastidiosa* (Figure 1B).

Pathogenicity of XfΔpopP and XfΔpopP-C on grapevine

In contrast to the grapevines inoculated with wild type *X. fastidiosa* and *XfΔpopP*-C, grapevines infected by *XfΔpopP* showed no symptom 12 weeks post inoculation (Figures 2A and 2B). Mock-inoculated control grapevines did not show any PD symptoms (Figures 2A and 2B). All diseased grapevines were detected positive for *X. fastidiosa* by PCR with *X. fastidiosa* specific primers Rst31/33 (data not shown), and grapevines in mock-inoculated group were detected negative. To further evaluate of the reduced virulence of *XfΔpopP*, the bacterial titers in infected grapevines with wild type *X fastidiosa*, *XfΔpopP*, and *XfΔpopP*-C were estimated by qPCR. The bacterial populations in grapevines infected with *XfΔpopP* showed about 66% less than those of grapevines inoculated with wild type *X. fastidiosa* (Figure 2C).

popP regulated gene expression

Most of selected genes were not differentially expressed among tested strains. However, TCS regulator *popP*, surface-exposed outer membrane gene *uspA1*, gum synthesized gene *gumB*, two-component system regulatory gene *algR*, type II secretory pathway protein-export membrane gene *secG*, and type V secretory pathway gene *mttC* were down regulated in *XfΔpopP* in comparison to wild type *X. fastidiosa* and *XfΔpopP*-C (Figure 3). Both housekeeping genes, *tapB* (temperature acclimation) and *dnaQ* (DNA polymerase III) were expressed almost equally in wild type *X. fastidiosa*, *XfΔpopP*, and *XfΔpopP*-C.

Discussion

TCS PhoP-PhoQ system of *Salmonella* and *Yersinia pestis* is responsible for the virulence within animal host cells [13-15]. PhoP-PhoQ of *X. oryzae* pv. *oryzae* (*Xoo*), *E. carotovora*, and *E. chrysanthemi* appears to be cell density-dependent in plant host [4-10]. In this study, while both wild type and *popP* mutant *X. fastidiosa* strains showed similar growth curves *in vitro*, grapevines infected with *popP* deletion strain have significantly decreased bacterial populations in comparison to the bacterial populations of grapevines inoculated with wild type. The severity of the disease has been known to be positively correlated with the titer of *X. fastidiosa*. Fritschi, et al. [36] investigated 18 grape genotypes in response to *X. fastidiosa* infection under greenhouse conditions and concluded that an inverse relationship was found between the level of PD resistance and bacterial populations in grapevines. No PD symptom was observed in the grapevines inoculated with *popP* deletion strain in compared with severe PD symptom developed by grapevines infected with wild type *X. fastidiosa* and complementary strain *XfΔpopP*-C consistently observed in repeated greenhouse inoculation tests. Similarity results were recently reported on the pathogenicity of *Xf popP* [37]. Our results demonstrate that *popP* is an important virulence factors and plays critical role for the survivals and the pathogenicity of *X. fastidiosa*.

Colonization, proliferation and biofilm formation of *X. fastidiosa* in water-conducting xylem vessels of host plants result in the blockage of water and nutrient in diseased plants [38,39]. In this study, the abilities of cell aggregation and biofilm formation of *popP* deficient strain were decreased while complementary strain *XfΔpopP*-C significantly

restored biofilm formation. *popP* is required for the regulation of density-dependent gene expression in *X. fastidiosa*, including genes responsible for biofilm formation. Biofilms are complex structures involving *X. fastidiosa* cells and an extracellular matrix. *X. fastidiosa* produces abundant multiple putative afimbrial adhesins, such as exopolysaccharide (EPS), lipopolysaccharide (LPS), and surface proteins, which contribute to the virulence of *X. fastidiosa* by attaching to the xylem wall in the plant and enhancing biofilm formation [33,40-43]. Previously study showed that *X. fastidiosa* EPS was coded by a cluster of nine *gum* genes closely related to the xanthan gum operon of *Xanthomonas campestris* pv. *campestris* [44]. The disruption of *gumB* in *X. fastidiosa* did not affect exopolysaccharide production, the *gumB* mutant, however, does show a reduced capacity to form biofilm [44]. The two-component system regulatory *algR* in *X. fastidiosa* was previously reported to regulate the synthesis of the LPS, which could play a role in biofilm formation and cell attachment as well [2,45-47]. Surface-exposed outer membrane gene *uspA1* was shown to be related to the virulence of *X. fastidiosa* [48]. Interestingly, in this study the expressions of *gumB*, *uspA1*, and *algR* were repressed in *popP* mutant strain in contrast to the wild type *X. fastidiosa*, the mutation of *popP* might reduce the production of LPS, resulting in the reduced capacity to form biofilm and cell-cell aggregation, both are important events of *X. fastidiosa* in successful colonization of the grapevine xylem. This might account for the significantly reduced disease severity of grapevines inoculated with *XfΔpopP*.

The ability of *X. fastidiosa* to colonize grapevines and to incite disease development is also dependent upon the capacity of this bacterium to produce a diverse set of virulence factors [33,49], which must be secreted to the bacterial cell surface or released them into the external environment. In *X. fastidiosa*, the important virulence determinants are delivered to the bacterial cell surface through type I, type II, and type V secretion systems [21,33,50,51]. It has been demonstrated that these actively secreted proteins are associated with bacterial pathogenicity or suppressing host defenses [52-54]. Matsumoto et al. [55] showed that XatA was one of six members of the AT-1 autotransporter family

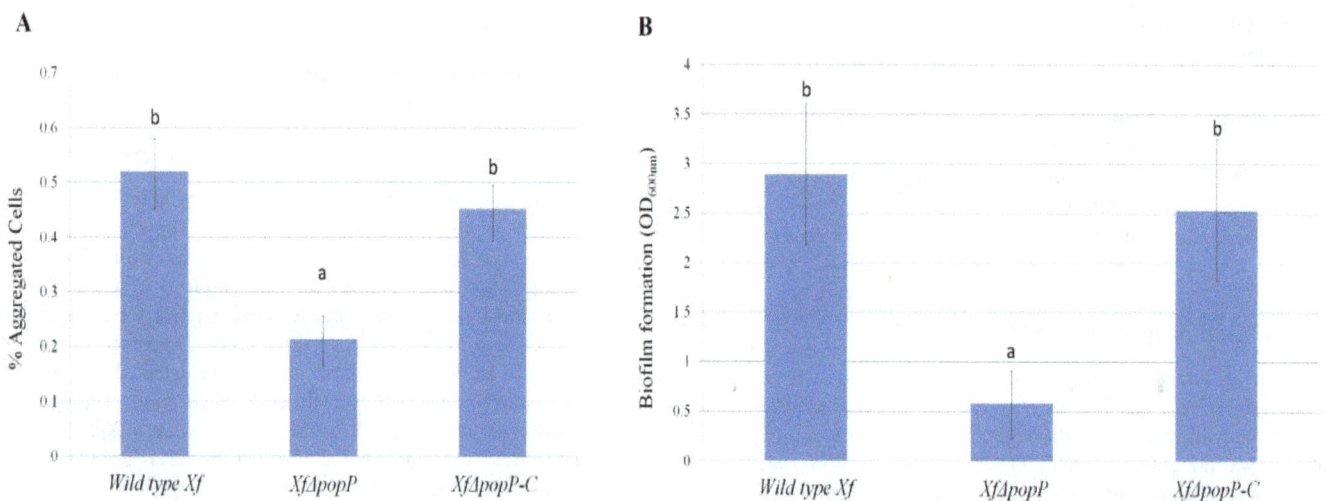

Figure 1: Quantitative assessment of cell-to-cell aggregation and biofilm formation of wild type *X. fastidiosa*, *XfΔpopP* and complement *XfΔpopP*-C. A) The percentages of aggregated cells of *X. fastidiosa* strains in PD2 broth [63]. B) Quantitative assessment of biofilm formation of all tested *X. fastidiosa* strains. Averages and standard deviations of three independent experiments are shown for all assessment experiments. Error bars indicate standard deviation. Letters indicate groups assigned by significant difference test (*t* test, *P*<0.05).

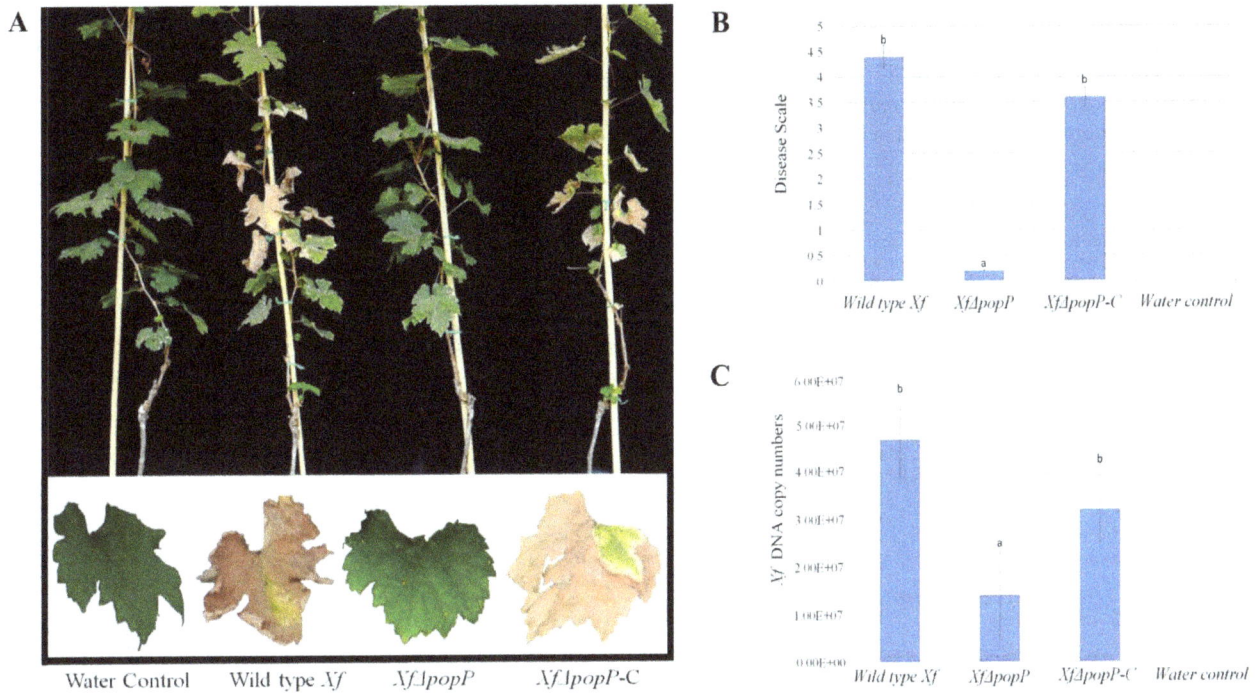

Figure 2: PD progression of Cabernet sauvignon grapevines inoculated with wild type *X. fastidiosa*, *XfΔpopP*, and *XfΔpopP*-C. A) Cabernet sauvignon grapevines infected with wild type *X. fastidiosa* and *XfΔpopP*-C, showing typical PD symptom of the scorched leaves, whereas no PD symptom was shown on grapevines infected with *XfΔpopP*. Controls are grapevines inoculated with water, which all shown no PD symptom. B) Disease severity of grapevine inoculated by all tested *X. fastidiosa* strains. Disease severity was based on a visual disease scale of 0 to 5 and was assessed 20 weeks after inoculation [28]. C) The genomic DNA copy number of wild type *X. fastidiosa*, *XfΔpopP*, and *XfΔpopP*-C in infected grapevines was estimated by qPCR. Averages and standard deviations of 5 independent replicates. Error bars indicate standard deviation. Letters indicate groups assigned by significant difference test (*t* test, *P*<0.05).

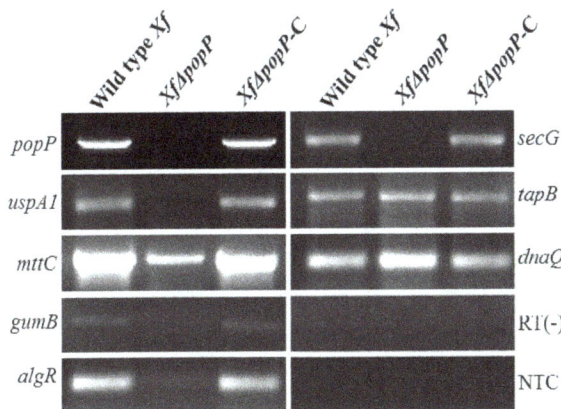

Figure 3: The detection of differential expression of virulence-related genes in wild type *X. fastidiosa*, *XfΔpopP*, and *XfΔpopP*-C by RT-PCR. Controls included positive control (the detectable expression of *dnaQ*) and negative controls [RT(-) and NTC], where RT(-) was a reaction with all components except RT reverse enzyme and NTC was a reaction with all components except RNA template. RT-PCR was conducted in three independent experiments.

secreted by type V secretion system in *X. fastidiosa* [55]. *xatA* mutant of *X. fastidiosa* shows a significant decrease in cells autoaggregation, biofilm formation, and disease symptoms on grapevines [55]. In this study type II sec-dependent secretion system *secG* and type V secretory pathway *mttC* were positively regulated by *popP*. The loss of *popP* in

X. fastidiosa might affect the synthesis of type II and type V secretory pathway, resulting in the blocking of the secretions of virulence factors, and the lost the pathogenicity on grapevine.

Although TCS PhoPQ is believed to be primarily involved in phosphate metabolism as it regulates the expression of a nonspecific acid phosphatase [56], it has been experimentally verified that the PhoPQ regulates a range of genes responsible for virulence via the sensing of extracellular signal, such as nutrient and antimicrobial peptides [3,57]. Previous studies indicated the PhoPQ was necessary for *X. oryzae* pv. *oryzae* to be able to tolerate a change in acidic conditions (pH 4.5 to 6.5) after invasion of its host rice [58]. Those results imply that PhoPQ signal transduction in phytopathogenic bacteria might be induced by fluctuations pH in host and the presence of extracellular molecule(s). Since grape xylem is a nutritionally limited environment, subject to a fluctuating pH and differentially available minerals [34,59-61], *X. fastidiosa* deficient in *popP* may be incapable of sensing the extracellular molecule(s) or fluctuations of pH in host or producing the biofilm formation-related factors, secrete virulence factors, resulting in complete loss of pathogenicity on grapevine. The preliminary studies identified several heat shock proteins, translation factors, and virulence regulator likely involving interactions with PopP, implicating that PopP in *X. fastidiosa* might regulate a wide variety of biological processes in sensing host xylem fluctuation environmental stimuli to regulate the virulence factor of *X. fastidiosa* in hosts [64,65].

While the exact molecular mechanisms by which PopP regulates in *X. fastidiosa* require further investigation, results from this study demonstrate that PopP plays a key role in controlling a variety of genes affecting biofilm formation and secreted virulence factors. These findings suggest that PopP is a critical component of the regulatory hierarchy governing the pathogenicity of *X. fastidiosa* in response to environmental fluctuations. These results constitute the first attempt in characterizing the pathogenicity role of *popP* to PD of grapevines, which provides new information for understanding the pathogenicity in *X. fastidiosa*.

Acknowledgments

Thanks to Guimei Qi for technical support. Our appreciation is also extended to Dr. Joseph Morse for comments on the manuscript. Funding for the project was supported by the United States Department of Agriculture, Agricultural Research Service. Trade names or commercial products in this publication are mentioned solely for the purpose of providing specific information and does not imply recommendation or endorsement by the United States Department of Agriculture. USDA is equal opportunity provider.

References

1. Wells JM, Raju BC, Hung HY, Weisburg WG, Mandelcopaul L, et al. (1987) Xylella fastidiosa gen. nov., sp. nov.: Gram negative, xylem limited, fastidious plant bacteria related to Xanthomonas spp. Int J Syst Bacteriol 37: 136-143.

2. Morris CE, Monier JM (2003) The ecological significance of biofilm formation by plant-associated bacteria. Annu Rev Phytopathol 41: 429-453.

3. Groisman EA (2001) The pleiotropic two-component regulatory system PhoP-PhoQ. J Bacteriol 183: 1835-1842.

4. Burdman S, Shen Y, Lee SW, Xue Q, Ronald P (2004) RaxH/RaxR: a two-component regulatory system in Xanthomonas oryzae pv. oryzae required for AvrXa21 activity. Mol Plant Microbe Interact 17: 602-612.

5. Lee SW, Jeong KS, Han SW, Lee SE, Phee BK, et al. (2008) The Xanthomonas oryzae pv. oryzae PhoPQ two-component system is required for AvrXA21 activity, hrpG expression, and virulence. J Bacteriol 190: 2183-2197.

6. Lee SW, Han SW, Bartley LE, Ronald PC (2006) From the Academy: Colloquium review. Unique characteristics of Xanthomonas oryzae pv. oryzae AvrXa21 and implications for plant innate immunity. Proc Natl Acad Sci USA 103: 18395-18400.

7. Lee SW, Han SW, Sririyanum M, Park CJ, Seo TS, et al. (2009) A Type 1-Secreted, Sulfated Peptide triggers XA21-Mediated Innate Immunity. Science 326: 850-853.

8. Flego D, Marits R, Eriksson AR, Koiv V, Karlsson MB, et al. (2000) A two-component regulatory system, pehRpehS, controls endopolygalacturonase production and virulence in the plant pathogen Erwinia carotovora subsp. carotovora. Mol Plant Microbe Interact 13: 430-438.

9. Llama-Palacios A, Lopez-Solanilla E, Poza-Carrion C, Garcia-Olmedo F, Rodriguez-Palenzuela P (2003) The Erwinia chrysanthemi phoP-phoQ operon plays an important role in growth at low pH, virulence and bacterial survival in plant tissue. Mol Microbiol 49: 347-357.

10. Llama-Palacios A, López-Solanilla E, Rodríguez-Palenzuela P (2005) Role of the PhoP-PhoQ system in the virulence of Erwinia chrysanthemi strain 3937: involvement in sensitivity to plant antimicrobial peptides, survival at acid Hh, and regulation of pectolytic enzymes. J Bacteriol 187: 2157-2162.

11. Alfano JR, Collmer A (2001) Mechanisms of bacterial pathogenesis in plants: familiar foes in a foreign kingdom. In: Groisman EA, editor. Principles of bacterial pathogensis. New York: Academic Press, pp. 180-211.

12. Scherer CA, Miller SI (2001) Molecular pathogenesis of salmonellae. In Groisman EA, editor. Principles of bacterial pathogenesis. New York: Academic Press, pp. 266-316.

13. Boyd AP, Cornelis GR (2001) Yersinia. In: Principles of bacterial pathogenesis. New York: Academic Press, pp. 228-253.

14. Marceau M, Sebbane F, Ewann F, Collyn F, Lindner B, et al. (2004) The pmrF polymyxin-resistance operon of Yersinia pseudotuberculosis is upregulated by the PhoP-PhoQ two-component system but not by PmrA-PmrB, and is not required for virulence. Microbiology 150: 3947-3957.

15. Oyston PC, Dorrell N, Williams K, Li SR, Green M, et al. (2000) The response regulator PhoP is important for survival under conditions of macrophage-induced stress and virulence in Yersinia pestis. Infect Immun 68: 3419-3425.

16. Moss JE, Fisher PE, Vick B, Groisman EA, Zychlinsky A (2000) The regulatory protein PhoP controls susceptibility to the host inflammatory response in Shigella flexneri. Cell Microbiol 2: 443-452.

17. Fields PI, Groisman EA, Heffron F (1989) A Salmonella locus that controls resistance to microbicidal proteins from phagocytic cells. Science 243: 1059-1062.

18. Galán JE, Curtiss R (1989) Virulence and vaccine potential of phoP mutants of Salmonella typhimurium. Microb Pathog 6: 433-443.

19. Miller SI, Kukral AM, Mekalanos JJ (1989) A two-component regulatory system (phoP phoQ) controls Salmonella typhimurium virulence. Proc Natl Acad Sci USA 86: 5054-5058.

20. Gunn JS, Hohmann EL, Miller SI (1996) Transcriptional regulation of Salmonella virulence: a PhoQ periplasmic domain mutation results in increased net phosphotransfer to PhoP. J Bacteriol 178: 6369-6373.

21. Simpson AJ, Reinach FC, Arruda P, Abreu FA, Acencio M, et al. (2000) The genome sequence of the plant pathogen Xylella fastidiosa. The Xylella fastidiosa Consortium of the Organization for Nucleotide Sequencing and Analysis. Nature 406: 151-159.

22. Shi XY, Dumenyo CK, Hernandez-Martinez R, Azad H, Cooksey DA (2007) Characterization of regulatory pathways in Xylella fastidiosa: genes and phenotypes controlled by algU. Appl Environ Microbiol 73: 6748-6756.

23. Matsumoto A, Young GM, Igo MM (2009) Chromosome-based genetic complementation system for Xylella fastidiosa. Appl Environ Microbiol 75: 1679-1687.

24. Kustu S, Santero E, Keener J, Popham D, Weiss D (1989) Expression of sigma 54 (ntrA)-dependent genes is probably united by a common mechanism. Microbiol Rev 53: 367-376.

25. Chuang SE, Daniels DL, Blattner FR (1993) Global regulation of gene expression in Escherichia coli. J Bacteriol 175: 2026-2036.

26. Shi XY, Dumenyo CK, Hernandez-Martinez R, Azad H, Cooksey DA (2009) Characterization of regulatory pathways in Xylella fastidiosa: genes and phenotypes controlled by gacA. Appl Environ Microbiol 75: 2275-2283.

27. Lin H, Doddapaneni H, Takahashi Y, Walker MA (2007) Comparative analysis of ESTs involved in grape responses to Xylella fastidiosa infection. BMC Plant Biol 7: 8.

28. Guilhabert MR, Kirkpatrick BC (2005) Identification of Xylella fastidiosa antivirulence genes: hemagglutinin adhesins contribute a biofilm maturation to X. fastidios and colonization and attenuate virulence. Mol Plant Microbe Interact 18: 856-868.

29. Francis M, Lin H, Rosa J, Doddapaneni H, Civerolo EL (2006) Genome-based PCR primers for specific and sensitive detection and quantification of Xylella fastidiosa. Eur J Plant Pathol 115: 203-213.

30. Minsavage GV, Thompson CM, Hopkins DL, Leite RMVBC, Stall RE (1994) Development of a polymerase chain reaction protocol for detection of Xylella fastidiosa in plant tissue. Phytopathology 84: 456-461.

31. Burdman S, Jurkevitch E, Soria-Díaz ME, Serrano AM, Okon Y (2000) Extracellular polysaccharide composition of Azospirillum brasilense and its relation with cell aggregation. FEMS Microbiol Lett 189: 259-264.

32. Leite B, Andersen PC, Ishida ML (2004) Colony aggregation and biofilm formation in xylem chemistry-based media for Xylella fastidiosa. FEMS Microbiol Lett 230: 283-290.

33. Van Sluys MA, de Oliveira MC, Monteiro-Vitorello CB, Miyaki CY, Furlan LR, et al. (2003) Comparative analyses of the complete genome sequences of Pierce's disease and citrus variegated chlorosis strains of Xylella fastidiosa. J Bacteriol 185: 1018-1026.

34. Bi JL, Dumenyo CK, Hernandez-Martinez R, Cooksey DA, Toscano NC (2007) Effect of host plant Xylem fluid on growth, aggregation, and attachment of Xylella fastidiosa. J Chem Ecol 33: 493-500.

35. Shi X, Bi J, Morse JG, Toscano NC, Cooksey DA (2010) Differential expression of genes of Xylella fastidiosa in xylem fluid of citrus and grapevine. FEMS Microbiol Lett 304: 82-88.

36. Fritschi FB, Lin H., Walker MA (2007) Xylella fastidiosa Population Dynamics in Grapevine Genotypes Differing in Susceptibility to Pierce's Disease. Am J Enol Vitic 3: 326-332.

37. Pierce BK, Kirkpatrick BC (2015) The PhoP/Q two-component regulatory system is essential for Xylella fastidiosa survival in Vitis vinifera grapevines. Phys Mol Plant Pathol 89: 55-61.

38. Osiro D, Colnago LA, Otoboni AM, Lemos EG, de Souza AA, et al. (2004) A kinetic model for Xylella fastidiosa adhesion, biofilm formation, and virulence. FEMS Microbiol Lett 236: 313-318.

39. Purcell AH, Hopkins DL (1996) Fastidious xylem-limited bacterial plant pathogens. Annu Rev Phytopathol 34: 131-151.

40. Kang Y, Liu H, Genin S, Schell MA, Denny TP (2002) Ralstonia solanacearum requires type 4 pili to adhere to multiple surfaces and for natural transformation and virulence. Mol Microbiol 46: 427-437.

41. Ojanen-Reuhs T, Kalkkinen N, Westerlund-Wikström B, van Doorn J, Haahtela K, et al. (1997) Characterization of the fimA gene encoding bundle-forming fimbriae of the plant pathogen Xanthomonas campestris pv. vesicatoria. J Bacteriol 179: 1280-1290.

42. Romantschuk M, Bamford DH (1986) The causal agent of halo blight in bean, Pseudomonas syringae pv. phaseolicola, attaches to stomata via its pili. Microb Pathog 1: 139-148.

43. van Doorn J, Boonekamp PM, Oudega B (1994) Partial characterization of fimbriae of Xanthomonas campestris pv. hyacinthi. Mol Plant Microbe Interact 7: 334-344.

44. Souza LC, Wulff NA, Gaurivaud P, Mariano AG, Virgílio AC, et al. (2006) Disruption of Xylella fastidiosa CVC gumB and gumF genes affects biofilm formation without a detectable influence on exopolysaccharide production. FEMS Microbiol Lett 257: 236-242.

45. Coyne MJ Jr, Russell KS, Coyle CL, Goldberg JB (1994) The Pseudomonas aeruginosa algC gene encodes phosphoglucomutase, required for the synthesis of a complete lipopolysaccharide core. J Bacteriol 176: 3500-3507.

46. de Souza AA, Takita MA, Pereira EO, Coletta-Filho HD, Machado MA (2005) Expression of pathogenicity-related genes of Xylella fastidiosa in vitro and in planta. Curr Microbiol 50: 223-228.

47. Martin DW, Schurr MJ, Yu H, Deretic V (1994) Analysis of promoters controlled by the putative sigma factor AlgU regulating conversion to mucoidy in Pseudomonas aeruginosa: relationship to sigma E and stress response. J Bacteriol 176: 6688-6696.

48. Caserta R, Takita MA, Targon ML, Rosselli-Murai LK, de Souza AP, et al. (2010) Expression of Xylella fastidiosa fimbrial and afimbrial proteins during biofilm formation. Appl Environ Microbiol 76: 4250-4259.

49. Chatterjee S, Wistrom C, Lindow SE (2008) A cell-cell signaling sensor is required for virulence and insect transmission of Xylella fastidiosa. Proc Natl Acad Sci U S A 105: 2670-2675.

50. Dautin N, Bernstein HD (2007) Protein secretion in gram-negative bacteria via the autotransporter pathway. Annu Rev Microbiol 61: 89-112.

51. Henderson IR, Navarro-Garcia F, Desvaux M, Fernandez RC, Ala'Aldeen D (2004) Type V protein secretion pathway: the autotransporter story. Microbiol Mol Biol Rev 68: 692-744.

52. Bruening G, Civerolo EL, Lee Y, Buzayan JM, Feldstein PA, et al. (2005) A major outer membrane protein of Xylella fastidiosa induces chlorosis in Chenopodium quinoa. Phytopathology 95: S14.

53. Nandi B, Nandy RK, Mukhopadhyay S, Nair GB, Shimada T, et al. (2000) Rapid method for species-specific identification of Vibrio cholerae using primers targeted to the gene of outer membrane protein OmpW. J Clin Microbiol 38: 4145-4151.

54. Pugsley AP (1993) The complete general secretory pathway in gram-negative bacteria. Microbiol Rev 57: 50-108.

55. Matsumoto A, Huston SL, Killiny N, Igo MM (2012) XatA, an AT-1 autotransporter important for the virulence of Xylella fastidiosa Temecula1. Microbiologyopen 1: 33-45.

56. Kier LD, Weppelman RM, Ames BN (1979) Regulation of nonspecific acid phosphatase in Salmonella: phoN and phoP genes. J Bacteriol 138: 155-161.

57. García Véscovi E, Soncini FC, Groisman EA (1996) Mg2+ as an extracellular signal: environmental regulation of Salmonella virulence. Cell 84: 165-174.

58. Grignon C, Sentenac H (1991) pH and ionic conditions in the apoplast. Annu Rev Plant Phys 42: 103-128.

59. Alves G, Ameglio T, Guilliot A, Fleurat-Lessard P, Lacointe A, et al. (2004) Winter variation in xylem sap pH of walnut trees: involvement of plasma membrane H+-ATPase of vessel-associated cells. Tree Physiol 24: 99-105.

60. Andersen PC, Brodbeck BV, Oden S, Shriner A, Leite B (2007) Influence of xylem fluid chemistry on planktonic growth, biofilm formation and aggregation of Xylella fastidiosa. FEMS Microbiol Lett 274: 210-217.

61. Basha SM, Mazhar H, Vasanthaiah HK (2010) Proteomics approach to identify unique xylem sap proteins in Pierce's disease-tolerant Vitis species. Appl Biochem Biotechnol 160: 932-944.

62. Kovach ME, Phillips RW, Elzer PH, Roop RM 2nd, Peterson KM (1994) pBBR1MCS: a broad-host-range cloning vector. Biotechniques 16: 800-802.

63. Davis MJ, French WJ, Schaad NW (1981) Axenic culture of the bacteria associated with phony peach disease of peach and plum leaf scald. Curr Microbiol 6: 309-314.

64. Krivanek AF, Famula TR, Tenscher A, Walker MA (2005) Inheritance of resistance to Xylella fastidiosa within a Vitis rupestris x Vitis arizonica hybrid population. Theor Appl Genet 111: 110-119.

65. Krivanek AF, Stevenson JF, Walker MA (2005) Development and Comparison of Symptom Indices for Quantifying Grapevine Resistance to Pierce's Disease. Phytopathology 95: 36-43.

15

Genetic Variability within *Xanthomonas axonopodis pv. punicae*, Causative Agent of Oily Spot Disease of Pomegranate

Chavan NP[1], Pandey R[1], Nawani N[1], Tandon GD[1] and Khetmalas MB[2]*

[1]*Dr. D.Y. Patil Biotechnology and Bioinformatics Institute, Dr. D.Y. Patil Vidyapeeth, Pune, India*
[2]*Rajiv Gandhi Institute of Information Technology and Biotechnology, Bharati Vidyapeeth Deemed University, Pune, India*

Abstract

Bacterial blight is one of the most devastating diseases of pomegranate in India. It is known to be caused by the strains of *Xanthomonas axonopodis pv. punicae (xap)*. As the control of disease varies with the use of pesticide agents, it becomes imperative to study the genetic variation among the pathogenic strains. Thirty-six strains of *Xanthomonas axonopodis* pv. *punicae*, were isolated from the diseased fruits of 3 varieties of pomegranate originating from 3 different provinces of Maharashtra, India. All the strains characterized phenotypically and genotypically were found to have diversity. The genetic diversity among the 36 Xanthomonas isolates was assessed using RAPD based techniques. A cluster dendrogram based on the random amplified polymorphic DNA (RAPD) showed that genetic diversity existed among the isolates of Xanthomonas. The genomic variation was found to be in the range of 0.55% to 0.95% among the isolates. The cluster analysis based upon band patterns formed two major clusters with 4 sub-groups. It can be concluded that this variability was genetic in nature irrespective of their location and the variety of pomegranate.

Keywords: Bacterial blight; *Xanthomonas axonopodis* pv. *punicae*; Genetic variability; RAPD

Introduction

Pomegranate is one of the most economically important fruit crops of India. Maharashtra region is the largest producer of pomegranate in the country. Pomegranate production suffers severely by the disease called oily spot disease which is predominant in the pomegranate cultivation in India. The quality and productivity of pomegranate crop is hampered by 70% to 80% due to this disease [1,2]. The bacterial blight was first reported in India from Rajasthan in 1952 [3]. It received a minor importance earlier but appeared as a serious threat in all pomegranate growing regions of Maharashtra, Northern Karnataka and Andhra Pradesh [4]. Solapur, Sangli, Nasik, Satara, Pune, Ahmadnagar and Wardha districts of Maharashtra state are major cultivators of Pomegranates, with minor plantations in other areas [5]. Since 1998, Oily spot disease has appeared as a major production problem in important pomegranate growing states in India [6]. The oily spot disease was first observed in the Maharashtra state at Mohol village of Pandharpur area in Solapur district in 2003 [7]. The disease continued to damage the crop for subsequent years, inspite of all possible and available protection measures adopted by the farmers. The disease could not be mitigated effectively due to rapid buildup of inoculums and spread of the disease widely. Many measures like biocontrol agents, combinations of antibiotics and pesticides along with phytochemicals and cultivation practices were not sufficient to remedy the problem [8,9]. There are reports of considerable amount of strain variation in the strains of *Xanthomonas axonopodis* pv. *punicae* from different parts of Karnataka [10]. Earlier reports based on physiological and biochemical tests have been used for characterization and study of phenotypic variation among *Xanthomonas* species [11-14] which are still much needed for identification of plant pathogenic bacteria to genus and species level [15]. Modern genetic techniques allow the exact differentiation of genetic variation within a population [16]. A rapid and specific identification test would be very useful to monitor the infection of pomegranate plants in order to develop strategies to control the disease in fields. A genetic variability study is essential to enhance the understanding of its taxonomy, epidemiology, and identification of the bacterial strain [17]. The similar control measures are not found to be very effective on all the varieties and their growing locations of pomegranate due to the existence of genetic variation among the strains responsible for the disease. Considering the seriousness of the disease, the present study was carried out to understand the diversity in *Xanthomonas axonopodis* pv. *punicae* prevalent among different pomegranate varieties and their various locations of Maharashtra in India.

Materials and Methods

Collection of diseased pomegranate fruit samples and isolation of organisms

Diseased fruit samples belonging to oily spot disease affected 3 varieties of pomegranate (Bhagawa, Ganesh and Mridula) were collected separately from 3 districts viz. Sangli, Solapur and Nasik of Maharashtra state. The basis used for sampling was the above three varieties are popular in the above mentioned three districts of Maharashtra. All the samples were carried to the laboratory in sterile polythene bags and stored at 4°C.

The infected fruit samples showing typical symptoms of bacterial blight were taken up for the isolation of the causal agent. The fruit samples were washed thoroughly with tap water and allowed to dry. The samples were surface sterilized with 0.1% mercuric chloride ($HgCl_2$) solution for one minute and washed three times serially in sterile distilled water to remove the traces of mercuric chloride. The infected portion was cut into small pieces and suspended in 10 mL of sterilized distilled water and squeezed gently with sterilized scalpel. When the water became slightly turbid due to bacterial cells, the suspension was serially diluted

*****Corresponding author:** Madhukar B Khetmalas, Rajiv Gandhi Institute of Information Technology and Biotechnology, Bharati Vidyapeeth Deemed University, Pune 411046, India, E-mail: madhukar.khetmalas@bharatividyapeeth.edu

up to 10^6 dilutions in 9 mL sterile water blanks. 0.1 mL of diluted bacterial cell suspension was spread on the sterilized nutrient agar petri plates. The inoculated plates were incubated at 30°C for 72 hours. The plates were observed for the development of well separated, typical, light yellow coloured bacterial colonies resembling *Xanthomonas* sp. Four colonies were chosen from each plate and purified by four quadrant streak method. The 36 bacterial isolates obtained upon isolation from the diseased fruit samples were designated as Xa1 to Xa36 and were stored at -80°C in glycerol stocks [15].

Morphological and biochemical characterization of the isolates for identification

The colony growth and morphological characteristics of the pathogen such as cell shape, size, color, Gram reaction characters were studied [15]. The biochemical characters such as hydrolysis of starch, gelatin liquefaction, hydrogen sulphide production, catalase, oxidase and acid production from different sugars *viz.*, glucose, lactose, fructose, sucrose, mannitol, maltose and dextrose by the isolates were studied [15,18,19]. The isolates were identified by comparing the characteristics with their description reported in Bergey's Manual of Systematic Bacteriology [19,20].

Pathogenicity testing

The freshly collected healthy pomegranate samples were used for pathogenicity test. The fruits samples were washed three times with sterilized water followed by surface sterilization with 10% sodium hypochlorite wash for 1 minute. The traces of sodium hypochlorite were removed by sterile water wash. Pomegranate samples were pricked by a sterile needle and sprayed with cell suspension of *Xanthomonas axonopodis* isolates as per their same fruit variety as well as other fruit varieties for cross infectivity (2×10^8 cfu/mL). In cross infectivity studies the isolates of Bhagawa variety were used to infect fruits of Ganesh and Mridula variety collected from same and different location. Likewise, the cross-infectivity studies were carried out with isolates from Ganesh and Mridula varieties. The inoculated fruits placed in sterile beakers, packed with sterile polythene bags for avoiding cross infectivity, were incubated at room temperature at the normal light. They were observed periodically for 4 weeks for the development of disease symptoms. The pathogens were reisolated from the diseased fruits and compared with their original cultures for their verification.

Susceptibility test of the isolates against broad spectrum antibiotics

Antibiotic susceptibility test was performed by agar well diffusion assay with all the 36 pathogenic isolates with 2.5 mg/mL of streptomycin sulfate (HiMedia) and Streptocycline (a broad-spectrum systemic antibacterial antibiotic product from Hindustan Antibiotics Ltd. 90% streptomycin sulfate + 10% tetracycline hydrochloride). The zones of inhibition were recorded.

Genetic variability

Extraction of total genomic DNA-The total genomic DNA was isolated for PCR-RAPD: Pure culture of each strain was streaked on the nutrient agar plates and incubated at 30°C for 72 h. Single colony from each plate was inoculated in 10 mL nutrient broth contained in 100 mL conical flasks. The flasks were incubated at 30°C for 72 h in

shaker incubator at a speed of 120 rpm. About 1.5 mL aliquots of broth from each flask were taken in 2.5 mL Eppendorf tubes and centrifuged at 13000 rpm for 5 min. The DNA extraction was followed as per the method of Yenjerappa [21]. The supernatant was poured off, 200 μL of lysis buffer was added to the tubes containing sediment pellet and mixed well. 166 μL 5M NaCl was added and vortexed for thorough mixing, the contents were again centrifuged at 13000 rpm for 10 min. The obtained supernatant was collected in a fresh tube to which 1 μL RNase A (10 mg/mL) was added, mixed well and incubated at 37°C for 30 min. Equal volume of phenol: chloroform: isoamyl alcohol (25:24:1) was added, mixed gently by inverting the tubes and centrifuged at 13000 rpm for 6 min. The upper aqueous phase was transferred to a fresh clean tube; 1.0 mL of cold 95% ethanol was added and mixed gently. The tubes were kept in deep freezer at -20°C for 1hr and centrifuged at 13000 rpm for 6 min. Ethanol was poured off, DNA pellet was air dried using speed vacuum for 5 min. The pellet was resuspended in 50 μL of 1X TE buffer, kept in the refrigerator at 4°C for overnight and stored in deep freezer at -20°C. The DNA was quantified using a spectrophotometer and electrophoresed on 0.8% agarose gel by comparison with control DNA samples of known concentration. The extracted DNA samples were taken up for RAPD analysis.

RAPD analysis

The total 10 random primers (OPA, OPB and OPF) were used in this study. The RAPD analysis was performed as per the procedure reported earlier with some modifications [22]. The PCR mixture was consisted of 2.5 μL 10X assay buffer with 15 mM $MgCl_2$, 1.0 μL dNTPs mix (2.5 mM each), 1.0 μL primer (5 pM μL^{-1}), 1.0 μL template DNA (25 ng μL^{-1}), 14.30 μL sterile nuclease free water and 2.0 μL Taq DNA polymerase (3.0 U μl^{-1}) (M/s Banglore Genei, Pvt. Ltd., Banglore, India).

The PCR amplifications were performed with Thermal Cycler (Thermo). The program was set for initial denaturation at 94°C for 4 min followed by 40 cycles at 94°C for 1 min, annealing at 35°C for 1 min, extension at 72°C for 2 min and final extension at 72°C for 7 min. The PCR products were analyzed by 0.8% agarose gel electrophoresis in 1X TAE buffer and stained with Ethidium bromide (0.5 μg mL^{-1}). The gel was visualized under UV transilluminator. The amplification patterns of all the isolates were compared with each other and bands of DNA fragment scored as '1' for presence and '0' for absence, generating '0' and '1' matrix. The DICE coefficient was used for the estimation of genetic variability. Pair-wise genetic similarities between strains were estimated by similarity coefficient. Clustering was done using the similarity coefficient and cluster obtained based on unweighted pair group method with arithmetic averaging (UPGMA) using cluster analysis of PyElph 1.4 PC software [23].

Results and Discussion

A total 36 (Xa1-Xa36) isolates causing oily spot disease of pomegranate were successfully isolated from diseased fruit samples of Bhagawa, Ganesh and Mridula varieties located in the pomegranate growing areas of Sangli, Solapur and Nasik districts of Maharashtra, India (Table 1). The isolated colonies emerged as shining yellow and mucoid on nutrient agar. They were Gram negative in nature and resembling with the *X. axonopodis* LMG 859 (the reference strain of *X. axonopodis* pv. *punicae* obtained from Shivaji university Kolhapur as a positive control). These isolates identified as *X. axonopodis* pv. *punicae*

Location	Sangli			Solapur			Nasik		
Variety	Bhagawa	Ganesh	Mridula	Bhagawa	Ganesh	Mridula	Bhagawa	Ganesh	Mridula
No. of Isolates	4	4	4	4	4	4	4	4	4

Table 1: The location of collection and variety of diseased fruits with the number of pathogenic isolates.

Isolate no.	Time in Days	Isolate no.	Time in Days
Bhagawa (Sangli)		Mridula (Solapur)	
Xa1	15	Xa21	16
Xa2	18	Xa22	22
Xa3	22	Xa23	21
Xa4	17	Xa24	22
Ganesh (Sangli)		Bhagawa (Nasik)	
Xa5	17	Xa25	17
Xa6	14	Xa26	17
Xa7	20	Xa27	19
Xa8	21	Xa28	20
Mridula (Sangli)		Ganesh (Nasik)	
Xa9	20	Xa29	20
Xa10	22	Xa30	22
Xa11	23	Xa31	18
Xa12	19	Xa32	21
Bhagawa(Solapur)		Mridula (Nasik)	
Xa13	23	Xa33	22
Xa14	24	Xa34	18
Xa15	20	Xa35	23
Xa16	23	Xa36	16
Ganesh (Solapur)			
Xa17	23		
Xa18	18		
Xa19	19		
Xa20	24		

Table 2: Oily spot disease causing isolates with time required for the complete development of disease.

Sr. No	Strain No.	Size of zone of inhibition	
		Streptomycin Sulfate	Streptocycline
1	Xa1	9	15
2	Xa2	12	21
3	Xa3	6	7
4	Xa4	0	9
5	Xa5	8	6
6	Xa6	16	16
7	Xa7	6	9
8	Xa8	0	6
9	Xa9	17	18
10	Xa10	6	7
11	Xa11	8	10
12	Xa12	15	16
13	Xa13	0	20
14	Xa14	7	8
15	Xa15	16	16
16	Xa16	6	6
17	Xa17	14	16
18	Xa18	10	12
19	Xa19	10	14
20	Xa20	15	6
21	Xa21	0	8
22	Xa22	8	20
23	Xa23	10	6
24	Xa24	6	13
25	Xa25	15	18
26	Xa26	8	9
27	Xa27	6	12
28	Xa28	12	6
29	Xa29	15	16
30	Xa30	8	10
31	Xa31	0	6
32	Xa32	6	19
33	Xa33	16	9
34	Xa34	9	14
35	Xa35	0	6
36	Xa36	0	10

Table 3: Inhibitory pattern of the isolates against wide spectrum antibiotics.

based on their cultural, morphological and biochemical characteristics [19,20].

All the 36 strains of *X. axonopodis* were found to be positive during *in vivo* pathogenicity test on fresh and healthy fruits as they displayed the symptoms similar to oily spot disease. All the strains were successfully reisolated from the infected fruits and found to be matching their parent strains. The variations were observed in the infectivity and pathogenicity pattern during *in vivo* pathogenicity tests. The difference was observed in time required for the complete development of disease. The time required for the development of disease was varied from 14 to 24 days with all the 36 isolates (Table 2). Some isolates infected immensely and exhibited early symptoms of the oily spot disease, the others were found to infect very slowly resulting in delayed appearance of disease symptoms.

Biochemical variability

A wide diversity was observed among the strains for the following biochemical tests:

Starch hydrolysis: Out of 36 all the 29 strains hydrolyzed starch except the 7 (Xa2, Xa5, Xa19, Xa24, Xa26, Xa29 and Xa36). The strains which were positive in starch hydrolysis varied in their degree of hydrolysis. The strains isolated from Ganesh variety, Xa6 (Sangli), Xa18 (Solapur) and Xa30 (Nasik) exhibited maximum zone of starch hydrolysis among all the 36 isolates.

Gelatin liquefaction: All the 36 strains found to be positive for gelatin liquefaction test.

Hydrogen sulfide production: Out of 36 strains 30 were positive for H_2S gas production test only 6 strains were not observed to be negative (Xa1, Xa17, Xa20, Xa23, Xa25, Xa35) showing variation in their ability to produce H_2S.

Catalase: All the 36 strains showed catalase test positive.

Oxidase: Out of 36, 27 strains showed positive to oxidase and 9 were negative.

Acids from carbohydrate: Except Xa2 and Xa26 all other strains produced acids from glucose as a carbon sources. When fructose was used as a carbon source all the 36 strains produced acid by utilizing it. When lactose, maltose and mannitol used as a carbon sources all the strains failed to produce acids. In case of sucrose except Xa6, Xa12, Xa16, Xa17, Xa21, Xa30, Xa33 all other strains utilized sucrose as a carbon source and produced acids. The results clearly indicate the variation in utilization of carbohydrate and production of acid.

There was no relation between the biochemical characteristics and disease progress or infectivity pattern during pathogenicity studies. Giri et al. was also recorded the biochemical variability within same species but not found any correlation with disease progress [10].

Susceptibility test

The strains were found to differ in susceptibility test against different wide spectrum antibiotics such as streptomycin sulfate and Streptocycline (Table 3). Due to their variation in susceptibility test against wide spectrum antibiotics all the strains were taken up for determining their genetic variations.

The results from the 36 *Xanthomonas axonopodis* pv. *punicae* strains isolated from infected fruits of pomegranate showed variations in biochemical tests and susceptibility test against the wide spectrum antibiotics.

Molecular variability

Genomic DNA extraction of Xanthomonas isolates: Thirty six isolates of *Xanthomonas axonopodis* causing oily spot disease were subjected to DNA extraction and the DNA samples were run on 0.8% agarose gel electrophoresis to check the integrity of DNA by observing under UV transilluminator.

RAPD analysis: Due to the existence of remarkable variation in susceptibility and pathogenicity the genetic diversity was assessed between *Xanthomonas axonopodis* isolates through RAPD. Main advantages of the RAPD technology include (i) suitability for work on anonymous genomes, (ii) applicability to problems where only limited quantities of DNA are available, (iii) high efficiency at low expense [24]. RAPD is the ideal technique in the absence of complete sequence information about genome of the pathogenic isolates, since it scans for sequence variation throughout the whole genome. The random amplification of polymorphic DNA (RAPD) is a quick method for developing species-specific probes and primers [25]. Molecular genetic markers have been developed into powerful tools to analyze genetic relationships and genetic diversity among various microbial groups. As an extension to the variety of existing techniques using polymorphic DNA markers, RAPD technique may be used in molecular ecology to determine taxonomic identity, assess kinship relationships, analyze mixed genome samples, and create specific probes.

Primer screening for RAPD analysis: Ten different decamer primers were screened out as mentioned in Table 4. The primers were diluted up to 10 pMole concentration with nuclease free water. During the RAPD analysis the amplification could be achieved only with two primers from operon primers series i.e. OPA-02 and OPB-03. Different annealing temperature conditions were checked across the set of Ten decamers i.e. 25°C, 30°C, 31°C, 32°C, 33°C, 34°C, 35°C, 36°C, 37°C and 38°C, the optimum annealing temperature was found to be 35°C for all the decamers used in the present study.

Amplification and dendrogram observed by decamer OPA-02: In RAPD analysis the amplification was observed using OPA-02. Other primers used in this study could not be amplified with all the 36 isolates of *Xanthomonas axonopodis*. The other primers were not

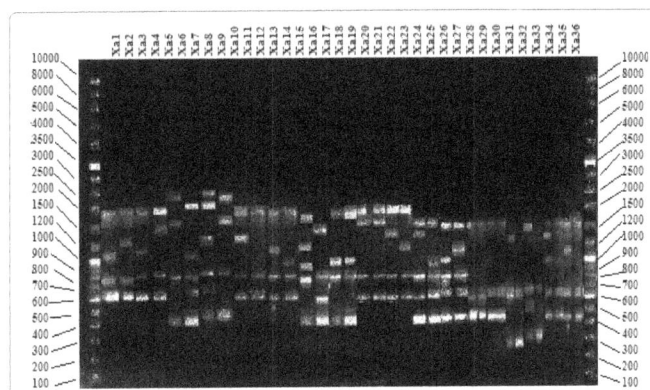

Figure 1: RAPD band pattern of amplified *Xanthomonas axonopodis pv. punicae* isolates.

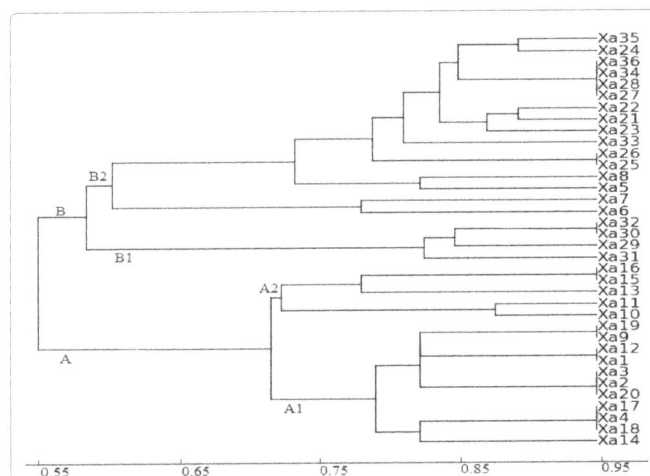

Figure 2: The dendrogram generated based on the RAPD data using the unweighted pair group method with arithmetic mean cluster analysis of genetic similarity coefficients.

amplified in all the isolates and banding pattern was not observed. Only OPA-02 primer exhibited good amplification with scorable bands. The amplification was observed in size range of 300 bp to 1800 bp.

Genetic variability among 36 isolates of *Xanthomonas axonopodis* pv. *punicae* was determined through RAPD analysis by using random primer OPA-02. The analysis resulted into 140 monomorphic and polymorphic bands. It revealed that maximum isolates exhibited monomorphic bands at 450, 650, 750 bp and polymorphic bands at 400, 900, 1000, 1100, 1500, 1700 bp (Figure 1). In OPA-02 primer *Xanthomonas axonopodis* pv. *punicae* showed 100% polymorphism. The dendrogram constructed based on unweighted pair group method with arithmetic averaging (UPGMA) analysis formed amplification with primer OPA-02 was comprised of two clusters A and B with a genetic similarity coefficient 0.58 (Figure 2 and Table 5). Cluster A and B were further divided in to two different sub-clusters A1, A2 and B1, B2. From these two clusters A and B (Xa17, Xa4, Xa18), (Xa19, Xa9), (Xa15, Xa16), (Xa30, Xa32), (Xa25, Xa26), (Xa27, Xa28, Xa34, Xa36) were very close with similarity coefficient ranged from 0.71 to 0.85. These isolates exhibited very minor variation so they grouped in the same clusters.

The Strains observed in cluster A were isolated from Bhagawa, Ganesh and Mridula varieties collected from Sangli and Solapur. The Strains observed in cluster B were also isolated from Bhagawa, Ganesh

Sr. No	Primer	Sequences
1	OPA-02	5'- TGC CGA GCT G3'
2	OPA-03	5'- AGT CAG CCA C 3'
3	OPA-08	5'- GTG ACG TAG G3'
4	OPA-20	5'- TCG GCG ATA G3'
5	OPB-02	5'- TGA TCC CTG G3'
6	OPB-03	5'- CAT CCC CCT G3'
7	OPB-04	5'- GGA CTG GAG T3'
8	OPB-05	5'- TGC GCC CTT C3'
9	OPF-06	5'- GGG AAT TCG G3'
10	OPF-07	5'- CCG ATA TCC C3'

Table 4: Random primers used for the amplification of *X. axonopodis pv. punicae* DNA.

Variety and Location of Pomegranate	Cluster A		Cluster B	
	A1	A2	B1	B2
Sangli				
Bhagawa	Xa1, Xa2, Xa3, Xa4	–	–	
Ganesh	–	–	–	Xa5, Xa6, Xa7 Xa8
Mridula	Xa9, Xa12	Xa10, Xa11	–	–
Solapur				
Bhagawa	Xa14	Xa13, Xa15, Xa16	–	–
Ganesh	Xa17, Xa18, Xa19, Xa20	–	–	
Mridula	–		–	Xa21, Xa22, Xa23, Xa24
Nasik				
Bhagawa	–	–	–	Xa25, Xa26, Xa27, Xa28
Ganesh	–	–	Xa29, Xa30, Xa31, Xa32	–
Mridula	–	–	–	Xa33, Xa34, Xa35, Xa36

Table 5: Grouping of 36 pathogenic isolates in different clusters on the basis of cluster analysis.

and Mridula varieties collected from Sangli, Solapur and Nasik. In the cluster A strains isolated from Ganesh, Bhagawa and Mridula varieties (Xa17, Xa4, Xa18), (Xa3, Xa2, Xa20), (Xa15, Xa16), (Xa1, Xa12), (Xa9, Xa19) with locations Sangli and Solapur districts exhibited very close relationship with each other. Similarly, in the cluster B strains isolated from above three varieties (Xa30, Xa32), (Xa26, Xa25) and (Xa27, Xa28, Xa34, Xa36) with location Nasik district revealed a very close relationship.

In cluster A none of the strain found to be present from Nasik district. The strains isolated from different varieties of pomegranates but with the same region like Bhagawa (Xa1 to Xa4), Ganesh (Xa5 to Xa8), Mridula (Xa9 to Xa12) from Sangli and Bhagawa (Xa13 to Xa16), Ganesh (Xa17 to Xa20), Mridula (Xa21 to Xa24) from Solapur, formed different cluster and sub clusters showing polymorphism. As in cluster B strains isolated from Nasik with Bhagawa, Ganesh and Mridula varieties formed different sub clusters as per the varieties as Bhagawa (Xa30, Xa32) and Ganesh (Xa25, Xa26) exhibiting polymorphism. This revealed the genetic variation between the strains isolated from same location but different varieties. The strains Xa27, Xa28, Xa34 and Xa36 formed one sub cluster showing very close relationship with a similarity coefficient of 0.80 but these were isolated with Bhagawa and Ganesh varieties only from Nasik.

A similar difference was also present within the same variety and same location (Mridula from Sangli and Bhagawa from Solapur), this further affirmed the existence of variation. Study of RAPD pattern has already been reported for the rapid, sensitive, and specific detection of genetic diversity among species and strains of Streptomyces [26]. Variability within the same species, X. axonopodis has also been reported earlier for various other plant diseases [10,27,28]. Mondal and Mani have also reported the genetic variability among the X. campestris strains isolated from oily spot infected pomegranates through Maharashtra and Delhi using ERIC-PCR method [29]. The present study exhibited a wide genetic variation existed between the 36 Xanthomonas axonopodis isolates in the range of 0.55% to 0.95%.

It confirmed that the variation among the strains was irrespective of their location and variety of pomegranate (Figure 1). Similar kind of diversity had been reported earlier in X. campestris pv. passiflorae from southern Brazil [30]. Kishun and Gupta have a similar finding from X. campestris pv. mangiferaeindicae population with a significant level of genetic diversity and the formation of 2 clusters in the phylogenetic tree [31]. The results from this study showed a high rate of genetic diversity with a 0.95 similarity coefficient. This was in contrast to a earlier study by Odipio et al. [32], who found very low genetic diversity among Ugandan isolates of X. campestris pv. musacearum as determined by

RAPD. This further confirms that a significant level of polymorphism was existing among the present 36 evaluated strains of Xanthomonas axonopodis pv. punicae [32].

The genetic diversity found among the 36 pathogenic isolates may developed due to the outcrossing behavior of pomegranate crop and the acquired antibiotic resistance due to overexposure to antimicrobial agents employed for the control of the disease.

It can be concluded that the biochemical, pathogenic, susceptibility and molecular techniques employed in the present study give strong evidence to the existence of variability among 36 X. axonopodis pv. punicae isolates regardless of variety and the geographical origin of oily spot disease in pomegranate. There is no correlation present between the biochemical, pathogenic, susceptibility and molecular variability patterns. The present study confirmed that the variability is existed between the 36 pathogenic isolates of X. axonopodis pv. punicae. This is the important finding providing information on the genetic structure of a pathogenic bacterial population. It might help in designing the strategy for the control of Oily spot disease of pomegranate.

Statistical analysis

The data was subjected to analysis of molecular variance (AMOVA) and the significance of the difference between the means was determined ($P<0.05$) using MedCalc. Values were expressed as means of 3 replicate determinations ± standard deviations (SD).

Acknowledgement

The author's express their sincere thanks to Dr. D. Y. Patil Vidyapeeth and Dr. D.Y. Patil Biotechnology and Bioinformatics Institute, Pune, India for providing the research facilities.

References

1. Sharma KK, Jadhav VT, Sharma J (2009) Present status of Pomegranate bacterial blight caused by Xanthomonas axonopodis pv. punicae and its management. ISHS Acta Horticulturae 890: II International Symposium on Pomegranate and Minor-including Mediterranean-Fruits, India.

2. Mondal KK, Singh D (2008) Bacterial blight of pomegranatea technical bulletin. Division of Plant Pathology, Indian Agricultural Research Institute. New Delhi, India.

3. Hingorani MK, Mehta PP (1952) Bacterial leaf spot of pomegranate. Indian Phytopathol 5: 55-56.

4. Yenjerappa ST, Ravikumar MR, Jawadagi RS, Khan NA (2004) In vitro and in vivo efficacy of bactericides against bacterial blight of pomegranate. Proceedings of National Symposium of Crop Surveillance: Disease Forecasting and Management, New Delhi, India.

5. Tiwari RK, Mistry NC, Sinhg R, Gandhi CP (2014) Indian Horticulture Database 2013.

6. Chavan NP. Pandey R, Nawani N, Nanda RK, Tandon GD, et al. (2016) Biocontrol potential of actinomycetes against *Xanthomonas axonopodis* pv. punicae, a causative agent for oily spot disease of pomegranate. Biocontrol Sci Technol 26: 351-372.

7. Dhandar DG, Nallathambi P, Rawal RD, Sawant DM (2004) Bacterial leaf and fruit spot: A new threat to pomegranate orchards in Maharashtra state. 26th Annual Conference and Symposium ISMPP, Goa, India.

8. Manjula CP, Khan ANA, Ravikumar MR (2002) Management of bacterial blight of pomegranate (*Punica granatum* L.) caused by *Xanthomonas axonopodis* pv. punicae. Ann. Meet. Symp. Plant Disease Scenario in Southern India, India.

9. Erayya LR, Kumaranag KM, Chandrashekar N, Khan AN (2014) *In vivo* efficacy of some antibiotics against bacterial blight of pomegranate caused by *Xanthomonas axonopodis* pv. punicae. Int Res J Biological Sci 3: 31-35.

10. Giri MS, Prashanthi SK, Kulkarni S, Benagi VI, Hegade YR (2011) Biochemical and molecular variability among *Xanthomonas axonopodis* pv. punicae isolates, the pathogen of pomegranate bacterial blight. Indian Phytopathol 64: 56-516.

11. Hayward AC (1964) Bacteriophage sensitivity and biochemical group in *Xanthomonas malvacearum*. J Gen Microbiol 35: 287-298.

12. Griffin DE, Dowler WM, Hartung JS, Bonde MR (1991) Differences in substrate utilization among isolates of *Xanthomonas oryzae* pv. oryzae and *X. campestris* pv. oryzicola from several countries. Phytopathol 81: 1222.

13. Verniere C, Pruvost O, Civerolo EL, Gambin O, Jacquemoud-Collet JP, et al. (1993) Evaluation of the biolog substrate utilization to identify and assess metabolic variation among strains of *Xanthomonas campestris* pv. Citri. Appl Environ Microbiol 59: 243-249.

14. Abdo-Hasan M, Khalil H, Debis B, MirAli N (2008) Molecular characterisation of Syrian races of *Xanthomonas axonopodis* pv. malvacearum. J Plant Pathol 90: 431-439.

15. Schaad NW (1992) Xanthomonas. In: Laboratory Guide for Identification of Plant Pathogenic Bacteria. (2ndedn), International Book Distributing Co. Charbagh, Lucknow, India.

16. Gabriel DW, DeFeyter R (1992) RFL Panalyses and gene tagging for bacterial identification and taxonomy. In: Gurr SJ, McPherson MJ, Bowles DJ. In 'Molecular plant pathology. Vol. 1, a practical approach'. IRL Press: Oxford, England.

17. Milgroom MG, Fry WE (1997) Contributions of population genetics to plant disease epidemiology and management. Adv Bot Res. 24: 1-30.

18. Salle AJ (1973) Laboratory manual on fundamental principles of bacteriology (7thedn). Mc Graw Hill Book Co., New York, USA.

19. Chand R, Kishun R (1991) Studies on bacterial blight of pomegranate. Indian Phytopathol 44: 370-372.

20. Garrity G, Brenner DJ, Krieg NR, Staley JR (2007) Bergey's Manual® of systematic bacteriology: Volume 2: The Proteobacteria. Springer, New York City.

21. Yenjerappa ST (2009) Epidemiology and management of bacterial blight of pomegranate caused by *Xanthomonas axonopodis* pv. punicae (Hingorani and Singh) Vauterin et al. Ph.D. thesis, University of Agricultural Sciences, Dharwad.

22. Williams JGK, Kubelik AR, Livak KJ, Rafalski JA, et al. (1990) DNA polymorphism amplified by arbitrary primers are useful as genetic markers. Nucleic Acids Res 18: 6531-6535.

23. Pavel AB, Vasile CI (2012) PyElph: A software tool for gel images analysis and phylogenetics. BMC Bioinformatics 13: 9.

24. Hadrys H, Balick M, Schierwater B (1990) Applications of random amplified polymorphic DNA (RAPD) in molecular ecology. Molec Ecol 1: 55-63.

25. Basagoudanavar SH, Rao JR, Omanwar S, Tiwari AK, Singh RK, et al. (2001) Identification of *Trypanosoma evansi* by DNA hybridization using a nonradioactive probe generated by arbitrary PCR. Acta Vet Hung 49: 191-95.

26. Martin P, Dary A, Andre A, Decaris B (2000) Identification and typing of Streptomyces strains: Evaluation of interspecific and intraclonal differences by RAPD fingerprinting. Res Microbiol 151: 853-864.

27. Ogunjobi AA, Dixon AGO, Fagade OE (2007) Molecular genetic study of cassava bacterial blight casual agent in Nigeria using random amplified polymorphic DNA. Electronic J Environ Agri Food Chem 6: 2364-2376.

28. Ogunjobi AA, Fagade OE, Dixon AGO (2010) Comparative analysis of genetic variation among *Xanthomonas axonopodis* pv. manihotis isolated from the western states of Nigeria using RAPD and AFLP. Indian J Microbiol 50: 132-138.

29. Mondal KK, Mani C (2009) ERIC-PCR-Generated genomic fingerprints and their relationships with pathogenic variability of *Xanthomonas campestris* pv. punicae, the incitant of bacterial blight of pomegranate. Curr Microbiol 59: 616-620.

30. Goncalves ER, Rosato YB (2000) Genotypic characterization of Xanthomonad strains isolated from passion fruit plants (*Passiflora* spp.) and their relatedness to different *Xanthomonas* species. Int J Sys Evol Microbiol 50: 811-821.

31. Kishun R, Gupta VK (2008) Detection of genetic diversity among Indian strains of *Xanthomonas campestris* pv. mangiferaeindicae using PCR-RAPD. Nat Preced 2403: 1.

32. Odipio J, Tusiime G, Tripathi L, Aritua V (2009) Genetic homogeneity among Ugandan isolates of *Xanthomonas campestris* pv. musacearum revealed by randomly amplified polymorphic DNA analysis. Afr J Biotechnol 8: 5652-5660.

16

Detection, Identification and Quantification of *Fusarium graminearum* and *Fusarium culmorum* in Wheat Kernels by PCR Techniques

Rabab Sanoubar[1]*, Astrid Bauer[2] and Luitgardis Seigner[2]

[1]*Department of Horticulture, Agriculture Faculty, Damascus University, Syria*
[2]*Iinstitute of Plant protection (Bayerische Landesanstalt für Landwirtschaft Institut für Pflanzenschutz, LfL), Freising, Germany*

Abstract

This study was carried out on 172 samples of winter wheat. The samples consisted of various cultivars that had been randomly collected from farmers' fields in different areas of Bavaria, South Germany. The objectives of this study were detecting the presence of *Tri-5* gene producing fungus that generates trichothecene mycotoxins, especially Deoxynivalenol (DON), by using conventional qualitative PCR; determining the correlation between the presence of *Tri-5* gene and DON content; evaluating the *Fusarium graminearum and Fusarium culmorum* infection by Real-Time PCR and estimating the correlation between DON content and the severity of *F. graminearum and F. culmorum* contamination. This study showed that 86% of all infected samples had a *Tri-5* gene and amplified a single 544bp fragment associated with a detectable amount of DON (ranged from 10 to 2990 µg kg^{-1}). This study demonstrated that *F. graminearum* is the predominant species associated with *Fusarium* head blight (FHB) and was considered as the predominant trichothecene producer that associated with FHB since there was a highly significant correlation (R^2=0.7) between DON and *F. graminearum* DNA content, compared to a weak correlation (R^2=0.03) between DON and DNA content of *F. culmorum* infected wheat kernels.

Keywords: Wheat; *Fusarium* head blight; *Tri-5* gene; Trichothecene DON; Conventional; Real-time PCR

Introduction

Fusarium head blight (FHB) of small grains was first described over a century ago and was considered as a major threat to wheat and barley during the early years of last century [1]. Head blight or scab of wheat caused epidemics in many wheat area worldwide [2,3]. The International Maize and Wheat Improvement Centre (CIMMYT) have considered FHB as a major factor limiting wheat production in many parts of the world [2]. FHB is also known as "tombstone" kernels of wheat because of the chalky and lifeless appearance of the infected kernels [4]. It has the capacity to destroy a potentially high-yielding crop within few weeks [5]. FHB is a significant disease of small-grain cereals throughout Europe [3], United States [6], Canada [7], South America [8], Asia [9] and Australia [10]. FHB was identified more than 120 years ago, in 1884, in England. The United States Department of Agriculture ranked FHB as the worst plant disease to appear since the 1950's [11]. It has increased worldwide [12] and it was considered as a major threat to wheat and barley during the early years of the twentieth century [13,14].

FHB is caused by a number of different fungal species of the genus *Fusarium* (*Fusarium* spp). However, *F. avenaceum, F. culmorum, F. graminearum* (teleomorph, *Gibberella zeae*), *F. poae*, and *Microdochium nivale* (teleomorph, *Monographella nivalis*) are the species which are most commonly associated with the FHB disease [15].

The threat posed by *Fusarium* spp. is multifaceted. It causes yield and quality losses due to sterility of the florets and formation of discolouration, which reduces kernel size and losses light weight kernels [9]. In addition, grain quality factors such as protein content and germination can be severely affected by the pathogen [16]. Several *Fusarium* species which cause FHB are able to produce trichothecene mycotoxin. *F. culmorum, F. graminearum, and F. poae* produce type B trichothecenes such as nivalenol (NIV), deoxnivalenol (DON), and fusaenon-X [17], while other species are not [18]. DON is the predominant Trichothecenes found in Europe and North America [19]. Trichothecene produced by this fungus pose a serious hazard to human and animal health [10] because these toxic materials are potent inhibitors of eukaryotic protein biosynthesis [20,21]. Acute adverse effects of the toxin in animals causes weight loss and feeding refusal in non-ruminant livestock, high rates of abortion, diarrhoea, emesis, alimentary haemorrhagy and contact dermatitis [22]. Human ingestion of grain contaminated with *F. graminearum* has been associated with alimentary toxicity as well as illness characterized by nausea, vomiting, anorexia, and convulsions [23]. Trichothecenes are also powerful modulators of human immune function and may promote neoplasms, cause autoimmune disease, or have long-term effects on resistance to infectious disease by altering immune response [24,25].

Several genes of *Fusarium* are involved in the biosynthesis of trichothecene and most of them are localized in a *Tri* gene cluster. The *Tri-5* gene encodes the enzyme trichodiene synthase [26], which catalyzes the first step in the trichothecene biosynthetic pathway in trichothecene-producing strains of *Fusarium* species. The development of *Tri-5* gene specific primers has allowed trichothecene-producing *Fusarium* spp. to be distinguished from nonproducing species using PCR-based assays [27]. The nucleotides sequence of the *Tri-5* gene has been characterized in several *Fusarium* species [28,29].

The main objectives of this study were detecting the presence of *Tri-5* gene producing fungus, which encodes the key enzyme in trichothecene production, especially DON, by using conventional

*Corresponding author: Rabab Sanoubar, Department of Horticulture, Agriculture Faculty, Damascus University, Syria, E-mail: rabab.sanoubar@unibo.it

PCR; determining the correlation between the presence of *Tri-5* gene and the DON content, which was analysed by the chemist Dr. Puttner. J. Lepschy; evaluating the amount of *F. graminearum and F. culmorum* infection through Real-Time PCR assay; investigating the relationship between DON content and the degree of *F. graminearum and F. culmorum* contamination and determining the aggressiveness of FHB towards plant host.

Materials and Methods

Fungal reference material

50 ng of extracted DNA from *F. graminearum* isolates were applied in tenfold serial dilutions (10^{-1} to 10^{-4}) as a quantitative standard in Real-Time PCR (RT-PCR) using a *F. graminearum* specific Taqman˙ hybridization probe for beta-tubulin gene. In parallel, 50 ng of DNA of *F. culmorum* strains were used in tenfold serial dilution also as a quantitative standard for RT-PCR using SYBR Green˙1.

Plant material

At harvest time, 172 winter wheat ears samples of various cultivars have been randomly collected from farmers' fields in different areas of Bavaria, South Germany. Directly after harvest, samples were sent to the Institute of Plant Protection, LfL and preserved at -20°C.

DNA extraction

DNA of infected wheat kernels was extracted by homogenising 10 mg of dried kernels in a mixer with the presence of 1 ml DNA extraction buffer (2% CTAB, 1.4 M NaCl, 100 mM Tris, 20 mM Na-EDTA, and 1% PVP-40). The mixture was vortexed and the flow was transferred to microcentrifuge tubes. 1 ml chloroform/isoamylalkohol (24:1) was added, well mixed and spined at 5000×g for 10 minute at 20°C. The aqueous phase containing DNA molecules was transferred into 2 ml fresh tubes where 100 μl Na-acetat (3M, pH 5.2) and 1 ml isopropanol (-20°C) were added and mixed by inverting the tubes many times. Tubes were placed in a freezer (-20°C) for at least 1 hour. For each sample, the lysate mixture was transferred to SV Minicolumn placed in 1.5 ml tubes and spined at 16.000×g for 1 minute at 4°C. The supernatant was discarded, and the SV Minicolumns were washed with ethanol many times as described by (30). Finally, 50 μl of distilled sterile water was added directly into SV Minicolumn which was placed in 1.5 ml microcentrifuge tube and incubated at room temperature for 5 minutes then spined at 16.000×g for 2 minutes at 4°C to collect the eluted DNA [30].

PCR assay

Two *Tri-5* specific primers have been used to detect the presence of *Tri-5* gene in *F.* spp. infected wheat kernels. 172 wheat samples were tested with a sample of *F. graminearum* used as a positive control. The sequences of these primers are: forward primer *Tri-5* F: (5'-AGCGACTACAGGCTTCCCTC-3') and reverse primer *Tri-5* R: (5'-AAACCATCCAGTTCTCCATCTG-3'). These primers were derived from the conserved region of *Tri-5* gene in *Fusarium* spp. *Tri-5* primers (Tr5F and Tr5R) amplified a single 544bp fragment in both DNA extracted of *F. graminearum* and *F. culmorum* and DNA of infected wheat grains. The total volume of reaction master mix was 22.7 μl. The PCR amplification was performed using (2.7 μl of 25 μg ml⁻¹) of both fungal *F. graminearum* DNA and DNA from 10 mg dry weight of wheat material, 0.5 μM of each of the *Tri-5*-specific primers, 0.8 mM concentration of nucleotides dNTPs, 0.5 unit of Taq polymerase and 2.27 μl of PCR buffer with 1.5 mM $MgCl_2$. The PCR negative control was a reaction master mix with 2.7 μl of distilled water instead of DNA template. Cycler programme was set as the following: one cycle at 95°C for 75 s then 32 cycles of 94°C for 20 s, 62°C for 17 s, 72°C for 45 s, and a final cycle at 72°C for 4 min and 15s. DNA banding were revealed by electrophoresis at 90 v on 2% agarose gels in 1x Tris-acetate EDTA buffer (TAE) (where 50X TAE contained: 2M Tris, 1M Acetic Acid, and 0.1M Na-EDTA x $2H_2O$ at PH 8.0) using ethidium bromide staining (30 μg of a ethidium bromide for 100 ml of 1X TAE buffer) and photographed under UV light using a camera and a photo print image visualizer.

Quantification of *F.* species by Real-Time PCR

Quantification of *F. graminearum* by RT-PCR using a TaqMan probe assay: Two primers, derived from the consensus beta-tubulin sequence which is associated with head blight in wheat, were used for *F. graminearum* quantification. The forward primer FGtubf: (5'-GTCTCGACAGCAATGGTGTT-3') and reverse primer FGtubr: (5'-GCTTGTGTTTTTCGTGGCAGT-3') specifically amplified a 111 bp fragment of the beta-tubulin gene of *F. graminearum* which was quantified by the TaqMan probe FGtubTM (FAM-5'ACAACGGAACGGCACCTCTGAGCTCCAGC3'-TAMRA). PCR was monitored on a Real-Time 7000 Sequence Detection System. PCR Master Mix contained: Hot Start Taq DNA polymerase and PCR buffer specifically adopted for quantitative PCR analysis using species-specific probes. The total volume of master mix reaction (23 μl) contained: optimal primer concentrations 0.3 μM of FGtubf and FGtubr primers, 1 x PCR buffer, 50 ng of template wheat DNA samples, and 0.2 μM of TaqMan probe, and 50 ng of *F. graminearum* dilution template DNA as a standard curve. There were four series of diluted standard curves, with 1:10 fold of dilution factor of *F. graminearum* DNA. The number of cycles in the PCR was set at 40, as the 40th cycle represented the extrapolated threshold cycle for a reaction with a theoretical single copy of the template DNA. PCR program was as the following: 95°C for 15 min, 40 cycles of 95°C for 15 s and 67°C for 1 min. All reactions were performed in triplicates. PCR efficiency was calculated from threshold cycles of the standard dilution curve.

Quantitation of *F. culmorum* by Real-Time PCR using DNA binding dye assay: Two specific primers were used for the detection of *F. culmorum* by amplifying 140bp fragment of *F. culmorum*. Forward primer sequence Fc03: (5'-TTCTTGCTAGGGTTGAGGATG-3') and reverse primer sequence Fc02: (5'-GACCTTGACTTTGAGCTTCTTG-3') were specifically amplified a 140bp fragment of *F. culmorum* genome that was quantified by the DNA binding dye, SYBR˙ Green 1. The SYBR Green 1 assay is similar to that of TaqMan assay except the presence of an intercalating agent such as fluorescent dye SYBR Green 1 instead of fluorescent probes TaqMan.

Results

Detection of *Tri-5* gene

The *Tri-5* specific PCR assay could provide a screening tool for detection of trichothecene-producing *Fusarium* species in plant tissues. In this study, 172 DNA samples of infected wheat kernels were analyzed using *Tri-5* gene primers in a PCR reaction to detect trichothecene producing *Fusarium* species. The separation of PCR products on agarose gels showed that 86% of DNA samples possessed a unique fragment of 544bp representing part of the *Tri-5* gene while 14% of samples didn't show any amplification product (Figure 1).

The same wheat samples were used to estimate the content of DON

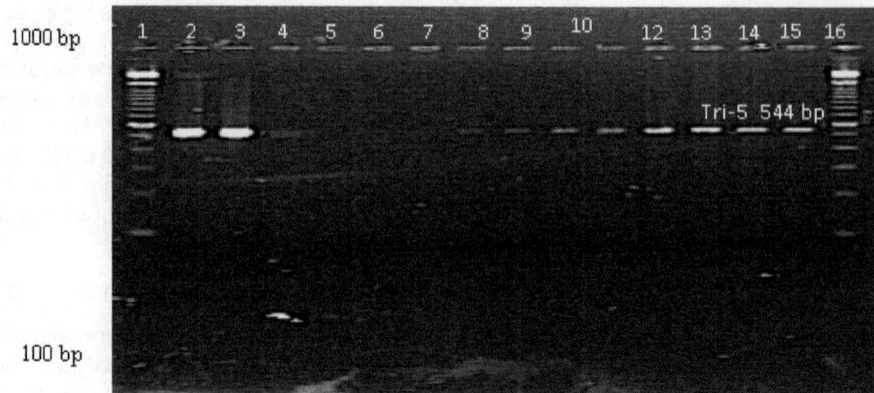

Figure 1: Analysis of DNA samples with specific primers (*Tri-5* gene) on 2% agarose gel.

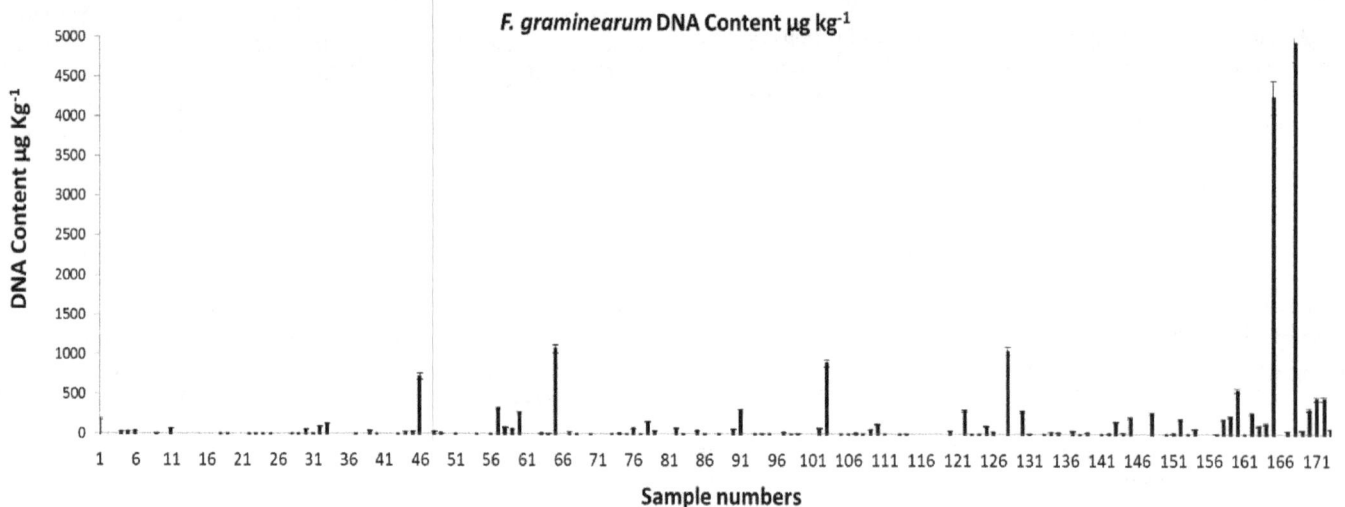

Figure 2: Infestation by *F. graminearum* in wheat samples harvest by Real-Time TaqMan probe. Samples were tested in triplicates.

in their tissues determined by HPLC (results provided by the chemist Dr. puttner J. Lepschy, LfL). The quantity of DON varied between the samples leading to regroup the samples into 3 categories according to their DON content.

The comparison between the amplification products produced with *Tri-5* gene primers and the DON contents demonstrated that the samples free of *Tri-5* gene products (14% of all samples) showed either absence of DON in their tissues (in 7% of all samples) or detected an amount of DON ranged from 11-226 µg kg^{-1} (in 7% of all samples).

For all samples possessing 544bp DNA fragments (86% of all samples), an amount of DON was detected in their tissues. Approximately 76% of them (59% of all samples) were considered DON high-producing strains (101-2990 µg kg^{-1}) and possessing an intensive DNA band on agarose gels, while 14% of them (27% of all samples) were considered DON low-producing strains (10-99 µg kg^{-1}) and showing a faint DNA band at 544bp.

Line 1, 16 standards 100 bp. Lines 2, 3 *F. graminearum* DNA used as positive control. Lines (4-15) some tested wheat samples: lines (5-7) free of DNA band, Lines (4, 8-11) faint bands, and lines (12-15) bright bands at 544bp.

Quantification of *F. graminearum* and *F. culmorum* using Real-Time PCR

The specific primers and TaqMan hybridization probe targeting the beta-tubulin gene amplifies DNAs from *F. graminearum* infected wheat. Based on determination of threshold cycle (**Ct**-values) in individual samples and known DNA standards during Real-Time PCR, amounts of target DNA present in the samples were calculated. The amplification of standard dilution curves of *F. graminearum* in Real-Time PCR gave linear and reliable results (R^2 values were between 0.997 and 0.989). The concentration of *F. graminearum* DNA ranged from 0.04 to 4945 µg kg^{-1} (mean of triplicate samples ranged from 9.19E-07 to 1.29E-01 µg ml^{-1}) of dry weight wheat kernels (Figure 2), while, *F. culmorum* DNA content was limited and ranged between 0.04 and 39.22 µg kg^{-1} (mean of triplicate samples ranged from 8.00E-07 to 7.13E-04 µg ml^{-1}) of dry weight wheat kernels over all Bavaria (Figure 3).

By comparing the results of *Tri-5* specific PCR assay, DON quantification by HPLC technique, and RT-PCR DNA quantification of *F. graminearum* and *F. culmorum* for the same wheat samples, we found that the same *Tri-5* DNA intensive bright samples that associated with the largest amounts of DON were contained also the highest

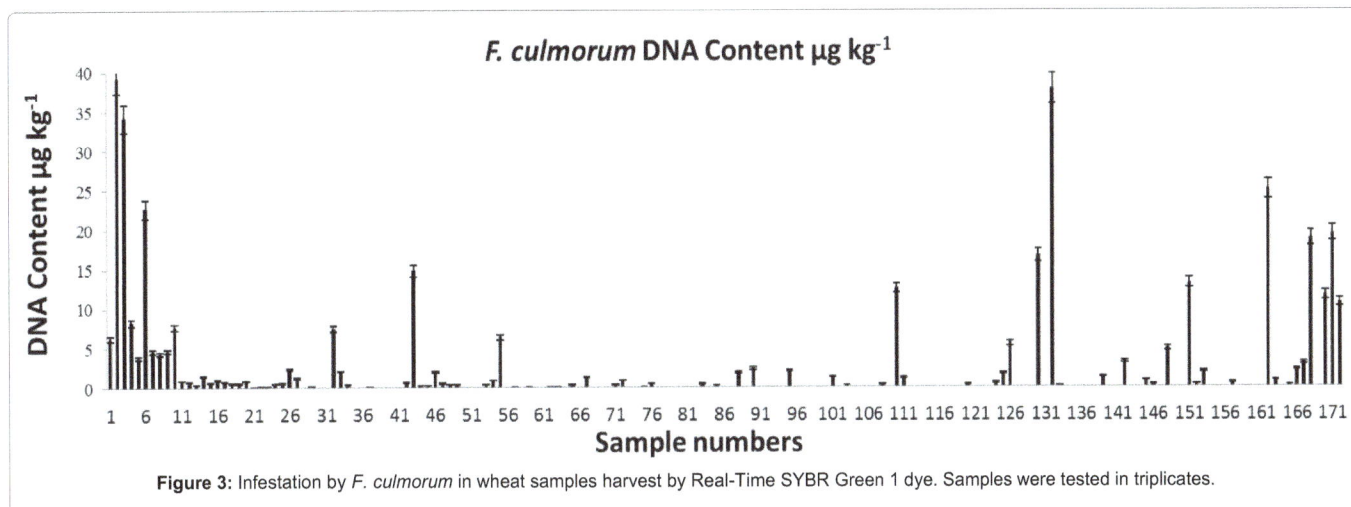

Figure 3: Infestation by *F. culmorum* in wheat samples harvest by Real-Time SYBR Green 1 dye. Samples were tested in triplicates.

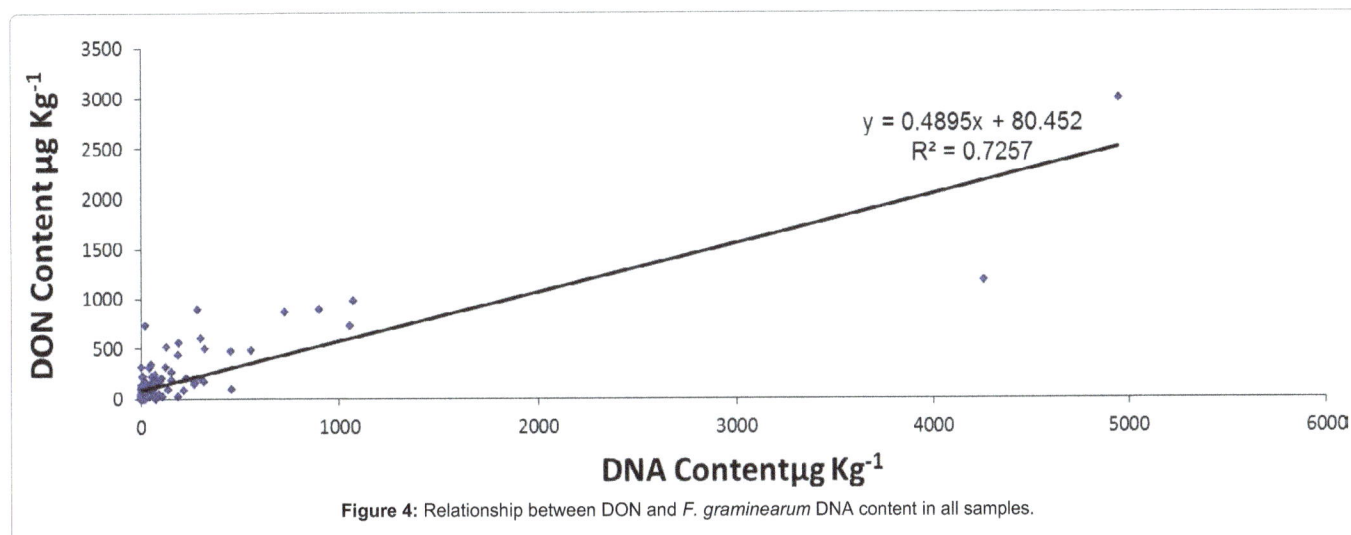

Figure 4: Relationship between DON and *F. graminearum* DNA content in all samples.

amount of *F. graminearum* DNA (4945-100.82 µg kg⁻¹) and trace amounts of *F. culmorum* DNA (39.22 to 10.67 µg kg⁻¹). Alternatively, faint *Tri-5* DNA samples which had a trace amount of DON (10-99 µg kg⁻¹) showed trace detectable amounts of both *F. graminearum* DNA ranged from 0.04 to 47.15 µg kg⁻¹ and *F. culmorum* DNA ranged from 0.04 to 1.99 µg kg⁻¹. In case of absence of *Tri-5* gene products which associated with free DON contents (7% of all samples), there was trace amounts of both *F. graminearum* and *F. culmorum* DNA (0.72 and 0.78 µg kg⁻¹, respectively). While in case of absence of *Tri-5* gene products on agarose gels but with presence of DON in their tissues (the other 7%), there was trace detectable amounts of both *F. graminearum* DNA (0.66-1.83 µg kg⁻¹) and *F. culmorum* DNA (0.08-39.22 µg kg⁻¹).

Correlation between *F. graminearum* DNA and DON content

The plot of *F. graminearum* DNA content in 172 wheat samples, determined by Real-Time TaqMan probe PCR and the DON content in their tissues determined by HPLC, showed a strong positive linear correlation between both parameters. Correlation coefficient was 0.725 (Figure 4). Moreover, the regression analysis of all data sets indicated a strong and highly significant correlation ($p < 0.05$) between DON contents in the plant tissues and *F. graminearum* DNA contents in wheat samples, and the regression equation was (y=0.4896x + 79.784; R^2=0.7252).

Correlation between *F. culmorum* DNA and DON content

The plot of DON content against *F. culmorum* DNA standard curve (Figure 5) showed a slight correlation between DON and *F. culmorum* DNA, whereas the linear correlation coefficient was ≈ 0.2 which is very far from +1. The regression analysis of all data set showed a weak correlation ($p < 0.01$) between *F. culmorum* DNA and DON content, and the regression equation was (y=8.9319x + 117.51; R^2=0.0353).

Moreover, comparison of the results of TaqMan Real-Time PCR for *F. graminearum* with analysis of DON content for the same samples showed that 73% of *F. graminearum* presence was associated with DON production. On the contrary, the results of SYBR Green 1 Real-Time PCR for *F. culmorum* with DON content showed that 56% of *F. culmorum* incidence was associated with DON production. However, in spite of the noticeable infections of wheat grains with *F. culmorum* 56%, its DNA content was low (0.04- 39.22 µg kg⁻¹) compared to that of *F. graminearum* (0.04 to 4945 µg kg⁻¹ Figure 6).

On the other hand, although the *F. graminearum* DNA was not detected in 20% of samples, but DON was found (10-320 µg kg⁻¹) and *F. culmorum* DNA was detected in some of these samples (2-39 µg kg⁻¹) (Figure 7).

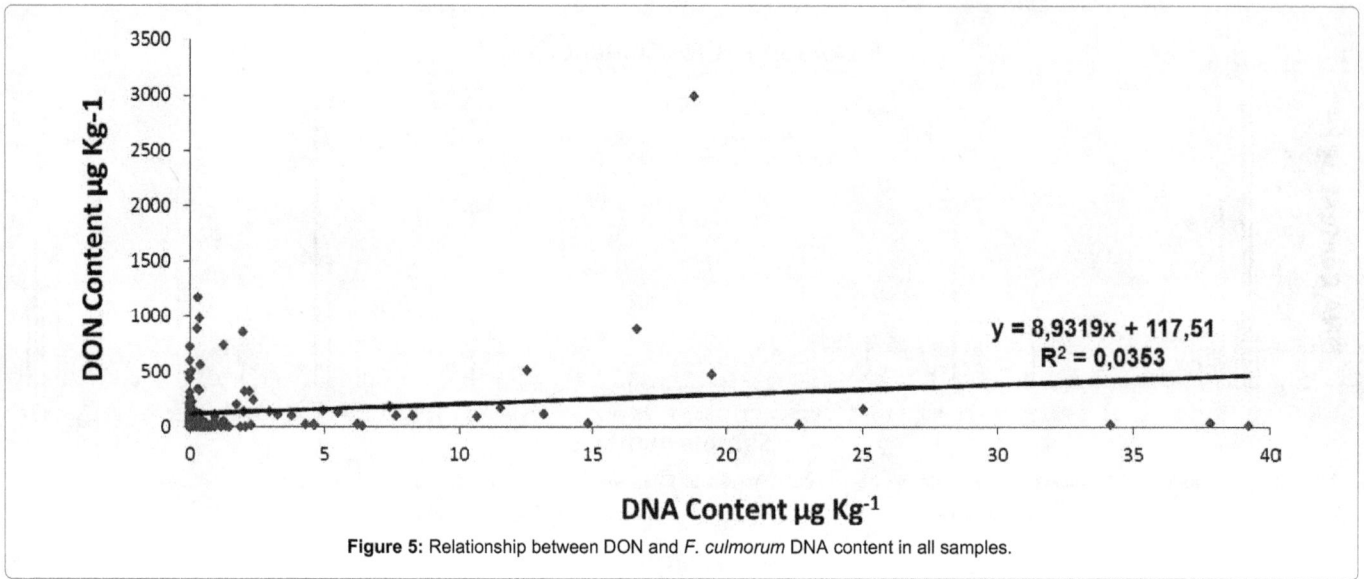

Figure 5: Relationship between DON and *F. culmorum* DNA content in all samples.

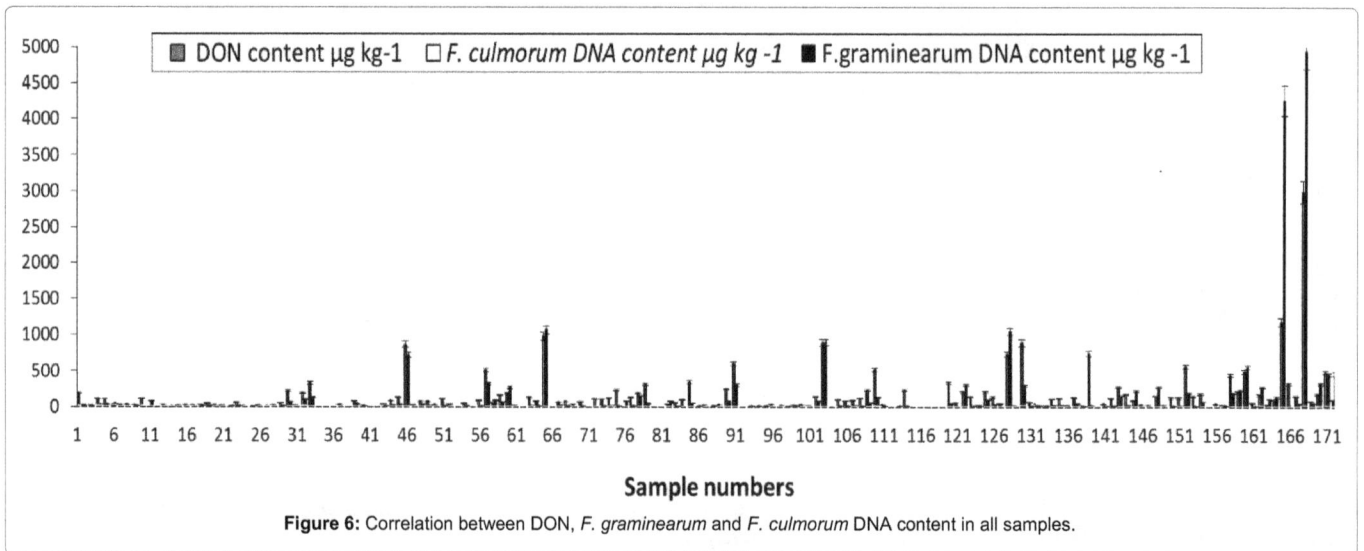

Figure 6: Correlation between DON, *F. graminearum* and *F. culmorum* DNA content in all samples.

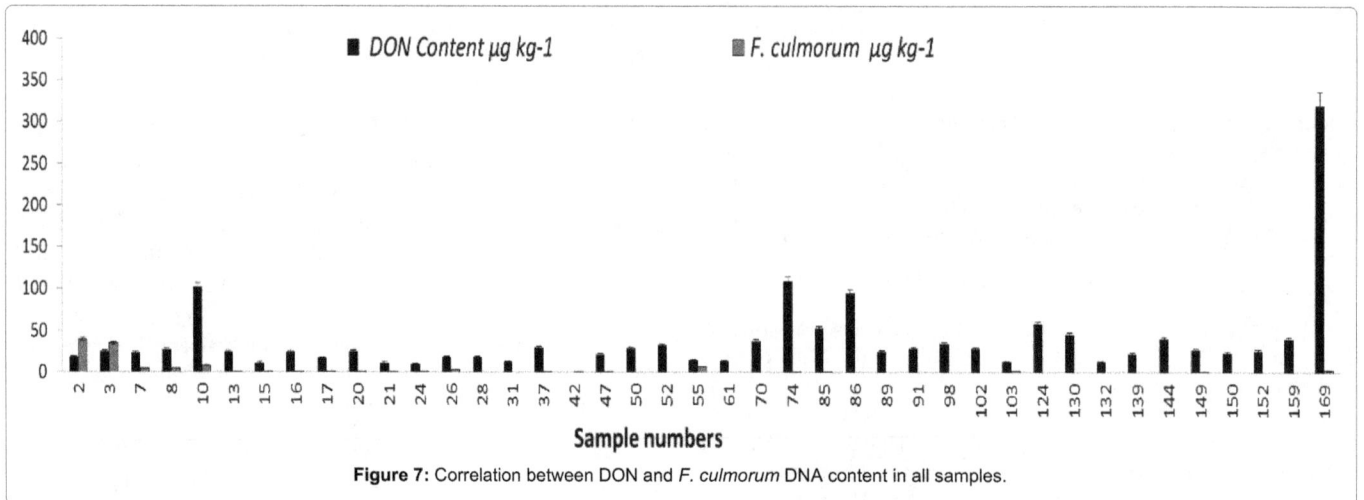

Figure 7: Correlation between DON and *F. culmorum* DNA content in all samples.

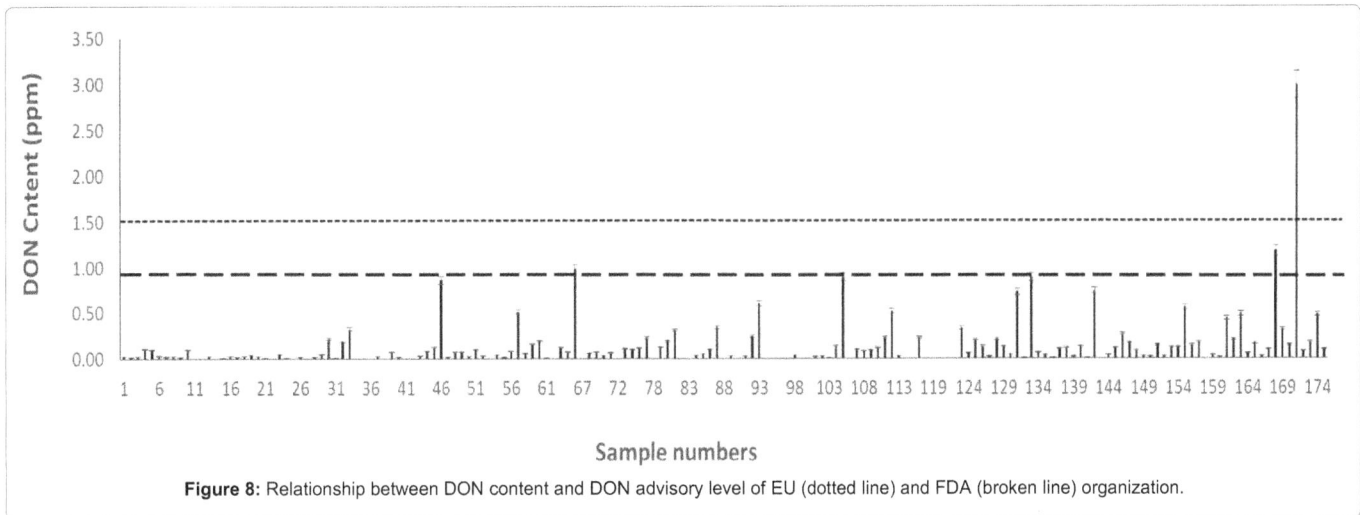

Figure 8: Relationship between DON content and DON advisory level of EU (dotted line) and FDA (broken line) organization.

Severity of *F. graminearum* and *F. culmorum* infection over Bavaria

The whole picture of DON contents in 172 infected wheat samples was highlighted in Figures 3-8 in comparsion with the DON established threshold by FDA organization and European Union. This figure showed that only two samples of all harvested wheat kernel lay above the FDA DON advisory level of 1 ppm, and only one sample is higher than the EU threshold of 1.25 ppm.

Discussion

Identification of trichothecene-producing *F.* spp. by PCR

Fusarium species are considered as a potential trichothecene-producing species [15]. The conserved region of *Tri-5* gene has been detected in *F. culmorum* [30,31], *F. graminearum* [32], *F. poae* [21], *F. sporotrichioides* [33], and *F. sambucinum* [7]. There was a direct relationship between *Tri-5* gene expression and the increase in deoxynivalenol production [34,35]. *Tri-5* primers were designed from highly conserved regions of the *Tri-5* gene of *Fusarium* spp. [17]. A *Tri5*-specific PCR assay has been developed to detect trichothecene-producing *Fusarium* species in contaminated wheat samples [36]. In our study, we used the qualitative *Tri-5* specific PCR assay to detect trichothecene-producing *Fusarium* species in contaminated wheat kernel samples (172 samples) collected from south Germany. 59% (101 samples) of 172 tested samples were positive in the *Tri-5* PCR assay and showed intensive bright DNA bands on agarose gels and were highly infected with one or more of *Fusarium* species containing *Tri-5* gene. Correspondingly, this result was significantly associated with the results of DON content where the same *Tri-5* DNA intensive bright samples had simultaneously the highest amount of DON ranged from 101 to 2990 µg kg^{-1}. This result was in accordance with some reports on *F. avenaceum* isolates that were positive in the *Tri-5* PCR assay and produced DON in culture [37]. In addition, the results of *F. graminearum* and *F. culmorum* DNA quantification by RT-PCR showed that the same *Tri-5* DNA intensive bright samples contained also the highest amount of *F. graminearum* DNA (4945-100.82 µg kg^{-1}) and *F. culmorum* DNA (39.22-10.67 µg kg^{-1}). Thus, according to our results we could say that there was a significant direct relationship between the density of *Tri-5* DNA bands on agarose gels and DON content. The increasing in DON content was also associated with higher concentration of *Fusarium* species, particularly, of *F. graminearum* DNA and slightly with *F. culmorum* DNA quantified by RT-PCR. In other words, we could say

the increased density of *Tri-5* DNA bands with the highest amounts of DON were related to the presence of high amounts of *F. graminearum* DNA rather than *F. culmorum*. For example, the highest amounts of *F. graminearum* DNA in some samples (4945, 4255, 1069, 1050 µg kg^{-1}) were associated with the largest amounts of DON (2990, 1174, 981, 728 µg kg^{-1}) while the concentration of *F. culmorum* for the same samples were very limited (18.81, 0.30, 0.36, 0.00 µg kg^{-1} respectively Figure 6). On the other hand, faint *Tri-5* DNA bands on agarose gels (27% of all tested samples) indicated samples containing a low concentration of *Tri-5* gene and as a result, the infection with *F.*spp. was also low. Accordingly, the amount of DON content in these samples was very low and ranged from 10 to 99 µg kg^{-1}. These results were in accordance with RT-PCR results, where low amounts of *F. graminearum* DNA (0.04-47.15 µg kg^{-1}) and *F. culmorum* DNA (0.07-1.99 µg kg^{-1}) were revealed.

Indeed, *F.*spp-infected wheat kernels that showed negative results in the *Tri-5* PCR assay and showing absence of DNA bands on agarose gels should be *Tri-5* gene free and should not be infected with *F.* spp. containing *Tri-5* gene and ,accordingly, these samples were DON free. In our results, 7% (12 samples) of all samples were negative in the *Tri-5* PCR assay and didn't produce DON and they were approximately free of *F. graminearum* DNA (0.72 µg kg^{-1}) and *F. culmorum* DNA (0.78 µg kg^{-1}). It has been demonstrated that within the same species and in the same cultural conditions toxin production by *Fusarium* strains may vary largely. Some strains produce large amount of trichothecene, whereas others produce small or undetectable amount of trichothecene [38-42]. However, there were other 7% of samples in which no *Tri-5* DNA was detected on agarose gels but an amount of DON ranged from 11 to 226 µg kg^{-1}, low detectable amounts of *F. graminearum* DNA ranged from 0.66 to 1.83 µg kg^{-1} and an amount of *F. culmorum* DNA (0.08-39.22 µg kg^{-1}) were detected. It might be possible that other genes involved in trichothecene biosynthesis have been identified outside the *Tri* biosynthetic gene cluster including *Tri*1 [43] and *Tri*101 [44] which requires more investigation in our samples. However, it is possible that the pathogenic isolates producing DON in very small amounts could produce other phytotxins instead of DON in the pathogenesis [45]. In our study, the same samples that produced *Tri-5* DNA intensive bright bands on agarose gels were containing the highest amount of DON revealed by HPLC analysis and had also the largest amounts of *F. graminearum* and *F. culmorum* DNA evaluated by RT-PCR. We conclude from the displayed results that there was a positive relationship between the three techniques used in this study as they provided us with similar results for the same samples.

Quantification of *F. graminearum* and *F. culmorum*

Molecular diagnostic of plant pathogenic fungi can be highly specific, very sensitive, and relatively fast [46]. We used in this study a fast and reliable method for the species-specific identification and absolute quantification of *F. graminearum* and *F. culmorum*. It is a RT-PCR assay using a TaqMan hybridization probe targeting the beta-tubulin gene for *F. graminearum* and SYBR Green 1 for *F. culmorum*. TaqMan method used in this study because of its sensitivity, selectivity, and reduction of fault signals due to primer-dimer formation [46] and allowed a fast species-specific identification and quantitation of plant infections by *F. graminearum* at very early stages where classical microbiological and toxin analysis methods fail to detect the pathogen [46]. The beta-tubulin gene of all non *F. graminearum* isolates failed to be amplified in the reaction while targeting DNA from all isolates yielded product in the PCR assay [47]. RT-PCR analysis confirmed that *F. graminearum* was more abundant in the infected grains than *F. culmorum* since the concentration of *F. graminearum* DNA ranged from 0.04 to 4945 µg kg^{-1} while *F. culmorum* DNA content ranged from 0.04 to 39.22 µg kg^{-1}. Consequently, *F. graminearum* infections were severe while the severity of *F. culmorum* infection was not high in wheat kernels. Comparison between the results of TaqMan Real-Time PCR analysis for *F. graminearum* and DON content showed that *F. graminearum* is an efficient DON producer where there was high positive significant correlation (R^2=0.7) between DON and *F. graminearum* DNA content. Therefore, *F. graminearum* was considered as the predominant trichothecene associated with FHB and produced the main part of DON in wheat crop. This is in accordance with former investigations [17]. In contrast, the slight weak correlation (R^2=0.03) between DON and *F. culmorum* DNA content may reflect that *F. culmorum* was the second important species in the DON producing *Fusarium* genus. Previous study suggested that *F. culmorum* along with *F. graminearum* were consistently the most pathogenic of the *Fusarium* species infecting cereal ears [12].

Moreover, the PCR analysis showed that in 20% of total infected samples, DON was found (10-320 µg kg^{-1}) where no *F. graminearum* DNA was detected and that was linked with only slight content of *F. culmorum* DNA (0.10- 39.22 µg kg^{-1} Figure 7). PCR analysis indicated that the presence of other *Fusarium* species within the field plots may account for the FHB disease and this result was consistent with the observation of ref. [48]. In these cases, DON content is probably attributable to the possibility that FHB infection in the samples is caused by a complex of *Fusarium* spp. which release DON mycotoxins, and other DON producing *Fusarium* spp. (like *F. pseudograminearum*, *F. poae* and/or *F. sporotrichoides*) might have been presented, that may require further investigation.

The U.S. Food and Drug Administration (FDA) recommend that DON levels in human foods should not exceed 1 ppm. Higher levels of DON are permitted in feed for poultry and ruminant animals. While the European Community supports the setting of European Union (EU) thresholds of trichothecenes as low as reasonably achievable in order to protect public health. For example, DON levels in human foods should not exceed 1.25 ppm. In general, the aggressiveness of *F. graminearum* and *F. culmorum* was relatively low over all Bavaria since the DON content was generally low (10-2990 µg kg^{-1} Figure 6). Indeed, the aggressiveness of *Fusarium* was not so high where only 2% of all harvested wheat kernel laid above the FDA DON advisory level of 1 ppm, and only one sample was higher than the EU threshold

of 1.25 ppm. Some authors reported that trichothecenes may play an important role in the aggressiveness of fungi towards plant host [49].

References

1. Dickson JG, Mains EB (1929) Scab of wheat and barley and its control. USDA Farmers Bulletin 1599. pp: 1-18.

2. Dubin HJ, Gilchrist L, Reeves J, McNab A (1997) Fusarium head scab: Global status and prospects. CIMMYT, Mexico, DF, Mexico, p. 130.

3. McMullen MP, Jones R, Gallenberg D (1997b) Scab of wheat and barley: A re-emerging disease of devastating impact. Plant Dis 81: 1340-1348.

4. Tuite J, Shaner G, Everson RJ (1990) Wheat scab in soft red winter in Indiana in 1986 and its relation to some quality measurements. Plant Dis 74: 959-962.

5. McMullen MP, Enz J, Lukach J, Stover R (1997a) Environmental conditions associated with Fusarium head blight epidemics of wheat and barley in the Northern Great Plains, North America. Cereal Res Commun 25: 777-778.

6. Liu ZZ, Wang ZY (1990) Improved scab resistance in China: Sources of resistance and problems. In: Saunders DA (ed) Wheat for the Non-traditional Warm Areas. Proc Int. Conf., CIMMYT, Mexico, D.F, pp. 178-188.

7. Hart LP, Ward R, Bafus R, Bedford K (1998) Return of an Old Problem: Fusarium Head Blight of Small Grains. Proceedings of the National Fusarium Head Blight Forum. Michigan State Univ., E. LansingThe American Phytopathological Society.

8. Hanson EW, Ausemus ER, Stakman EC (1950) Varietal resistance of spring wheats to fusarial head blight. Phytopathology 40: 902-914.

9. Mathre DE (1997) Compendium of barley diseases. (2nded) The Am Phytopathological Soc Press, St. Paul, MN.

10. Bechtel DB, Kaleikau LA, Gaines RL, Seitz LM (1985) The effects of Fusarium graminearum infection on wheat kernels. Cereal Chem 62: 191-197.

11. Wood M, Comis D, Harden D, McGraw L, Stelljes KB (1999) Fighting Fusarium. Agricultural Research. June issue. USDA-ARS, Beltsville, MD.

12. Parry DW, Jenkinson P, McLeod L (1995) Fusarium ear blight (scab) in small grains -a review. Plant Pathol 44: 207-238.

13. Muriuki JG (2001) Deoxynivalenol and nivalenol in pathogenesis of Fusarium head blight in wheat. Thesis, University of Minnesota.

14. Stack RW (2003) History of Fusarium head blight with emphasis on North America. In: Leonard KJ, Bushnell WR (ed) Fusarium head blight of wheat and barley. APS Press, St. Paul, MN, pp: 1-34.

15. Edwards SG, Pirgozliev SR, Hare MC, Jenkinson P (2001) Quantification of trichothecene-producing Fusarium species in harvested grain by competitive PCR to determine efficacies of fungicides against Fusarium head blight of winter wheat. American Society for Microbiology 67: 1575-1580.

16. Schwarz PB, Casper HH, Barr JM (1995) Survey of the occurrence of deoxynivalenol (vomitoxin) in barley grown in Minnesota North Dakota and South Dakota during 1993. MBAA Tech Q 32: 190- 194.

17. McMullen MP, Schatz B, Stover R, Gregoire T (1997c) Studies of fungicide efficacy, application timing, and application technologies to reduce Fusarum head blight and deoxynivalenol. Cereal Res. Commun. 25: 779-780.

18. Marasas WFO, Nelson PE, Toussoun TA (1984) Toxigenic Fusarium species: identify and mycotoxicology. Pennsylvania State University Press, University Park. Pa.

19. Bottalico A, Perrone G (2002) Toxigenic Fusarium species and mycotoxins associated with head blight in small grain cereals in Europe. European Journal of Plant Pathology 108: 611-624.

20. Boyacioglu D, Hettiarachchy NS, Stack RW (1992) Effect of three systemic fungicides on deoxynivalenol (vomitoxin) production by Fusarium graminearum in wheat. Can J Plant Sci 72: 93-101.

21. van Eeuwijk FA, Mesterhazy A, Kling CI, Ruckenbauer P, Saur L, et al. (1995) Assessing non-specificity of resistance in wheat to head blight caused by inoculation with European strains of Fusarium culmorum, F. graminearum and F. nivale using a multiplicative model for interaction. Theor Appl Genet 90: 221-228.

22. Bennett JW, Klich M (2003) Mycotoxins. Clin Microbiol Rev 16: 497-516.

23. Murphy M, Armstrong D (1995) Fusariosis in patients with neoplastic disease. Infect Med 12: 66-67.

24. Berek L, Petri IB, Mesterházy A, Téren J, Molnár J (2001) Effects of mycotoxins on human immune functions in vitro. Toxicol In Vitro 15: 25-30.

25. Lindsay JA (1997) Chronic sequelae of foodborne disease. Emerg Infect Dis 3: 443-452.

26. Bai G, Kolb FL, Shaner G, Domier LL (1999) Amplified fragment length polymorphism markers linked to a major quantitative trait locus controlling scab resistance in wheat. Phytopathology 89: 343-348.

27. Niessen ML, Vogel RF (1998) Group specific PCR-detection of potential trichothecene-producing Fusarium-species in pure cultures and cereal samples. Syst Appl Microbiol 21: 618-631.

28. Fekete C, Logrieco A, Giczey G, Hornok L (1997) Screening of fungi for the presence of the trichodiene synthase encoding sequence by hybridization to the Tri5 gene cloned from Fusarium poae. Mycopathologia 138: 91-97.

29. Hohn TM, Desjardins AE (1992) Isolation and gene disruption of the Tox5 gene encoding trichodiene synthase in Gibberella pulicaris. Mol Plant Microbe Interact 5: 249-256.

30. Bauer A, Seigner L, Battner P, Tischner H (2004) Monitoring of FHB using PCR for qualitative and quantitative detection of Fusarium spp. Proceedings of the 2nd International Symposium on Fusarium Head Blight 2: 553.

31. Snijders CHA, Krechting CF (1992) Inhibition of deoxynivalenol translocation and fungal colonization in Fusarium head blight resistant wheat. Can J Bot 70: 1570-1576.

32. Moschini RC, Fortugno C (1996) Predicting wheat head blight incidence using models based in meteorological factors in Pergamino, Argentina. Eur J Plant Pathol 102: 211-218.

33. Hart LP, Pestka JJ, Liu MT (1984) Effect of kernel development and wet periods on production of deoxynivalenol in wheat infected with Gibberella zeae. Phytopathology 74: 1415-1418.

34. Fernando WG, Paulitz TC, Seaman WL, Dutilleul P, Miller JD (1997) Head Blight Gradients Caused by Gibberella zeae from Area Sources of Inoculum in Wheat Field Plots. Phytopathology 87: 414-421.

35. Doohan FM, Weston G, Rezanoor HN, Parry DW, Nicholson P (1999) Development and use of a reverse transcription-PCR assay to study expression of Tri5 by Fusarium species in vitro and in planta. Appl Environ Microbiol 65: 3850-3854.

36. Neissen ML, Vogel RF (1997) A molecular approach to the detection of potential trichothecene producing fungi. In: Mesterhazy A (ed) Cereals research communication. Proceeding of the Fifth European Fusarium Seminer, Szeged, Hungary-1997.Cereals Reasearch Institute, Szeged, Hungary. pp. 245-249.

37. Abramson D, Clear RM, Smith DM (1993) Trichothecene production by Fusarium spp. isolated from Manitoba grain. Can J Plant Pathol 15: 147-152.

38. Mesterhazy A (1995) Types and components of resistance to Fusarium head blight of wheat. Plant Breeding 114: 377-386.

39. Mesterhzy A (1997) Breeding for resistance to Fusarium head blight of wheat. Proceedings of the 5th European Fusarium Seminar. Cereal Res Commun 25: 231-866.

40. Schroeder HW, Christensen JJ (1963) Factors affecting resistance of wheat to scab caused by Gibberella zeae. Phytopathology 53: 831-838.

41. Snijders CHA, Perkowski J (1990) Effects of head blight caused by Fusarium culmorum on toxin content and weight of wheat kernels. Phytopathology 80: 566-570.

42. Walker S, Leath S, Hagler W, Murphy J (2001) Variation among isolates of Fusarium graminearum associated with Fusarium head blight in North Carolina. Plant Dis 85: 404-410.

43. McCormick SP, Harris LJ, Alexander NJ, Ouellet T, Saparno A, et al. (2004) Tri1 in Fusarium graminearum encodes a P450 oxygenase. Appl Environ Microbiol 70: 2044-2051.

44. Kimura M, Kaneko I, Komiyama M, Takatsuki A, Koshino H, et al. (1998) Trichothecene 3-O-acetyltransferase protects both the producing organism and transformed yeast from related mycotoxins. Cloning and characterization of Tri101. J Biol Chem 273: 1654-1661.

45. Hestbjerg H, Felding G, Elmholt S (2002) Fusarium culmorum infection of barley seedling: correlation between aggressiveness and deoxynivalenol content. Journal of Phytopathology 4: 308-312.

46. McCartney HA, Foster SJ, Fraaije BA, Ward E (2003) Molecular diagnostics for fungal plant pathogens. Pest Manag Sci 59: 129-142.

47. Reischer GH, Lemmens M, Farnleitner A, Adler A, Mach RL (2004) Quantification of Fusarium graminearum in infected wheat by species specific real-time PCR applying a TaqMan Probe. J Microbiol Methods 59: 141-146.

48. Doohan M F, Parry DW, Jenkinson P, Nicholson P (1998) The use of species-specific PCR based assays to analyse Fusarium ear blight of wheat 47: 19-205.

49. Proctor RH, Hohn TM, McCormick SP (1995) Reduced virulence of Gibberella zeae caused by disruption of a trichothecene toxin biosynthetic gene. Mol Plant Microbe Interact 8: 593-601.

17

Isolation and Sequencing of *Actin1*, *Actin2* and *Tubulin1* Genes Involved in Cytoskeleton Formation in *Phytophthora cinnamomi*

Ivone M Martins[1,2], M Carmen López[3], Angél Dominguez[3] and Altino Choupina[1,2]*

[1]*Department of Biology and Biotechnology, Polytechnic Institute of Bragança, Apartado 1172, 5301-854 Bragança, Portugal*
[2]*Mountain Research Center, Polytechnic Institute of Bragança, Apartado 1172, 5301-854 Bragança, Portugal*
[3]*Department of Microbiology and Genetics, CIETUS-IBSAL, University of Salamanca / CSIC, Plaza de Drs. Queen s/n, 37007 Salamanca, Spain*

Abstract

Oomycetes from the genus *Phytophthora* are fungus-like plant pathogens that are devastating for agriculture and natural ecosystems. On the Nordeste Transmontano region (northeast Portugal), the *Castanea sativa* chestnut culture is extremely important. The biggest productivity and yield break occurs due to the ink disease, caused by *Phytophthora cinnamomi* which is one of the most widely distributed *Phytophthora* species, with nearly 1000 host species. The knowledge about molecular mechanisms responsible for pathogenicity is an important tool in order to combat associate diseases of this pathogen. Complete open reading frames (ORFs) of *act1*, *act2* and *tub1* genes who participate in cytoskeleton formation in *P. cinnamomi* were achieved by high-efficiency thermal asymmetric interlaced (HE-TAIL) polymerase chain reaction (PCR). *act1* gene comprises a 1128 bp ORF, encoding a deduced protein of 375 amino acids (aa) and 41,972 kDa. *act2* ORF comprises 1083 bp and encodes a deduced protein of 360 aa and 40,237 kDa. *tub1* has a total length of 2263 bp and encodes a 453 aa protein with a molecular weight of 49.911 kDa. Bioinformatics analyses shows that actin1 is ortholog to the *act1* genes of *Phytophthora infestans*, *Phytophthora megasperma* and *Phytophthora melonis*; actin2 is ortholog to the *act2* genes of *P. infestans*, *Phytophthora brassicae*, *P. melonis* and *Pythium splendens* and tubulin1 shows the highest orthology to *P. infestans* and *P. capsici* α-tubulin genes.

Analysed 3D structure of the three putative proteins revealed a spatial conformation highly similar to those described for orthologous proteins obtained by X-ray diffraction.

Keywords: Actin; *Castanea sativa*; Cytoskeleton; Ink disease; Phytophthora cinnamomi; Tubulin

Introduction

Phytophthora cinnamomi is a destructive and widespread soil-borne oomycete that infects woody plant hosts [1]. On the Nordeste Transmontano region (northeast Portugal), this pathogen is the responsible by the ink disease affecting Castanea sativa chestnut. The most common symptoms are root necrosis and reduction in root growth, which invariably lead to tree death [2]. Due to their particular physiological characteristics, no efficient treatments against diseases caused by these microorganisms are presently available [3]. In order to develop such treatments appeared essential to dissect the molecular mechanisms of the interaction between *Phytophthora* species and host plants.

Actin and tubulin are highly abundant conserved proteins in eukaryotic cells, which participate in more protein-protein interactions than any other proteins [4,5]. These properties, together with the actin capacity to carry out the transition between monomeric (G-actin) and filamentous (F-actin) states under the control of nucleotide hydrolysis, ions, and a large number of actin-binding proteins, make actin a critical player in many cellular functions, ranging from cell motility and the maintenance of cell shape and polarity to the regulation of transcription [6,7]. In vertebrates there are three groups of actin isoforms: alpha, beta and gamma. The alpha actins are found in muscle tissues and are a major constituent of the contractile apparatus. The beta and gamma actins co-exist in most cell types as components of the cytoskeleton and act as mediators of internal cell motility. In plants there are many isoforms which are probably involved in a variety of functions such as cytoplasmic streaming, cell shape determination, tip growth, graviperception, cell wall deposition, etc.

A moderate-sized protein consisting of approximately 375 residues, actin is encoded by a large, highly conserved, gene family. Some single-celled eukaryotes like yeasts and amoebae have a single actin gene, whereas many multicellular organisms contain multiple actin genes. Actin in *Phytophthora infestans* is encoded by at least two genes, *actA* and *actB*, in contrast to unicellular and filamentous fungi *Saccharomyces cerevisiae*, *Kluyveromyces lactis* and *Yarrowia lipolytica*, where a single gene have been detected [8,9]. On the other hand microtubules are major constituents of the cell cytoskeleton participating in a wide range of cellular functions, such as motility, division, maintenance of cell shape, and intracellular transport. Also play an essential role in nuclear division as components of the mitotic spindle and dimeric tubulin is their primary component. A key property of tubulin is its ability to assemble into microtubules via interaction between polymerized α- and β-tubulin monomers (heterodimer), and to undergo disassembly at appropriate times in the cell cycle. However, microtubule role is variable depending on the organism, cell type and other factors.

*Corresponding author: Altino Choupina, Department of Biology and Biotechnology, Polytechnic Institute of Bragança, Apartado 1172, 5301-854 Bragança, Portugal, E-mail: albracho@ipb.pt

The aim of this study was to isolate and sequence three genes involved in cytosqueleton formation in *P. cinnamomi*.

Materials and Methods

Biological material

P. cinnamomi isolate Pr120 was isolated from soil samples from sites of *C. sativa* affected by the ink disease on the Nordeste Transmontano region (northeast Portugal). The strain was grown in the dark for 4-6 days at 22-25°C in PDA (Potato-Dextrose Agar) medium. This isolate is preserved at the Laboratory of Molecular Biology of the Polytechnic Institute of Bragança (IPB).

Total genomic DNA from *P. cinnamomi* mycelium was isolated according to the reported methods [10,11].

Fluorescence microscopy

For nuclei staining, *P. cinnamomi* mycelium was grown in PDA medium and resuspended in PBS 10X containing 1 mg/ml of Dapi (4-6-Diamidino-2-phenylindole) (Sigma-Aldrich), and visualized with the appropriate UV filter. For cell wall staining, *P. cinnamomi* mycelium was grown in PDA medium and resuspended in PBS 10X containing 10 mg/ml of Calcofluor white (Sigma-Aldrich), using the appropriate UV filter. For actin visualization, *P. cinnamomi* mycelium was grown in PDA medium and stained with 0.1 μg/ml rhodamine-conjugated phalloidin (Invitrogen) [12]. Images were obtained using a Leica HC (Germany) fluorescence microscope, type 020-523.010.

Amplification of the actin and tubulin genes

PCR was used to amplify *act1*, *act2* and *tub1* genes from *P. cinnamomi*. Degenerated primers Act1 (5'- GYMATGGASGAC-GAYATTCARGC-3') and Act2 (5'-GYMGYCTTAGAAGCACTT-GCGRTG) to amplify *act1*, Act3 (5'-CAWTCAAGATGGCTGAC-GAWGAYG) and Act4 (5'-CARCTTAGAAGCACT TGCGGTGC) to amplify *act2*, were designed based on actin A and B sequences alignment from *P. infestans*. Degenerated primers Tub1 (5'-GGYAATGC-STGTTGGGAAYTMTAT) and Tub2 (5'-CATMCCYTCWCCSAC RTACCAGTG) to amplify *tub1*, were designed based on α-tubulin sequences alignment from *S. cerevisiae* and *P. palmivora*. The thermal program used for PCR reactions consisted of one cycle of 94°C for 5 min, 30 cycles of 94°C for 30 sec; 57°C for 30 sec; 72°C for 1.5 min and 1 cycle of 72°C for 7 min. Each 25 μl PCR contained 0.8 mM dNTPs, 0.2 mM of each primer, 100 ng genomic DNA, and 2.5 U Taq DNA polymerase in the appropriate buffer. Aliquots of the PCR reactions were separated on 0.8% w/v agarose gels and stained with ethidium bromide, to check for the presence of the expected amplicon. PCR fragments were purified with "DNA and Gel Band Purification" kit (GE Healthcare), following the manufacturer's instructions. Amplified DNA was cloned into pGEM-T Easy (Promega) which is a linearized high-copy-number vector with a single 3'-terminal thymidine at both ends.

Bacterial transformation and DNA extraction

Plasmids were propagated in *Escherichia coli* (DH5α cells, [13]) and plasmid DNA was extracted-purified with the Wizard® Plus SV Minipreps DNA Purification System (Promega), following the manufacturer's instructions. Both DNA fragments obtained were sequenced in a capillar automatic sequencer 3100 Genetic Analyzer (Applied Biosystem).

Amplify unknown genomic DNA sequence of *act1*, *act2* and *tub1* genes

HE-TAIL PCR is an efficient method to amplify unknown genomic DNA sequences adjacent to short known regions by flanking the known sequence with asymmetric PCR. In this procedure gene specific primers, Act1.1 (5'-GCCGTTYTCCTTGATCAGCGG), Act1.2 (5'-AGGCGTTGTCGCCCCAGACC), Act2.1 (5'-CGGCCGCGGT-GACGCTGACG) and Act2.2 (5'-GGTCTGGGGCGACAACGCCT) were used to amplify unknown genomic DNA sequences of *act1* and *act2* genes. Gene specific primers Tub1.1 (5'-GCGTTGAACACCAG-GAAACCCTG) and Tub1.2 (5'-CGAGATCACCAACAGCGCCTTC-GA) were used to amplify unknown genomic DNA sequence of *tub1* gene. Degenerated primers R1 (5'-NGTCGASWGANAWGAA), R2 (5'-GTNCGASWCANAWGTT), R3 (5'-WGTGNAGWANCANA-GA) and R4 (5'-NCAGCT WSCTNTSCTT) were applied [14]. Three rounds of PCR were performed using the product of the previous PCR as a template for the next (Table 1).

The primary PCR was performed in a 50 μl volume containing 80 ng of genomic DNA, 0.2 mM of primers M1 or M3 2 mM of a random primer (R1, R2, R3, R4), 0.2 mM of each dNTP and 1U Taq DNA polymerase in the appropriate buffer. The secondary PCR was performed with primers M2 or S4 (0.2 mM) and the same random primer R (2 mM) as used in the primary reaction. 1 μl of 1/50 dilution of the primary PCR was used as a template. Single-step annealing-extension PCR consisting of a combined annealing and extension step at 65°C or 68°C was used in primary and secondary PCR reactions. The tertiary reaction was carried out with 1 μl of 1/10 dilution of the secondary reaction, 0.2 mM of primers S1 and S2, 0.2 mM of random primer R (the same as used in the previous cycles), 0.2 mM of each dNTP, 1U DNA Taq polymerase in the appropriate buffer. To exclude nonspecific amplification, a tertiary control reaction R-R was set up without adding gene-specific primers.

Amplification of promoter and terminator sequences of *act1*, *act2* and *tub1* genes

In this procedure gene-specific primers were designed and, in combination with those, the degenerated primer R2 was applied [14]. Three rounds of PCR were performed using the product of the previous PCR as a template for the next. A detailed cycler program is given in Table 1.

Reaction	Number of cycles	Thermal settings
Primary	1	93°C (1 min); 95°C (5 min)
	5	94°C (30 sec); 62°C (1 min); 72°C (2 min 30 sec)
	1	94°C (30 sec); 25°C ramping 72°C (3 min); 72°C (2 min 30 sec)
	15	94°C (20 sec); 65°C (3 min 30 sec);
		94°C (20 sec); 65°C (3 min 30 sec);
		94°C (30 sec); 42°C (1min); 72°C (2 min 30 sec)
	1	72°C (5 min); 4°C Hold
Secondary	12	94°C (20 sec); 65°C (3 min 30 sec);
		94°C (20 sec); 65°C (3 min30sec);
		94°C (30 sec); 42°C (1 min); 72°C (2 min 30 sec)
	1	72°C (5 min); 4°C Hold
Tertiary	30	94°C (30 sec); 42°C (1 min); 72°C (2 min 30 sec)
	1	72°C (5 min); 4°C Hold

Table 1: HE-TAIL PCR cycle settings. Primary, secondary and tertiary nested PCR reactions were performed sequentially. The primary PCR reaction consists of 15 TAIL cycles, while the secondary reaction contains 12 TAIL cycles.

The primary PCR was performed in a 50 µl volume containing 80 ng of genomic DNA, 0.1 mM of gene-specific primers, 2 mM of primer R2, 0.2 mM of each dNTP and 1U Taq DNA polymerase in the appropriate buffer. The secondary PCR was performed with gene-specific primers (0.2 mM) and the same primer R (2 mM) as used in the primary reaction. 1 µl of 1/50 dilution of the primary PCR was used as a template. Single-step annealing-extension PCR consisting of a combined annealing and extension step at 65°C or 68°C was used in primary and secondary PCR reactions. The tertiary reaction was carried out with 1 µl of 1/10 dilution of the secondary reaction, 0.2 mM of gene-specific primers, 0.2 mM of primer R2, 0.2 mM of each dNTP, 1U DNA Taq polymerase in the appropriate buffer. To exclude nonspecific amplification, a tertiary control reaction R2-R2 was set up without adding gene-specific primers.

Protein structure analysis

The 3D structure of the proteins was achieved by using the PyMOL 1.3 r1 edu program and Swiss Model [15-17].

Results and Discussion

Microscopy visualization

In order to visualize the morphology of *P. cinnamomi* a staining and microscopic observation of three cell structures was performed (Figure 1): nucleus, through DAPI staining (Figure 1A). Each mycelium presents more than one individualized nucleous. Cell wall was observed by staining with calcofluor white (Figure 1B) who indicated the lack division septa and actin cytoskeleton and according to [1]. Actin citoskeleton was visualized with rhodamin-phalloidin, although actin distribution can be observed throughout all the mycelium, appeared more concentrated at the hyphae extremities and adjacent to the plasma membrane (Figure 1C), according to [7]. In oomycetes actin forms a cap immediately after the apical plasma membrane, whereas in other organisms like *P. cinnamomi*, the actin appears in small plates in the tips of hyphae [7].

Hypha of *P. cinnamomi* also shows strong punctate staining in the sub-apical region, and strong staining bands along the walls of the hypha. Some evidences suggest that microfilaments reinforce the hypha tip of oomycetes. Disruption of the F-actin cap and reduction of F-actin at the tips of oomycetes hypha could, along with cell wall softening, increase tip yielding and thus playing a potential role in invasive hypha growth [7].

act1, *act2* and *tub1* amplification

Performing a PCR using genomic DNA from *P. cinnamomi* as template and degenerated primers designed based on actin A and B sequences alignment from *P. infestans*, and the α-tubulin sequences alignment from *S. cerevisiae* and *P. palmivora*, we were able to amplify three fragments of about 1200 bp.

After sequencing and bioinformatics analysis, we found that our sequences presented identity with several sequences of the database with the characteristic actin and tubulin domains. We named our sequences Actin1, Actin2 and Tubulin1, EMBL database, accession numbers AM412175.1, AM412176.1 and AM412177.1, respectively.

After performing a HE-TAIL PCR, the fragments of interest were selected and sequenced. By using BioAlign and SeqMan programs we were able to align and analyse the sequences. Through this method we could amplify both promoter (+113 bp) and terminator (+640 bp) of the *act1* gene. We were not able to amplify the promoter and terminator of the *act2* gene because the sequences that we isolated and should flank this gene did not align with the original fragment. For the *tub1* gene we could amplify both promoter (+352 bp) and terminator (+549 bp) regions.

A BLAST analysis (blast.ncbi.nlm.nih.gov) of both genes was performed against *Phytophthora*, *Saccharomyces cerevisiae*, *Kluyveromyces lactis* and *Aspergillus nidulans*. act1 and act2 of *P. cinnamomi* present an identity of 88%. act1 present an identity of 100% against *P. infestans*, *P. megasperma* and *P. melonis* and act2 is ortholog to the *act2* genes of *P. infestans*, *P. brassicae*, *P. melonis* and *Pythium splendens* (100% identity). act2 present higher identity to the ascomycetous genes 91%, rather than 78% of *act1*.

tub1 show 100% identity to *P. infestans* and *P. sojae* and 70% to *Y. lipolytica*, *K. lactis* and *S. cerevisiae*.

3D Structure of Actin1, Actin2 and Tubulin1

After obtaining the deduced amino acid sequence of the *act1*, *act2* and *tub1* genes we have determined the 3D structure of the proteins using the InterProScan program.

act1 (Figure 2A) presents 82.3% of identity with the 3D structure of the reference protein 3ci5A and *act2* (Figure 2B) presents 86.9% of identity with the 3D structure of the reference protein 3eksA (described in the data bases). *tub1* encodes for a α-tubulin protein and presents

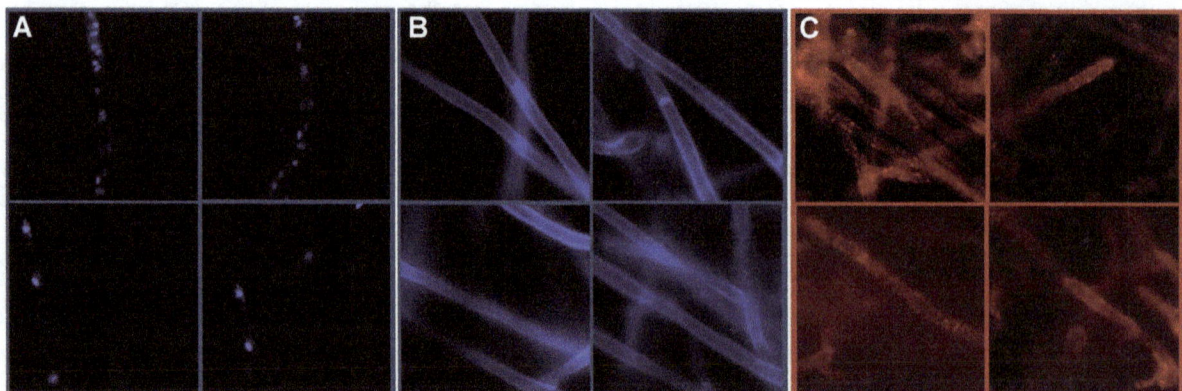

Figure 1: Fluorescence microscopy of stained *P. cinnamomi* mycelium at different locations. **A)** Each mycelium presents more than one individualized nucleous nuclei when stained with DAPI (1 mg/ml); **B)** A lack of division septa and actin cytoskeleton is observed in a cell wall stained with Calcofluor white (50 mg/ml); **C)** Actin distribution appears more concentrated at the hyphae extremities and adjacent to the plasma membrane when actin cytoskeleton was stained with Rhodamin-Phalloidin (0.1 µg/ml).

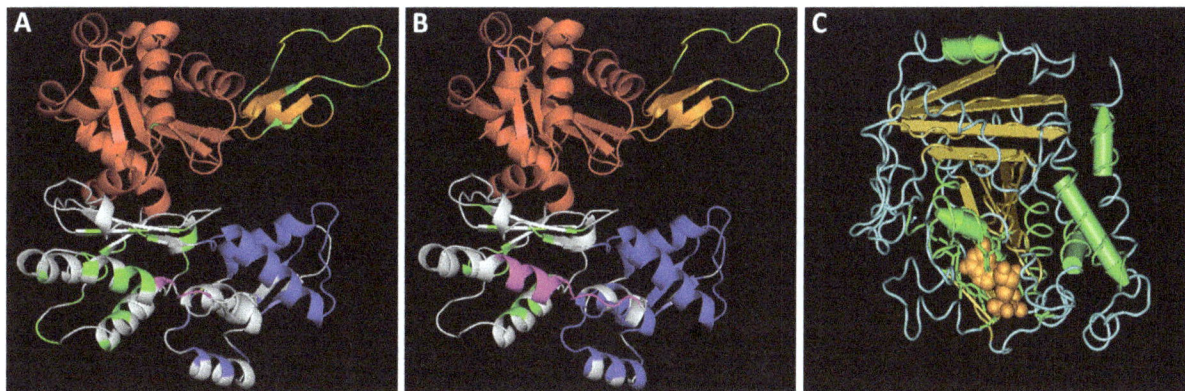

Figure 2: 3D structures of Actin1 Actin2 and Tubulin1. A) 3D structure of the protein encoded by *act1*, subdomain-1 is represented in red, subdomain-2 in orange, subdomain-3 in white subdomain-4 in blue; **B)** 3D structure of the protein encoded by *act2*, subdomain-1 is represented in red, subdomain-2 in orange, subdomain-3 in white subdomain-4 in blue. At subdomain-2 is represented the DNase I chain in yellow (as a loop); **C)** 3D structure of the protein α-tubulin encoded by the gene *tub1*, in yellow we can observe the GTP subdomain.

76.3% of identity with the 3D structure of the reference protein 1Z2B of the data base. The high identity percentage with the reference protein sequences on the database, allowed us to determine the 3D structure of the proteins encoded by *tub1* (Figure 2C).

Looking at the 3D structure we can observe the actin proteins subdomains, represented in different colors (Figures 2A and B). Subdomain-1 is represented in red, subdomain-2 in orange, subdomain-3 in white subdomain-4 in blue. At the subdomain-2 is represented the DNase I chain in yellow (as a loop). Actin from many sources forms a tight complex with deoxyribonuclease (DNase I). The formation of this complex results in the inhibition of DNase I activity, and actin loses its ability to polymerize [18]. Also, subdomain-2 plays an important role in the movement of actin filaments *in vivo* [19]. In vertebrates, the four α-actin isoforms present in various muscle cells and the β- and γ-actin isoforms present in no muscle cells differ at only four or five positions. Although these differences among isoforms seem minor, the isoforms have different functions: α-actin is associated with contractile structures, and β-actin is at the front of the cell where actin filaments polymerize [20]. Actin exists as a monomer in low salt concentrations, but filaments form rapidly as salt concentration rises, with the consequent hydrolysis of ATP. Each actin protomer binds one molecule of ATP and has one high affinity site for either calcium or magnesium ions, as well as several low affinity sites. It has been shown that an ATPase domain of actin shares similarity with ATPase domains of hexokinase and hsp70 proteins.

Regarding the 3D structure of the observe the α-tubulin protein, we can observe the α-tubulin protein encoded by the gene *tub1* as well as the GTP domain (aa residues D82, T129, V161, S162, E167, N190, Y208). At microtubules the α-tubulin subunit contains GTP while the β-tubulin subunit contains ATP. Microtubules assemble by polymerization of α-β dimers of tubulin. Polymerization is a polar process that reflects the polarity of the tubulin dimer, which in turn dictates the polarity of the microtubule [21]. To form microtubules, the dimers of α- and β-tubulin bind to GTP and assemble onto the (+) ends of microtubules [5]. The energy to drive the microtubule machine comes from GTP hydrolysis. Tubulin is a GTPase whose activity is stimulated by polymerization. A crucial observation is that tubulin polymerizes in the presence of non-hydrolysable GTP to form stable microtubules. Thus, polymerization is driven by the high affinity of the tubulin–GTP dimer for the end of the microtubule [22].

Because of the extremely high diversity and adaptability of the oomycetes lifestyle it is imperative to dissect the multiple roles of the cytoskeleton and is functional interplay in a range of as many as possible species. Over the next few years more and more structural and regulatory components of the oomycete cytoskeleton will be localized and their dynamic interactions that ultimately generate function and determine cell form will be studied *in vivo*.

Acknowledgments

This study was supported by the project COMBATINTA/SP2.P11/02 Interreg IIIA–Cross-Border Cooperation Spain-Portugal, financed by The European Regional Development Fund.

References

1. Erwin DC, Ribeiro OK (1996) *Phytophthora* diseases worldwide, American Phytopathological Society Press, St. Paul, Minnesota, USA.

2. Abreu CG (1996) Doença da tinta: causas e consequências do declínio do castanhal. Estudos Transmontanos 6: 269-289.

3. Hardham AR (2005) Phytophthora cinnamomi. Mol Plant Pathol 6: 589-604.

4. Dominguez R, Holmes KC (2011) Actin Structure And Function. Annu Rev Biophys 40: 169-186.

5. Heald R, Nogales E (2002) Microtubule dynamics. J Cell Sci 115: 3-4.

6. Kabsch W, Holmes KC (1995) The actin fold. Faseb J 9: 167-174.

7. Walker SK, Chitcholtan K, Yu Y, Christenhusz GM, Garrill A (2006) Invasive hyphal growth: an F-actin depleted zone is associated with invasive hyphae of the oomycetes Achlya bisexualis and *Phytophthora cinnamomi*. Fungal Genet Biol 43: 357-365.

8. Ng R, Abelson J (1980) Isolation and sequence of the gene for actin in *Saccharomyces cerevisiae*. Proc Natl Acad Sci U S A 77: 3912-3916.

9. Unkles SE, Moon RP, Hawkins AR, Duncan JM, Kinghorn JR (1991) Actin in the oomycetous fungus *Phytophthora infestans* is the product of several genes. Gene 100: 105-112.

10. Raeder U, Broda P (1985) Rapid preparation of DNA from filamentous fungi. Letters in Applied Microbiology 1: 17-20.

11. Specht CA, DiRusso CC, Novotny CP, Ullrich RC (1982) A method for extracting high-molecular-weight deoxyribonucleic acid from fungi. Anal Biochem 119: 158-163.

12. Alfa C, Fantes P, Hyams J, McLeod M, Warbrick E (1993) Experiments with fission yeast: a laboratory course manual. Cold Spring Harbor Laboratory Press, Cold Spring Harbor, New York, USA.

13. Hanahan D (1983) Studies on transformation of *Escherichia coli* with plasmids. J Mol Biol 166: 557-580.

14. Michiels A, Tucker M, Van Den Ende W, Van Laere A (2003) Chromosomal walking of flanking regions from short known sequences in GC-rich plant genomic DNA. Plant Molecular Biology Reporter 21: 295-302.

15. Arnold K, Bordoli L, Kopp J, Schwede T (2006) The SWISS-MODEL workspace: a web-based environment for protein structure homology modelling. Bioinformatics 22: 195-201.

16. Guex N, Peitsch MC (1997) SWISS-MODEL and the Swiss-PdbViewer: an environment for comparative protein modeling. Electrophoresis 18: 2714-2723.

17. Schwede T, Kopp J, Guex N, Peitsch MC (2003) SWISS-MODEL: An automated protein homology-modeling server. Nucleic Acids Res 31: 3381-3385.

18. Lazarides E, Lindberg U (1974) Actin is the naturally occurring inhibitor of deoxyribonuclease I. Proc Natl Acad Sci U S A 71: 4742-4746.

19. Schwyter DH, Kron SJ, Toyoshima YY, Spudich JA, Reisler E (1990) Subtilisin cleavage of actin inhibits in vitro sliding movement of actin filaments over myosin. J Cell Biol 111: 465-470.

20. Lodish H, Berk A, Zipursky L, Matsudaira P, Baltimore D, et al. (2000) Molecular Cell Biology. (4thedn), WH Freeman and Company, New York, USA.

21. Howard J, Hyman AA (2003) Dynamics and mechanics of the microtubule plus end. Nature 422: 753-758.

22. Howard J, Hyman AA (2009) Growth, fluctuation and switching at microtubule plus ends. Nat Rev Mol Cell Biol 10: 569-574.

18

Stimulating of Biodegradation of Oxamyl Pesticide by Low Dose Gamma Irradiated Fungi

Abd El-Moneim MR Afify[1]*, Mohamed A Abo-El-Seoud[2], Ghada M Ibrahim[1] and Bassam W Kassem[2]

[1]*Department of Biochemistry, Faculty of Agriculture, Cairo University, Giza, Egypt*
[2]*Department of Plant Research, Nuclear Research Center, Egyptian Atomic Energy Authority, Abu-Zabal 13759, Egypt*

Abstract

This investigation has been conducted to study the possibility of stimulating *Trichoderma spp* with low dose gamma radiation for biodegradation of Oxamyl pesticides. Fungi strains capable for biodegradation of oxamyl are identified as *Trichoderma spp.*, including *T. harzianum, T. viride, Aspergillus niger, Fusarium oxysporum* and *Penicillium cyclopium*. The results indicated that *Trichoderma* spp. used Oxamyl as source of carbon and nitrogen and possesses enzyme(s), which acts on amide and ester bond in Oxamyl structure. Degradation of oxamyl was 72.5% within 10 days of incubation by *T. harzianum* strain. It is very important to note that degradation of oxamyl 82.05% within 10 days of incubation by *T. viride* strain. This indicated that the isolates of *Trichoderma* spp. were potentially useful for oxamyl bioremediation. The biomass of *Trichoderma spp* strain were increased and reached its maximum at 250 Gy by 21.97 and 40.0 when using *Trichoderma spp.*, as well as *T. viride*, respectively. As a general trends the gamma radiation over than 0.25 KGr reduce the growth of *Trichoderma spp* by 50.27 and 38.13, using *Trichoderma spp.* as well as *T. viride*, respectively.

Keywords: Oxamyl; *Trichoderma harzianum; Trichoderma viride*; Biodegradation; Gamma-irradiation

Introduction

Contamination of surface water by organophosphate and carbamate compounds is of concern because of the potential toxicity to aquatic organisms, especially those at lower trophic levels. Many organophosphate and carbamate compounds have acute and chronic toxicity to fish and aquatic invertebrates. *Trichoderma spp* strain improvement usually by exposing the microorganism to radiation that produces the enzyme by techniques, such as classical mutagenesis, which involves γ-rays, UV-rays [1].

Pesticides must persist long enough to control biological targets, but should not become a pollution problem [2,3]. Environmental of Oral and dermal exposure of rats to Pesticide cyanophos was characterized by studying acetylcholine esterase, aspartate transaminase, alanine transaminase and alkaline phosphatase (ALP) enzyme activities were studied by Afify and El-Beltagi [4], as biomarker for pesticides pollution. The toxicity of some pesticides, such as organophosphates and carbamate insecticides, is mainly caused by the inhibition of ChE activity of vertebrates and invertebrates. This inhibition leads to the accumulation of acetylcholine in the synaptic terminals, and therefore, to a change in the normal transmission of the nervous impulse. This interference may result in neurological manifestations, such as irritability, restlessness, muscular twitching and convulsions that may end in the respiratory failure and death of the animal [5]. Oxamyl and carbofuran, a broad-spectrum carbamate pesticide, has been used extensively in agriculture as a soil-incorporated to control a variety of insect pests of crops, including canola, corn, alfalfa, potatoes and strawberries [6]. Studies on microbial degradation are useful in the development of bioremediation strategies for the detoxification of these insecticides by microorganisms. Bioremediation is defined as the process, whereby organic wastes are biologically degraded under controlled conditions to an innocuous state, or to levels below concentration limits established by regulatory authorities [7].

A number of isolates capable of carrying out some form of degradation of pesticide have been isolated from soils and several fungi are have been isolated and studied, including *Aspergillus niger* [8],

Fusarium graminearum [9], *Mucor ramannianus* [10] and *Gliocladium* sp. [11]. Recently, a strain of *Trichoderma harzianum* has been shown to degrade carbofuran [12] and organochlorines through an oxidative system [13].

Low dose of ionizing radiation on microorganisms is responsible of accelerated enzyme activity [14]. The lowest dose of gamma irradiation (1 MCi for 10 min) enhanced three isolates of *Aspergillus niger*, investigated to produce more biomass and polygalactronase, pectinmethylglacturonase, cellulase and protease [15]. *Trichoderma harzianum, T. viride* and *T. knoingii* irradiated with 0.5 KGy dosage resulted in the highest percentage of pathogen growth reduction by producing highly active exo-enzymes [16]. The two thermophilic isolates, *Streptomyces albaduncus* and *S. erythogresius*, were exposed to increasing doses of gamma radiation up to 5 KGy. All radiation did not affect the physiological properties, but relativity higher doses enhanced the utilization of carbon sources and increased their sodium chloride tolerance from 8 to 10%. Dose level of 2 KGy enhanced the antimicrobial activity of both isolates, either at first or second generation against bacteria, moulds and yeasts among them [17]. The low doses of gamma ray (10 and 20 Gy) significantly increased the alcohol-dehydrogenase enzyme activity of *Saccharomyces cerevisiae* [18].

The present work aimed to apply gamma radiation on *Trichoderma spp*, and to enhance effective hydrolytic enzymes in their bio-control abilities for biodegradation of oxaml pesticides. Low dose of gamma

*Corresponding author: Abd El-Moneim MR Afify, Department of Biochemistry, Faculty of Agriculture, Cairo University, Giza, Egypt
E-mail: abdelmoneimafify@yahoo.com

radiation used to enhancement *T. viride* and *T. harzianum*. Several successful attempts had been made to increase the bio-control potential of *Trichoderma* spp by exposing them to gamma radiation.

Materials and Methods

Soil sampling and characterization

Soil sample were collected from 10 different sub-samples and taken from the areas of 25 m², (0-20 cm) depth, from heavy clay soil had a previous history of treatment with Oxamyl in the last 10 years at field located in El-Fayoum governorate, Egypt.

Detailed physical and chemical properties of the soil are presented in Table 1. In the laboratory, the soil was gently air-dried to the point of soil moisture suitable for sieving. After sieving to a maximum particle size of <2 mm, the soil kept in a plastic bag at 4°C for 7 days before use.

Chemicals and reagents

Technical grade Oxamyl (99.1% purity) was purchased from Sigma Aldrich Co., Nasr city, Egypt. All other chemicals and solvents were ultra pure grade and obtained from El-Gomhouria CO. for Trading Chemicals and Medical Appliances, Egypt.

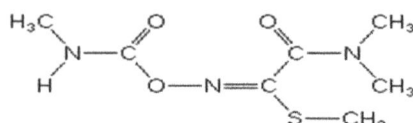

Chemical structure of Oxamyl (Methyl 2-(dimethylamino)-*N*-[(methylamino) carbonyl] oxy]-2-oxoethanimidothioate).

Enrichment procedure and isolation of microorganisms

Soil contamination: Prepared soil (200 g) was supplemented with Oxamyl at concentration of 50 mg/kg soil, introduced in a form of methanol solution. After mixing and solvent evaporation, the soil was incubated in the dark at 30 ± 1°C, in a thermostatic chamber for 90 days. The water content of the soil was adjusted to 50%. Throughout the incubation period, water losses exceeding 5% of the initial values were compensated by the addition of deionized water. After 30 and 60 days of incubation, the soil was contaminated again with the same dosage of Oxamyl

Identification of isolates: The pure isolated fungi were identified according to the most documented keys in fungi identification [19,20]. The morphological identification of isolated fungal strains is based on the morphology of the fungal culture colony or hyphae and the characteristics of spores.

Exposure of *T. harzianum, T. viride* to gamma radiation

The most effective Oxamyl degrading fungi (*Trichoderma spp.,* including *T. harzianum, T. viride*) selected and exposed to different doses of gamma radiation. Slants of 7 days old culture were irradiated with doses of 0.0, 0.02; 0.05; 0.1; 0.25; 0.5; 1.0; 2.0 and 5.0 KGy, and three replicates were used for each dose. Radiation treatments were carried out at Atomic Energy Authority, Abu-Zabal at dose rate of Egypt's Mega-gamma-1 type, J 6600-Cobalt-60 Irradiator.

Chemical analyses

Extraction and purification of oxamyl: Oxamyl was analyzed by high performance liquid chromatography (HPLC) at Atomic Energy Authority, Abu-Zabal, Egypt. In order to extract Oxamyl from the soil and liquid phases, the soil was slurry centrifuged at 6000 rpm at 25°C for 15 min to separate the liquid from the soil. The liquid phase was filtered through cellulose acetate paper (Whatman- number 1, England), prior to the liquid–liquid partitioning extraction procedure. Briefly, 2 mL of methanol were added to 2 mL of liquid sample, and then the mixture was sonicated twice for 10 min on a 50/60 voltage cycle. After sonication, Oxamyl was extracted in a separation funnel with dichloromethane. For the method of high-performance liquid chromatography (HPLC), the supernatant was dissolved in the same volume of pure grade methanol and filtrated by membrane filters (0.45 μm). An aliquot of the residue in a 20 μL sample size was injected into a HPLC. The analytical column was Zorbax SB-C18 column (250×4.6 mm, 5 μm), and the solutes were detected using PDA detector with gradient UV-VIS detection ranging from 200 to 600 nm. The mobile phase consisted of 70% methanol and 30% water at a flow rate of 1.0 mL min^{-1}.

Data analysis: Data was analyzed by SPSS program Version 11.5.0. The significance of treatments was set at p-value less than or equal to 0.05 by the one-way ANOVA test.

Results and Discussion

The biochemical and genetic basis of microbial degradation has received considerable attention. Several genes/enzymes which provide microorganisms with the ability to degrade organo-pesticides, have been identified and characterized. The ability of these organisms to reduce the concentration of xenobiotics [21] is directly linked to their long-term adaptation to environments, where these compounds exist. Gamma irradiation may be used to enhance the performance of such microorganisms that have the preferred properties, essential for biodegradation. Therefore, gamma irradiation was used to activate several fungi and determine the activities of their growth under this condition, as well be discussed in our investigation.

Oxamyl concentration (mg/L)	Dry weight biomass mg/100 ml*				
	T. harzianum	*T. viride*	*Aspergillus niger*	*Fusarium oxysporum*	*Penicillium cyclopium*
0 (control)	133.2 ± (0.40414)	169.0 ± (0.57735)	154.2 ± (0.46188)	95.6 ± (0.51961)	160.6 ± (0.05773)
20	168.6 ± (0.28867)	178.8 ± (0.51961)	154.6 ± (0.11547)	121.8 ± (0.46188)	157.0 ± (0.28867)
50	171.2 ± (0.11547)	185.6 ± (0.057735)	156.2 ± (0.17320)	145.8 ± (0.63508)	151.8 ± (0.11547)
100	176.0 ± (0.92376)	199.2 ± (0.17320)	159.4 ± (0.69282)	137.8 ± (0.11547)	146.0 ± (0.23094)
200	182.0 ± (0.28867)	175.0 ± (0.40414)	166.0 ± (0.28867)	120.2 ± (0.86602)	141.6 ± (0.80829)
250	169.3 ± (0.17320)	160.0 ± (0.63508)	145.0 ± (0.51961)	108.7 ± (0.17320)	122.0 ± (0.40414)
300	152.8 ± (0.11547)	149.1 ± (0.51961)	132.0 ± (0.92376)	96.2 ± (0.40414)	105.8 ± (0.17320)

* The values are the means of three replicates with the standard error (in parentheses) which was within 5% of the mean.

Table 1: Effect of oxamyl on the dry weight biomass of isolated fungi in MSM containing different concentrations of oxamyl within 7 days of incubation.

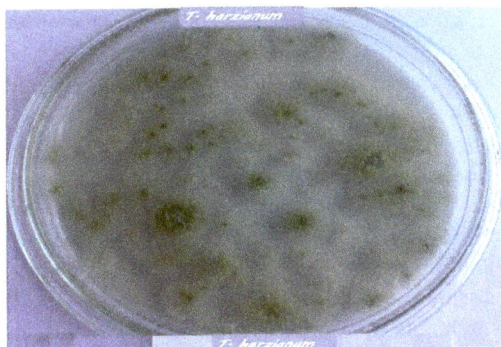

Figure 1: *Trichoderma harzianum* strain.

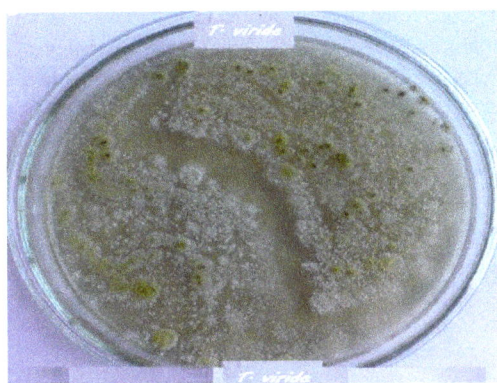

Figure 2: *Trichoderma viride* strain.

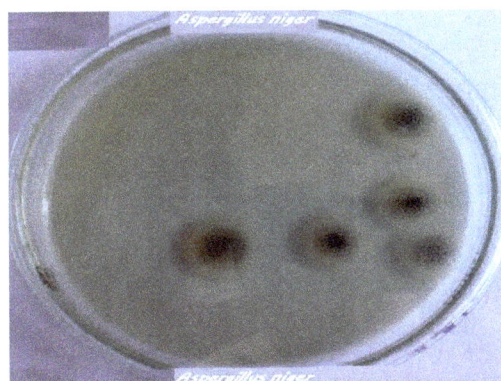

Figure 3: *Aspergillus niger* strain.

Isolation and identification of fungi

Fungi were isolated from soil collected from El-Fayoum governorate treated with different concentration of oxamyl (20-300 mg/L). After several tests of culture on synthetic medium containing the Oxamyl, five fungi were selected and tested for their ability to degrade Oxamyl, pesticide. The fungi strains were identified as *Trichoderma spp.*, including *T. harzianum, T. viride, Aspergillus niger, Fusarium oxysporum* and *Penicillium cyclopium* (Table 1) (Figures 1-5). The results showed that *Trichoderma spp.*, including *T. harzianum* and *T. viride,* reach its maximum growth using oxamyl concentration of 200 and 100 mg/L and yielded 182.0 and 199.2, respectively. On the other hand, *Aspergillus niger* reach its maximum yield 166.0 with 200 mg/L,

while the growth of *Fusarium oxysporum* and *Fusarium oxysporum* were weak (Figures 2 and 6). From the above results, *Trichoderma* spp. used Oxamyl as source of carbon and nitrogen and possesses enzyme(s) which acts on amide and ester bond in Oxamyl. Degradation of oxamyl was 72.5% within 10 days of incubation by *T. harzianum* strain. While the degradation of oxamyl was 82.05% within 10 days of incubation by *T. viride* strain, this indicated that the isolates of *Trichoderma* spp. were potentially useful for oxamyl bioremediation (Table 2) (Figure 7).

These results agree with those of Rajagopal et al. [22], who isolated *Bacillus* sp., *Micrococcus* sp., *Arthrobacter* sp. and *Azospirillum* sp. capable of using pesticides as source of carbon and nitrogen. This increase in the biomass of the culture with Oxamyl could be explained by the fact that this product constitutes, an additional carbon and nitrogen contribution, which allows the synthesis of new secondary metabolites favoring the production of microbial biomass, and in consequence, support a faster use of Oxamyl. The increase in biomass (mycelial dry weight) was reported when *T. viride* strain incubated with pesticides [12]. Therefore, *Trichoderma spp* have been selected for enhancement by gamma radiation for better biodegradation of oxamyl pesticide.

Effect of gamma radiation on *Trichoderma spp*

Results in Table 3 and Figures 8 and 9 indicated the effect of different gamma-radiation doses 0.0, 0.02; 0.05; 0.1; 0.25; 0.5; 1.0; 2.0 and 5.0 KGy, on biomass of *Trichoderma spp.*, as well as *T. viride* grown on MSM with oxamyl at concentration of 200 mg L^{-1} within 7 days of incubation. The biomass of *Trichoderma spp* strain were increased

Figure 4: *Fusarium oxysporum* strain.

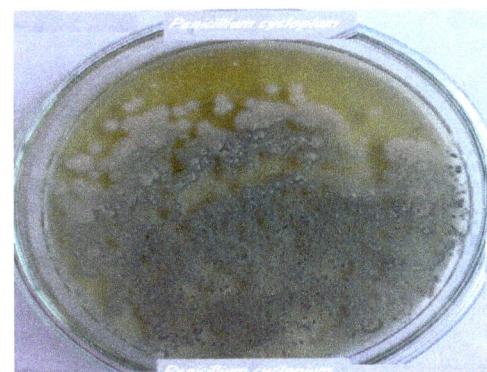

Figure 5: *Penicillium cyclopium* strain.

Fungal strain	Time of incubation (d)	Oxamyl control	Oxamyl inoculation	Oxamyl loss (%)
T. harzianum	1	198.0	175.9	12.05
	2	196.2	148.6	25.7
	3	193.2	129.0	35.5
	4	190.3	111.0	44.5
	5	189.3	101.2	49.4
	6	184.9	90.6	54.7
	7	178.6	81.9	59.05
	8	172.1	72.0	64
	9	167.2	64.0	68
	10	160.0	55.0	72.5
T. viride	1	198.0	181.8	9.1
	2	196.2	168.2	15.9
	3	193.2	142.0	29
	4	190.3	122.2	38.9
	5	189.3	95.7	52.15
	6	184.9	76.5	61.75
	7	178.6	61.0	69.5
	8	172.1	52.1	73.95
	9	167.2	44.2	77.9
	10	160.0	35.9	82.05

Table 2: Biodegradation of oxamyl (200 mg L^{-1}) in mineral salt medium by *Trichoderma spp.*

Gamma irradiation dose (Gy)	Mycelial dry weight (mg/100 ml)*			
	T. harzianum		T. viride	
	Growth mg/100 ml	% of Change	Growth mg/100 ml	% of Change
0	182.0 ± (0.057735)	0.0	165.0 ± (0.57735)	0.0
20	198.5 ± (0.02886)	9.07	174.0 ± (0.14433)	5.45
50	204.0 ± (0.37527)	12.08	189.1 ± (0.23094)	14.60
100	217.8 ± (0.23094)	19.67	209.5 ± (0.40414)	26.96
250	222.0 ± (0.40414)	21.97	231.0 ± (0.51961)	40.0
500	175.2 ± (0.23094)	-3.74	151.0 ± (0.40414)	-8.84
1000	129.6 ± (0.17320)	-28.79	129.0 ± (0.46188)	-21.81
2000	108.0 ± (0.63508)	-59.34	119.0 ± (0.11547)	-27.87
5000	91.50 ± (0.11547)	-50.27	102.0 ± (0.28867)	-38.13

*The values are the means of three replicates with the standard error (in parentheses) which was within 5% of the mean.

Table 3: Mycelial dry weight (mg/100 ml) of gamma irradiated *Trichoderma* spp. grown on MSM with 200 mg L^{-1} of oxamyl at 30ºC for 7 days of incubation.

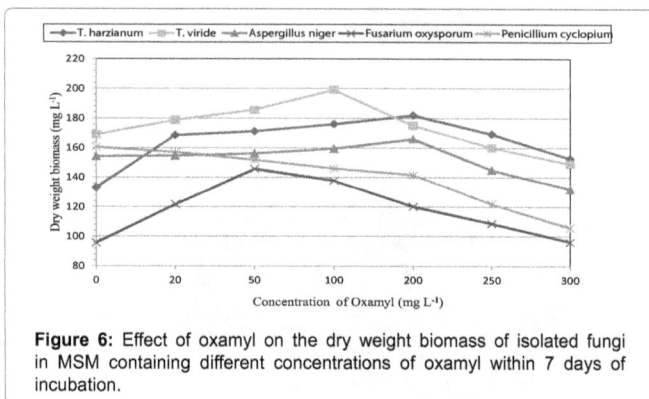

Figure 6: Effect of oxamyl on the dry weight biomass of isolated fungi in MSM containing different concentrations of oxamyl within 7 days of incubation.

respectively. Previous studies have shown that relatively low dose of ionizing radiation on microorganisms is responsible of accelerated enzyme activity [14]. The low doses of gamma ray (10 and 20 Gy) significantly increased the alcohol-dehydrogenase enzyme activity of *Saccharomyces cerevisiae* [18]. *Trichoderma harzianum, T. viride* and *T. knoingii* irradiated with 0.5 KGy dosage resulted in the highest percentage of pathogen growth reduction by producing highly active exo-enzymes (cellulosae and chitinase isoenzymes), as confirmed by Haggag and Mohamed [16]. These results are in agreed with stated that growth of *T. viride* was increased at 0.5 KGy of gamma-radiation. Mycelial dry weight increased in isolates of *Aspergillus tamaru, A. flavus* and *A. niveus,* when exposed to gamma-irradiation doses of 0.2 and 0.5 KGy [23]. Previous studies have shown that relatively low dose of ionizing radiation on microorganisms is responsible of accelerated enzyme activity [14]. The low doses of gamma ray (10 and 20 Gy) significantly increased the alcohol-dehydrogenase enzyme activity of *Saccharomyces cerevisiae* [18]. *Trichoderma harzianum, T. viride* and *T. knoingii* irradiated with 0.5 KGy dosage resulted in the highest percentage of pathogen growth reduction by producing highly active

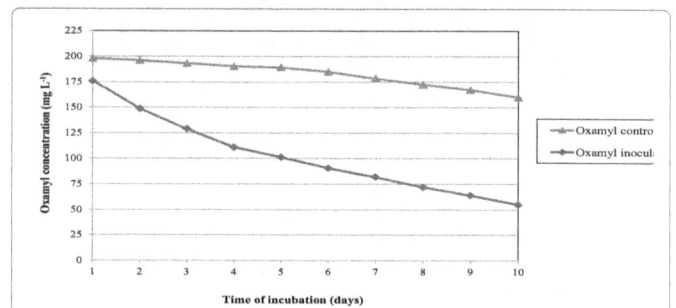

Figure 7: Biodegradation of Oxamyl (200 mg L^{-1}) in MSM by *T. harzianum strain* within 7 days of incubation at 30°C.

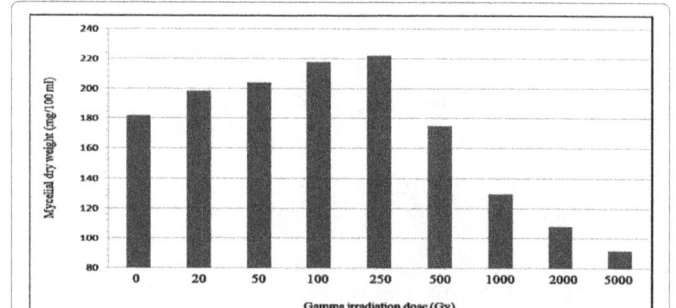

Figure 8: Mycelial dry weight (mg/100 ml) of gamma irradiated *T. harzianum* grown on MSM with 200 mg L^{-1}of Oxamyl at 30ºC for 7 days of incubation.

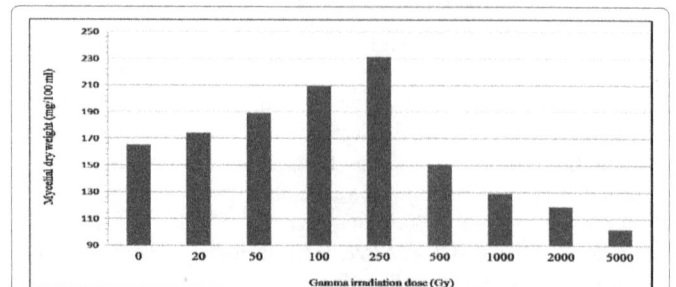

Figure 9: Mycelial dry weight (mg/100ml) of gamma irradiated *T. viride* grown on MSM with 200 mg L^{-1} of Oxamyl at 30ºC for 7 days of incubation.

and reached its maximum at 250 Gy by 21.97% and 40.0% when using *Trichoderma spp.,* as well as *T. viride,* respectively. As a general trend, the gamma radiation over than 0.25 Kg reduce the growth of *T.* spp by 50.27% and 38.13% using *Trichoderma spp.,* as well as *T. viride,* exo-enzymes. The results showed that *Trichoderma* sp. presented a good growth in the presence of DDD and 21% of the pesticide was degraded. In the experiments where DDD was added after 5 days of *Trichoderma* sp. growth, and with the addition of H₂O₂, the total biodegradation occurred [24]. Haggag and Mohamed [16] found that mutagenesis of three *Trichoderma* species by gamma irradiation exhibited high capabilities to produce efficient antibiotics, enzymes and phenols. On the other hand The tested UV-induced mutants were higher in their production of enzymes (cellulases, chitinases and β-1,3-glucanases) than their parental wild type strain (*T. viride*). Cellulase was the greatest enzyme production by the tested *T. viride* strains, followed by β-1,3-glucanase then chitinase. Therefore, the enhancement of *Trichoderma spp* by gamma radiation induce the activation of the main enzymes cellulases, chitinases and β-1,3-glucanases, which depend mainly on the dose of radiation [25]. The enzyme cellulase, a multi enzyme complex made up of several proteins, catalyzes the conversion of cellulose to glucose in an enzymatic hydrolysis. This agrees with previous reports that the amylolytic potential of *T. viride* was increased at 0.5 KGy of gamma-radiation [26]. *Trichoderma harzianum*, *T. viride* and *T. knoingii* irradiated with 0.5 KGy dosage resulted in the highest percentage of pathogen growth reduction by producing highly active exo-enzymes. Therefore, *Trichoderma* spp. mutants were effective in reducing the pathogen growth in rhizophere soil, as compared to the wild type strains.

From the results above, *Trichoderma* strains is attributable to increase biodegradation of the oxymyl pesticide according to one or more complex mechanisms, including nutrient competition, antibiosis, the activity of cell wall-lytic enzymes, induction of systemic resistance and increased plant nutrient availability, as confirmed by Ene and Alexandru [27].

Conclusion

The proposal that our enhancement *Trichoderma* spp. will involve in the biodegradation of oxamyl pesticides well be belongs to produce of several active enzymes by the two spp of *Trichoderma*, as confirmed by fungi of *Aspergillus niger* and *Fusarium graminearum* [9]. Data in this work indicate the possibility of applying gamma-radiation doses to increase oxamyl degradation by enhancement *T. harzianum* and *T. viride* with low dose gamma radiation. An enrichment procedure allowed isolating of two effective fungal strains belonging to *T. harzianum* and *T. viride*, that may participate in efficient degradation of the oxamyl. Obtained results have implicated for the development of a bioremediation strategy of oxamyl-polluted soils. However, use of pesticide-degrading microbial systems for removal of pesticide compounds from the contaminated sites requires an understanding of ecological requirements of degrading strains. There is a need for further research on the biochemical and genetic aspects of oxamyl degradation by the isolated fungi. Therefore, this point needs further investigation to study the activity of enzymes needed to degrade oxamyl to its main metabolite and further biodegradable products. In the future, it could even apply *Trichoderma* spp. directly under special condition or synthetize these enzyme and applied in the field for biodegradation of oxamyl pesticides to clean environments from pollutants.

Acknowledgment

This work was supported by Scientists Next Generation Program, Academy of Scientific Research and Technology, Egypt, and Faculty of Agriculture Biochemistry Department, Cairo University.

References

1. Parekh S, Vinci VA, Strobel RJ (2000) Improvement of microbial strains and fermentation processes. Appl Microbiol Biotechnol 54: 287-301.

2. Kaufman DD (1987) Accelerated biodegradation of pesticides in soil and its effect on pesticide efficacy. Proc Br Crop Prot Conf Weeds 2: 515-522.

3. Negro CL, Senkman LE, Collins PA (2011) Metabolic responses of pleustonic and burrowing freshwater crabs exposed to endosulfan. Fresenius Environ Bull 21.

4. Afify AMR, El-Beltagi HS (2011) Effect of the insecticide cyanophos on liver function in adult male rats. Fresenius Environ Bull 20: 1084-1088.

5. WHO (World Health Organization) (1986) Organophosphorus Insecticides: A General Introduction. Environmental Health Criteria 63, Geneva. Wei JC (1979) Handbook of Fungi Identification. Technology Press, Shanghai, China.

6. Chapalamadugu S, Chaudhry GR (1992) Microbiological and biotechnological aspects of metabolism of carbamates and organophosphates. Crit Rev Biotechnol 12: 357-389.

7. Vidali M (2001) Bioremediation. An overview. Pure Appl Chem 73: 1163-1172.

8. Zhang Q, Liu Y, Liu YH (2003) Purification and characterization of a novel carbaryl hydrolase from Aspergillus niger PY168. FEMS Microbiol Lett 228: 39-44.

9. Salama AK (1998) Metabolism of carbofuran by Aspergillus niger and Fusarium graminearum. J Environ Sci Health B 33: 253-266.

10. Seo J, Jeon J, Kim SD, Kang S, Han J, et al. (2007) Fungal biodegradation of carbofuran and carbofuran phenol by the fungus Mucor ramannianus: identification of metabolites. Water Sci Technol 55: 163-167.

11. Slaoui M, Ouhssine M, Berny E, El-Yachioui M (2007) Biodegradation of the carbofuran by a fungus isolated from treated soil. Afr J Biotechnol 6: 419-423.

12. Wootton MA, Kremer RJ, Keaster AJ (1993) Effects of carbofuran and the corn rhizosphere on growth of soil microorganisms. Bull Environ Contam Toxicol 50: 49-56.

13. Katayama A, Matsumura F (2009) Degradation of organochlorine pesticides, particularly endosulfan, by Trichoderma harzianum. Environl Toxicol Chem 12: 1059-1065.

14. Chakravarty B, Sen S (2001) Enhancement of regeneration potential and variability by gamma-irradiation in cultured cells of Scilla indica. Biol Plant 44: 189-193.

15. Gherbawy YA (1998) Effect of gamma irradiation on the production of cell wall degrading enzymes by Aspergillus niger. Int J Food Microbiol 40: 127-131.

16. Haggag WM, Mohamed HAA (2002) Enhancement of antifungal metabolites production from gamma-rays induced mutants of some Trichodema sp. for control onion white rot disease. Plant Pathol Bull 11: 45-56.

17. Moussa LLA, Mansour FA, Serag MS, Abou El-Nour SAM (2005) Effect of gamma radiation on the physiological properties and genetic materials of Streptomyces albaduncus and S. erythogresius. Int J Agri Biol 7: 197-202.

18. Ben-Akacha N, Zehlila A, Mejri S, Jerbi T, Gargouri M (2008) Effect of gamma-ray on activity and stability of alcohol-dehydrogenase from Saccharomyces cerevisiae. Biochem Eng J 40: 184-188.

19. Carmichael JW, Kendrick BW, Conners IL, Lynne S (1980) Genera of hyphomycetes. The University of Alberta, CA, USA.

20. Barnett HL, Hunter BB (1998) Illustrated genera of imperfect fungi. APS Press, St. Paul, Minnesota, USA.

21. Afify AMR (2009) Biological function of xenobiotics through protein binding and transportation in living cells. Int J Agric Res 5: 562-575.

22. Rajagopal BS, Rao VR, Nagendrappa G, Sethunathan N (1984) Metabolism of carbaryl and carbofuran by soil –enrichment and bacterial cultures. Can J Microbiol 30: 1458-1466.

23. Younis NA (1999) A comparison study on protease, alpha-amylase and growth of certain fungal strains of Aspergillus sp. After exposure to gamma-rays. Arab J Nucl Sci Appl 32: 257-264.

24. Ortega SN, Nitschke M, Mouad AM, Landgraf MD, Rezende MO, et al. (2011) Isolation of Brazilian marine fungi capable of growing on DDD pesticide. Biodegradation 22: 43-50.

25. Shafique S, Bajwa R, Shafique S (2009) Cellulase biosynthesis by selected trichoderma species. Pak J Bot 41: 907-916.

26. Gbedemah CM, Awafo V (1990) Inter Atomic Energy Agency. Vienna, Austria 149: 77-83.

27. Ene M, Alexandru M (2008) Microscopical examination of plant reaction in case of infection with Trichoderma and Mycorrhizal fungi. Roumanian Biotechnol Lett 13: 13-19.

19

Pathogenic and Genetic Characterization of *Xanthomonas campestris* Pv. *campestris* Races Based on Rep-PCR and Multilocus Sequence Analysis

Priyanka Singh Rathaur[1,2], Dinesh Singh[1*], Richa Raghuwanshi[2] and Yadava DK[3]

[1]Division of Plant Pathology, Indian Agricultural Research Institute, New Delhi-110012 India
[2]Department of Botany, Mahila Mahavidyalya ,Banaras Hindu University, Varanasi-221005, U.P, India
[3]Division of Genetics, Indian Agricultural Research Institute, New Delhi-110012, India

Abstract

Xanthomonas campestris pv. *campestris* (Pammel) Dowson (Xcc) is the causal agent of black rot disease of crucifers worldwide. Seventy five isolates of Xcc were collected from 12 agro-climatic regions of India to determine the distribution pattern of races and diversity of the population. Based on pathogenic reaction on seven standard differential crucifers, race 1, 4 & 6 were found to be prevalent. For assessing the pathogenic diversity, forty one cultivars of crucifers comprising seven *Brassica* and coeno species were inoculated artificially under field conditions. *Brassica juncea* cultivars (Pusa Bold, Varuna, Pusa Vijay, Pusa Mustard 21 and Pusa Mustard 25) showed resistance against all the strains of Xcc, whereas the *Brassica olerecea* cultivar Pusa Ageti was found to be resistant to races 1 and 4. Genetic characterization of these 75 strains of Xcc was carried out using rep-PCR (ERIC, REP and BOX-PCRs) followed by phylogenetic analysis. The strains of Xcc clustered into 6 groups at 50% similarity coefficient and among these groups, 28 strains of Xcc belonging to races 1, 4 & 6 were clustered together under Group 5. Sequences of the 16S rRNA, *hrp*F and *ef*P genes of five strains representing the races 1, 4 and 6 were used for multilocus sequence analysis. Based on sequence analysis of 16S rRNA and *hrp*F genes, the Indian strains were found to be very closely related to the strain Xcc ATCC 33913 (race 3, UK), whereas based on *ef*P sequences, they were found to be closely related to strains race 1 Xcc B100 (Italy) and race 9 Xcc 8004 (UK).

Keywords: Race; DNA fingerprinting; Diversity; 16S rRNA; *hrp*F; *ef*P gene

Introduction

Black rot caused by *Xanthomonas campestris* pv. *campestris* (Pammel) Dowson (Xcc) is an important disease of crucifers, which substantially damages the crops by 10–50% under favourable environmental conditions. The pathogen infects a large number of cruciferous plants, including agriculturally important crops such as cole crops (broccoli, cabbage, cauliflower and knoll khol), turnip, radish, oliferous *Brassica* crops, ornamental plants, and weeds [1,2]. Based on the interaction of various strains of Xcc with different *Brassica* species, a total of nine pathogenic races of Xcc were established by using set of seven cultivars of crucifer [3-6]. Various molecular techniques such as RFLP patterns, repetitive sequence based PCR (rep-PCR), 16S rRNA gene analysis, *hrp* (hypersensitive response and pathogenicity) gene and amplified fragment length polymorphism have been used to study the genetic variability among various bacteria [7-10]. Although, the PCR-based DNA-fingerprinting is a fast, reliable and comparatively low cost technique and its effectiveness depend on the primers chosen for analysis and the quality of the DNA. Three families of repetitive sequences (Rep) including repetitive extragenic palindromic (REP) sequences, Enterobacterial Repetitive Intergenic Consensus (ERIC) sequences, and the BOX element have been identified [11]. Studies of the 16S rRNA gene and the 16S-23S intergenic region are generally used to identify Xcc strains at the genus and species level. In Xcc, the *hrp* cluster consists of 26 genes extending from *hpa*2 to *hrp*F [12] and among them the *hrp*F gene is suitable for the differentiation of pathovars and used for diagnosis of *Xanthomonas* species [9,13,14]. Another gene *ef*P encodes the elongation factor P protein that stimulates the peptidyl transferase activity of fully assembled 70 S prokaryotic ribosomes and enhances the synthesis of certain dipeptides initiated by N-formylmethionine [15]. This gene has been used by [16] to study population genetic structure and diversity of *Helicobacter pylori*.

In order to design effective control strategies, especially seed health tests and development of resistant varieties against black rot disease, knowledge of the extent of genetic variability in the pathogen population is important. Thus, the variation within Xcc needs to be determined. In the present study, we collected 75 isolates of Xcc from black rot-infected cole crop and other crucifer crops from diverse agro-climatic regions of the country for assessing their race profile, virulence pattern of Xcc strains and their genetic diversity by using rep-PCR and multilocus sequence analysis using *hrp*F, 16S rRNA and *ef*P genes of representative strains of Xcc races.

Materials and Methods

Isolation of *X. campestris* pv. *campestris*

Leaf samples (one or two plants from each field) exhibiting typical black rot symptoms of V-shaped lesions with blackened veins were collected from major cole crops notably, cauliflower (*Brassica oleracea* var. *botrytis* L.), cabbage (*B. oleracea* var. *capitata* L.), knol-khol (*B. oleracea* var. *gongylodes* L., and broccoli (*B. oleracea* var. *italica* Plenck), radish (*Raphanus sativus* L.), turnip (*Brassica rapa* var. *rapa*), Indian mustard [*Brassica juncea* (L.) Czerna and cross] and vegetable

*Corresponding author: Dinesh Singh, 1Division of Plant Pathology, Indian Agricultural Research Institute, New Delhi-110012, India
E-mail: dinesh_iari@rediffmail.com

mustard [*B. juncea* (L.) Czernajew] from 12 agro-climatic regions of India which covers major 18 states of India (Table 1). One or two leaf samples, representing one or two plants from each field, were selected for isolation of the causal agent from the typical V-shaped lesions. The diseased samples were dried between sheets of paper at room temperature before isolation of Xcc. From each leaf sample, 0.5 to 1 cm² of leaf tissue was excised with a sterilised blade from the margin of a lesion. Then the excised leaf sections were sequentially passed through 70% alcohol for 20 seconds, 0.01% $HgCl_2$ for 20 seconds and twice in sterile distilled water for 40 seconds. Leaf sections were chopped with a sterile blade over sterile glass slides with few drops of distilled water and allowed to infuse for about 3 min. Then a loopful of suspension was streaked on nutrient sucrose agar medium containing 23 g of nutrient agar, 20 g of sucrose and 5 g of agar powder per liter and incubated at 28°C for 48 h [17]. Cultures were maintained in slants on YGCA medium for short times and for longer period cultures were stored at -80°C with 50% glycerol in nutrient broth until further study.

S.No.	Agroclimatic Zones	States	Host	Isolates
1.	Western Himalaya Region	Punjab	Cabbage	**Race 1**- Xcc-C189
			Cauliflower	**Race 1**- Xcc-C186, Xcc-C187, Xcc- C188
		Jammu Kashmir	Radish	**Race-4**-Xcc-C113
			Cauliflower	**Race-4**- Xcc-C112, **Race 1**- Xcc-C125, Xcc-C126
		Himachal Pradesh	Cabbage	**Race 1**-Xcc-C118, Xcc-C205, Xcc-C206, Xcc-C207
			Cauliflower	**Race 1**- Xcc-C114, Xcc-C128, Xcc-C210
			Radish	**Race 1**- Xcc-C208, Xcc-C209
2.	Eastern Himalaya Region	Meghalaya	Cabbage	**Race 1**- Xcc-C8
			Broccoli	**Race 1**- Xcc-C9
			Cauliflower	**Race 1**- Xcc-C119
3.	Lower Gangatic Plains Regions	West Bengal	Cauliflower	**Race 4**- Xcc-C197, Xcc-C198 **Race 1**- Xcc-C199, Xcc-C200,
4.	Middle Gangatic Plains Regions	Uttar Pradesh	Indian Mustard	**Race 1**- Xcc-C162
			Cabbage	**Race 4**- Xcc-C117,Xcc-C170 **Race 1**- Xcc-C167
			Cauliflower	**Race 4**- Xcc-C157, Xcc-C195, Xcc-C196 **Race 1**- Xcc-C161, Xcc-C178
			Turnip	**Race 4**- Xcc-C171 **Race 1**- Xcc-C166
		Bihar	Cabbage	**Race 1**- Xcc-C216
			Cauliflower	**Race 1**- Xcc-C217, Xcc-C218
		Jharkhand	Cauliflower	**Race 4**- Xcc-C111, **Race 1**- Xcc-C131
			Cabbage	**Race 1**- Xcc-C130
5.	Upper Gangatic Plains Regions	Uttrakhand	Cauliflower	**Race 1**- Xcc-C116, Xcc-C120, **Race 4**- Xcc-C227
			Knolkhol	**Race 4**- Xcc-C219, Xcc-C220
			Cabbage	**Race 4**- Xcc-C221, Xcc-C228
			Indian Mustard	**Race 4**- Xcc-C226
6.	Eastern Plateaus & Hill Region	Orissa	Cauliflower	**Race 4**- Xcc-C127, Xcc-C115
7.	Trans Gangatic Plains Regions	Haryana	Cabbage	**Race 1**- Xcc-C129
			Cauliflower	**Race 4**-,Xcc-C110
		Delhi	Cauliflower	**Race 1**- Xcc-C23 **Race 4**- Xcc-C147
			Broccoli	**Race 1**- Xcc-C4
			Cabbage	**Race 1**- Xcc-C6 **Race 6**- Xcc-C278,
			Knol Khol	**Race 1**- Xcc-C14,
			Vegetable Mustard	**Race 1**- Xcc-C16
			Turnip	**Race 1**- Xcc-C19
			Indian Mustard	**Race 1**- Xcc-C21
8.	Central Plateau & Hill Region	Madhya Pradesh	Cauliflower	**Race 1**- Xcc-C230, Xcc-C231
9.	Southern Plateau & Hill Region	Karnataka	Cabbage	**Race 1**- Xcc-C124
			Cauliflower	**Race 1**- Xcc-C132, Xcc-C211, Xcc-C212,
10.	West Coast plains & Hill Region	Maharashtra	Cauliflower	**Race 1**- Xcc-C248
		Goa	Cauliflower	**Race 1**- Xcc-C247
11.	Gujarat plains & Hill Region	Gujarat	Cabbage	**Race 4**- Xcc-C121,Xcc-C123
			Cauliflower	**Race 1**- Xcc-C122
12.	Western Dry Region	Rajasthan	Cabbage	**Race 1**-Xcc-C190

Table 1: List of *Xanthomnas campestris* pv. *campestris* strains isolated from crucifer crops, from different states, Host and races of India.

Pathogenicity tests

Seventy five isolates of Xcc were tested for their pathogenicity on seedlings of a susceptible cultivar of cauliflower (cv. Pusa Sharad) which were grown under field conditions at the Indian Agricultural Research Institute, New Delhi. The inoculum of Xcc was prepared by culturing the bacteria on nutrient sucrose medium for forty-eight hours at 28°C, then the bacterial growth was scraped from the Petri plates and suspended in sterile distilled water and inoculum of each isolates was maintained 0.1 OD at 600 nm (U-2900 US/VJS spectrophotometer, model no. 2J1-004, Hitachi) before inoculation as described by Singh et al. [2]. This bacterial suspension was inoculated with small scissors by dipping in the bacterial suspension and transferring onto the three youngest leaves of 30 days old plants with three replications as described by Vicente et al. [2]. The black rot disease reaction was recorded at 15 days after inoculation.

Race characterisation

Race characterisation of 75 strains of Xcc representing the diverse agro-climatic regions of India was achieved using a set of seven cultivars of *Brassica* species i.e., turnip (*B. rapa* var. *rapa*) 'Just Right Turnip F1' and 'Seven Top Turnip', Indian mustard (*B. juncea* (L.) Czernajew) 'Florida Broad Leaf', Ethiopian mustard (*Brassica carinata* L. Braun) 'PI 199947', rapeseed mustard (*Brassica napus* L) 'Cobra 14R', cauliflower (*B. oleracea* var. *botrytis* L.) 'Miracle F1' and cabbage (*B. oleracea* var. *capitata* L.) 'Wirosa F1' (Vicente et al. 2001; 2006). The seeds of these different lines were obtained from the University of Warwick, UK and from Otis S. Twilley Seed Co. Inc. (121 Gary Rd, Hodges SC 29653-9168 [*B. juncea, B. rapa* var. *rapa* EC732033 to EC732035]). Forty eight hour old cultures of the 75 strains of Xcc were pelleted and resuspended in sterilised distilled water to maintain an OD 0.1 at 600 nm. The cultures were inoculated on to 35 days old plants of the cultivars Just Right Turnip F1, Seven Top Turnip, Florida Broad Leaf, PI 199947 and Cobra 14R; and plants of cultivars Miracle F1 and Wirosa F1 30 days after planting. Leaves were inoculated by clipping secondary veins, near the margins, with small scissors dipped in bacterial suspension. Ten points of inoculation were made in the youngest leaves on each plant with three replications. The number of infected points per leaf and the severity of symptoms were assessed after 15 and 30 days of inoculation. Based on the relative size of the largest lesion on the leaf, they were rated on a scale of 0-9 as described by Vicente et al. [18].

Pathogenic variability

The pathogenic variability was studied during the winter season (November to March) for two consecutive years i.e., 2012–2013 and 2013–2014 at the Indian Agricultural Research Institute, New Delhi under field conditions by scoring disease reaction on the leaves of crucifers as given in Table 1. A total of 41 representatives of the Crucifereae family including 7 economically important *Brassica* species i.e., *B. campestris*, *B. carinata*, *B. juncea*, *B. napus*, *B. nigra*, *B. oleracea*, and *B. rapa* and three *Brassica* coeno species (*R. sativus*, *Eruca sativa* and *Sinapis alba*) were tested against the 75 strains of Xcc, which are collected from diverse agroclimatic regions of India. All the 75 strains of Xcc were inoculated on 35 days old plants of turnip, radish and oliferous *Brassica* spp. and 30 days after planting of cole crops. Later the severity of symptoms was assessed as by described by Vicente et al. [18] In the present study, screening was carried out in the field during winter seasons, where temperature ranged from 7.2 to 30.1°C and rainfall ranged from 0 to 6.5 mm (2012–2014). Based on disease scores, the inoculated plants were grouped into four categories and the percentage of inoculated points showing symptoms as resistant, partially resistant, susceptible and highly susceptible were recorded as described by Vicente et al. [18].

DNA extraction and molecular characterisation

The strains of Xcc were grown in nutrient broth medium for 24 h at 28°C. The total genomic DNA of the bacteria was extracted using the CTAB method [19]. Molecular characterisation of 75 isolates of Xcc was achieved using a *hrp*F gene based PCR as described by Singh et al. [20].

Genetic diversity by rep-PCR

Genetic diversity of the 75 strains of Xcc was assessed by rep-PCR using the BOX, ERIC and REP. The primers and conditions were as described by Schaad et al. [17]. The data from rep-PCR (BOX, ERIC, REP-PCR) fingerprinting profiles were combined together to generate a similarity matrix by using the SIMQUAL module of NTSYSpc 2.02e. The similarity matrix was used for cluster analysis by unweighted

pair group method of arithmetic average (UPGMA) using sequential, agglomerative, hierarchical, nested clustering module of NTSYSpc 2.02e [21]. The output data are graphically presented as a phylogenetic tree.

Multilocus sequence analysis

Out of 75 strains, five strains representing three races of Xcc i.e., Xcc-C18 (race 1,turnip, Laxminagar, Delhi), Xcc-C131 (race 1, cauliflower, Ranchi, Jharkhand), Xcc-C168 (race 4, cabbage, Uttar Pradesh), Xcc-C106 (race 4, vegetable mustard, Laxminagar, Delhi) and Xcc-C278 (race 6, cabbage, Laxminagar, Delhi) were selected to study genetic variability by using 16S rRNA, *hrp*F and *ef*P gene sequences. The primers for 16S rRNA and *ef*P gene amplification were designed using Primer 3 (www.frodo.wi.nit.edu) (Table 2) and checked for specificity *in silico* using www.insilico.ehu.es.The primers were validated for their universality across *Xanthomonas* and related bacteria by primer blasting using www.ncbi.nlm.nih.gov. About 25.0 µl of PCR reaction mixture containing $5 \times$ Taq buffer (5.0 µl), 10 mM dNTPs (0.5 µl), 25 mM $MgCl_2$ (1.5 µl), 10 µM forward and reverse primers (0.5 µl each), Taq polymerase (0.24 µl), molecular grade water and 1.0 µl DNA (100 ng) was used for PCR amplification. PCR was carried out under the following conditions using a Gradient thermocycler (C-1000TM, BIORAD): 95°C for two min followed by 30 cycles of 95°C for 30 seconds, annealing temperature and duration was varied according to gene (Table 2), 72°C for 1 min and terminated by a final elongation at 72°C for five min. The *hrp*F gene primers and protocol used were as described by Singh et al. [20]. The PCR product was mixed with 1.0 µl of loading dye. Electrophoresis was carried out using 1.2% agarose gel, prepared in $1 \times$ Tris-acetate-EDTA (TAE) buffer containing ethidium bromide (0.5 µg ml⁻¹). Electrophoresis was carried out at 60V for 1.5 hours, visualised on a gel documentation system (BIORAD model Gel DocTM XR+) under UV light (300 nm) and photographed using image Lab version 2.0.1 software (BIORAD) for gel analysis. PCR products of the five strains were purified using Gel and PCR Clean-Up kit (Promega). Sequencing was performed by Sanger's method (Applied Biosystem Machine-3130, Chromas Biotech, Bangalore, India). Sequence data of isolates were submitted to NCBI Genbank. For the *hrp*F, 16S rRNA and *ef*P analysis, the sequences obtained were aligned pairwise. Multiple alignments compared to those of the type/ reference strains were performed with ClustalW (1.7) software [22] 994). Phylogenetic trees were generated using MEGA version 6.0 Tamura et al. [23] with default parameters, K2P distance model and the neighbour-joining algorithm (Saitou and Nei, [24]). Statistical support for tree nodes was evaluated by bootstrap (Felsenstein, [25,26]) analyses with 1,000 samplings.

Locus	Sequences of Primer 5' ------------------------- 3'	Base pair (bp)	Annealing temperature and duration	Reference
16srDNA	Xcc16S-F:5'-GCAAGCGTTACTCGGAATTA-3' Xcc16S-R: 5'-TACGACTTCACCCCAGTCAT-3'	959	56°C for 30 Sec	Used in this study
*ef*P gene	DXEP1F: 5'-TCATCACCGAGACCGAATA-3' DXEP1R: 5'-TCCTGGTTGACGAACAGC-3'	434	54°C for 30 Sec	Used in this study
hrp gene	DhrpXccF:5'-GTGGCCATGTCGTCGACTC-3' DhrpXccR:5'-GGAATAAACTGTTTCCCCAATG-3'	769	60°C for 40 Sec	Singh *et al.*, 2014
BOX	BOXA1R: 5'-CTACGGCAAGGCGACGCTGACG-3'	Multiple bands	53°C for 1min	Schaad *et al.* 2001
ERIC	ERIC1R: 5'-ATGTAAGCTCCTGGGGATTCAC-3' ERIC2: 5'-AAGTAAGTGACTGGGGTGAGCG-3'	Multiple bands	46°C for 1min	Schaad *et al.* 2001
REP	REP1R: 5'-IIIICGICGICATCIGGC-3' REP21: 5'- ICGICTTATTATCIGGCCAC-3'	Multiple bands	48°C for 1 min	Schaad *et al.* 2001

Table 2: Primer used in this study.

Results

Isolation of *X. campestris* pv. *campestris*

Seventy-five strains of Xcc were isolated from eight major crops from diverse agro-climatic regions of India (Table 1). These isolates of Xcc produced yellow, translucent, raised, mucoid colonies on nutrient sucrose agar medium. They were found to be Gram-negative, rod shaped, aerobic with monotrichous flagella. Pathogenicity of all the isolates was tested on cauliflower (*B. oleracea* var. *botrytis*) cv. Pusa Sharad by artificial inoculation and all the isolates produced typical black rot disease symptoms of yellow or dead tissue at the edge of leaves, similar to tip burn, which frequently progressed into a V-shape followed by blackening of the veins within 6-15 days of inoculation. These isolates were considered to be Xcc on the basis of these morphological and pathogenicity tests. For further confirmation, these isolates were tested by PCR amplification using the primers Dhrp_Xcc_F and Dhrp_Xcc_R, which generate a 769 bp product and they were specific to Xcc. The strains of Xcc caused black rot, are morphologically indistinguishable and yielded a 769 bp PCR amplification product. These data confirmed them as the causal agent of black rot disease prior to further pathogenic and genomic diversity analysis.

Race characterisation

In the present study, a set of seven selective cultivars of *Brassica* species were used for the race profiling of 75 strains of Xcc. Three races i.e., races 1, 4 and 6 were found in the Indian Xcc strains. Xcc strains which showed positive disease reactions in 5 cultivars [Wirosa F1 (*B. oleracea*), Just Right Hybrid Turnip (*B. rapa*), line 14R of Cobra (*B. napus*), PI 199947 (*B. carinata*) and Florida Broad Leaf Mustard (*B. juncea*)] were designated as race 1 strain. Race 4 showed positive disease reactions on only 2 cultivars i.e., Wirosa F1 (*B. oleracea* var. *capitata*) and Miracle F1 (*B. oleracea* var. *botrytis*)), whereas race 6 showed positive reactions in all seven cultivars. Out of 75 strains of Xcc, 51 strains belonged to race 1 (68%), 23 strains belonged to race 4 (31%) and a single strain Xcc-C278 belonged to race 6 (1 %). The Race 6 Xcc strain was isolated from infected leaves of cabbage (Delhi), while race 1 and race 4 strains were isolated from infected leaves of cauliflower, cabbage, broccoli, knol khol, turnip, radish, Indian mustard, vegetable mustard and black mustard. Besides these, race 1 and race 4 Xcc also infected kale and Brussels sprouts respectively.

Pathogenic variability

Pathogenic variability of Xcc was tested in two cropping seasons on 41 cultivars of crucifers including seven *Brassica* spp, namely *B. campestris, B. carinata, B. juncea, B. napus, B. nigra, B. oleracea* and *B. rapa,* and three *Brassica*s coeno species, namely *R. sativus, E. sativa* and *S. alba*. Based on the pathogenicity profiles of the 75 strains of Xcc, *B. juncea* cultivars (Cv. Pusa Bold, Varuna, Pusa Vijay, Pusa Mustard 21 and Pusa Mustard 25) showed resistance to all the races of Xcc (Table 3). However, Pusa Ageti (cabbage) showed resistant to races 1 and race 4 only. Pusa Swarnim, a cultivar of *B. carinata,* and Pusa Swarnima, a cultivar of *B. rapa var. rapa* were moderately resistant to race1, whilst the Bhawani cultivar of *B. rapa* cv. *toria* was moderately resistant to races 1 and 4 (Table 3). Among these races, race 1 dominated in all 12 agro-climatic regions of India and occurred in most of the indian states except Orissa, whereas race 4 did not occur in Punjab, Himachal Pradesh, Bihar, Madhya Pradesh, Meghalaya, Maharashtra, Goa and Rajasthan states.

Genetic diversity by rep-PCR

The patterns from the rep-PCR of all investigated strains consisted

Species of crucifers	Common name (cultivars)	Disease Reaction of *X. campestris* pv. *campestris* races on crucifer host		
		Race 1 (51)	Race 4 (23)	Race 6 (1)
Brassica juncea	Indian Mustard (Pusa Bold)	R	R	R
B. juncea	Varuna	R	R	R
B. juncea	Pusa Vijay	R	R	R
B. juncea	Pusa Mustard-21	R	R	R
B. juncea	Pusa Mustard-25	R	R	R
B. napus	GSL-1	S	S	S
B. napus	GSL-2	S	S	S
B. napus	GSL-5	S	S	S
B. napus	PAC-401	S	S	S
B. carinata	(Pusa Swarnim) IGC-01	MR	S	S
B. carinata	(Pusa Aditya) NPC-9	S	S	S
B. carinata	HC-2	S	S	S
B. carinata	Kiran	S	S	S
B.rapa cv.toria	Bhawani	S	S	S
B.rapa cv.toria	Pt-303	MR	S	S
B.rapa cv.toria	TL-15	S	S	S
B.rapa cv.toria	T-9	S	S	S
B.rapa subsp. sarson	Pusa Kalyani	S	S	S
B.rapa subsp. sarson	BSH-1	S	S	S
B.rapa subsp. sarson	BSH-9	S	S	S
B.rapa subsp. tricularis	PBT-37	S	S	S
B.rapa subsp. sarson	Pusa Gold	S	S	S
B.rapa subsp. sarson	YSH-0401	S	S	S
B.rapa subsp. sarson	NIC-394	S	S	S
Sinapis alba	IC-390162	S	S	S
Eruca sativa	RTN-314	S	S	S
E. sativa	T-27	S	S	S
B. nigra	IC-247	S	S	S
B.nigra	SRB-9	S	S	S
B.oleracea var. capitata	PusaAgeti (cabbage)	R	R	S
B.oleracea var. capitata	B. oleracea (cabbage)	S	S	S
B. oleracea var. botrytis	Pusa Sharad (cauliflower)	S	S	S
B. oleracea var. botrytis	B.o (cauliflower)	S	S	S
B. oleracea var. gongylodes	PalamTender (knol khol)	S	S	S
B. rapa var. rapa	Pusa Swarnima (Turnip)	MR	S	R
B. rapa var. rapa	Pusa Swati (Turnip)	S	S	S
R. sativus	Radish(Pusa Deshi)	S	S	S
R. caudatus	*R. caudatus*	S	S	S
Brassica oleracea var. gemmifera	Brussel Sprout (E-220 NANO)	S	S	S
B. juncea	Pusa Sag	R	R	R
Camellia sativa	*Camellia sativa*	S	S	S

Group A: (Race 1) Xcc-C189, Xcc-C186, Xcc-C187, Xcc-C188, Xcc-C125, Xcc-C126,Xcc-C118, Xcc-C205, Xcc-C206, Xcc-C207,Xcc-C114, Xcc-C128, Xcc-C210 ,Xcc-C208, Xcc-C209, Xcc-8, Xcc-C9,Xcc-C119, Xcc-C199, Xcc-C200, Xcc-C162 ,Xcc-C167 , Xcc-C161, Xcc-C178, Xcc-C166,Xcc-C216, Xcc-C217, Xcc-C218,Xcc-C131,Xcc-C130, Xcc-C116, Xcc-C120, Xcc-C129, Xcc-C23, Xcc-C4 Xcc-C6 , Xcc-C14, Xcc-C16, Xcc-C19 ,Xcc-C21,Xcc-C230, Xcc-C231, Xcc-C124, Xcc-C132, Xcc-C211, Xcc-C212, Xcc-C248 ,Xcc-C247, Xcc-C122, Xcc-C190
Group B: (Race 4) Xcc-C113, Xcc-C112,Xcc-C197, Xcc-C198 ,Xcc-C117,Xcc-C170,Xcc-C157, Xcc-C195, Xcc-C196, Xcc-C171, Xcc-C111 ,Xcc-C227 Xcc-C219, Xcc-C220,Xcc-C221, Xcc-C228 Xcc-C226, Xcc-C127, Xcc-C115, Xcc-C110, Xcc-C147, Xcc-C121,Xcc-C123
Group C: (Race 6) Xcc-C278

Table 3: Pathogenic reaction of crucifer cultivars against *X. campestris* pv. *campestris* strain under field condition. R-Resistant Reaction, MR-Moderately Resistant, S- Susptible Reaction.

of 3 to 10 fragments ranging in size from approximately 0.3 to 7 kb. Amplification profiles of each strain of Xcc varied between all three methods of PCR as 3-9 amplicons were obtained in the REP-PCR, 8-14 in ERIC and 10-16 in BOX-PCR (Figure 1).

Computer assisted analysis of rep-PCR profiles showed a very high level of diversity among the Xcc strains. All the strains of Xcc could be clustered into 6 groups at 50 % similarity coefficient. In these groups, Group 1, 2 and 5 contained 21, 12 and 28 strains of Xcc and these strains were further clustered at 75% similarity co-efficient into 20 subgroups (Figure 2). Group I was further divided in to six subgroups (IA, IB, IC, ID, IE and IF), group II divided into five subgroups (IIA, IIB, IIC, IID, IIE) and group V in to three subgroups (VA, VB and VC). Group 3 and 4 contained 10 and 3 strains respectively. Group III was further clustered into two subgroups (IIIA, IIIB) and group IV was divided into three subgroups (IVA, IVB and IVC). Group 6 contained the single isolate Xcc-C1. The 28 strains of Xcc found in group 5 were found to belong to races 1, 4 and 6. This reveals the existence of diversity among Xcc strains, which represent the diverse agro-climatic regions of India.

Multilocus sequence analysis

Three different genetic loci, i.e., hrpF, 16S rRNA and efP genes, were used separately to study the genetic diversity among five strains of Xcc representing races 1, 4 and 6. DNA sequence analysis of 16S rRNA and the 16S-23S intergenic regions has revealed a much greater diversity than previously recognised among strains leading to important revisions in the taxonomy and systematics of xanthomonads [8,25]. The partial nucleotide sequences of the 16S rRNA of four strains of Xcc i.e., Xcc-C18 (race1, turnip), Xcc-C106 (race 4, vegetable mustard), Xcc-C131 (race 1, cauliflower) and Xcc-C168 (race 4, cabbage) and Xcc-C278 (race 6, cabbage) showed more than 99.8% identity to each other and they were very closely related to Xcc strain ATCC 33913 (Race 3, Brussels sprout, UK as earlier reported by Popovic et al. [27]. These strains also showed more than 99.8% identity with *X. arboricola* pathovars *glycines, punicae, poinsettiicola* LMG 8676, *X. campestris* pv. *viticola* and *X. campestris* LMG 5793, *X. citri* subsp. *citri* Aw12879, *X. gardneri* DSM 19127 (bacterial leaf spot disease of tomato), *X. meloni, X. dye* 12167 and *X. populi.* (Figure 3). Our result finding was corroborative with the result obtained by Popovic et al. [27].

The *ef*P gene sequence analysis of the representative Indian Xcc isolates showed high similarity identity matrix (>98%) among each other and were very closely related to the sequence of strains Xcc B100 (race 1) and Xcc 8004 (race 9). However, the sequence of Xcc strain ATCC 33913 (race 3), was distinct from those of other strains of Xcc (Figure 4) but showing more than 95% with identity *X. campestris* pv. *rapahani* 756C, *X. campestris* pv. *vesicatoria* and *X. axonopodis* pv. *citrumelo.* For genetic diversity assessment, the *ef*P gene was earlier used by Achtman et al. [16] to study the population genetic structure and diversity of *H. pylori,* which revealed about 85% relatedness for strains from different geographical regions. Based on their *ef*P sequences, all the Xcc isolates analysed here grouped in the same cluster (Group 1 and Figure 4), except Xcc ATCC 33913 so this phylogenetic analysis

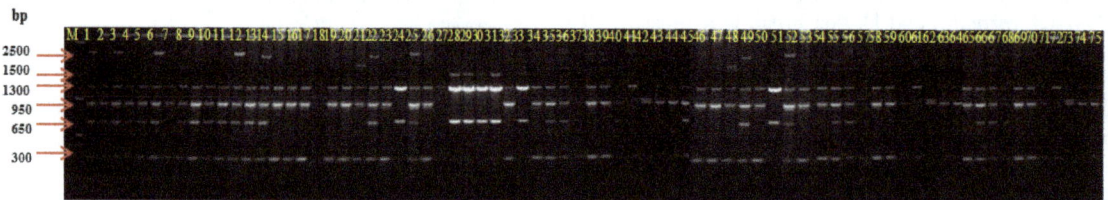

Fig. 1a. Box Fingerprints

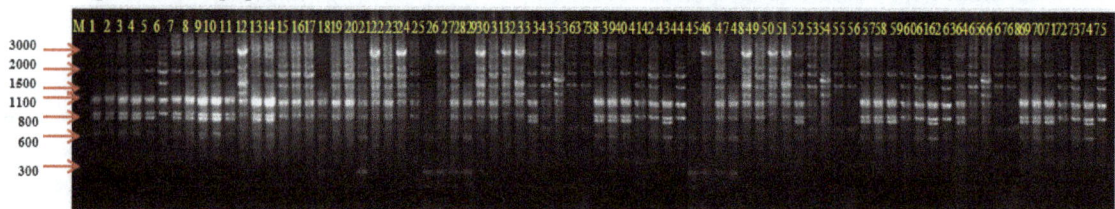

Fig. 1b. ERIC Fingerprints

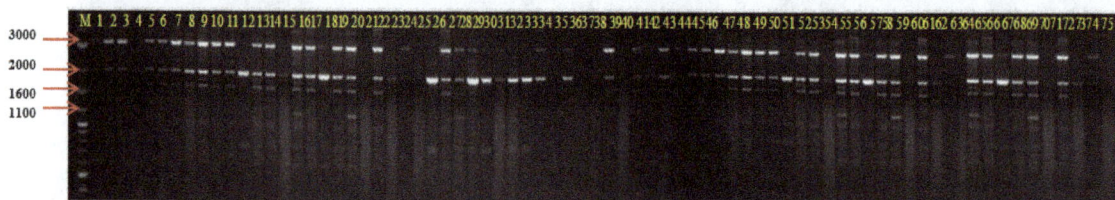

Fig. 1c. REP Fingerprint

Lane M: 1.0 Kb DNA ladder, lanes 1-75: Xcc-C4, Xcc-C6, Xcc-C8, Xcc-C9, Xcc-C14, Xcc-C16, Xcc-C19, Xcc-C21, Xcc-C23, Xcc-C110, Xcc-C111, Xcc-C112, Xcc-C113, Xcc-C114, Xcc-C115, Xcc-C116, Xcc-C117, Xcc-C118, Xcc-C119, Xcc-C120, Xcc-C121, Xcc-C122, Xcc-C123, Xcc-C124 Xcc-C125, Xcc-C126, Xcc-C127, Xcc-C128, Xcc-C129, Xcc-C130, Xcc-C131, Xcc-C132, Xcc-C147, Xcc-C157, Xcc-C161, Xcc-C162, Xcc-C166, Xcc-C167, Xcc-C170, Xcc-C171, Xcc-C178, Xcc-C185, Xcc-C187, Xcc-C186, Xcc-C187, Xcc-C188, Xcc-C189, Xcc-C190, Xcc-C195, Xcc-C196, Xcc-C197, Xcc-C198, Xcc-C199, Xcc-C200, Xcc-C205, Xcc-C206, Xcc-C207, Xcc-C208, Xcc-C209, Xcc-C210, Xcc-C211, Xcc-C212, Xcc-C216, Xcc-C217, Xcc-C227, Xcc-C228, Xcc-C230, Xcc-C231, Xcc-C247, Xcc-C248, Xcc-C278. Xcc-C218, Xcc-C219, Xcc-C220, Xcc-C221, Xcc-C221, Xcc-C226,

Figure 1: Larvicidal activity of *Callicarpa americana* leaf extracts against fourth instar larvae of *Aedes albopictus* at various concentrations. Percent mortality was determined 24 hours after treatment. Mosquitoes that were non-responsive to physical stimulation were counted as dead. Error bars represent the standard error of the mean of three independent assays.

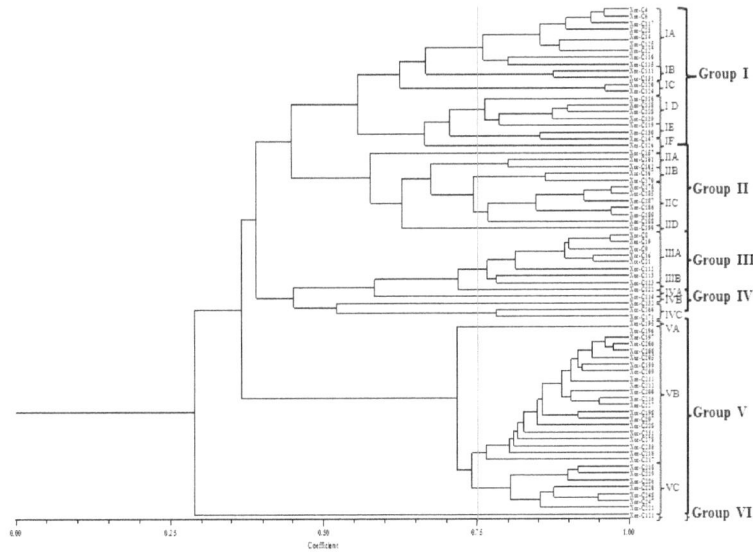

Figure 2: Combined Dendrogram representing genetic relationships between the strains of *X. campestris* pv. *campestris* based on BOX, ERIC and REP-PCR fingerprints. Similarity (%) between patterns was calculated by using the Jaccard's coefficient. The data were sorted by using the UPGMA clustering method.

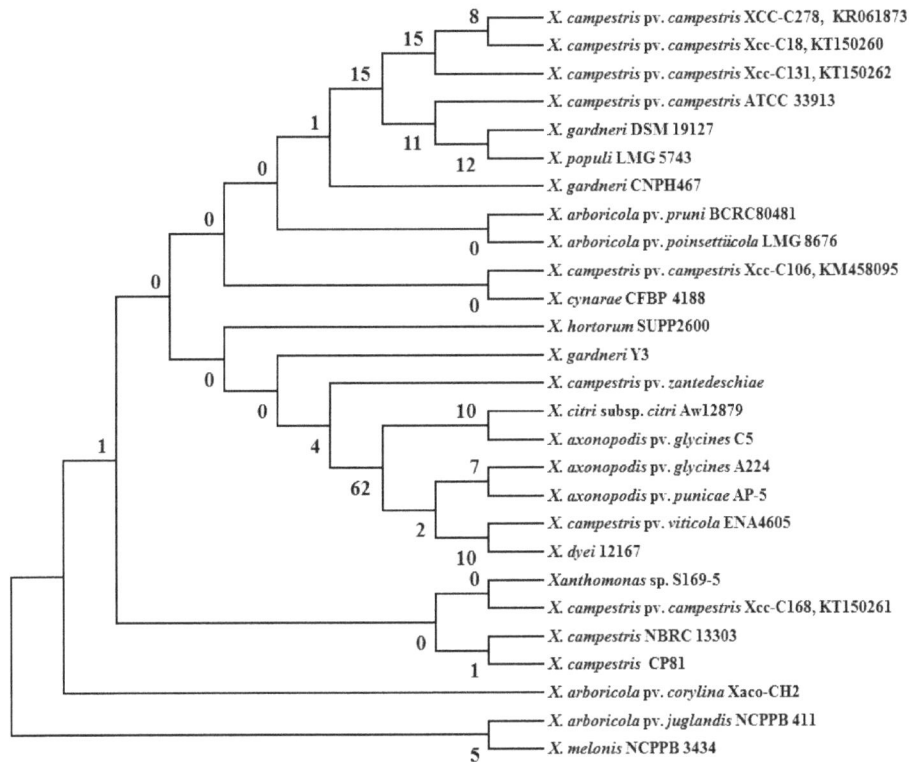

Figure 3: Phylogenetic tree of 16S rRNA sequences of 5 Indian isolates of *X. campestris* pv. *campestris* and 22 nucleotide sequences data obtained from NCBI database. The evolutionary history was inferred using the Neighbor-Joining method by using MEGA 6.0.

showed very clear differentiation at the pathovar level. However, we could not find any relation of races with sequence similarity of the *efP* gene.

Phylogenetic sequevar analysis of *hrp*F genes demonstrated that the 5 representative Indian isolates formed a distinct cluster from the strains of Xcc from other countries, obtained from NCBI database. Out

of 8 strains of Xcc and one strain of *X. campestris* pv. *raphani*, 5 were clustered in Group 1 and 3 strains in Group 2 (Figure 5). The highly similar sequences of Xcc B100 and Xcc 8004 taken from NCBI Genbank were clustered in Group 2. The sequence from the five representative Indian Xcc strains showed more than 90% homology with each other and also with Xcc strain ATCC33913 and *X. campetris* pv. *raphani* 756C in group 1 (Figure 5).

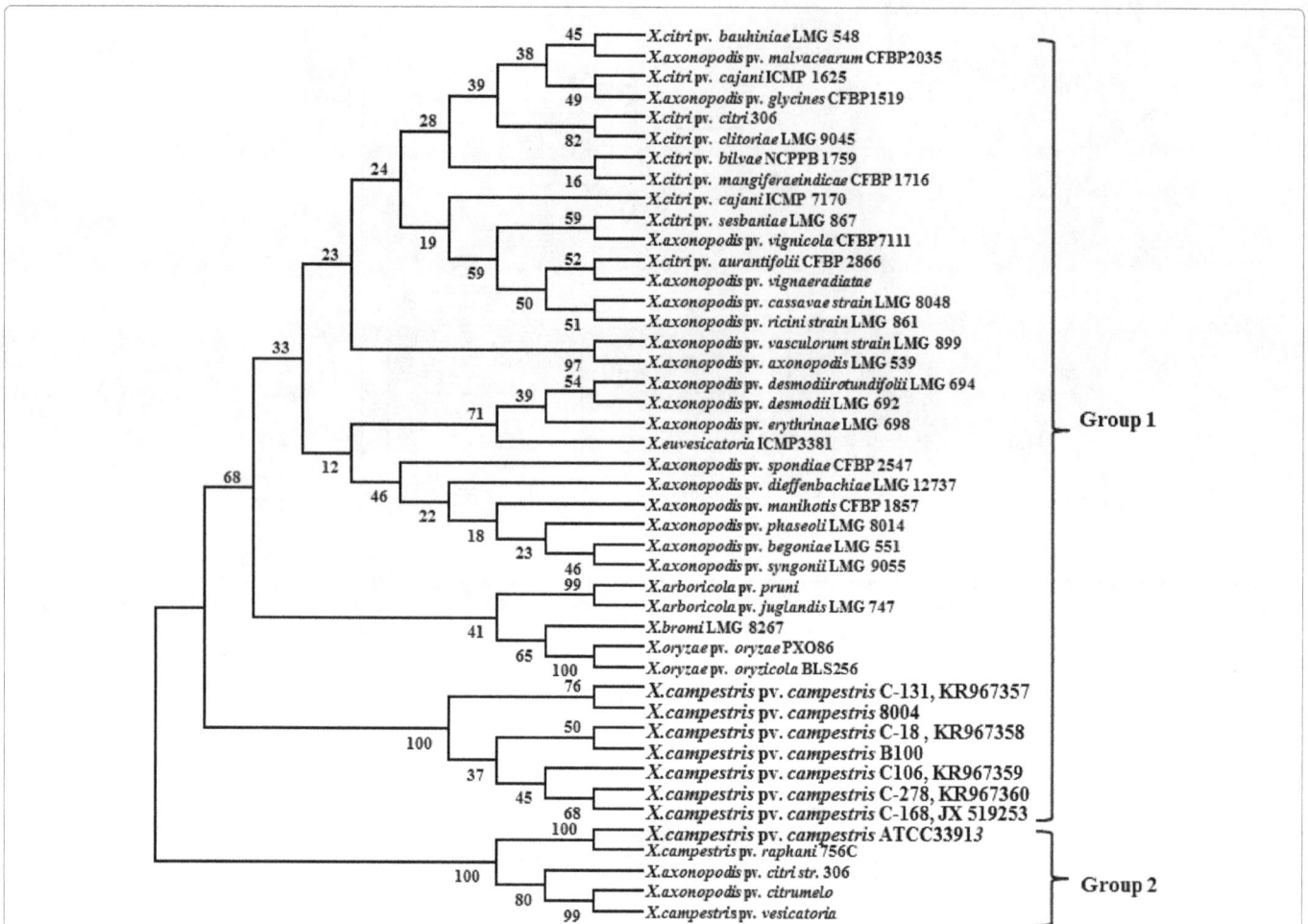

Figure 4: Phylogenetic tree of *efP* gene sequences of 5 Indian isolates of *X. campestris* pv. *campestris* and 44 nucleotide sequences data obtained from NCBI database. The evolutionary history was inferred using the Neighbor-Joining method by using MEGA 6.0.

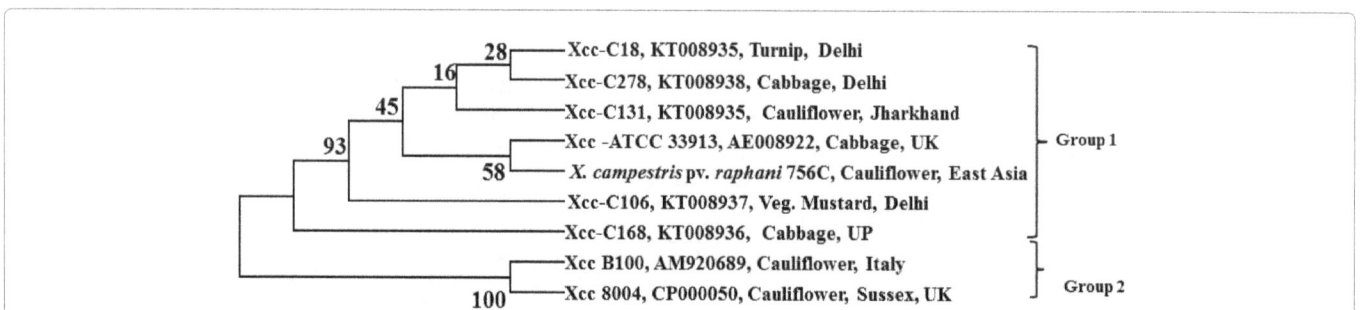

Figure 5: Taxonomic analysis of partial *hrpF* gene sequences of *X. campestris* pv. *campestris* strains using neighbor-joining phylogenetic distribution. The dandrogram was generated by Mega (version 6.06) software using UPGMA mean applied to the distance matrix of nucleotide differences. Number at branch point indicates percent bootstrap support for 1000 interactions.

Discussion

Black rot disease is widely distributed in all agro-climatic regions of India. The isolates of Xcc were collected from 12 agro-climatic regions under 18 states of India from 8 crucifer crops. In the present studies, black rot producing isolates of Xcc were characterized by pathogenicity test and advanced technique like PCR based molecular technique using 16srRNA, *efP*, *hrp* gene [13,14] which is to highly conserved to enable differentiation of the pathovars [20]). All the isolates were identical in classical tests and produced typical black rot disease symptom as

V- shaped yellowing and necrosis of tissue and blackening of veins. Crucifers especially cole crops, tubers and oliferous crops belonging to *Brassica* and *Raphanus* species are affected by black rot incited by Xcc.

In the present work, race characterization had been done in India. Presently, nine races of Xcc are known in the world and among them races 1 and 4 are predominant [18], which is corroborated by our findings. Race typing of 51 strains of Xcc by Vicente [28] found three races (1, 4 and 6) in Portugal. While, Jensen et al. [1] reported races 1, 3, 4, 5, 6 and 7 in cabbage from Nepal and among them 1, 4 and 6 were the most common. In our study also race 1 was found to be predominant followed by race 4 in *B. oleracea* cultivars in the all agro-climatic regions in India, whereas race 6 was found only once on cabbage in Delhi state. In contrast Lema et al. [29] isolated race 6 from *B. rapa*. However, races 2, 3, 5, 7, and 9 were not found in India. In the present study, all isolates of Xcc produced typical V- shaped symptoms of black rot disease on the crucifer crops.

In recent studies, using defined races, race specific resistance in different collections of *B. oleracea*, *B. rapa* and *B. napus* have been identified [30]. All the tested cultivars of crucifers were challenged with concentrated suspensions of Xcc, using leaf clipping of young plants to establish the extent and type of resistance in all tested species of *Brassica* and *R. sativus* which clearly favours the pathogen and the rapid development of symptoms. The black rot symptoms appeared within six days in susceptible cultivars, while symptom development was delayed up to15 days in resistant cultivars.

Genetic diversity of plant pathogenic bacteria has been studied by DNA based approaches to generate evidence of genome plasticity, ecological distribution, dispersal and evolution. Knowledge of the existence of genetic variability in the pathogen population is important for plant breeding and consequently in crop improvement program. For genomic variability, rep-PCR fingerprinting using primer sets (BOX, REP and ERIC) are highly conserved repetitive sequences [31], showed different banding pattern among Xcc isolates of different crucifer hosts in India. In our studies, similar levels of resolution were observed compared with previous rep-PCR studies of *Xanthomonas* [1,2,32]. Rep-PCR has proven to be a reliable fingerprinting tool reinforcing its applicability to track Xcc strains in culture collections and for quality control of commercial inoculants. It is a rapid, low-cost, and reliable method that has been extensively used to assess the genetic diversity of Xcc strains [2,29] and has potential to determine the diversity within Xcc similar to that of other techniques such as pulse-field gel electrophoresis or amplified fragment length polymorphism. Moreover, the method is adequate for grouping or defining species or even pathovars and strains [29]. Rep-PCR has also been used successfully to characterise a large number of bacteria and differentiated closely related strains of bacteria [1,31]. The data obtained from the pathogenicity and genetic diversity analysis in this study confirmed previous findings describing the heterogeneity within Xcc strains. The strains within the groups revealed complex but polymorphic bands resulting in grouping of closely related strains based on fingerprinting patterns. However, the strains recovered from samples from the same host and states/agro-climatic region with the same race did not fall into one group but into subgroups spread over the dendrogram. Similar results were also reported by Jensen et al. [1] from Nepal, Valverde et al. [10] from Israel and Zaccardelli et al. [9] from Italy. Although, a previous study had demonstrated a correlation of races with their geographical origin, [33] but we found that the strains of Xcc could not be separated on the basis of races and their geographical origin. Here, three different genes were used separately to study genetic diversity of five strains of Xcc representing races 1, 4 and 6. The results from 16S rRNA and *hrp*F gene analyses revealed diversity at the genus and species level and *ef*P gene can be used to differentiate at the pathovar level. None of these three genes are able to differentiate at the race level. This study revealed the existence of variations within Indian isolates of Xcc.

The evolutionary processes occurring in the pathogen and how pathogen populations adapt to changes in host resistance or susceptibility are equally important in the evolutionary arms race. Race study is depend on simpler gene-for-gene model has been proposed by He et al. [34] based on the interactions between Xcc isolates and cultivars of *Brassica* (*B. juncea*, *B. oleracea*, *B. rapa*), radish (*Raphanus sativus*) and pepper (*Capsicum annuum*). Arms races are ultimately determined at the phenotypic level, but examining individual antagonistic genetic systems, such as resistance (R) genes and their corresponding avirulence (Avr) genes, and cell-wall attacking enzymes and their inhibitors, can provide new insights. In our study we found three races, which are distinguished based pathogenic reactions on the hosts and they are directly related to effector proteins. Sometime very similar results were found between plant and human pathogenic bacteria. A current study about the evolution linkage between human pathogenic bacteria and *Xanthomonas* AvrRxo1-ORF1 has been reported by Qian Han et al. [35]. However, evolution of races in Xcc is required further study. Evolutionary analysis can suggest how arms races are produced, and to what extent genetic variation are shaped by co-evolutionary outcomes.

Conclusion

Seventy five Xcc isolates were characterized into three races, i.e., races 1, 4 and 6. Pathogenic study will helpful to develop resistant cultivar of crucifers. This study revealed the existence of variations within Indian isolates of Xcc. In this study, although some isolates obtained from the same source produced similar BOX, ERIC, REP-PCR profiles, other strains showed different profiles despite belonging to the same race. On the other hand, some isolates had similar profiles despite belonging to different races. This suggests that rep PCR can be used for rapid initial screening of isolates to select non-identical ones for further analysis, including race typing. However, our results showed that rep-PCR, 16S rRNA, *hrp*F and *ef*P sequence analysis were not suitable for differentiation of Xcc races, presumably because the races are dependent on which different combinations of avirulence genes in the pathogen population will be recognised by different combinations of resistance genes in the host (the 'gene for gene' hypothesis).

Acknowledgements

The authors are thankful to SERB, Department of Science and Technology, New Delhi for providing financial support under project entitled "Characterization and identification of *Xanthomonas campestris* pv. *campestris* races causing black rot disease of crucifers" to conduct the experiments. The authors are also thankful to Heads, Division of Plant Pathology, IARI, New Delhi and Department of Botany, Mahila Mahavidyalya, Banaras Hindu University, Varanasi-221005 (India) for their keen interest and help throughout the course of these investigations.

References

1. Jensen BD, Vicente JG, Manandhar HK, Roberts SJ (2010) Occurrence and diversity of *Xanthomonas campestris* pv. *campestris* in vegetable brassica fields in Nepal. Plant Disease 94: 298-305.

2. Singh D, Dhar S, Yadava DK (2011) Genetic and pathogenic variability of Indian strains of *Xanthomonas campestris* pv. *campestris* causing black rot disease in crucifers. Curr Microbiol 63: 551-560.

3. Kamoun S, Kamdar HV, Tola E, Kado CI (1992) Incompatible interactions between crucifers and *Xanthomonas campestris* involve a vascular hypersensitive response: role of the hrpx locus. Mol. Plant Microbe Interact 5: 22-23.

4. Ignatov A, Hidam K, Kuginuki Y (1999) Pathotypes of *Xanthomonas campestris* pv. *campestris* in Japan. Acta Phytopathologica et Entomologica Hungarica 34: 177-182.

5. Taylor JD, Conway J, Roberts SJ, Astley D, Vicente JG (2002) Sources and Origin of Resistance to *Xanthomonas campestris* pv. *campestris* in Brassica Genomes. Phytopathology 92: 105-111.

6. Vicente JG, Taylor JD, Sharpe AG, Parkin IA, Lydiate DJ, et al. (2002) Inheritance of Race-Specific Resistance to *Xanthomonas campestris* pv. *campestris* in Brassica Genomes. Phytopathology 92: 1134-1141.

7. Louws FJ, Fulbright DW, Stephens CT, DeBruijn FJ (1994) Specific genomic fingerprints of phytopathogenic *Xanthomonas* and Pseudomonas pathovars and strains generated with repetitive sequences and PCR. Appl Environ Microbiol 60: 2286-2295.

8. Gonçalves ER, Rosato YB (2002) Phylogenetic analysis of *Xanthomonas* species based upon 16S-23S rDNA intergenic spacer sequences. Int J Syst Evol Microbiol 52: 355-361.

9. Zaccardelli M, Francesco C, Annalisa S, Massimo M (2007) Detection and identification of the crucifer pathogen, *Xanthomonas campestris* pv. *campestris*, by PCR amplification of the conserved Hrp/type III secretion system gene hrc C. Euro J Plant Pathol 118:299-306.

10. Valverde A, Hubert T, Stolov A, Dagar A, Kopelowitz J, et al. (2007) Assessment of genetic diversity of *Xanthomonas campestris* pv. *campestris* isolates from Israel by various DNA fingerprinting techniques. Plant Pathology 56: 17-25.

11. Higgins CF, Ames GF, Barnes WM, Clement JM, Hofnung M (1982) A novel intercistronic regulatory element of prokaryotic operons. Nature 298: 760-762.

12. da Silva AC, Ferro JA, Reinach FC, Farah CS, Furlan LR, et al. (2002) Comparison of the genomes of two *Xanthomonas* pathogens with differing host specificities. Nature 417: 459-463.

13. Berg T, Tesoriero L, Hailstones DL (2006) A multiplex real-time PCR assay for detection of *Xanthomonas campestris* from brassicas. Lett Appl Microbiol 42: 624-630.

14. Singh D, Dhar S (2011) Bio-PCR based diagnosis of *Xanthomonas campestris* pv. *campestris* in black rot of infected leaves of crucifers. Indian Phytopathology 64: 7-11.

15. Aoki H, Dekany K, Adams SL, Ganoza MC (1997) The gene encoding the elongation factor P protein is essential for viability and is required for protein synthesis. J Biol Chem 272: 32254-32259.

16. Achtman M, Azuma T, Berg DE, Ito Y, Morelli G, et al. (1999) Recombination and clonal groupings within Helicobacter pylori from different geographical regions. Mol Microbiol 32: 459-470.

17. Schaad NW, Jones J B, Lacy GH (2001) *Xanthomonas*. Laboratory guide for identification of plant-pathogenic bacteria, American Phytopathological Society Press, St. Paul.

18. Vicente JG, Conway J, Roberts SJ, Taylor JD (2001) Identification and Origin of *Xanthomonas campestris* pv. *campestris* Races and Related Pathovars. Phytopathology 91: 492-499.

19. Murray MG, Thompson WF (1980) Rapid isolation of high molecular weight plant DNA. Nucleic Acids Res 8: 4321-4325.

20. Singh D, Raghavendra BT, Rathaur PS, Singh H, Raghuwanshi R, et al. (2014) Detection of black rot disease causing pathogen *Xanthomonas campestris* pv. *campestris* by bio-PCR from seeds and plant parts of cole crops. Seed Sci Technol 42: 36-46.

21. Rohlf FJ (2000) NTSYS-pc: Numerical Taxonomy and Multivariate Analysis System, Version 2.2. Exeter Software. Setauket, New York.

22. Thompson JD, Gibson TJ, Plewniak F, Jeanmougin F, Higgins DG (1997) The clustal X windows interface: flexible strategies for multiple sequence alignment aided by quality analysis tools. Nucleic Acids Res 25: 4876-7882

23. Tamura K, Stecher G, Peterson D, Filipski A, Kumar S (2013) MEGA6: Molecular Evolutionary Genetics Analysis version 6.0. Mol Biol Evol 30: 2725-2729.

24. Saitou N, Nei M (1987) The neighbor-joining method: a new method for reconstructing phylogenetic trees. Mol Biol Evol 4: 406-425.

25. Felsenstein J (1985) Confidence limits on phylogenies: an approach using the bootstrap. Evolution 39: 783-791.

26. Hauben L, Vauterin L, Swings J, Moore ER (1997) Comparison of 16S ribosomal DNA sequences of all *Xanthomonas* species. Int J Syst Bacteriol 47: 328-335.

27. Popovic T, Josic D, Starovic M, Milovanovic P, Dolovac N, et al. (2013) Phenotypic and genotypic characterization of *Xanthomonas campestris* strains isolated from cabbage, kale and broccoli. Arch Biol Sci Belgrade 65: 585-593

28. Vicente JG (2004) A podridaonegra das cruciferas. (Lopes, G., ed) Alcobaca: COTHN Centro Operativo e Tecnologico Hortofruticola.

29. Lema M, Cartea ME, Sotelo T, Velasco P, Soengas P (2012) Discrimination of *Xanthomonas campestris* pv. *campestris* races among strains from northwestern Spain by Brassica spp. genotypes and rep-PCR. Eur J Plant Pathol 133: 159-169.

30. Ignatov A, Kuginuki Y, Hida K (2000) Distribution and inheritance of race-specific resistance to *Xanthomonas campestris* pv.*campestris* in Brassica rapa and B. napus. J Russian Phytopathol Soc 1: 89-94

31. Versalovic J, Koeuth T, Lupski JR (1991) Distribution of repetitive DNA sequences in eubacteria and application to fingerprinting of bacterial genomes. Nucleic Acids Res 19: 6823-6831.

32. Vicente JG, Everett B, Roberts SJ (2006) Identification of Isolates that Cause a Leaf Spot Disease of Brassicas as *Xanthomonas campestris* pv. raphani and Pathogenic and Genetic Comparison with Related Pathovars. Phytopathology 96: 735-745.

33. Massimo Z, Francesco C, Annalisa S, Massimo M (2007) Detection and identification of the crucifer pathogen, *Xanthomonas campestris* pv. *campestris*, by PCR amplification of the conserved Hrp/ type III secretion system gene hrc C. Eur J Plant Pathol 118: 299-306.

34. He YQ, Zhang L, Jiang BL, Zhang ZC, Xu RQ et al. (2007) Comparative and functional genomics reveals genetic diversity and determinants of host specificity among reference strains and a large collection of Chinese isolates of the phytopathogen *Xanthomonas campestris* pv. *campestris*. Genome Biology 8: R218.

35. Han Q, Zhou C, Wu S, Liu Y, Triplett L, et al. (2015) Crystal Structure of *Xanthomonas* AvrRxo1-ORF1, a Type III Effector with a Polynucleotide Kinase Domain, and Its Interactor AvrRxo1-ORF2. Structure 23: 1900-1909.

20

Screening for Resistance to Crown Rust in Oat Genotypes through Morphological and Molecular Parameters

Yogesh Ruwali, Lalan Kumar* and JS Verma

G.B. Pant University of agriculture and technology, Pantnagar, U.S.Nagar, Uttarakhand, India

Abstract

A collection of 20 oat genotypes from different sources were evaluated in small isolated field plots for crown rust severity in natural epidemics with virulent *P. coronata* isolates and further screened with linked molecular markers. Large variation was observed for disease severity under field conditions in spreader plot. Genotypes with partial resistance due to a reduction of disease severity in spite of a compatible interaction (rust score 3) and moderate susceptibility were identified. The twenty genotypes displaying the variable disease severity with visible necrosis were selected for further studies regarding presence or absence of major genes (Pc91 and Pc68) for resistance to crown rust. In field nurseries, on the basis of latency period and disease severity (DS) none of the twenty genotypes fell into the resistant pool (score 1). Most of them showed a prolonged latency period, reduced infection frequency and colony size, and increased percentage of early aborted colonies not associated with host cell necrosis. Result on screening for crown rust resistance in field screening and by linked markers showed the absence of major resistance gene in the target population, several advance lines were identified as moderately resistant to crown rust reaction.

Keywords: Partial resistance; *Puccinia coronate*; Oat; SCAR; NCBI

Introduction

Despite being high fed fodder crop, oat is now gaining importance due to its unique and important quality characteristics, particularly lipid and protein. Oat is predominantly used as green fodder in north India where crown rust has started making its appearance in oat fields. Crown rust, caused by *Puccinia coronata*, is one of the most destructive diseases of oat (*Avena sativa* L.) in major oat growing countries. Over the past 10 years, yield losses of 10 to 20% in oat due to crown rust were reported for various American states from 1991 to 1993 [1]. Crown rust is most important where dews are frequent and temperatures are mild (15-25°C) during the oat growing season, which is a characteristic climate of Pantnagar region. More than 100 race-specific resistance genes to crown rust have been identified out of which 96 were defined as Pc, with the majority considered to be dominant genes [2]. The resistance caused by these genes is typically race-specific, expressed as a hypersensitive reaction, and of limited durability. The non-durability of this resistance has caused breeders to look for more durable types of resistance such as partial resistance (PR). PR has been identified in barley and oat and is expressed as a reduced rate of epidemic development despite a compatible interaction i.e. high infection type [3]. Pc91 is a major crown rust resistance gene effective at all stages of plant development [4] also Pc68, on the other hand, which was introgressed in *A. sativa* from *Avena sterilis* L. [5], is considered to be one of the most effective genes against this disease. The objectives of this work were to screen the oat genotypes for crown rust resistance, to deduce the status of major genes for resistance and to identify useful material for resistance breeding programs.

Materials and Methods

Field studies

The observation on crown/leaf rust reaction was carried out in a spreader plot at the Instructional Dairy Farm of G. B. Pant University of Agriculture & Technology, Pantnagar, Uttarakhand, India where seeds from Kent and UPO 270 were mixed in equal proportion and then sown as spreader rows on 15th October 2010 in two replications leaving the space for test entries (Figure 1). Row to row distance was kept 20 cm and the experimental test material was sown 20 days after the planting of spreader rows i.e. on 5th November 2010 in between the spreader lines (Figure 2) and separately in an isolated normal plot. Each test entry was represented by 40-45 plants in a 2 m long single row in spreader plot. Artificial inoculation was done at 20 days of planting of spreader rows, by spraying the solution of crown rust spores, cultured from previous year's infected oat leaves, over the

Figure 1: Advance spreader row planting.

*Corresponding author: Lalan Kumar, G.B. Pant University of agriculture and technology, Pantnagar, U.S. Nagar, Uttarakhand, India
E-mail: Lalan53singh@gmail.com

Figure 2: Spreader row along with test entries.

No.	Genotype	Source	No.	Genotype	Source
1.	D. Sel.-1	Pantnagar	11.	UPO-275	Pantnagar
2.	D. Sel.-5	Pantnagar	12.	Kent	Australia
3.	D. Sel.-6	Pantnagar	13.	UPO-212	Pantnagar
4.	Wright	U.S.A	14.	No.-1	Pantnagar
5.	HFO-114	Haryana	15.	OS-6	Haryana
6.	OL-125	Ludhiana	16.	EC-605833	Exotic
7.	UPO-265	Pantnagar	17.	EC-605836	Exotic
8.	EC-246199	Exotic	18.	EC-605838	Exotic
9.	UPO-271	Pantnagar	19.	UPO-260	Pantnagar
10.	UPO-273	Pantnagar	20.	UPO-270	Pantnagar

Table 1: List of oat genotypes evaluated for crown rust resistance.

spreader rows. Appropriate moisture was maintained in the spreader plot, for optimum growth conditions of the pathogen throughout the crop growth period, by partial irrigation (Table 1) [6].

Laboratorial studies

CTAB procedure was used for isolation of DNA [7]. Further PCR amplification by using the 25 µl reaction mixture containing 1X KCl buffer (Fermentas) containing 0.2 mM dNTPs, 30 ng of each forward and reverse primer, 1.5 mM $MgCl_2$, 0.8 U Taq DNA polymerase (Fermentas) and 100 ng of DNA. Thermal cycler reaction were carried out according to the following temperature profile 4 min initial denaturation at 94°C; 37 cycles of 94°C for 1 min, varying annealing Tm according to primer for 45 s, 72°C for 1 min and final extension of 7 min at 72°C; and final hold 4°C. All amplifications were performed twice and independently to make sure that the results were correct. Electrophoresis was done at 50 V for 4 h in 1 X TBE electrophoresis buffer for SCAR. Gels were documented using Gel Doc system (Bio-Rad) and electrophoresis of amplified product was done sequentially. All the primers used in the study were synthesized by Merck Specialties Pvt. Ltd. In laboratory screening through molecular markers, three primers were used. Primer 1 and primer 2 are SCAR markers linked to Pc91 reported by McCartney et al. [4] while primer 3 was designed by using software primer 3+ available at NCBI website, using the sequence of Pc68 gene submitted by Satheeskumar et al. [8] (Table 2).

Observation and scoring assessment

All the twenty entries were evaluated for crown rust reaction separately. To estimate latency period (LP) the number of days was

recorded since the seedling emergence till the first disease symptom manifestation for all entries. Disease severity (DS) was estimated two times during the growing season in terms of the percentage of leaves covered by the lesions (orange-yellow spores exposed by rupture of the leaf epidermis) at 30 days after sowing of test entries and then at 50% heading stage. The observation was taken by counting the number of lesions with its size of 1.0 cm and above. Based on the frequency of lesions and number of infected leafs, the 0 to 9 scale was used and the entries were scored accordingly. At maturity biological yield of all test entries was measured by weighing 5 plants randomly selected from the spreader plot and normal irrigated plot. The genotype with lowest LP, maximum number of infected leaves, lesion number at both stages of observation and maximum decrease in biological yield was given the score 9 (most susceptible), and rest were accordingly scored.

Results

Field reaction of genotypes

Disease severity (DS) ranged from very high to low, and the frequency distribution was markedly skewed towards high DS (Figure 3). The observation on leaf rust reaction was recorded two times in the spreader row plot first at 30 days after planting of the entries and again at 50% heading stage. Rust scoring was done as per ICARDA rust scale (0-9) depending upon the number of infected leaves, the number of lesions and decrease in biological yield in stressed conditions. Four of selected genotypes viz. D. Sel.-1, D. Sel.-5, EC-605838 and UPO-260 showed a significantly longer relative latency period (RLP) than rest of the genotypes evident from the late appearance of rust spores in them. The relative infection frequency (RIF) of these genotypes was significantly lower than the susceptible pool genotypes as visible leaf necrosis in them was less and also showed least decrease in biological yield in stressed conditions (Table 3).

Laboratory results

The three crown rust linked primers did not gave amplification

Sr.		Primer seq 5'- 3'	Phase	%GC	Tm
1	F	GGACTATCTAGTTTATGGAGGAG	Dominant-coupling	43.0	56.6
	R	AGGCAAAACGAGCAGTGTAA		45.0	62.3
2	F	CTTGTATTGTGCGTTGGAA	Dominant-repulsion	45.0	59.2
	R	CTTGTATTGTGCGTTGGAA		42.0	59.7
3	F	CGAGGGCTACATTCAAGAGC	Designed by primer 3+ (NCBI)	55.0	60.0
	R	TCCAAGGGTCTTCTCGGTAA		50.0	59.7

Table 2: Details of rust resistance linked primers used in the study of oat genotypes.

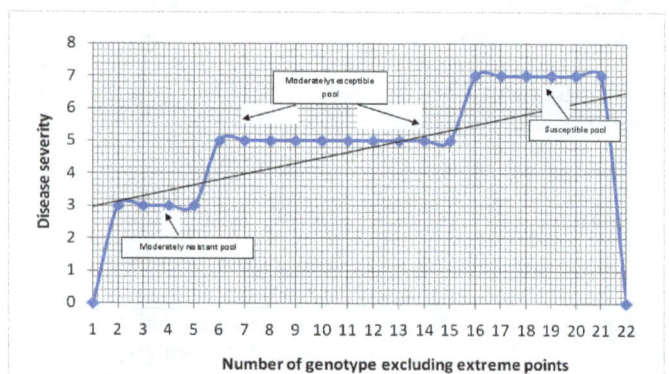

Figure 3: Frequency distribution of test genotypes against disease score/severity.

Screening for Resistance to Crown Rust in Oat Genotypes through Morphological and Molecular...

129

S.No	Genotype	Biological yield/ plant (gm)		Rust scores*	Amplification with linked primer	
		Normal Plot	Spreader Plot		Pc91	Pc68
	D. Sel.-1	76	71.8	3	nil	nil
2	D. Sel.-5	55.7	55.1	3	nil	nil
3	D. Sel.-6	53.9	40.8	7	nil	nil
4	Wright	52	48.4	5	nil	nil
5	HFO-114	54	53.3	5	nil	nil
6	OL-125	60.1	49.6	7	nil	nil
7	UPO-265	54	48.4	7	nil	nil
8	EC- 246199	61	56.7	5	nil	nil
9	UPO-271	55.7	51	5	nil	nil
10	UPO-273	46	42.5	5	nil	nil
11	UPO-275	60.3	56.3	5	nil	nil
12	KENT	51.7	43.1	7	nil	nil
13	UPO-212	46.7	44	5	nil	nil
14	No.- 1	62.2	57.1	5	nil	nil
15	OS- 6	45.1	36.4	7	nil	nil
16	EC-605833	54.7	51.6	5	nil	nil
17	EC-605836	54.6	49	5	nil	nil
18	EC-605838	50.1	50.5	3	nil	nil
19	UPO-260	52.7	51.8	3	nil	nil
20	UPO-270	62.7	39.8	9	nil	nil

*1=resistant, 3=moderately resistant, 5=moderately susceptible, 7=susceptible, 9=highly susceptible

Table 3: Details of selected oat genotypes in the field and with SCAR markers.

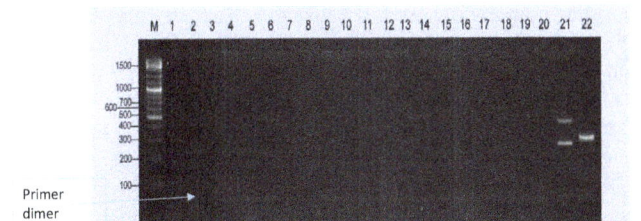

Figure 4: PCR based amplification of all 20 genotypes along with control genotype by SCAR Pc68 and Pc91 crown resistance gene marker.

bands with any of the test genotype even after three precautious reaction procedures. Only primer-dimer were visible in the gel-doc visualization; however the control crown rust resistance genotype Amalgam and starter which were donor of Pc91 and Pc68 gives the particular band with all three primers (Figure 4) thus its confirmed that no complimentary sequence was available in the target genotype DNA. Based on the results obtained by screening of the oat genotypes with linked primer (Table 3) it can be stated that none of the two major gene viz., Pc91 and Pc68 responsible for crown rust resistance, were present in the oat genotype evaluated. It was further supported by the field scoring data where none of the genotypes has fallen in the resistant pool (rust score 1).

Discussion

Puccinia coronata, is one of the most destructive diseases causing agent of Crown rust in (*Avena sativa* L.) in major oat growing countries. The infection behaviour of this disease was varied widely amongst the susceptible and resistance genotype as well as stressed conditions. This may be due to unfavourable growth behaviour of *Puccinia coronata*. It was further supported by the field scoring data where none of the genotypes has fallen in the resistant pool (rust score 1). Thus, the resistance difference among the 20 test genotypes can be generalized by

acknowledging the absence of the two major resistance genes among them and other minor resistance genes at play, i.e. for the genotypes showing moderate resistance reaction with rust reaction score 3 it can be generalized that this type of partial resistance could be by virtue of genes with minor effect for crown rust resistance [9] and for genotypes with moderate susceptibility and susceptibility reactions for crown rust it can be generalized that the pathogen has evolved with matching virulent genes.

Absence of the two major gene i.e. Pc91 and Pc68 for resistance to crown rust in the experimental material of present study may be compensated by the fact that several advance generation improved lines (D. Sel.-1 and D. Sel.-5, UPO-260), and exotic material (EC-605838) have been found partially resistant to crown rust reaction, thus can be used against crown rust in more affected regions. The observation in the present investigation for crown rust reaction may have been confounded by presence of other pathogens in the field, causing induced resistance/susceptibility reaction [10,11], but the authenticity been provided by SCAR markers for absence of major resistance gene in the test genotypes. As the population of suitable genotype (matching races) of pathogen, increases in the field location where the cropping of single oat genotype may prove fatal because of the rapid buildup of infection [12]. Thus there is an urgent requirement for the incorporation of major genes for resistance in different oat genotypes and subsequently finding new genes for crown rust resistance, so that growing oats in field remains economical in future.

Acknowledgements

The authors gratefully acknowledge the HOD, Genetics and Plant Breeding and Incharge Molecular Marker Laboratory, PCPGR, Pantnagar for providing required laboratory facilities. Financial support received from the Directorate of Experimental Station, Pantnagar is also duly acknowledged.

References

1. Long DL, Hughes ME (2001) Small Grain Losses Due to Rust. USDA-ARS Cereal Dis. Lab., University of Minnesota. On-line publication CDL-EP#007.

2. Zhu S, Kaeppler HF (2003) Identification of quantitative trait loci for resistance to crown rust in oat line MAM17-5. Crop Sci 43: 358-366.

3. Parlevliet JE (1975) Partial resistance of barley to leaf rust, *Puccinia hordei*. I. Effect of cultivar and development stage on latent period. Euphytica 24: 21-27.

4. McCartney CA, Stonehouse RG, Rossnagel BG, Eckstein PE, Scoles GJ, et al. (2011) Mapping of the oat crown rust resistance gene Pc91. Theor Appl Genet 122: 317-325.

5. Franceli RK, Felipe AS, Graichen, Jose AM, Ana BL, et al. (2010) Molecular mapping of Pc68, a crown rust resistance gene in *Avena sativa*. Euphytica 175: 423-432.

6. Smith RCG, Heritage AD, Stapper M, Barrs HD (1986) Effect of stripe rust (*Puccinia striiformis* west.) and irrigation on the yield and foliage temperature of wheat. Field Crop Res 14: 39-51.

7. Doyle JJ, Doyle JL (1987) A rapid DNA isolation procedure for small quantities of fresh leaf tissue, Phytochem. Bull 19: 11-15.

8. Satheeskumar S, Sharp PJ, Lagudah ES, McIntosh RA, Molnar SJ (2011) Genetic association of crown rust resistance gene Pc68, storage protein loci, and resistance gene analogues in oats. Genome 54: 484-497.

9. Parlevliet JE (1978) Further evidence of polygenic inheritance of partial resistance in cultivar and development stage on latent period. Euphytica 24: 21-27.

10. Adhikari KN, McIntosh RA (1998) Susceptibility in oats to stem rust induced by coinfection with leaf rust. Plant Pathol 47: 420-426.

11. Jackson EW, Obert DE, Menz M, Hu G, Bonman JM (2008) Qualitative and quantitative trait loci conditioning resistance to Puccinia coronata pathotypes NQMG and LGCG in the oat (Avena sativa L.) cultivars Ogle and TAM O-301. Theor Appl Genet 116: 517-527.

12. Webster J (1980) Introduction to fungi. (2ndedn), Cambridge University Press, UK.

21

Serine Exoproteinases Secreted by the Pathogenic Fungi of *Alternaria* Genus

Tatiana A Valueva[1]*, Natalia N Kudryavtseva[1], Alexis V Sofyin[1], Boris Ts Zaitchik[1], Marina A Pobedinskaya[2], Lyudmila Yu Kokaeva[2] and Sergey N Elansky[2,3]

[1]*A.N. Bach Institute of Biochemistry of the Russian Academy of Sciences, Leninsky prospect 33, Moscow 119071, Russia*
[2]*Department of Mycology, M.V. Lomonosov Moscow State University, Moscow 119991, Russia*
[3]*A.G.Lorkh Potato Research Institute of the Russian Academy of Sciences, Moscow region, 140051, Kraskovo-1, Lorkh Street 23, Russia*

Abstract

Fungi of different species of the *Alternaria* genus isolated from potato or tomato plants in several Russian regions excreted proteolytic enzymes during the growth in the medium containing thermostable proteins of pea and carrot. The growing of fungi in such medium can be considered as a model system for studying the infection process. It has been shown increased production of proteolytic enzymes that included serine proteases belonging to the subtilisin and trypsin families. In most isolates, the proteolytic enzymes production has been observed during the exponential phase of growth. The data obtained demonstrated that the exoproteinase activity depended on the natures of both the isolate and its host plant, but it was defined by the isolate's genotype predominantly. The data also clearly demonstrated a phenomenon of exceeding of the exoprotease activity, especially trypsin-like, in the tomato isolates in comparison with that in the potato isolates. They indirectly indicate the possibility of pathogenic specialization of *Alternaria* spp. in the Solanaceae that is in accordance with the existence of intra- and inter-specific variations in species of *Alternaria*.

Keywords: *Alternaria* genus; Subtilisin-like proteinases; Trypsin-like proteinases; Potato and tomato plants

Introduction

Genus *Alternaria* Ness is ubiquitous, including species found worldwide in association with a large variety of plants. Many species are saprophytes, animal/plant pathogens or postharvest pathogens. Species of the genus are reported to occur in different ecosystems and geographic regions [1-3]. As a plant pathogen, *Alternaria* has a wide range of hosts ranking 10th in terms of their total number [4]. As postharvest pathogens, *Alternaria* species contribute to extensive losses of our agricultural output due to spoilage [5].

Early blight is a dangerous fungal disease caused by *Alternaria* spp., which affects tomato (*Solanum lycopersicum* L.) and potato (*Solanum tuberosum* L.), two of the most important vegetable crops. Epidemics of early blight occur with high intensity in almost all areas where these crops are grown and it is an important disease worldwide [1,6,7]. In Russia the potato and tomato pathogens of the early blight can be a lot of species of the *Alternaria* genus, among which are both the large-spore species, such as *A. solani* Sorauer and *A. tomatophila* Simmons, and some small-spore species, such as *A. alternata* (Fr.) Keissl., *A. tenuissima* (Kunze) Wiltshire, *A. infectoria* E.G. Simmons, and *A. arborescens* [8]. However, the population of the pathogens was poorly investigated. The taxons listed represent heterogeneous groups including species, which differ in morphological, ecological, physiological and biochemical characteristics. Different species may have their own biological features and vary in such practically important indicators as aggressiveness, virulence to different potato or tomato cultivars, resistance to fungicides, toxigenicity, temperature of growth optimum in winter. Moreover, the isolates of the same species can significantly vary mycologically.

Fungi utilise extracellular enzymes not only for nutrition, but also in pathogenesis: they can function in overcoming the natural host resistance as well as in providing soluble compounds that are then assimilated [9]. In *Alternaria* spp. the proteases excretion has been shown and it is considered that these enzymes may function as pathogenic factors in some fungus-plant interactions [10,11]. It has been also suggested that proteases can facilitate the local penetration of a plant cell wall by breaking down the fibrous glycoproteins that contribute to its stability [12]. Phytopathogenic fungi, such as *Fusarium*, *Alternaria*, and *Rhizoctonia*, produce alkaline serine proteases that are indispensable to their growth. Probably these proteases are nutrient-mobilizing enzymes whose main function is the support of fungal growth after the host cell death has occurred [9,13].

Study of enzymes excreted by phytopathogenic fungi is complicated by the presence of a host-plant, particularly by the plant enzymes as well as inhibitors of microbial enzymes that occur in plants. Therefore the most practical way to study the production by fungi of exozymes is by studying these enzymes in artificial growth media that don't include mentioned disturbing factors. Taking this into account, the aim of the paper was to get an overview of the proteolytic activities that are excreted by different *Alternaria* isolates growing in submerged culture. We hypothesized that the findings could clarify a possible contribution of exoproteinases to the pathogenicity of the fungi studied.

Materials and Methods

Samples

In the study there were analysed early blight strains isolated from

---

*Corresponding author: Tatiana A. Valueva, A.N. Bach Institute of Biochemistry of the Russian Academy of Sciences, Leninsky prospect 33, Moscow 119071, Russia
E-mail: valueva@inbi.ras.ru

infected potato (leafs) and tomato (leafs or fruits) plants in Leningrad, Moscow, Astrakhan, and Kostroma regions, Mariy El and Tatarstan Republics, the Stavropol and Primorsky (another name is Far East) territories.

Strain isolation

Isolation of strains into the pure culture was carried out using the wet chambers. After the appearance of conidia on the surface of the growth medium, an infected sample was analysed by using MBS-10 stereo binocular microscope (Russia) with 70-100x magnifications. For this purpose the fungus conidia were transferred by a sterile needle onto Petri dish with Wort Agar medium supplemented with penicillin (1000 U/ml) and fungus was grown at 25°C until the colony diameter reached 4-5 cm. Then a piece of mycelium from the edge of the colony was transferred onto a fresh Petri dish with Wort Agar medium.

Species identification

Identification of species was performed according to morphological criteria [14]. Isolates were grown on Petri dishes with potato carrot agar (PCA) under fluorescent lamps at 25°C. After 7-10 days of growth, colonies were analysed by microscope to register the features of the formation of conidial chains and the spores' morphology.

Material preparing for the enzyme activity analysis

The fungi maintenance and their growing as a submersed culture in the liquid medium (pea-carrot broth) were carried out as described earlier [11]. After 2, 5, 7, 10, 14, and 18 days of the growth, the mycelium was harvested on a weighed Whatman No. 41 filter paper. It was washed with a small quantity of warm distilled water, heated overnight in an oven at about 90°C, cooled in a desiccator, and weighed. Further loss in weight was not obtained by longer periods of drying. Crude culture filtrate obtained after mycelium harvesting was used in enzyme activity assays.

Enzyme assays

The substrates, such as azocasein, N,α-benzoyl-L-Arg-pNa (BAPNA) and N-carbobenzyloxy-L-Ala-L-Ala-L-Leu-pNa (Z-AALPNA), were purchased from Sigma-Aldrich (USA). All other reagents were of the highest grade commercially available.

The proteolytic enzyme activity was determined by method [15], using 0.5% azocasein in 0.1 M Tris-HCl buffer, pH 7.5, as a substrate. One unit of the proteolytic activity (U) was the amount of enzyme that increased optical density at 366 nm by 0.1 per min in the supernatant after precipitation by TCA of the reaction mixture proteins.

The amidase enzyme activity was determined by the Erlanger's method [16], using p-nitroanilide substrates: BAPNA in the assay of the trypsin-like activity and Z-AALPNA in the assay of the subtilisin-like activity. The substrate concentration was 0.5 mM. One unit of the amidase activity (AU) was the amount of enzyme that hydrolyzed 1 nmol of the substrate in 1 min.

Statistical analysis

All experiments were carried out at least in 3-fold repetition. The data presented in the figures are the averages; standard deviations were also calculated, but they are not shown in the figures to avoid overloading. Significant difference was defined as p<0.05.

Results and Discussion

Strategies of plant disease management should be established on an epidemiological basis. Epidemiology is the study of the spatial and temporal dynamics of epidemics [17], which in turn are the result of an interaction between host and pathogen populations [18]. Thus, the study of plant diseases epidemics should take the population approach [7]. Under this paradigm, defining the genetic structure of populations is the first logical step of studying the pathogen population, because the genetic structure reflects the evolutionary history and the potential of a population to evolve [19,20].

In this study we used the microorganisms that are pathogens of early blight: small-spore species *A. alternata* and *A. infectoria* and large-spore species *A. solani*. These samples have been collected on the diseased Solanaceae plants, such as the potato (and called the potato's isolates) or tomato (and called the tomato's isolates).

According to the results obtained during reconstruction of taxonomic relations between the studied isolates they were divided into three groups. The first group included small-spore isolates *A. alternata* such as PL14, PL3b (KF 998555), PL44 (KG 998557), PL18a (KF 998556), TL106-031 (KF 998558), TL49 (2) (KF 998552), TL12d (KF998554) and TL5

The second group contained all the studied large-spore isolates *A. solani*: PL043-021 (KF998549), TL14e/2 (KF 998551), TL125 and TF11a (KF998551).

Lastly, the third group was formed by the *A. infectoria* isolates such as one *A. infectoria* 5a (KF 998559).

As we have shown previously [11], the phytopathogenic *Alternaria* fungi can be grown in the medium containing thermostable proteins of pea and carrot, because the combination of these plant components provides the optimum mycelium growth level of the isolates under study. Simultaneously, such medium would be a model for studying the infection process. It was also demonstrated that during *Alternaria* growth in this medium, fungi produced extracellular proteinases with the slightly alkaline activity optimum [11].

Figure 1 shows the results of the study of proteinases excretion by the mentioned 13 isolates. The increase of the exoproteolytic activity in both large-spore and small-spore isolates of *Alternaria* was observed in the exponential growth phase of the culture. Both in potato and in tomato isolates the accumulation of the exoproteolytic activity was initiated in the early time of the fungi growth, and the character of the process for both types of isolates was identical in general. At the same time, it should be pointed out that the maximum of the exoproteolytic activity in potato isolates was attained some earlier (in 10 days) than in the tomato ones (in 14 days) (Figure 1). Attention is drawn to the fact that the only fungus isolated on the tomato fruit (*A. solani* 11a) excreted very low-level proteolytic activity into the culture fluid (Figure 1).

If to compare large- and small-spore isolates, one can note that the maximum of the exoproteolytic activity was achieved on 10-14 day of the culture growth in large-spore isolates, while the exoproteolytic activity in small-spore isolates could either increase during the whole period under study or markedly decline after 14 days (Figure 1).

A gradual increase of the subtilisin-like activity during the fungi growth was typical for most isolates studied (Figure 2). Only in two small-spore isolates (5 and 44) the process of excretion of the subtilisin-like proteases had the maximum in 10 days and then it decreased gradually (Figure 2). This activity was not detectable in isolates 11a (large-spore, tomato fruit) and 3b (small-spore, potato).

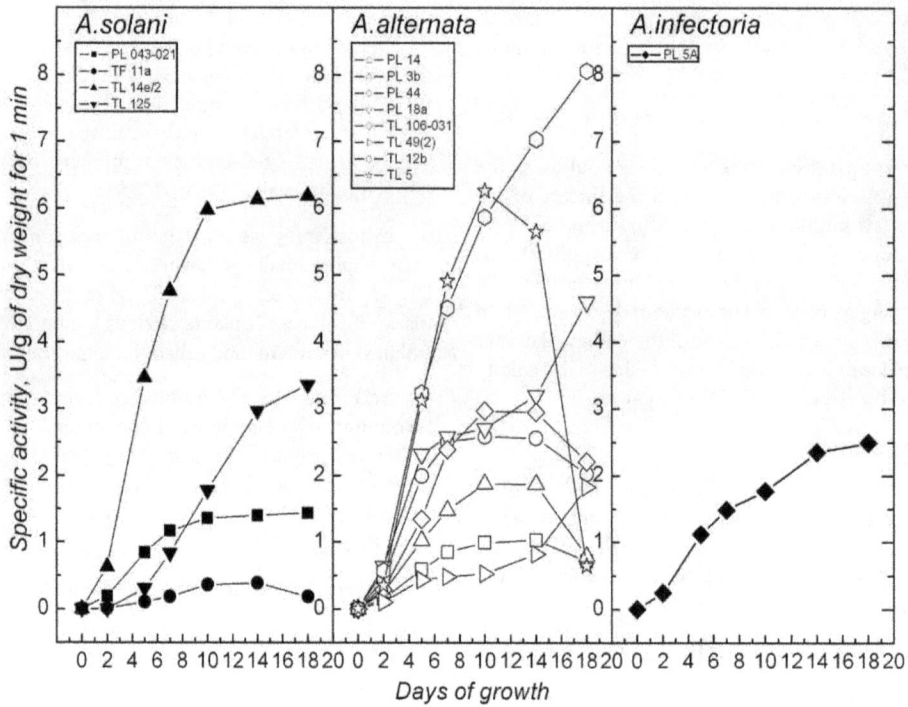

Figure 1: Dynamics of the specific exoproteolitic activity during the growth of isolates representing three *Alternaria* clades. Insets show the isolates names with their corresponding symbols. The standard deviation of activity for each of the values did not exceed 5% and are not represented to avoid overloading of the figure.

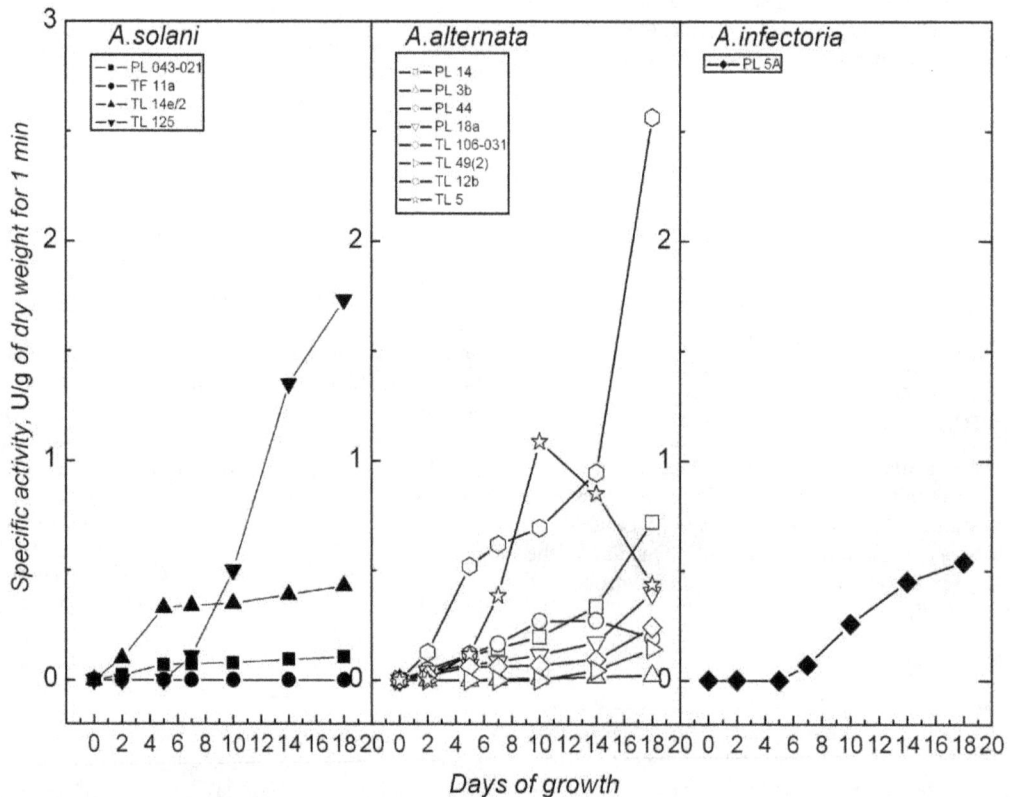

Figure 2: Dynamics of the specific extracellular subtilisin-like activity during the growth of isolates representing three *Alternaria* clades. Insets show the isolates names with their corresponding symbols. The standard deviation of activity for each of the values did not exceed 5% and are not represented to avoid overloading of the figure.

The characteristic feature of the trypsin-like activity dynamics in small-spore isolates was the achievement of maximum in 12 days of growth followed by a decrease of the activity level that, to a greater or lesser degree, depended on the isolate nature (Figure 3). The highest trypsin-like activity was observed in small-spore isolate 12b with the activity increased sharply from the first day of the culture development. The trypsin-like activity was almost completely absent in two tomato large-spore isolates 11a and 125 (Figure 3).

The group, which *A. infectoria* belongs to, occupies a particular place. The exoproteolytic activity of the *A. infectoria* 5a isolate increased uniformly throughout 18 days of the fungus development cycle, whereas the subtilisin-like activity was only measurable after 6 days of growth with the increase in the subsequent growth of the fungus. It should be pointed out that the trypsin-like activity of the isolate considerably exceeded the subtilisin-like one and was practically equal to general exoproteolytic activity (Figures 1-3). This does suggest that *A. infectoria* 5A exhibits high pathogenicity.

The high variability of *Alternaria* populations can be attributed to the interesting features of these fungi, taking into account their strictly asexual reproduction with unknown sexual or reproductive stage [10]. In addition, recombination (meiotic or mitotic) and other evolutionary forces, such as genetic drift, gene flow, and natural mutations, can affect the observed high variability. This process may also include host plant material distribution in broad geographical areas.

The previous data [11] indicate the absence of the clear correlation of the serine exoproteases composition with the nature of host plants.

Apparently this is due to high variability of the isoenzymes excreted by fungi due to their high reproductive capacity (these species are capable of producing very large amounts of spores in a short period) [19]. As one can see from the above data, the measured levels of activities of the isolates studied were markedly different from each other, and to analyse the data their average values for all members of each of three groups were calculated for the fungi's growth period and then the charts of these trends of the activities were plotted.

Figure 4 shows that there were no essential differences in the dynamics of changes of both the exoproteinase and subtilisin-like activities for all the three groups. At the same time, the trypsin-like activity was markedly higher in the members of small-spore isolates and was characterised by the presence of the explicit maximum by the 10th day of growth (for *A. infectoria* it was by the 14th day). In contrast, the chart of the trypsin-like activity of the large-spore isolates was a plateau-like shape.

However, as it has emerged, it is more interesting to compare the calculated trends of the studied activities depending on the host plants of isolates. Figure 5 presents such data for the large-spore *A. solani*. One can see that the exoproteinase activity was about three times higher in tomato than in potato isolates. The same trend was characteristic of the trypsin-like activity. The subtilisin-like activity is practically absent in the potato isolates, while this activity in the tomato isolates is proved to be higher than the trypsin-like one by the end of the growth period.

The similar trends for small-spore *Alternaria* isolates are shown in Figure 6. In the case the accumulation dynamics of both the exoprotease

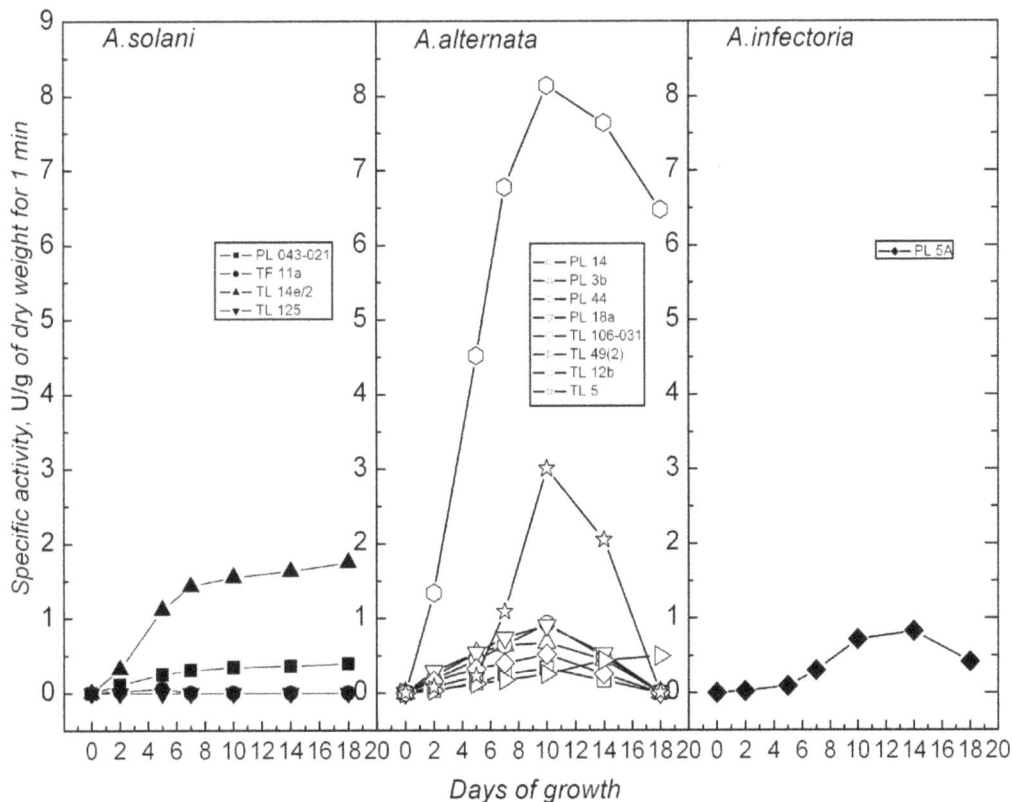

Figure 3: Dynamics of the specific extracellular trypsin-like activity during the growth of isolates representing three *Alternaria* clades. Insets show the isolates names with their corresponding symbols. The standard deviation of activity for each of the values did not exceed 5% and are not represented to avoid overloading of the figure.

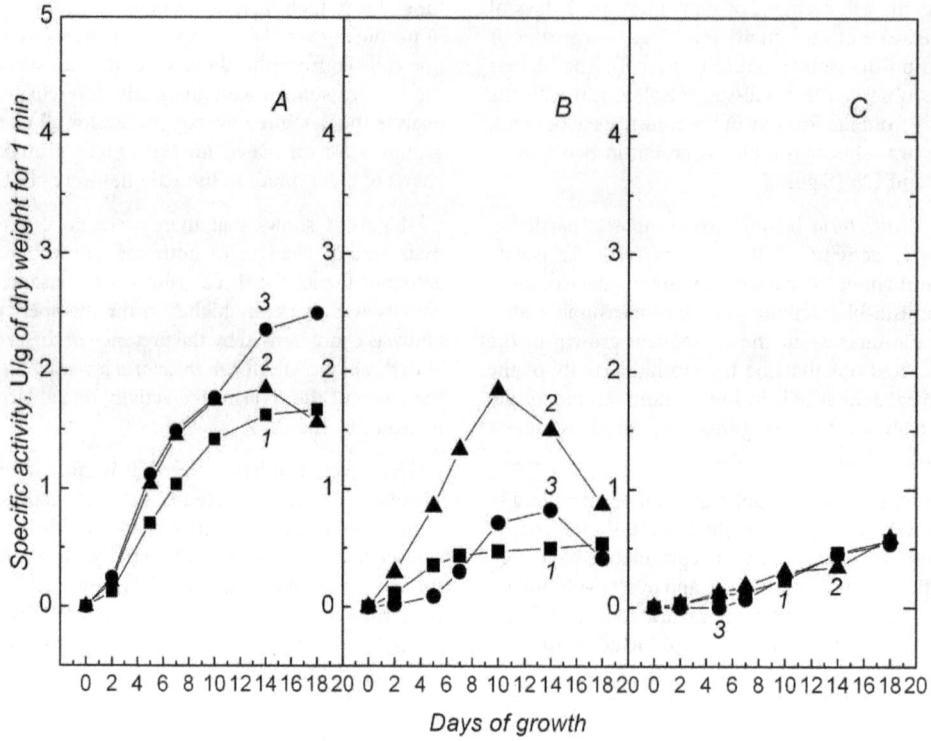

Figure 4: Calculated trends of the specific exoproteolitic (A), extracellular trypsin-like (B), and extracellular subtilisin-like (C) activities in the three *Alternaria* clades. 1. – large-spore *A. solani* clade; 2 – small-spore *A. alternata* clade; 3 – small-spore *A. infectoria* clade.

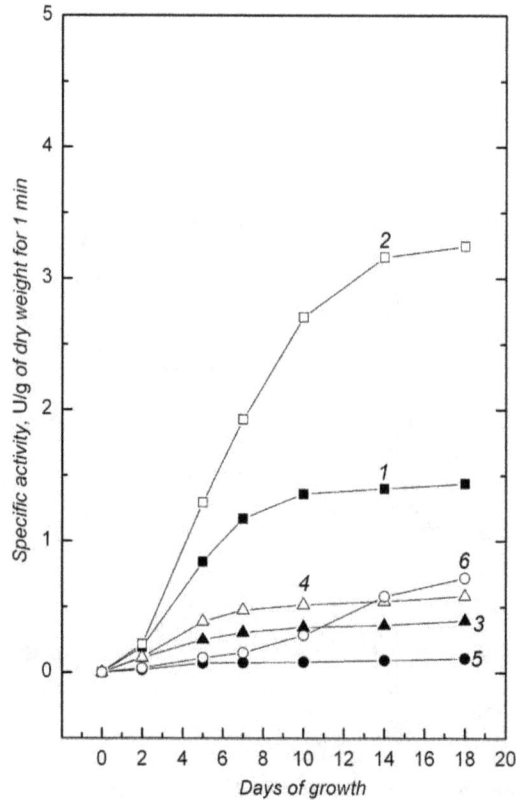

Figure 5: Calculated trends of the specific exoproteolitic (1, 2), extracellular trypsin-like (3, 4), and extracellular subtilisin-like (5, 6) activities in the large-spore *A. solani* clade depending on the host plant (potato – 1, 3, 5; tomato – 2, 4, 6) of isolates.

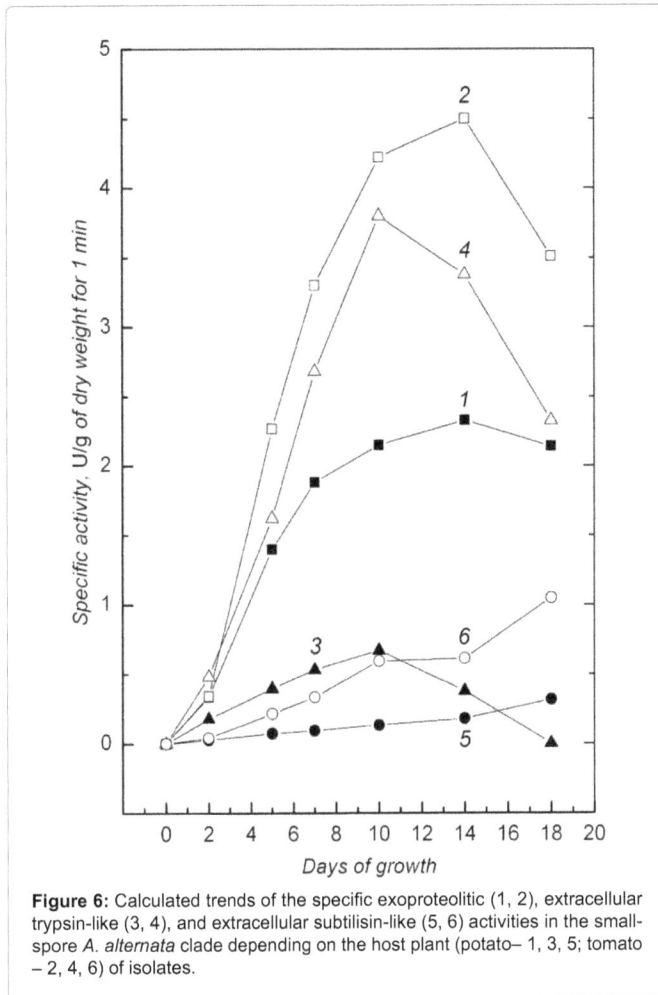

Figure 6: Calculated trends of the specific exoproteolitic (1, 2), extracellular trypsin-like (3, 4), and extracellular subtilisin-like (5, 6) activities in the small-spore *A. alternata* clade depending on the host plant (potato– 1, 3, 5; tomato – 2, 4, 6) of isolates.

and trypsin-like activity was also almost identical. And similarly to the discussed above, in the tomato isolates the exoproteinase activity was significantly higher than that in the potato isolates. At the same time, it is noteworthy that the charts for subtilisin-like activities had absolutely other shapes. It should be emphasized that this activity was practically zero in all potato isolates studied, whereas in tomato ones it, firstly, was significantly lower than the trypsin-like activity at the initial growth phase and, secondly, there was a tendency to its accumulation towards the end of the fungi's growth period.

It is well known that various types of *Alternaria* can exist as both saprophytic and pathogenic fungi, infecting solanaceous plants. There is a speculation that the less the restriction of host range is, the less evolutional selection pressure on a microorganism is [21]. In other words, such microorganisms are able to adapt to the wider range of hosts. Julian and Lucas [22] have suggested that microorganisms with low pathogenicity may be less susceptible to the host plant's developmental pressure than more pathogenic species. Therefore, less pathogenic organisms should be more adaptogenic. Our data do not contradict this because the large-spore species that have already been evolutionarily formed, are characterized by the presence of general regularities in displaying the exoproteinase activities, whereas such regularity is absent, as a rule, in the isolates belonging to the small-spore species. This may indicate a continuation of the processes of evolution and specialization in relation to the host plant in them. For that matter it is important to note that the data obtained in this paper

clearly demonstrate a phenomenon of exceeding the exoproteases and, especially, trypsin-like activities in the tomato isolates in comparison with that in the potato isolates.

Based on the data available, one can suggest that the mechanism of plant destruction is the same both for small-spore and large-spore *Alternaria* species. At present there is no evidence of pathogenic specificity of the *A. solani* isolates collected on the tomato or potato plants, however, the potato isolates were less aggressive to tomato plants of different genotypes than vice versa [23]. The data obtained may also indirectly indicate that in the studied *Alternara* isolates pathogenic specialization is possible in the Solanaceae, which is consistent with the presence of intra- and inter-specific variation in the *Alternaria* species. Thus, there are several mechanisms available for *Alternaria* isolates to adapt its enzyme activities to their specific needs on their particular hosts.

Acknowledgments

This work was financially supported by a grant of the Ministry of Education and Science (project number RFMEFI62114X0002).

References

1. Rotem J (1994) The Genus Alternaria: Biology, Epidemiology and Pathogenicity. The American Phytopathological Society, St Paul, Minnesota USA.

2. De Hoog GS, Horré R (2002) Molecular taxonomy of the *Alternaria* and *Ulocladium* species from humans and their identification in the routine laboratory. Mycoses 45: 259-276.

3. Hong SG, Pryor BM (2004) Development of selective media for the isolation and enumeration of Alternaria species from soil and plant debris. Can J Microbiol 50: 461-468.

4. Farr DF, Bills GF, Chamuris GP, Rossman AY (1989) Fungi on Plant and Plant Products in the United States. The American Phytopathological Society, St Paul, Minnesota USA.

5. Wilson CL, Wisniewski ME (1994) Biological control of postharvest diseases, theory and practice. CRC Press, Inc., Boca Raton, FL 182.

6. Pscheidt JW, Stevenson WR (1986) Early blight of potato and tomato: A Literature Review, College of Agriculture & Life Sciences, Madison, WI, USA.

7. Milgroom MG, Peever TL (2003) Population biology of plant pathogens: the synthesis of plant disease epidemiology and population genetics. Plant Dis 87: 608-617.

8. Orina AC, Gannibal PhB, Levitin MM (2010) Specific diversity, biological characters and geography of Alternaria fungi associated with Solanaceous plants. Mycol Phytopath (Russia) 44: 150-9.

9. Griffin DH (1994) Fungal physiology Wiley-Liss John Wiley & Sons, Inc. Publ.

10. Valueva TA, Kudryavtseva NN, Sofyin AV, Revina TA, Gvozdeva EL, Ievleva EV (2011) Comparative analyses of exoproteinases produced by three phytopathogenic microorganisms. Journal of Pathogens 2011: 1-9.

11. Valueva TA, Kudryavtzeva NN, Gvozdeva EL, Sof'in AV, Il'ina NYu, Kladnitskaya GV, Pobedinskaya MA, Elansky SN (2013) Serine proteinases secreted by two isolates the fungus Alternaria solani. J Basic Applied Sci 9: 105-115.

12. Carpita NC, Gibeaut DM (1993) Structural models of primary cell walls in flowering plants: consistency of molecular structure with the physical properties of the walls during growth. Plant J 3: 1-10

13. Pekkarinen A, Mannonen L, Jones BL, Niku-Paavola M-L (2000) Production of proteases by Fusarium species grown on barley grains and in media containing cereal proteins. J Cereal Sci 31: 253-261.

14. Simmons EG (2007) *Alternaria* an identification manual. Utrecht the Netherlands: CBS Fungal Biodiversity Center theory and practice CRC Press, Boca Raton, FL.

15. Charney J, Toarelli RM (1947) A colorimetric method for the determination of the proteolytic activity of duodenal juice. J Biol Chem 171: 501-505

16. Erlanger DF, Kokowsky N, Cohen W (1961) The preparation and properties of two new chromogenic substrates of trypsin. Arch Biochem Biophys 95: 271-278.

17. Campbell CL, Laurence V (1990) Madden introduction to plant disease Epidemiology. John Wiley&Sons, New York.

18. Zadoks JC, Schein RD (1979) Epidemiology and plant disease management. Oxford University Press, New York, USA.

19. Leung H, Williams PH (1986) Enzyme polymorphism and genetic differentiation among geographic isolates of the rice blast fungus. Phytopathology 76: 778-783.

20. McDermott JM, McDonald BA (1993) Gene flow in plant pathosystems. Annu Rev Phytopathol 31: 353-73.

21. Pryor BM, Michailides TJ (2002) Morphological, pathogenic and molecular characterization of Alternaria isolates associated with Alternaria late blight of pistachio. Phytopatology 92: 406-416.

22. Julian AM, Lucas JA (1990) Isozyme polymorphism in pathotypes of *Pseudocercosporella herpotrichoides* and related species from cereals. Plant Pathol 39: 178-190.

23. Lavrova OI, Elansky SN, Dyakov YT (2003) Selection of *Phytophthora infestans* isolates in asexual generations. J Russian Phytopathol Soc 4: 1-7.

22

Morphological and Molecular Screening of Turmeric (*Curcuma longa* L.) Cultivars for Resistance against Parasitic Nematode, *Meloidogyne incognita*

Swatilekha Mohanta[1], Swain PK[1], Sial P[1] and Rout GR[2]*

[1]*Department of Nematology, College of Agriculture, OUAT, Regional Research Technology Transfer Station, Pottangi, Bhubaneswar-751003, Odisha, India*
[2]*Department of Biotechnology, College of Agriculture, OUAT, Bhubaneswar-751003, India*

Abstract

Turmeric (*Curcuma longa*) is a high value export oriented important commercial crop among the spices. The production was declined due to several biotic and abiotic stresses. Among biotic stresses, root-knot nematode, *Meloidogyne incognita* is a major threat to turmeric cultivation. Seventy cultivars were screened to identify the resistance to root-knot nematode, *Meloidogyne incognita*. The result revealed that cultivars 'Dugirala', 'PTS-31', 'Ansitapani', 'PTS-42', 'PTS-47'noted as fully resistant;'361 Gorakhpur', '328 Sugandham', 'PTS-21' rated as moderately resistant and rest other cultivars were susceptible. The cultivar '328 Sugandham' was moderately resistant to root-knot nematode. This was further confirmed through DNA amplification studies with ISSR markers. The similarity matrix was obtained after multivariate analysis using Nei and Li's coefficient and the matrix value was ranged from 0.35 to 0.89, with a mean value of 0.62. The two cultivars 'Dugirala' and '361 Gorakhpur' with 48% similarity with other 21 cultivars. Both the cultivars were resistance to root knot nematode (RKN) having indexed ranged from 2.0 to 3.0. The five cultivars i.e. 'Tu No.4', 'Tu No.1', 'Erode local', 'TC-4' and ' 'Phulbani Wild' were 78% similarity and susceptible to RKN having index from 4.0 to 5.0. Cultivars 'Dugirala', '328 Sugandham' and 'PTS-47' exhibited resistance to both root knot nematodes. This investigation as an understanding of the level and partitioning of genetic variation within the cultivars with resistant/susceptible to root knot nematode disease would provide an important input into determining efficient management strategies for breeding program.

Keywords: *Meloidogyne incognita*; Turmeric; Screening; Resistance; ISSR marker

Introduction

Turmeric (*Curcuma longa*), an herbaceous plant is native to tropical south East Asia. It is a high value export oriented important commercial crop among spices in India. The tuber crops represent the most important food commodity in many subtropical and tropical countries [1]. The rhizome has 1.8 to 5.4 percent curcumin, the pigment and 2.5 to 7.2 percent of essential oil. It is used as a dye with varied application in drug and cosmetic industries. In India, it is grown in an area of 104,500 ha producing annually 3,28,800 tones. Although, India is leading in its production (75% of world output), the average productivity and quality are not satisfactory for which the export value is reduced dramatically. Annually 18 to 20 crores worth of turmeric are exported. In India, Andhra Pradesh is the leading state followed by Maharashtra, Tamil Nadu, Orissa, Kerala and Bihar. However, the production and productivity of this high value cash crop is declining day by day because of several biotic and abiotic stresses. Among biotic stresses root-knot nematode, *Meloidogyne incognita* is a major threat to turmeric cultivation [2,3]. Nematodes causes' serious yield performance and quality reduction in most of the tuber crops [4-6]. Root-knot nematodes (*Meloidogyne* spp), first identified as a potential threat to yam production [7] (Bridge) and also in sweet potato [8]. This extensive polyphagous species is a sedentary endo-parasitic nematode that induces multinucleated modified transfer cells inside the vascular bundles of roots through a series of physiological and biochemical changes thereby resulting in galling root dysfunction, reduced water flow and photosynthesis [9]. Management of this important phytophagous nematode through conventional tactics has become a difficult task because of limited availability of nematicides in the world market as well as environmental concern. Few reports are available on molecular screening on tomato, cotton, peanut with regard to root knot nematode resistance [10-13]. Molecular markers have now come up as the most desirable tool for detecting and characterizing variation among the resistance and susceptible at the DNA level Among the different molecular markers, inter simple sequence repeats (ISSR) techniques have proven to be a reliable, reproducible, easy to generate, inexpensive and versatile set of markers that relies on repeatable amplification of DNA sequences using single primers. Therefore, the present study was undertaken to identify some resistant turmeric cultivars as an ecofriendly alternative to nematicides based on physical markers and to correlate these findings with molecular investigation through ISSR marker assisted DNA amplification studies.

Materials and Methods

Seventy cultivars of turmeric were collected from Regional Research Technology Transfer Station (RRTTS), Pottangi (Odisha) for screening their resistance against root-knot nematode, *Meloidogyne incognita*. These cultivars were planted in the 8″ diameter surface sterilized earthen pots containing 3 kg steam sterilized soil and kept in the experimental garden of the Department of Nematology, College of Agriculture, Orissa University of Agriculture and Technology,

***Corresponding author:** Rout GR, Department of Biotechnology, College of Agriculture, OUAT, Bhubaneswar, India, E-mail: grrout@rediffmail.com

Bhubaneswar. One month after the planting freshly hatched second stage juveniles of *M. incognita* were inoculated @ 3000 J2/pot around the root zone of the plant for infection and development. Sixty days after inoculation, the plants in pots were uprooted carefully and roots were evaluated for resistance against *M. incognita* by following 1-5 point scale [14] on the basis of development of galls and egg masses on the root as follow (Table 1).

Data on root-knot indices were subjected to statistical analysis

Sl. No.	Cultivar	Root-knot Index	Reaction
V1	Dugirala	2.00	R
V2	Tu. No.4	4.00	S
V3	Erode local	4.00	S
V4	PTS-53	3.50	S
V5	Sudarsan	4.50	HS
V6	PTS-31	2.00	R
V7	CLS-33	5.00	HS
V8	TC-4	3.50	S
V9	Phulbani Wild	4.00	S
V10	361 Gorakhpur	3.00	MR
V11	Ansitapani	2.00	R
V12	Tu. No.1	4.50	HS
V13	PTS-34	4.50	HS
V14	Bataguda	4.00	S
V15	PTS-17	5.00	HS
V16	PTS-8	4.50	HS
V17	PTS-42	2.00	R
V18	Ethamkalam	5.00	HS
V19	328 Sugandham	2.50	MR
V20	PTS-47	2.00	R
V21	PTS-21	2.50	MR
V22	Kasturi Manjari	4.50	HS
V23	PCT-7	3.50	S
V24	Black turmeric	5.00	HS
V25	Chayapusupu-1	5.00	HS
V26	CAS-15	5.00	HS
V27	CAS-51	5.00	HS
V28	CAS-53	5.00	HS
V29	CLS-3	3.50	S
V30	CLS-21	5.00	HS
V31	Florescent	4.00	S
V32	GL-Puram	5.00	HS
V33	Kuchipudi	4.00	S
V34	K. Local	5.00	HS
V35	Lakadong	4.00	S
V36	Mydukur	4.00	S
V37	Mundapadar	5.00	HS
Sl. No.	Cultivar	Root-knot Index	Reaction
V38	NB-60	5.00	HS
V39	NB-6206	5.00	HS
V40	No.38	4.00	S
V41	PCT-9	5.00	HS
V42	PTS-1	4.00	S
V43	PTS-20	4.00	S
V44	PTS-3	5.00	HS
V45	PTS-4	4.00	S
V46	PTS-11	5.00	HS
V47	PTS-12	4.00	S
V48	PTS-13	4.00	S
V49	PTS-27	4.50	S
V50	PTS-30	4.00	S
V51	PTS-33	5.00	HS
V52	PTS-43	4.00	S
V53	PTS-44	5.00	HS
V54	PTS-48	4.00	S
V55	PTS-50	5.00	HS
V56	PTS-51	5.00	HS
V57	PTS-54	4.00	S
V58	PTS-55	5.00	HS
V59	PTS-57	4.00	S
V60	PTS-62	5.00	HS
V61	Rajpuri local	5.00	HS
V62	Rajendra Sonia	5.00	HS
V63	Ranga	5.00	HS
V64	Raikia	5.00	HS
V65	Roma	5.00	HS
V66	Surama	5.00	HS
V67	Tu. No.-6	5.00	HS
V68	VK-9	5.00	HS
V69	VK-154	4.00	S
V70	Wynad local	5.00	HS
	Mean	4.27	
	Sem (0.05)	0.21	
	CD (0.05)	0.58	
	CV	6.85	

R: Resistance, MR: Moderate Resistance, S: Susceptible, HS: Highly Susceptible.

Table 1: Screening of turmeric cultivars against *M. incognita*.

by following analysis of variance through complete randomized block design. Leaf samples of selected resistant, moderately resistant, susceptible and highly susceptible cultivars were collected for DNA extraction and amplification by ISSR marker and these selected cultivars were planted in raised beds (1 m × 3 m) at RRTTS, Pottangi with three replications. Yield was recorded after fully maturation of the cultivars. Data were subjected to statistical analysis through analysis of variance in a randomized block design. The results of the pot culture studies are pertaining to resistance by *M. incognita*, plant growth and yield were correlated to confirm the resistance based on morphological features with that of molecular investigation.

Genomic DNA extraction and quantification

DNA was extracted from fresh leaves using the cetyl-trimethyl ammonium bromide (CTAB) method [15,16]. Approx. 200 mg of fresh leaves were ground to a powder in liquid nitrogen using a mortar and pestle. The powder was transferred to a 50-ml falcon tube with 10 ml of CTAB buffer [2% (w/v) CTAB, 1.4 M NaCl, 20 mM EDTA, 100 mM Tris (tris(hydroxymethyl) amino methane)-HCl, pH 8.0, and 0.2% (v/v) β-mercaptoethanol. The homogenate was incubated at 60°C for 2 h, extracted with an equal volume of chloroform/isoamyl alcohol (24:1, v/v), and centrifuged at 9838 × *g* for 20 min. DNA was precipitated from the aqueous phase by mixing it with unequal volume of isopropanol. After centrifugation at 9838 × *g* for 10 min, the resultant DNA pellet was washed with 70% (v/v) ethanol, air-dried, and re-suspended in TE (10 mM Tris-HCl, pH 8.0, and 0.1 mM EDTA) buffer. DNA quantifications were performed by visualizing under UV light, after electrophoresis on 0.8% (w/v) agarose gel at 50 V for 45 min and comparing with a known amount of lambda DNA marker

(Emerk Bioscience, India). The resuspended DNA was then diluted in TE buffer to 5 µg/µl concentration for use in polymerase chain reaction (PCR).

Primer screening

Twenty synthesized inter simple sequence repeat(ISSR) primers (M/S Emerk Bioscience, Bangalore, India) were initially screened to determine the suitability of each primer for the study. Primers were selected for further analysis based on their ability to detect distinct, clearly resolved, and polymorphic amplified products within the varieties. To ensure reproducibility, the primers generating no, weak, or complex patterns were discarded.

ISSR assay

PCRs with a single primer were carried out in a final volume of 25 µl containing 20 ng template DNA, 100 µM of each deoxyribonucleotide triphosphate, 20 ng of oligonucleotides synthesized primer (M/S Bangalore Genei, Bangalore, India), 1.5 mM $MgCl_2$, 1X Taq buffer (10 mM Tris-HCl, pH 9.0, 50 mM KCl, 0.001% gelatin), and 0.5 U Taq DNA polymerase (M/S Emerk Bioscience, India). Amplification was performed in a thermal cycler (Peqlab, United Kingdom) programmed for a preliminary 2 min denaturation step at 94°C, followed by 40cycles of denaturation at 94°C for 20s, annealing at required temperature for 30s, extension at 72°C for 1 min, and finally amplification at 72°C for 10 min. Amplification products were separated alongside a molecular weight marker (3.0 Kb plus ladder, M/S Emerk Bioscience, India) by 1.5% (w/v) agarose gel. Electrophoresis in 1X TAE (Tris acetate/EDTA) buffer. The gel was prestained with ethidium bromide and visualized under UV light. Gel photographs were scanned through a Gel Documentation System (Gel Doc., UVITECH, UK), and the amplification product sizes were evaluated using the software Quantity one (Bio-Rad) (Rohlf).

Data analysis

During data analysis, only reproducible polymorphic bands in amplification reactions were considered as present. Each band was treated as a separate putative locus, and scored as present (1) or absent (0) in each cultivar. The binary data of the ISSR fingerprints were used further for population genetic analyses. The numbers of monomorphic and polymorphic bands were derived from the binary data, and their percentages were calculated

Bands with similar mobility to those detected in the negative control, if any, were not scored. Similarity index was estimated using the formula, $S = 2 N_{AB}/N_A + N_B$ [17].

Where, N_{AB} is the number of amplified products common to both A and B.

Results and Discussion

The screening of seventy cultivars against the *M. incognita* on the basis of varying degree of galling in the plant roots as indicated by root-knot indices. None of the seventy tested cultivars reacted highly resistant to *M. incognita* (Table 1). The root-knot indices of all the cultivars ranged between 2.0-5.0. Statistical analysis of data indicated that there were significant differences among the cultivars. 'Dugirala', 'PTS-31', 'Ansitapani', 'PTS-42' and 'PTS-47' with root-knot index 2.0 and resistant to *M. incognita*, which were significantly different from other cultivars. The root knot indices of cultivars '361-Gorakhpur', '328-Sugandham' and 'PTS-21' were 3.0, 2.5 and 2.5 respectively and rated as moderately resistant. Rest other cultivars were susceptible

to highly susceptible to root-knot indices ranging between 4.0-5.0. Eapen et al. [18] reported that cultivars like 'Erode', 'Cls. No.4' were rated as highly resistant and 'C 11.320', 'Kattapana', 'Cls. No.21' as a moderate resistant to *M. incognita*. In the present study, the cultivar 'Dugirala' showed resistance to nematode which was conformity with Mani et al. [19]. There were significant differences among the cultivars on the basis of plant growth, yield performance and root-knot indices. 'PTS-21' rated moderately resistant to *M. incognita* has shown highest plant height, leaf length, leaf width and rhizome yield. The rest two moderately resistant cultivars '361-Gorakhpur' and '328-Sugandham' have shown moderate plant growth and rhizome yield. Some of the cultivars resistant to *M. incognita* exhibited moderate to low plant growth, rhizome yield and low root-knot indices. Among these, 'PTS-47' (6.9 Kg/3m²) was the highest yielder followed by 'PTS-42' (6.46 Kg/3m²). The cultivars like '361-Gorakhpur' and 'PTS-21'exhibiting moderately resistant and the cultivars like 'Ansitapani' and'PTS-42' exhibiting resistant to *M. incognita*. Similarly, cultivars like 'Erode local', 'PTS-53', 'Sudarsan', 'CLS-33', 'Phulbani Wild', 'PTS-17' and 'Kasturi Manjari' were susceptible to *M. incognita*. Two cultivars 'Dugirala' and 'PTS-47' were found resistant to *M. incognita*.

On the basis of root-knot indices, out of seventy cultivars, 23 cultivars were selected to compare the resistance, moderate resistance, susceptible and highly susceptible on the basis of ISSR markers. The present study offers an optimization of primer screening for evaluation of genetic relationship among twenty three cultivars of *Curcuma longa* through ISSR analysis (Table 2). The cultivar 'PTS-53' was used initially for screening of synthesis primers for amplification by using polymerase chain reactions. The results showed some primers produced relatively more amplification fragments compared to other primers. The reproducibility of the amplification product was tested on DNA from three independent extractions of the cultivars. Most of the amplification reactions were duplicated. Only bands that were consistently reproduced across amplifications were considered for the analysis. Bands with the same mobility were considered as identical fragments, receiving equal values, regardless of their staining intensity. When multiple bands in a region were difficult to resolve, data for that region of the gel was not included in the analysis. Among the twenty primers tested, only eleven of them produced unambiguous DNA fragments. All the twenty three cultivars of *Curcuma longa* extensively amplified using these eleven ISSR primers (Table 3) and produced 66 fragments ranging from 100bp to 2500bp. The minimum size fragment of 100bp was amplified by the primer USB-835 and the maximum size fragment of 2500bp was amplified by primer USB-807, USB-708, USB-810, and USB-837 and USB-840. Out of 66 fragments, only 50 fragments (75%) were polymorphic. The pattern of ISSR produced by the primers USB-810, USB-841, USB-807 and USB-835 are shown in Figure 1. The genetic variation through molecular markers has been highlighted in a number of medicinal plants [2,20-22]. The present results have shown the narrow variation within some of the cultivars. The similarity matrix was obtained after multivariate analysis using Nei and Li's coefficient and is presented in Table 4. The matrix value was ranged from 0.35 to 0.89, with a mean value of 0.62. The high matrix values indicated that there were distantly related to each other. The similarity matrix obtained in the present study was used to construct a dendrogram with the unweight UPGMA method and resulted in their distant clustering in the dendrogram (Figure 2). The dendrogram shows two major clusters. The first major cluster (A) had only two cultivars 'Dugirala' and '361 Gorakhpur' with 48% similarity with other major cluster (B) having 21 cultivars. Both the cultivars of major cluster -1 were resistance to root knot nematode (RKN) having indexed ranged

Sl. No.	Turmeric cultivars	Plant Ht. (cm)	Leaves/Tiller	Tillers/Plant	Leaf Length (cm)	Leaf Width (cm)	Yield of fresh rhizome (kg/3m²)	Root-knot Index	Reaction to Root-knot nematode	Reaction to Taphrina leaf blotch
V1	Dugirala	87.8	6.6	3.4	42.4	9.7	2.1	2.00	R	R
V2	Tu. No.4	102.4	6.2	3.2	50.4	11.2	8.6	4.00	S	HS
V3	Erode local	96.4	6.2	3	46.8	10.8	8.5	4.00	S	R
V4	PTS-53	82.8	6.8	3	41	12.5	6.9	3.50	S	R
V5	Sudarsan	76.8	7.2	2.2	36.8	13	2.9	4.50	HS	R
V6	PTS-31	87.2	6.4	3.2	25	10.7	1.8	2.00	R	S
V7	CLS-33	73.4	6	1.6	36.6	11.4	5	5.00	HS	R
V8	TC-4	71.4	6.8	2.4	37.4	11.4	5.4	3.50	S	S
V9	Phulbani Wild	72.8	5.6	3.4	38	8.9	3.8	4.00	S	R
V10	361 Gorakhpur	78.6	6	1.8	37.6	10.9	7.2	3.00	MR	S
V11	Ansitapani	87.6	5.6	3.4	46.2	11.8	5.4	2.00	R	S
V12	Tu. No.1	83	6	2.2	40.2	12.9	6.7	4.50	HS	HS
V13	PTS-34	82.2	5.8	3.2	43.4	12.9	4.6	4.50	HS	HS
V14	Bataguda	84	6	2.8	39.6	12	8.5	4.00	S	S
V15	PTS-17	92.4	6.4	2	44.4	11.4	6.5	5.00	HS	R
V16	PTS-8	96.6	6.2	3.2	51.6	13.4	4.1	4.50	HS	HS
V17	PTS-42	73.8	6.2	2.6	35.2	10.4	6.4	2.00	R	HS
V18	Ethamkalam	80.2	6.8	3	43.6	9.7	2.2	5.00	HS	S
V19	328 Sugandham	92	5.8	2.8	45.2	12.5	5.9	2.50	MR	R
V20	PTS-47	89.8	6.4	2.8	42.6	13.2	6.9	2.00	R	R
V21	PTS-21	121.8	6.6	2.2	60.6	13.8	12.3	2.50	MR	HS
V22	Kasturi Manjari	85.4	5.2	3	43.4	11.8	5	4.50	HS	R
V23	PCT-7	90.8	5.2	3.2	49.6	12.7	4.9	3.50	HS	S
	Sem (0.05)	0.49	0.29	0.10	0.46	0.30	0.15	0.33		
	CD (0.05)	1.45	0.85	0.30	1.35	0.89	0.43	0.96		
	CV	0.99	8.1	6.45	1.88	4.51	4.49	13.08		
	Mean	86.5	6.17	2.76	42.5	11.7	5.72	3.57		

R: Resistance, MR: Moderate Resistance, S: Susceptible, HS: Highly Susceptible

Table 2: Morphological characteristics of 23 cultivars of *C.longa* and reaction of turmeric cultivars to *M. incognita* and *T. maculans*.

Name of Primer	Sequence of the primer	Total No. amplification products	No. of polymorphic products	Size range (Kb)
USB-807	5'-AGAGAGAGAGAGAGAGAGT-3'	07	07	200-2500
USB-808	5'- AGAGAGAGAGAGAGAGAGC-3'	07	06	100-2500
USB-810	5'-GAGAGAGAGAGAGAGAT -3'	06	05	600-2500
USB-811	5'-GAGAGAGAGAGAGAGAC-3'	05	04	200-2000
USB-815	5'-CTCTCTCTCTCTCTCTG-3'	07	05	200-1500
USB-835	5'-AGAGAGAGAGAGAGAGTC-3'	04	01	500-1500
USB-836	5'-AGAGAGAGAGAGAGAGT-3'	06	06	200-2000
USB-837	5'-AGAGAGAGAGAGAGAGCC-3'	07	05	500-2500
USB-840	5'-GAGAGAGAGAGAGAGACTT-3'	06	03	200-2500
USB-841	5'-GAGAGAGAGAGAGAGACTC-3'	05	02	500-1500
USB-842	5'-AGAGAGAGAGAGAGAGCA-3'	06	06	500 -2000

Table 3: Total number of amplified fragments and number of polymorphic bands generated by PCR using selected ISSR primers in 23 cultivars of *Curcuma longa*.

from 2.0 to 3.0. Second major cluster (B) having 21 cultivars and again divided into two minor clusters (B1 and B2). One minor cluster (B1) having five cultivars i.e. 'Tu No.4', 'Tu No.1', 'Erode local', 'TC-4' and 'Phulbani Wild'. Among the five cultivars, two cultivars i.e. 'Tu No.4', 'Tu No.1' were making one group with 78% similarity and susceptible to RKN having index from 4.0 to 5.0. Another cultivar 'Erode local' making one group with 71% similarity with other two cultivars were also susceptible to RKN. Second minor cluster (B2) again subdivided into two sub-minor clusters i.e. C1 and C2. First sub-minor cluster (C1) having 4 cultivars with 63% similarity and all are highly susceptible to

RKN. Second sub-minor cluster (C2) having 12 cultivars and making two groups. One group having two cultivars ('Ansitapami' and 'PST-31') with 75% similarity and other group having 10 cultivars with 89% similarity. The cultivars 'PTS-21', '328 Sugandham','PTS-42' and 'PTS-47' were resistance to RKN with root knot indices ranged from 2.0 -2.50. Chu et al. [12] identified RAPD based markers to select for nematode resistance in *Arachis hypogaea*. In another study, Tahery [23] revealed that the identification of ISSR markers associated with root knot nematode resistance of *Hibiscus cannabinus*. He found 13 polymorphic ISSR markers between the resistant and susceptible

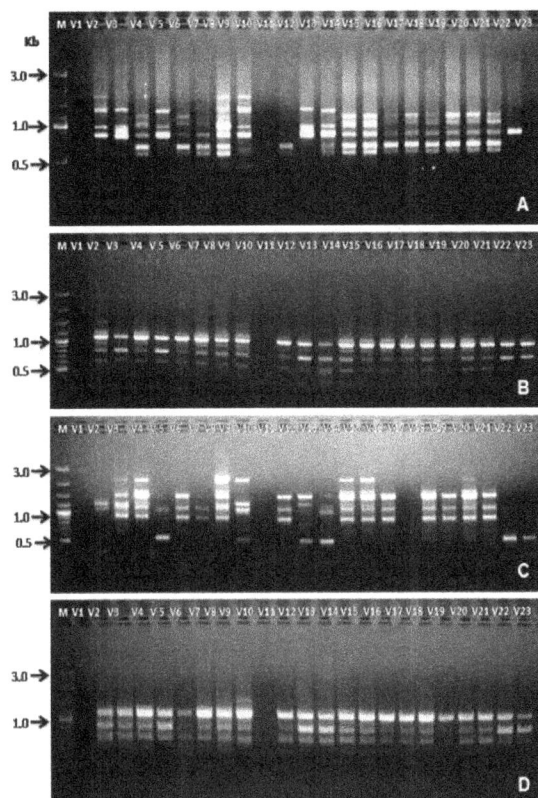

Figure 1: ISSR patterns of 23 varieties of *Curcurma longa* generated by primer USB-810(A), USB-841 (B), USB-807(C) and USB-835(D) MKb molecular weight ladder, V1-V23 assigned as cultivars indicated in Table 2.

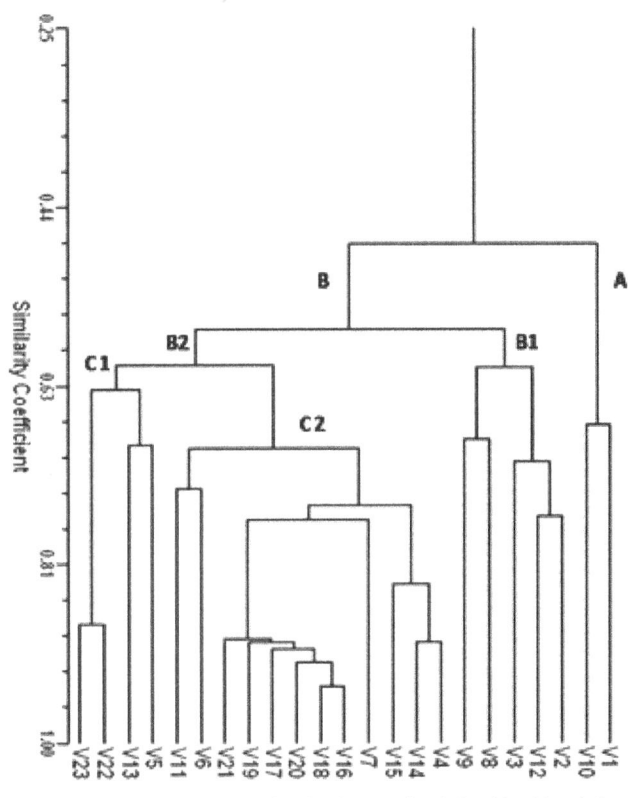

Figure 2: UPGMA dendrogram showing the genetic relationships 23 varieties of *Curcurma longa*. V1-V23 assigned as cultivars indicated in Table 2.

	V1	V2	V3	V4	V5	V6	V7	V8	V9	V10	V11	V12	V13	V14	V15	V16	V17	V18	V19	V20	V21	V22	V23
V1	1.00																						
V2	0.48	1.00																					
V3	0.43	0.70	1.00																				
V4	0.45	0.52	0.68	1.00																			
V5	0.60	0.65	0.52	0.48	1.00																		
V6	0.56	0.48	0.50	0.78	0.59	1.00																	
V7	0.50	0.50	0.52	0.72	0.63	0.69	1.00																
V8	**0.35**	0.60	0.70	0.81	0.50	0.62	0.57	1.00															
V9	0.40	0.68	0.50	0.59	0.67	0.55	0.50	0.68	1.00														
V10	0.67	0.38	0.33	0.42	0.47	0.53	0.57	0.38	0.44	1.00													
V11	0.44	0.52	0.55	0.67	0.67	0.73	0.67	0.60	0.61	0.50	1.00												
V12	0.56	0.76	0.71	0.59	0.67	0.55	0.50	0.61	0.55	0.37	0.61	1.00											
V13	0.56	0.55	0.50	0.60	0.69	0.65	0.59	0.55	0.63	0.53	0.73	0.72	1.00										
V14	0.40	0.48	0.64	0.94	0.72	0.76	0.76	0.72	0.62	0.44	0.71	0.55	0.63	1.00									
V15	0.44	0.52	0.62	0.83	0.56	0.71	0.75	0.67	0.60	0.50	0.80	0.60	0.71	0.88	1.00								
V16	0.47	0.54	0.57	0.75	0.58	0.72	0.76	0.61	0.55	0.44	0.61	0.55	0.55	0.79	0.68	1.00							
V17	0.53	0.46	0.55	0.74	0.56	0.71	0.75	0.59	0.45	0.50	0.59	0.60	0.61	0.78	0.67	0.88	1.00						
V18	0.50	0.50	0.59	0.79	0.61	0.76	0.81	0.64	0.50	0.47	0.65	0.57	0.58	0.83	0.72	0.94	0.94	1.00					
V19	0.44	0.46	0.55	0.74	0.56	0.71	0.75	0.59	0.45	0.41	0.59	0.52	0.53	0.78	0.67	0.88	0.88	0.94	1.00				
V20	0.47	0.48	0.64	0.84	0.58	0.72	0.76	0.68	0.55	0.44	0.61	0.55	0.55	0.89	0.78	**0.89**	0.88	0.94	0.88	1.00			
V21	0.53	0.52	0.55	0.74	0.65	0.81	0.75	0.59	0.52	0.50	0.69	0.60	0.61	0.78	0.76	0.88	0.88	0.94	0.88	0.88	1.00		
V22	0.63	0.59	0.48	0.57	0.65	0.61	0.65	0.46	0.52	0.50	0.50	0.60	0.61	0.60	0.58	0.68	0.76	0.72	0.67	0.68	0.76	1.00	
V23	0.63	0.67	0.55	0.50	0.65	0.53	0.65	0.40	0.45	0.50	0.50	0.68	0.61	0.52	0.58	0.60	0.67	0.63	0.58	0.60	0.67	0.88	1.00

Table 4: Similarity matrix for Nei and Li's coefficient of a total of twenty three variety of turmeric. V1-V23 assigned as name of the cultivars indicated in Table 2.

parents. The marker analysis showed that the ISSR primers were significantly associated with the root knot nematode gall index. Jenkins et al. [10] identified SSR markers for marker assisted selection of root-knot nematode resistant to cotton. They found that the chromosome 11 and 14 of cotton genotype have been associated with root knot nematode resistance which opening the way for marker assisted selection in applied breeding. This investigation as an understanding of the level and partitioning of genetic variation within the cultivars with resistant/susceptible to root knot nematode disease would provide an important input into determining efficient management strategies [24]. The genetic variability in a gene pool is normally considered as the major resource for turmeric improvement program.

Acknowledgement

The authors wish to acknowledge to Department of Agril. Biotechnology, College of Agriculture, Bhubaneswar for providing laboratory facility.

References

1. Food and Agriculture Organization (2001) Production year book for 2000. FAO, Rome.

2. Bai H, Sheela MS, Jiji T (1995) Nemic association and avoidable yield loss in turmeric, Curcuma longa L. Pest Management in Horticultural Ecosystems. 1: 105-110.

3. Ray S, Mohanty KC, Mahapatra SN, Patanaik PR, Ray P (1995) Yield loss in ginger (Zingiber officinale Rosc.) and turmeric (Curcuma longa L.) due to root knot nematode (Melodogyne incognita). JOSAC, 4: 67-69.

4. Caveness FE, Wilson JE, Terry R (1981) Root-knot nematodes on tannia (Xanthosoma sagitifolium).Nematologia Mediterranea, 9: 201-203.

5. McSorley R, O'Hair SK, Parrado JL (1983) Nematodes of cassava, Manihot esculenta Cranz. Nematropica, 13: 262-264.

6. Coyne DL (1994) Nematode parasites of cassava. African Crop Science Journal 2: 355-359.

7. Bridge J (1973) Nematode as pests of yams in Nigeria. Mededelingen Faculteit Landbouwwetenschappen Ghent 38: 841-852.

8. Gapasin RM, Valdez RB, Mendoza EMT (1988) Phenolic involvement in sweet potato resistance to Meloidogyne incognita and M. javanica. Annals of Tropical Research 10: 63-72.

9. Mohanta S, Swain PK, Mishra BK (2013) Root-knot nematode induced physiological and biochemical changes in turmeric under enhanced CO2 gradient. Journal of Plant Protection and Environment, 10: 74-77.

10. Jenkins JN, McCarty JC, Wubben M., Hayes R, Gutierrez OA, et al. (2012)

11. Danso Y, Akromah R, Osei K (2011) Molecular marker screening of tomato (Solanum lycopersicum L.) germplasms for root knot nematodes (Meloidogyne species) resistance. AJB, 10: 1511-1515.

12. ChuY, Holbrook CC, Timper P, Ozias-Akins P (2007) Development of a PCR based molecular marker to select for nematode resistance in Peanut. Crop Sci., 47: 841-845.

13. Skupinova S, Vejli P, Sedlaki P, Bardovai M, Srbek L, et al. (2004) Using DNA markers for characterization of tomato resistance against root nematode Meloidogyne incognita. Plant Soil, Environment 50: 59-64.

14. Adegbite AA, Amusa NA, Agbaje GO, Taiwo LB (2005) Screening of cowpea varieties for resistance to root-knot nematode under field conditions. Nematropica 35: 155-159.

15. Bai D, Brandle J, Reeleder R (1997) Genetic diversity in North American ginseng (Panax quinquefolius L.) grown in Ontario detected by RAPD analysis. Genome 40: 111-115.

16. Doyle JJ, Doyle JL (1990) Isolation of plant DNA from fresh tissue. Focus, 12: 13-15

17. Nei M, Li WH (1979) Mathematical model for studying genetic variation in terms of restriction endonucleases. ProcNatlAcadSci., USA76: 5269-5273

18. Eapen SJ, Ramana KV, Sasi Kumar B, George JK (1998) Resistant to Meloidogyne incognita in ginger and turmeric germplasms. Proceedings of national symposium on rational approaches in nematode management for sustainable agriculture, Anand, India, 22-25 November: 106-109.

19. Mani A, Naidu PH, Madhavachari S (1987) Occurrence and control of Meloidogyne incognita on turmeric in Andhra Pradesh, India. International Nematology Network Newsletter, 4: 13-18.

20. Williams JG, Kubelik AR, Livak KJ, Rafalski JA, Tingey SV (1990) DNA polymorphisms amplified by arbitrary primers are useful as genetic markers. Nucleic Acids Res 18: 6531-6535.

21. Rout GR, Das P, Goel S, Raina SN (1998) Determination of genetic stability of micropropagated plants of ginger using Random Amplified Polymorphic DNA (RAPD) markers. Bot.Bull. Acad. Sin.,39: 23-27

22. Rohlf FJ (1995) NTSYS-PC–Numerical Taxonomy and Multivariate Analysis System, Version 2.0, Exter Software,Setauket, New York, USA

23. Tahery Y (2012) Identification of ISSR markers associated with root knot nematode resistance of Hibiscus cannabinus. Annals of Biological Research, 3: 259-269.

24. Rout GR (2006) Identification of Tinospora cordifolia (Willd.) Miers ex Hook F & Thomas using RAPD markers.Z Naturforsch C 61: 118-122.

SSR markers for marker assisted selection of root-knot nematode (M.incognita) resistant plants in cotton (Gossypium hirsutum L.). Euphytica, 183: 49-54.

23

The Comparison of Antibodies Raised Against PLRV with Two Different Approaches - Viral Particles Purification and Recombinant Production of CP

Aseel DG* and Hafez EE

Department of Plant Protection and Biomolecular Diagnosis, Arid Lands Cultivation Research Institute (ALCRI), Egypt

Abstract

Serology is one of the most important techniques which extensively used in different fields especially the agriculture one. Serological methods usually which are extensively used because of their specificity in disease diagnosis and relative ease of completion. The methods are widely used in plant virology include enzyme-linked immunosorbent assay, dot-blot immunoassay, immunospecific electron microscopy, and tissue-blot immunoassay. The main stone of the serology test is antiserum which is either mono or poly. Due to the high cost of the mono-antisera production and its low sensitivity of the ones immunized by the purified virus particles, the recombinant protein could be good substitution. In this study, we aimed to produce polysera using the viral particles and with the recombinant coat protein of the Potato Leaf Roll Virus as well and used both in rabbit immunization separately. The sensitivity, specificity, and reactivity of the two produced sera were tested in comparing with the real-time polymerase chain reaction. Results revealed that polysera produced by recombinant coat protein are more specific and sensitive than the sera produced by the purified particles. Moreover, the results obtained by dot and tissue blot confirmed the results obtained by enzyme-linked immunosorbent assay techniques. The real time-polymerase chain reaction results were similar to that obtained by serological methods except the time and the substrate. In conclusion, production of the polysera using recombinant protein is test easy, cheap, and sensitive test for on line vial detection.

Keywords: Potato Leafroll Virus (PLRV); Real-Time Quantitative Reverse Transcription Polymerase Chain Reaction (qRT-PCR); rCP; Antiserum production; Dot Blot Immunoassay (DBIA); TBI

Introduction

Potato (*Solanum tuberosum* L.) is one of the most important vegetable crops in human nutrition having potential of vital food security. It is the fourth largest food crop cultivated in more than 100 countries throughout the world and has gained a status of globally traded commodity [1].

Potato is infected by at least 40 viruses and 2 viroids [2] and mixed viral infection is frequent [3]. The primary infection triggers the rolling of young leaves with upright growth pattern appearing pale yellow, tinged purple, pink or red with many cultivars. Whereas, secondary symptoms turned out to be severe with overall rolled leaves with leathery texture, stunted growth and tuber necrosis [4].

PLRV is a major menace for the potato production all over the world [5]. PLRV is the typical member of the genus *Polerovirus* of the Family *Luteoviridae* [6]. PLRV has a monopartite, single stranded RNA genome, transmitted by aphids in a circulative non-propagative manner and is mainly restricted to phloem tissues of infected plants [7]. PLRV forms 25 to 30 nm diameter isometric particles that encapsulate genomic RNA of about 5.9kb that contains six large open reading frames (ORF) [8]. In addition, the 3' end proximal ORFs are expressed via a sub genomic RNA synthesized in host cells during the infection process. The ORF3 encodes the major capsid protein (CP) of about 23 kDa and ORF4 encodes a 17 kDa putative movement protein. The ORF5 encodes the carboxy terminal region of the minor capsid component expressed by translational readthrough of the ORF 3 amber stop codon. The resulting full-length protein has a MW of about 74 kDa but, in preparations of purified virus, it is present in a C-terminally truncated form of about 54 kDa [9].

Recombinant coat protein (CP) was used as an immunogen to produce monoclonal antibodies (MAbs) and polyclonal antibodies (PAbs) against many plant viruses. It's an alternative approach to produce structural proteins of viruses, in particular CP, in *E. coli*, overcoming

difficulties associated with the development of antibodies of good quality. It will be useful when purified viruses or virus proteins are not available especially with viruses are phloem-limited viruses, such as *Luteoviruses*, are present at very low concentrations in their hosts, resulting in low yield of purified virions. Also, the filamentous particles of *Potyviruses* are relatively low stable and they tend to aggregate with plant debris by Souiri et al. [10]. Therefore, polyclonal antibodies raised against purified virions contaminated with host tissue components, cross react with host antigens and often give variable background reactions, thus limiting their use in ELISA based diagnostic methods. Recombinant polyclonal antibodies (Pab) specific to *Tomato spotted wilt virus* (TSWV) [11], *Potato virus Y* [12], *Alfalfa mosaic virus* (AMV) Khatabi et al. [13] have been generated and used for serological detection of cited viruses [14].

Recombinant antisera for Egyptian isolates of both PVX [15] and PLRV [16] were induced using denatured CP technology. These antisera were reactive in I-ELISA but not in DAS-ELISA. Abdel-Salam et al. [8] described a modified technique involving the use of a mixture of native and denature CP for each virus in the antiserum production to enhance the binding capacity of the produced antibodies with their corresponding antigens in DAS-ELISA.

Serological methods widely used in plant virology include enzyme-linked immunosorbent assay (ELISA), dot-blot immunoassay (DBIA), immunospecific electron microscopy, and tissue-blot immunoassay (TBIA). Serologically based tests commonly are employed today because

*Corresponding author: Dalia Gamil Aseel, Department of Plant Protection and Biomolecular Diagnosis, Arid Lands Cultivation Research Institute (ALCRI), City of Scientific Research and Technological Applications, Alexandria, Egypt
E-mail: daliagamil52@gmail.com

of their specificity in disease diagnosis and relative ease of completion [17-19].

We report in this study the sensitivity and specificity of the antisera produced by the recombinant protein in comparing with those raised by purified virus particles. Moreover, the advantages of the recombinant antisera are, rapidity, easy, and inexpensive for the diagnosis.

Material and Methods

qReal-Time PCR for detection PLRV-CP gene

The total RNA of *potato leafroll virus* was extracted from one gram of the infected leaf tissues by using RNeasy® Plant Mini Kit (QIAGEN, Germany) according to manufacturer's instructions. The total RNAs were reverse transcribed using reverse primer of PLRV-CP. In each reaction, 3 µL RNA (30ng) were added to 17 µL of reaction mixture (2.5 µL of 5x RT reaction buffer; 2.5 µL of 25 mM dNTPs; 1 µL of PLRV-CP R primer (100pmol); 0.2 µL of reverse transcriptase (200 u) and 10.8 µL of H_2O). The program was performed at 42°C for 1 h; enzyme stopped at 65°C for 20 min and final step at 4°C for 10 min. Consequently, qPCR SYBR`Green Kit (Thermo, USA) was used to quantify expression of coding region of coat protein gene in infected potato samples by leaf roll virus in assay after the normalization of certain concentrations. The specific primer DAF-CP sense 5`-AGTACGGTCGTGGTTAAAGG3-3 ` and DAR-CP antisense 5`-CTATTTGGGGGTTTTGCAAAG3-3 ` were designed according to Presting et al. [20], were used to target the specific gene (CP for PLRV). A 18S rRNA- sense 5`-TACCTGGTTGATCCTGCCAGTAG-3` and 18S rRNA antisense 5`-CCAATCCCTAGTCTGCATCGT-3` were used as a housekeeping gene for normalizing RNA levels of the target gene. The qPCR SYBR Green based real-time PCR was performed in a total volume of 10µL followed by 1 µL cDNA 10 ng, 5 µL SYBER Green 1x, 0.7 µL DAF 10 ppm, 0.7 µL DAR 10 ppm up to 2.6 µL water. The reaction was performed with a pre-denaturation at 95°C for 15 min, and 40 cycles of denaturation at 95°C for 15 secs, at annealing at 60°C for 1 min and elongation at 72°C for 30 sec. Fluorescent signal measurements were carried out during the elongation step. The qRT-PCR reactions were carried out in thermo piko qRT-PCR apparatus. All qRT-PCRs were performed in duplicate. The qRT-PCR reactions were carried out in 10 µL into Thermo picko qRT-PCR plate 96 well.

Data analysis

Delta Delta Threshold cycle (ΔΔCT) expression values were calculated for RNA samples of PLRV to determine gene expressions using 18S rRNA (reference gene) and the other PLRV-CP gene. $\Delta\Delta C_T$ expression = $2^{(-\Delta\Delta CT)}$, the equations show the mathematical model of the relative expression ration for the real time PCR. The ratio of the target gene is expressed in sample versus control in comparison to reference gene [21].

Virus purification

According to the method described by Gooding & Hebert [22] with some modification, the following processes were applied: Using blander for 5 min, 100g of infected leaves were homogenized in 0.5M Na_2HPO_4-KH_2PO_4 buffer (1:1 W/V) pH 7.2, containing 1% 2-β-mercaptoethanol. The homogenate was squeezed through cheesecloth. An 8-ml n-butanol/100 ml extract was added to the homogenate and stirred for 2 h at 4°C. Then, the emulsion was centrifuged at 12.000 rpm for 30 min. The supernatant was placed on the stirrer at 4°C and 4g polyethelen glycol (PEG, mo. Wt. 6000)/100 ml were added. The mixture was incubated on the stirrer for 2 hours at 4°C and centrifugation was performed for 15 min at 12.000 rpm. The small clear glassy pellet was resuspended in 20 ml of 0.01 M phosphate buffer, (pH 7.2)/100 ml of initial extract and

centrifuged at 12.000 rpm for 15 min. Further purification was obtained by a second precipitation with PEG. A 0.4 g NaCl and 4 g PEG were added for each 10 ml virus suspension and incubated on the stirrer for 2 h at 4°C. Pellet was collected by ultracentrifugation for 2 h at 30.000 rpm. Finally, the pellet was immediately resuspended in 2 ml of 0.01M phosphate buffer, pH 7.2, mixed well; centrifuged at 12.000 rpm for 5 min and was stored at -20°C.

PLRV-CP, cloning and protein expression

This is done by using the protocol described by Dalia et al. [23].

Antisera production

The virus purified and the purified recombinant PLRV-CP fusion protein were emulsified with an equal volume of Freund's complete adjuvant for first injection and with Freund's incomplete adjuvant for subsequent three injections intramuscularly into a New Zealand white rabbit. The rabbit was first bled two weeks after the last immunization. The whole blood was kept for one hour at room temperature for clotting then the clot was released and the blood was heated at 37°C for 30 min then stored at 4°C overnight. The serum was decanted from the clot and centrifuged at 2,000 rpm to remove cell debris. The serum was filtered and brought to 0.025% sodium azide. Aliquots of serum, mixed with equal volumes of glycerol, were stored at -20°C. The serum fractions were collected and stored at -20°C until required according to El-Attar et al. [16].

SDS-Polyacrylamide Gel Electrophoresis (SDS-PAGE)

Proteins were separated by Sodium Dodecyl Sulphate Polyacrylamide Gel Electrophoresis (SDS-PAGE) using the protocol described by Sambrooke and Russell [24]. Gel electrophoresis was performed using the Mini-PROTEIN II vertical gel electrophoresis system (BioRad); 5% staking gel and 12% resolving gel, The SDS gel electrophoresis was carried out at about 80 V in IX Tris/glycine-SDS running buffer. After electrophoresis, the gel was stained by shaking for 1 h in Coomassie brilliant blue R-250 stain and de-stained with de-staining solution overnight until the bands were clearly defined.

Serological detection of PLRV

The produced PLRV-antiserum was tested using I-ELISA, DAS-ELISA, DBIA and TBIA. Two antisera were used for comparison: Antiserum raised against virus particles (Viral antiserum) and Antiserum raised against PLRV coat protein (r CP antiserum).

I-ELISA and Double antibody sandwich-ELISA: The indirect-ELISA was performed according to Koenig [25] and modified by Fegla et al. [26]. Whenever, Double antibody sandwich-ELISA was applied according to Clark and Adams, [27]: The reactions were read visually (yellow color) by a Micro ELISA reader (Stat fax -2100), absorbance at 405 nm (ELISA value).

DBIA and TBIA: Both Dot Blot Immunoassay (DBIA) and Tissue Blot Immunoassay (TBIA) were approached according to Kamenova and Adkins [28] with some modifications. A nitrocellulose membranes (Amersham Biosciences Corp., Piscataway, NJ) was pre-wetted in 100% methanol for 10 secs and then washed in distilled water for 1 min. For DBIA, the membranes were marked into 1-cm squares and a 5 µL sample (homogenized in carbonate coating buffer used in ELISA) was spotted in the center of each square. For TBIA, leaves first were rolled into a tight bundle before being cut with a sterile razor blade into small pieces. Tuber and stem were cut transversely. Freshly cut edges of all tissues were pressed firmly against the membrane. Membranes for both assays were placed in petri dishes blocked with 5% (wt/vol) bovine

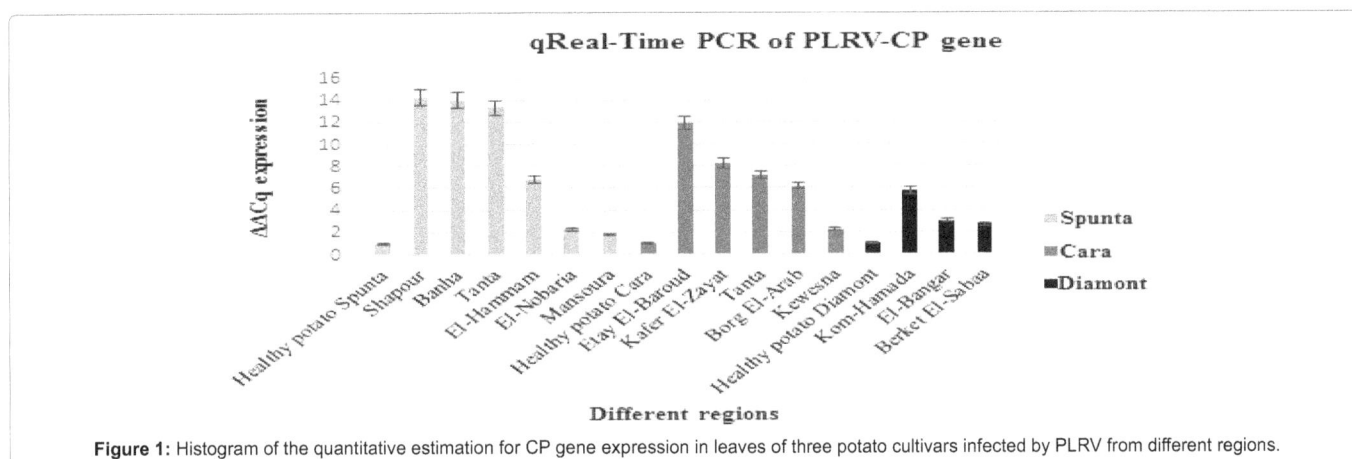

Figure 1: Histogram of the quantitative estimation for CP gene expression in leaves of three potato cultivars infected by PLRV from different regions.

serum albumin (BSA) in PBS buffer for 30 min with gentle shaking at 37°C. After a brief rinse in PBS-T for three times, the membranes were incubated with polyclonal antibodies in serum buffer with incubated at 37°C for 30 min. Membranes were washed in three changes of PBS-T, including 0.5% (wt/vol) Nitrocellulose membranes for 5 min each, then incubated with alkaline phosphatase-conjugated goat anti-rabbit IgG (Sigma-Aldrich, Oakville, ON, Canada), at 1:7500 dilution in serum buffer at 37°C for 30 min then membrane was washed 3 times with PBST for 5 min. Membranes were then washed as before and incubated with freshly prepared substrate NBT/ BCIP sodium salt solution reagent with gently agitation for 1 min. The reaction was stopped by washing the membrane in deionized water for several minutes. The membrane was air dried on a filter paper and photographed.

Results and Discussion

qReal-Time PCR for detection PLRV-CP gene

PLRV is one of the most destructive viruses in potatoes [29] and in seed-production schemes, the absence or very low incidence of the virus is a prerequisite [30]. While most of the viruses infecting potatoes can be detected easily in leaf, stem and tuber tissues by Dot and Tissue blot. On the other hand, quantitative Real-Time PCR (qRT-PCR) is possibly the best method to analyze gene expression because of the large dynamic range; high sensitivity and reproducibility [31-35]. Thus, in order to estimate the relative gene expression of some viral genes understudy, the relative amounts of viral RNA for target PLRV-CP gene for three potato cultivars (Spunta, Cara and Diamont) were compared with the amount of viral RNA in healthy leaves samples. Results revealed that, gene expression in all examined infected samples was higher than that for the control samples. The highest expression values were; 14.32 (Spunta, komhamada), 11.93 (Cara, Etay El-Baroud) and 5.71 (Diamont, komhamada). But the lowest expressions of CP gene were observed with Spunta (Mansoura) is (1.75), with Cara (kewesna) 2.21 followed by 2.64 with Diamont (Berket El-Sabaa). Similarly, Arif et al. [36] tested 22 plants of 14 lines harbored both PVY-CP and PLRV-Replicase genes. They found sixteen plants of 11 double transgenic lines showed high level of the expressions for both genes. In addition, in the present study, the expression PLRV-CP gene varied from region to another for the same cultivar. We assume that expression variability of the viral genes in host cells due to time and place of sampling collection. Also, the relative environmental conditions; degree of infection; virus type and stage of viral life cycle; plant genotype and plant-virus interactions affect the expression level for specific genes Pallas and García, [37]. In some studies, Real Time PCR has been described for efficient detection of PLRV in dormant tubers but due to high costs

of reagents and equipment involved in real time PCR, the method described herein is more applicable and cost-effective (Figure 1).

Purified of virus particles and r-coat protein

The purified virus protein was separated on SDS-PAGE and size of band 80kDa with RT about 53kDa respectively as shown in Figure 2A. This result are in agreement with Brault et al. [38]; Filichkin et al. [39]; Jolly and Mayo, [40]; Wang et al. [41] found within the read through protein (RTD) there is a highly conserved N-terminal region and a variable C-terminal region. The full length RTP can be detected readily in infected tissue, but in purified virus preparations a significant portion of the C-terminus of the RTD is proteolytically processed yielding a 51–58 kDa RTP. This phenomenon has been seen among other members of the family *Luteoviridae* and despite such truncations, the virus is still aphid transmissible [41]. Also, this result are similar with Bahner et al. [42] that reported the major protein component detected by staining with Coomassie blue was the 23 K protein but small amounts of a 53K protein were always present. A 53K polypeptide has been detected in protein from particles of PLRV in preparations purified from potato tissue and *Physalis floridana* using either Celluclast or Driselase Waterhouse, [43] to macerate the tissue prior to extracting the virus particles. Moreover, the purified *luteovirus* particles contain two types of proteins: a major capsid protein (CP) of ~22 kDa and a minor capsid component of 54 kDa, which is a truncated form of a translation read-through protein of the CP gene termination codon. The read-through domain (RTD) contains determinants responsible for virus transmission according to Gonçalves et al. [44].

The purified protein for rCP-PLRV was separated on SDS-PAGE analysis and an enriched expected size of band (approximately 23kDa) was observed (Figure 2). Also, Hossain et al. [45] amplified and cloned 346 bp amplicon of PLRV-CP gene. Our purified PLRV-CP fragments were then sub-cloned into expression vector and transformed into *E. coli* cells. The expressed proteins were purified and one band of ~23 KDa was detected on SDS-PAGE (Figure 2B). According to Mayo and Miller [46], PLRV virions are assembled mainly from the 23 kDa coat protein (CP), but contain minor amounts of readthrough protein (RTP) translated when the stop codon of the CP is suppressed.

Antiserum production for both viral particles and recombinant coat protein

The polyclonal antibodies were obtained from rabbit bleeding after two weeks from the last injection, as presented in Figure 3. The titers of antisera raised against PLRV were 1: 6400, 1/25600 and 1/51200 as determined by indirect ELISA. For experimental point of view, the

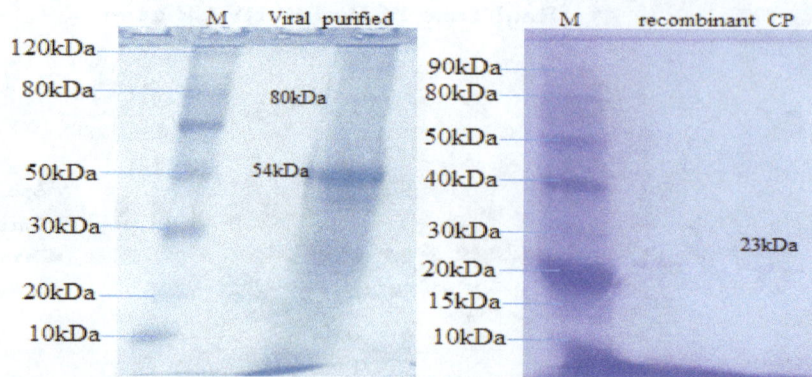

Figure 2: SDS-PAGE analysis showing, (A): virus purification of PLRV, (B): the purified PLRV-CP (23kDa) as a result of the recombinant vector (PLRV-CP gene) expression, where M, 200 kDa protein marker.

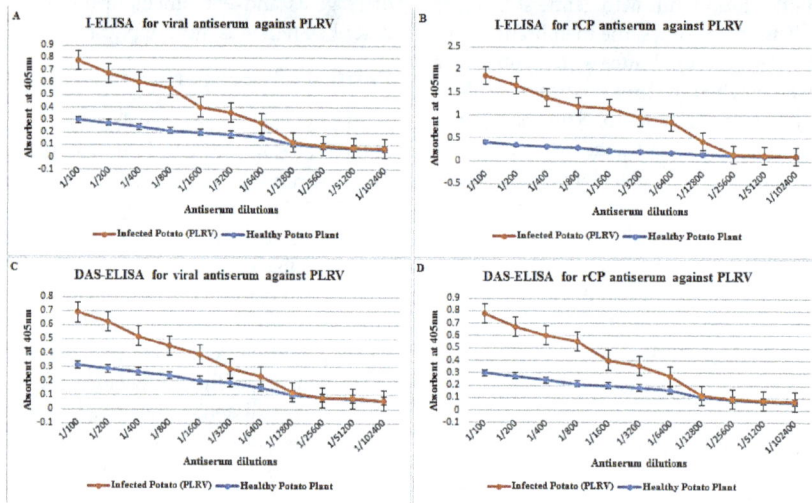

Figure 3: Indirect ELISA and DAS-ELISA using different dilutions of crude PLRV-CP rabbit polyclonal antibody. Healthy and PLRV (1:10, w/v) were tested. Goat-anti rabbit AP-conjugate was used as secondary antibody.

antiserum titer decreased gradually, as it reacts at 1:3200 and 1: 6400 for virus particles and rCP respectively (Figures 3A and 3B respectively). The obtained results were in agreement with that obtained by El-Sharkawy [47], who reported that the antiserum produced against *cow pea aphid borne virus* (CABMV) had titer of (1: 1024) by indirect ELISA. Antiserum obtained after first bleeding (1: 1024) was used in the subsequent experiments. The antiserum produced against PLRV-rCP was applied successfully for the sensitive detection of PLRV in potato plant more than that produced from viral particles. While DAS-ELISA found of virus purified and r CP results were 1/400 and 1/800 respectively (Figures 3C and 3D). Also, we observed sensitivity of Indirect than DAS-ELISA and absorbent value was highest in Indirect-ELISA than DAS-ELISA. In addition, Abdel-Salam et al. [8] noticed that the r-antisera for PVX and PLRV were induced through immunizing the animal with equal concentrations of virus CP prepared under denatured and native conditions. This would expose both epitopes (outer epitopes) and cryptopes (hidden epitopes) to the animal immune system and therefore enhance immunogeneictiy. The similar result detected by the Indirect ELISA had a greater sensitivity than either DAS- or cocktail-ELISA. The observed high sensitivity of indirect ELISA compared with DAS-ELISA could be due to the binding of more virus antigen on plates that were not percolated with specific antibodies, as suggested by Hsu and Aebig and cited by Hsu

and Lawson [48]. Moreover, when evaluating the sensitivity of indirect and DAS-ELISA, their relative ability for accurate quantitation of viral antigens in crude extracts should be considered. For instance, those observed an increase in absorbance values by indirect ELISA with increasing dilutions of *C. quinoa* leaves and hibiscus roots. This likely reflects competition between viral and plant proteins for the finite number of binding sites on the microtiter plate, as previously observed by Lommel et al. [49]. Also, El-Attar et al. [16] addressed the possibility of using recombinant PLRV-CP to produce PLRV specific antisera and to test their suitability for use in serological diagnostic assays for surveys or in certification programs. This investigation suggested that the recombinant virus coat proteins expressed in bacterial cells have great potential as an alternative source of antigens for raising specific antibodies to plant viruses. Such recombinant virus coat proteins can be produced in large quantities and can be manipulated or modified as needed for specific uses. A similar suggestion was reported by Soliman et al. [15].

Dot blot immunoassay (DBIA) found to be sensitive enough for detection of PLRV in infected potato plants (leave, stem and tuber) as presented in Figure 4. A purple colour was obtained from tissues in positive reactions, whereas extracts from healthy plants remained green in representing negative reactions and colour less in blank. DBIA, the virus could be detected in sap extracted from leaves, stems and tubers of

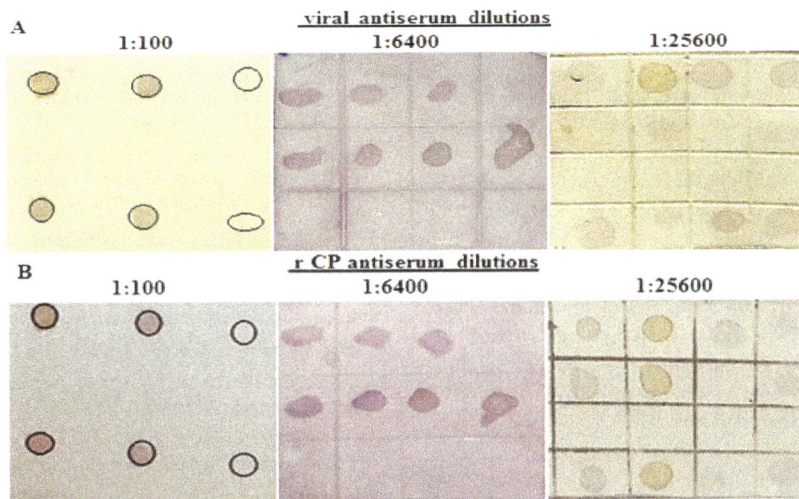

Figure 4: Detection of PLRV in potato leaf, stem and tuber samples using PLRV-antiserum produced for both (A) virus particles and (B) recombinant CP by Dot blot.

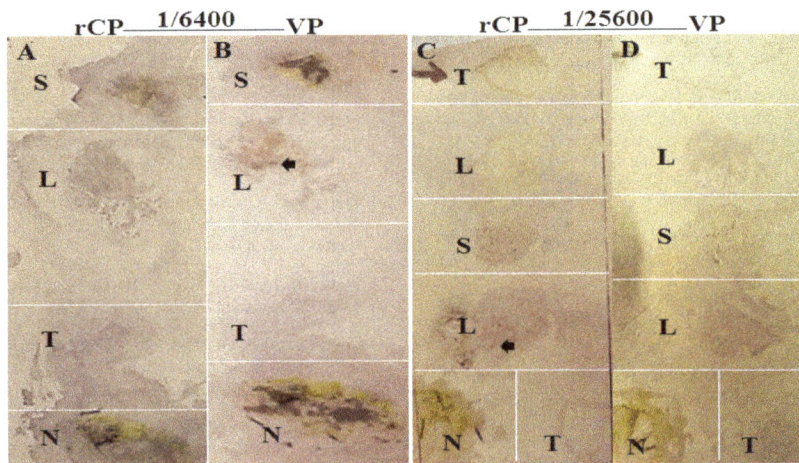

Figure 5: Tissue blotting immunoassay: (A and B); rCP and virus particles used 1/6400 dilution from antiserum, (C and D); rCP and virus particles used 1/25600 dilution from antiserum respectively.

Dilutions of polyclonal antibodies	DAS-ELISA		I-ELISA		DBIA		TBIA		RT-PCR	
	rCP	Vp	rCP	Vp	rCP	Vp	rCP	Vp	rCP	Vp
1/100	+	+	+	+	+	+	+	+	+	-
1/200	+	+	+	+	+	+	+	+	-	-
1/400	+	+	+	+	+	+	+	+	-	-
1/800	+	-	+	+	+	+	+	+	-	-
1/1600	-	-	+	+	+	+	+	+	-	-
1/3200	-	-	+	+	+	+	+	+	-	-
1/6400	-	-	+	-	+	+	+	+	-	-
1/12800	-	-	-	-	+	+	+	+	-	-
1/25600	-	-	-	-	+	+	+	+	-	-
1/51200	-	-	-	-	-	-	-	-	-	-
1/102400	-	-	-	-	-	-	-	-	-	-

Table 1: Comparison for both r CP and viral particles Polyclonal antibodies used detection of PLRV in potato samples by I-ELISA, DAS-ELISA, DBIA, TBIA and RT-PCR.

infected plants and used rCP antiserum and viral antiserum at dilutions up to 1×10^2, 1×10^3 and 1×10^5, respectively. Such results generally are in line with those reported by Chaicharoen et al. [50] and El-Sharkawy [47]. The rate of false DBIA positives was estimated. Since, DBIA has

been found to be more sensitive than ELISA. The immunogenicity of the r-antisera for PLRV was also examined for detecting these viruses in commercial potato plants in the field. Similar results showed the high antigenicity of r CP of *Potato Virus Y* and PLRV immunogenicity of its r-antiserum upon evaluation with DBIA [8]. Also, Fulladolsa Palma et al. [51] investigated that the unlike ELISA, DBIA is not quantitative. The dot-blot immunoassay was successfully used to detect several genera of viruses in different hosts, including *Potyvirus* (PVY and PVA in potato, CYVV in clover), *Potexvirus* (PVX in potato, HVX in hosta), *Carlavirus* (PVS in potato), *Luteovirus* (PLRV in potato), *Cucumovirus* (CMV in pumpkin), *Tobamovirus* (TMVin Nicotiana benthamiana), and *Nepovirus* (TRSV in N. benthamiana). Signals from virus-infected samples were clearly visible for all of the samples tested, including those infected with PLRV, from which it is often difficult to detect a signal after ELISA due to low virus titer in the host plant.

Sensitivity of PLRV detection in leaves (L), stems(S) and tubers (T) extracts using TBIA was examined for the presence of PLRV in potato sap extracts. For PLRV was abundantly present in all tissues tested. Signals obtained from potatoes of PLRV infected plants are shown in Figure 5. Strongest blots detected in leaves and stems and sharper images were obtained from antiserum dilution 1/6400 produced from rCP than Virus particles. While, the antiserum dilution 1/25600 high blot concentration was observed in leaves, stems than tubers, but, healthy potato plant was used as a negative control (N).

Results demonstrated in Table 1 revealed that, the dilution end point of the produced antiserum against PLRV was used at dilution 1/6400 and 1/3200 of rCP and Virus respectively by indirect ELISA and 1/25600 used in Dot-blot and Tissue blot for both antiserum but 1/800 and 1/400 of rCP and Virus particles respectively by DAS-ELISA.

Conclusion

We studied and compared the sensitivity, specificity and reactivity of antiserum produced from r CP more specific and sensitive than the antiserum produced by the virus purified particles. Serological methods such as DBIA and TBIA easy, rapidly, inexpensive, sensitivity and reactivity than I-ELISA, DAS-ELISA and RT-PCR.

Acknowledgements

We wish to thank Saad Hammad for Specialist Lab, Plant Protection and Biomolecular Diagnosis Department, (SRTA, City). for preparation of the samples collection from different localities in Egypt.

References

1. He Z, Larkin RP, Honeycutt W (2012) Sustainable potato production. In: Global case studies. (1st edn), Springer, Dordrecht, Heidelberg, New York, London.

2. Jeffries C, Barker H, Khurana SM (2005) Potato viruses (and viroids) and their management. In: Potato production, improvement and post-harvest management. The Haworth's Food Products Press, New York, USA.

3. Kerlan C, Moury B (2008) Potato virus Y. Encyclopedia of virology. In: BWJ Mahy, Regenmortel V, (eds), (3rdedn). 287-296 p.

4. Douglas DR, Pavek JJ (1972) Net necrosis of potato tubers associated with primary, secondary and tertiary infection of leafroll. Am Potato J 49: 330-333.

5. Ehrenfeld N, Romano E, Serrano C, Arce-Johnson P (2004) Replicase mediated resistance against potato leafroll virus in potato desirée plants. Biol Res 37: 71-82.

6. Pringle CR (1998) Virus taxonomy - San Diego. Report of the 27th Meeting of the Executive Committee of the International Committee on Taxonomy of Viruses. Arch Virol. 143: 1449-1459.

7. Mayo MA, Ziegler-Graff V (1996) Molecular biology of luteoviruses. Adv Virus Res 46: 416-460.

8. Abdel-Salam AM, El-Attar AK, Soliman AM (2013) The use of native and denatured recombinant coat protein forms for induction of good quality antisera for Potato virus X and Potato leaf roll virus. Am J Res Commun 1: 70.

9. Bahner I, Lamp J, Mayo M, Hay RT (1990) Expression of the genome of potato leafroll virus: readthrough of the coat protein termination codon in vivo. J Gen Virol 71: 2251- 2256.

10. Souiri A, Zemzami M, Amzazi S, Ennaji MM (2014). Polyclonal and monoclonal antibody-based methods for detection of plant viruses. Eur J Sci Res 123: 281-295.

11. Vaira AM, Vecchiati M, Masenga V, Accotto GP (1996) A polyclonal antiserum against a recombinant viral protein combines specificity with versatility. J Virol Methods 56: 209-219.

12. Folwarczna J, Plchová H, Moravec T, Hoffmeisterová H, Dedic P, et al. (2008) Production of polyclonal antibodies to a recombinant coat protein of potato virus Y. Folia Microbiol (Praha) 53: 438-442.

13. Khatabi B, He B, Hajimorad MR (2012) Diagnostic potential of polyclonal antibodies against bacterially expressed recombinant coat protein of Alfalfa mosaic virus. Plant Dis 96: 1352-1357.

14. Kapoor R, Mandal B, Paul PK, Chigurupati P, Jain RK (2014) Production of cocktail of polyclonal antibodies using bacterial expressed recombinant protein for multiple virus detection. J Virol Methods 196: 7-14.

15. Soliman AM, Barsoum BN, Mohamed GG, El-Attar AK, Mazyad HM (2006) Expression of the coat protein gene of the Egyptian isolate of potato virus X in Escherichia coli and production of polyclonal antibodies against it. Arab J Biotech 9: 115-128.

16. El-Attar AK, Riad BY, Saad A, Soliman AM, Mazyad HM (2010) Expression of the coat protein gene of potato lafroll virus in Escherichia coli and development of polyclonal antibodies against recombinant coat protein. Arab J Biotech 13: 85-98.

17. Derrick KS (1973) Quantitative assay for plant viruses using serologically specific electron microscopy. Virol 56: 652-653.

18. Lin NS, Hsu YH, Hsu HT (1990) Immunological detection of plant viruses and a mycoplasmalike organism by direct tissue blotting on nitrocellulose membranes. Phytopathol 80: 824-828.

19. Powell CA (1987) Detection of three plant viruses by dot-immunobinding assay. Phytopathol 77: 306-309.

20. Presting GG, Smith OP, Brown CR (1995) Resistance to Potato leafroll virus in potato plants transformed with the coat protein gene or with vector control constructs. Phytopathol 85: 436-442.

21. Schmittgen TD, Livak KJ (2008) Analyzing real-time PCR data by the comparative CT method. Nat Protoc 3: 1101-1108.

22. Gooding GV, Hebert TT (1967) A simple technique for purification of tobacco mosaic virus in large quantities. Phytopathology 57: 1285.

23. Dalia GA, Riad SA, Fegla GI, Hafez EE (2015) Molecular characterization of some viruses infecting potatoes in north Egypt using advanced techniques. Ph.D. Thesis, University of Alexandria, egypt.

24. Sambrook J, Russell DW (2001) Molecular cloning - A laboratory manual. (3rd edn), Cold Spring Harbor Laboratory Press: Cold Spring Harbor, NY.

25. Koenig R (1981) Indirect ELISA method for broad specificity detection of plant viruses. J Gen Virol 55: 53-62.

26. Fegla GI, EL-Samra IA, Nouman K, Younes HA (1997) Host range, transmission and serology of an isolate of tomato yellow leaf curl virus from tomato of plastic houses in northern Egypt. Proc1st Sci Confe Agric Sci Assiut University, Egypt.

27. Clark FM, Adams AM (1977) Characteristics of the microplate method of enzyme–linked immunosorbent assay for the detection of plant viruses. J Virol Methods 34: 475-483.

28. Kamenova I, Adkins S (2004) Comparison of detection methods for a novel tobamovirus isolated from Florida hibiscus. Plant Dis 88: 34-40.

29. Beemster ABR, De Bokx JA (1987) Survey of properties and symptoms. Viruses of Potatoes and Seed- Potato Production. In: De Bokx JA, Van der Want JPA, (eds). The Netherlands, p: 84-113.

30. Van der Zaag DE (1987) Growing seed potatoes. Viruses of Potatoes and Seed-Potato Production. In: de Bokx JA, Van der Want JPA, (eds). The Netherlands, p: 176-203.

31. Bustin SA (2000) Absolute quantification of mRNA using real-time reverse transcription polymerase chain reaction assays. J Mol Endocrinol 25: 169-193.

32. Bustin SA (2002) Quantification of mRNA using real-time reverse transcriptase PCR (RT-PCR): Trends and problems. J Mol Edndocrinol 29: 23-39.

33. Bustin SA, Benes V, Nolan T Pfaffl MW (2005) Quantitative real-time RT-PCR – A perspective. J Mol Endocrinol 34: 597-601.

34. Huggett J, Dheda K, Bustin S, Zumla A (2005) Real-time RT-PCR normalization; Strategies and considerations. Genes Immun 6: 279-284.

35. Van Guilder HD, Vrana KE, Freeman WM (2008) Twenty-five years of quantitative PCR for gene expression analysis. Biotechniques 44: 619-626.

36. Arif M, Thomas PE, Crosslin JM, Brown CR (2009) Development of molecular resistance in potato against potato leaf roll virus and potato virus y through agrobacterium-mediated double transgenesis. Pak J Bot 41: 945-954.

37. Palls V, Garcia JA (2011) How do plant viruses induce disease? Interaction and interference with host components. J Gen Virol 92: 2691-2705.

38. Brault V, Van den Heuvel JF, Verbeek M, Ziegler-Graff V, Reutenauer A, et al. (1995) Aphid transmission of beet western yellows luteovirus requires the minor capsid read-through protein P74. Embo J 14: 650-659.

39. Filichkin SA, Lister RM, McGrath PF, Young MJ (1994) In vivo expression and mutational analysis of the barley yellow dwarf virus readthrough gene. Virol 205: 290-299.

40. Jolly CA, Mayo MA (1994) Changes in the amino acid sequence of the coat protein readthrough domain of potato leafroll luteovirus affect the formation of an epitope and aphid transmission. Virol 201: 182-185.

The Comparison of Antibodies Raised Against PLRV with Two Different Approaches - Viral Particles...

149

41. Wang JY, Chay C, Gildow FE, Gray SM (1995) Readthrough protein associated with virions of barley yellow dwarf luteovirus and its potential role in regulating the efficiency of aphid transmission. Virol 206: 954-962.

42. Bahner I, Lamp J, Mayo M, Hay RT (1990) Expression of the genome of potato leafroll virus: Read through of the coat protein termination codon in vivo. J Gen Virol 71: 2251-2256.

43. Waterhouse PM (1981) Purification, properties and relationships of carrot red leaf virus, and its interaction with carrot mottle virus. Ph.D. Thesis, University of Dundee, UK.

44. Gonçalves MC, Van der Wilk F, Dullemans AM, Verbeek M, Vega J, et al. (2005) Aphid Transmission and Buchnera sp. GroEL Affinity of a Potato leafroll virus RTD Deficient Mutant. Fitopatol Bras 30.

45. Hossain M, Idrees B, Bushra ANT, Tayyab H (2013) Molecular characterization, cloning and sequencing of coat protein gene of a Pakistani potato leaf roll virus isolate and its phylogenetic analysis. Afr J Biotechnol 12: 1196-1202.

46. Mayo M, Miller WA (1999) The structure and expression of luteovirus genomes. In the Luteoviridae. 23-42.

47. El-Sharkawy MM (2005) Biological and serological studies on certain viruses affecting leguminous crops. M.Sc. Thesis, Faculty of Agriculture, Kafr El-Sheikh, Tanta University, Egypt. p: 110.

48. Hsu HT, Lawson RH (1991) Direct tissue blotting for detection of tomato spotted wilt virus in Impatiens. Plant Dis 75: 292-295.

49. Lommel SA, Mccain AH, Morris TJ (1982) Evaluation of indirect enzyme-linked immunosorbent assay for the detection of plant viruses. Phytopathology 72: 1018-1022.

50. Chaicharoen A, Honyproayoon R, Adcharapum C, Ratchanee H (2003) Comparison of indirect ELISA, DIBA and DTBI assays for detection of cowpea aphid-born mosaic virus. Proc of 41st-Kasetsart Univ. Annu Conf. pp: 423-431.

51. Palma ACF, Kota R, Charkowski AO (2013) Optimization of a chemiluminescent dot-blot immunoassay for detection of potato viruses. Am J Potato Res 90: 306-312.

24

Specific Detection of *Klebsiella variicola* and *K. oxytoca* by Loop-Mediated Isothermal Amplification

Jarred Yasuhara-Bell[1], Caleb Ayin[2], April Hatada[2], Yonghoon Yoo[2], Robert L. Schlub[3] and Anne M. Alvarez[2*]

[1]*Department of Molecular Biosciences and Bioengineering, College of Tropical Agriculture and Human Resources, University of Hawai'i at Mānoa, 3190 Maile Way, St. John Room 315, Honolulu, HI 96822, USA*

[2]*Department of Plant and Environmental Protection Sciences, College of Tropical Agriculture and Human Resources, University of Hawai'i at Mānoa, 3190 Maile Way, St. John Room 315, Honolulu, HI 96822, USA*

[3]*Cooperative Extension Service, University of Guam, Agriculture and Life Sciences Building Room 105E, Mangilao, Guam 96923, USA*

Abstract

Klebsiella spp. are opportunistic pathogens with clinical, veterinary and plant-associated isolates. A previous study showed that bacterial ooze from wetwood of severely declined ironwood trees in Guam contained *Ralstonia solanacearum*, *Klebsiella variicola* and *K. oxytoca*. In this study, Loop-Mediated Isothermal Amplification (LAMP) detected *K. variicola* and *K. oxytoca* specifically, using unique primer sets designed individually for each organism. Each LAMP detected its target specifically, while showing negative results for non-target bacteria and negative controls. LAMP detected *Klebsiella* in inoculated-ironwood stem tissues and bacterial ooze. Due to the presence of plant inhibitors, different sampling protocols were tested. Soaking plant tissue samples to allow diffusion of bacteria into solution, followed by boiling, provided optimum detection of *Klebsiella* directly from plant samples. False negatives obtained when using crushed plant samples were eliminated by including an enrichment step, involving plating and 12-h incubation. DGGE (denaturing gradient gel electrophoresis) and a colony-blot immunoassay using a *Klebsiella*-specific antibody also detected *Klebsiella* in inoculated ironwood. DGGE bands and antibody cross-reactions from closely related enterobacters showed the potential for false positive results. The nature of LAMP makes it ideal for point-of-care testing, and when combined with the specificity of the LAMP primers developed in this study, demonstrates its potential as a routine field test for *Klebsiella* in ironwood in Guam, as well as clinical and veterinary diagnosis of *Klebsiella* infection. Additionally, the regions targeted for detection in this study have application across all forms of molecular-based diagnostics.

Keywords: *Klebsiella*; Detection; Loop-mediated amplification; *Casuarina equisetifolia*; ironwood; Decline

Introduction

Ironwood (*Casuarina equisetifolia*) is a common forest tree in the Pacific islands and is widely propagated in Australia, India and China for firewood, wood for construction, land reclamation and windbreaks. Ironwood is salt tolerant and grows on nutrient-poor sandy soils common on coastlines [1]. For the past twelve years, ironwood trees on Guam have been undergoing a slow decline and thousands have died throughout the island [2-4]. Ironwood decline is associated with thinning foliage, dieback of branches, and wetwood formation. Various causal agents, including the root and butt rot fungus *Ganoderma* spp., termites, tree-care practices, and most recently bacteria, have been implicated [5]. However, the disease etiology has not been clearly established [2,4,6,7]. A disease showing similar symptoms was first reported in Mauritius [8], and later in India [9]. The disease symptoms include defoliation, yellowing, wilt, dieback and finally death of the trees.

In a 2011 survey of ironwood decline in Guam, involvement of *Ralstonia solanacearum* (*Rs*), a bacterial pathogen that causes wilt in a large number of plant hosts including ironwood, was suspected when ooze from diseased trees produced positive reactions with *Rs*-specific immunostrips (Agdia, Inc.). Following the initial observation by Putnam in 2011 [10], numerous samples from symptomatic plants were sent to Hawai'i for isolation and identification. However, *Rs* could not be cultured, resulting in an additional survey in Guam in 2012. Three complimentary approaches were used to isolate and identify the pathogen: growth on modified *Rs*-semiselective medium (mSMSA), *Rs*-specific immunostrip assays, and confirmation with a *Rs*-specific loop-mediated isothermal amplification (LAMP) reaction [11,12]. The

identity of isolates was determined in Hawai'i and the two predominant types of bacteria associated with ironwood decline were characterized [13,14]. Bacteriological tests, along with multiplex PCR, BOX-PCR, and phylogenetic analysis of the *dnaA* RIF marker [15], were performed to further characterize and evaluate the genetic diversity among the *Rs* isolates [13,14]. Non-*Rs* isolates were presumptively identified as *Klebsiella* spp. using bacteriological tests [13,14,16,17]. Most strains showed closest similarity to *K. variicola* following sequence analysis of the 16S rRNA gene, while one showed closest match to *K. oxytoca*. Strains identified as *Rs* were capable of wilting tomato seedlings within 7-10 days and ironwood seedlings within 10-30 days. *Klebsiella* strains did not produce symptoms in inoculated plants, but were re-isolated from stem tissues as far as 10 cm above the inoculation site, indicating possible colonization of the plant. Results indicated, for the first time, that *Rs* is involved in the decline of ironwood in Guam and *Klebsiella* spp. are associated with wetwood, a phenomenon associated with various bacteria and reported in other forest ecosystems [18,19].

*Corresponding author: Anne Alvarez, Department of Plant and Environmental Protection Sciences, College of Tropical Agriculture and Human Resources, University of Hawai'i at Mānoa, 3190 Maile Way, St. John Room 315A, Honolulu, HI 96822, USA, 808), E-mail: alvarez@hawaii.edu.

Ironwood decline has not yet been observed in Hawai'i, or even Saipan, which is much closer to Guam, even though *Rs* is present throughout the Pacific. Molecular and immunodiagnostic tools for rapid pathogen identification are needed to obtain relevant data in future field studies to better understand ironwood decline and its potential to spread to other islands of the Pacific. Immunostrips (Agdia, Inc.) and LAMP technology using a hand-held SMART-DART device [11,12] are available for *Rs* detection, but similar tests are needed to determine the possible role of *Klebsiella* spp. in ironwood decline. Once developed and validated, field-ready diagnostic tools can be used to determine the association of these two bacteria within naturally infected symptomatic plants, and lack of association with healthy ones. These assays could also be used to screen nursery trees for latent and active infections, thereby providing tools to produce healthy nursery stocks for the replenishment of dead ironwood trees and reforestation in Guam. This study compares several diagnostic approaches, including a LAMP assay, for detecting *Klebsiella* spp. in ironwood, as well as mixed cultures with *Ralstonia solanacearum* and other bacterial genera.

Materials and Methods

Bacterial strains and culture conditions

The strains used in this study (Table 1) were from the Pacific Bacterial Collection at the University of Hawai'i at Mānoa and included three *Klebsiella variicola* strains (A6126, A6127, and A6128), one *K. oxytoca* strain (A6125) and two *Ralstonia solanacearum* strains (A6123 and A6124), which were all isolated from ironwood in Guam, one *K. pneumoniae* strain (A6133, aka ATCC 13883), as well as *Agrobacterium tumefaciens* (A2961, aka C58), *Enterobacter aerogenes* (A3131, aka ATCC 13048), *E. cloacae* (A5149), *Escherichia coli* (A6091, aka LJH 1947), *Pseudomonas fluorescens* (A3275, aka A811-1), and *Ralstonia solanacearum* (GMI 1000). Bacteria were removed from storage, plated onto TZC medium (17 g/L agar, 10 g/L peptone, 5 g/L glucose and 0.001% 2,3,5-triphenyl-tetrazolium chloride (TZC)) and then incubated at 26°C (±2°C).

Study design

DNA from *Klebsiella* spp. and *Ralstonia solanacearum* strains were tested individually, as well as in mixtures of two different genera or species. Individual and mixed DNA were further tested in the presence of plant tissue. Molecular tests of DNA included PCR followed by DGGE (denaturing gradient gel electrophoresis) analysis and LAMP. DNA from non-*Klebsiella* and non-*Ralstonia* bacteria were also tested with LAMP to determine specificity.

Cultures of *Klebsiella* spp. and *Ralstonia solanacearum* strains were used individually, as well as in mixtures of two different genera or species, and tested with a colony-blot assay, in the absence and presence of plant tissue. Cultures were also inoculated into plants individually and as mixtures. Samples were taken from plants and tested with LAMP, DGGE, and colony-blot assay.

Plant inoculation

Plant inoculations were performed according to a previously established protocol [13,14]. Briefly, bacterial suspensions at 10^8 CFU/ml were prepared from 48 hr cultures of test strains from TZC plates. Seeds of *C. equisetifolia* trees were collected from the University of Hawai'i at Mānoa campus, germinated on damp filter paper, placed in the dark and then transplanted to community pots. Established seedlings were transplanted to 6.7 cm pots containing Sunshine Mix #4 (Sun Gro Horticulture, Agawam, MA). Roots of 15-week-old seedlings

were wounded by drawing a sterile scalpel through the soil at four sides of the pot in the root zone. Then, 10 ml of the bacterial suspension were pipetted into the soil after the wounding process. Older ironwood seedlings, ~17 weeks old, were inoculated with 50 ml of bacterial suspension in the same manner. Two plants were co-inoculated with *Ralstonia solanacearum* and *Klebsiella variicola*, but one was an older, 10-month old ironwood (Plant 1) that was used to try to produce bacterial ooze. The ironwood seedlings were then grown under the same conditions as described previously [14].

Plant sampling

Plant samplings were performed according to a previously established protocol [14]. Briefly, stem sections of plants were surface sterilized in 10% Clorox (final concentration: 0.8% sodium hypochlorite) for 30 seconds and triple-rinsed in sterile de-ionized water. Stem sections were then macerated in ~200 μl of sterile de-ionized water and the subsequent suspension was streaked onto TZC. Following a 24 h enrichment phase on TZC, the identity of the re-isolated bacteria was presumptively determined by appearance, and then confirmed using the various tests described hereafter. Tests were also performed directly on the extracted plant samples [15-19]. Bacterial ooze from plant samples was also collected by soaking stem sections in water to allow bacteria to diffuse from the plant material. For co-inoculated plants, five sample sections were taken from Plant 1, while eight were taken from a younger one (Plant 2), with sections being numbered sequentially away from the inoculum site. Tests were performed directly on bacterial ooze, as well as isolated colonies following a 12-h and 24-h enrichment.

DNA extraction

A Chelex DNA extraction was performed on isolated bacteria, as well as plant samples. Briefly, 0.75-1.0 ml of 40% Chelex 100 resin (Bio-Rad, Hercules, CA) in 1X TE buffer (10 mM Tris HCl and 1 mM EDTA at pH 8) with 10% Triton X-100 (Sigma-Aldrich, St. Louis, MO) was added to each sample. Samples were mixed by pipeting vigorously and vortexing and then heated to 95°C for 10 minutes on a digital heat block. Samples were stored at 4°C. Various other DNA extraction protocols were used on plant samples to remove inhibitors [20-24], including the PowerPlant Pro DNA isolation Kit (MO BIO Laboratories Inc., Carlsbad, CA).

DGGE

DGGE primers (341F(-GC)/907RA) used in this study were reported previously to amplify the variable V3 region of the 16S rRNA gene [25-27]. PCR reactions for 341F(-GC)/907RA were performed in

Organism	Strain	Other ID	Hcp 1 LAMP	UGH LAMP
Agrobacterium tumefaciens	A2961	C58	-	-
Enterobacter aerogenes	A3133	ATCC 13048	-	-
Enterobacter cloacae	A5149	B193	-	-
Escherichia coli	A6091	LJH 1947	-	-
Klebsiella oxytoca	A6125	S-3	-	+
Klebsiella pneumoniae	A6133	ATCC 13883	-	-
Klebsiella variicola	A6126	S-4	+	-
Klebsiella variicola	A6127	S-8	+	-
Klebsiella variicola	A6128	S-19	+	-
Pseudomonas fluorescens	A3275	A811-1	-	-
Ralstonia solanacearum	A6123	S-25	-	-
Ralstonia solanacearum	A6124	S-26	-	-
Ralstonia solanacearum	A3292	GMI-1000	-	-

Table 1: Bacterial strains used in this study and corresponding LAMP results.

a 25 µl reaction volume containing 1 µl bacterial DNA and 24 µl PCR reaction master mix [1 µl of each primer (10 µM), 0.13 µl GoTaq DNA polymerase (5 U/ µl) (Promega, Madison, WI), 5 µl 5X GoTaq Reaction Buffer (Promega, Madison, WI), 0.5 µl dNTP mix (10 mM) (Promega, Madison, WI) and 16.37 µl DNase/RNase free water]. PCR reaction conditions were as follows: an initial denaturing at 94°C for 5 min, followed by 25 cycles of denaturing at 94°C for 30 s, annealing at 57°C for 1.5 min, and elongation at 72°C for 1 min, with a final elongation at 72°C for 10 min. PCR products were resolved using DGGE. PCR primer sequences are listed in (Table 2).

DGGE analysis was performed with a DGGE-2000 system (C.B.S Scientific Co.), using a 0.75 mm-thick 6% polyacrylamide gel (ratio of acrylamide to bis-acrylamide, 37.5:1) with a linear denaturing-agent gradient of 35-65% (100% denaturant agent was defined as 7 M urea with 40% formamide) that was submerged in 0.5X TAE buffer (40 mM Tris, 40 mM acetic acid, 1 mM EDTA; pH 7.4) at 60°C. PCR samples were mixed with 6 µl of dye solution [0.1% bromphenol blue (w/v), 70% glycerol (v/v)] and applied to the gels. Electrophoresis conditions were 5 h 30 min at 75 V. Gels were stained for 30 min in 1X TAE buffer with ethidium bromide and visualized using the Foto/Analyst Express System (Fotodyne Inc., Hartland, WI). Products were excised from gels and resuspended in sterile water [28]. Products were reamplified using 341F/907RA primers (Table 2), under the same reaction conditions, and then sequenced for identification. Sequences were queried in the National Center for Biotechnology Information (NCBI) database using the Basic Alignment Search Tool (BLAST) [29].

DNA sequencing

PCR products were cleaned for sequencing using ExoSAP-IT (Affymetrix, Santa Clara, CA), according to the manufacturer's instructions. Cleaned PCR products were sequenced at the Greenwood Molecular Biology Facility at the University of Hawai'i sequencing facility, using each forward and reverse primer, according to specifications.

Colony-blot immunoassay

The colony blot immunoassay was performed according to methods established previously [30,31], with modifications. Briefly, 1 µl of bacterial suspension (OD_{A600}=0.1), or plant extract, were spotted onto 0.45 µm nitrocellulose membranes (NitroBind, Osmonics, Inc., Minnetonka, MN) and allowed to dry completely. Dried nitrocellulose membranes were blocked in 5% PBS-milk for 1 h and then boiled in PBS for 10 min to eliminate endogenous enzymatic activity. Membranes were then washed three to five times using 0.05% PBS-Tween 20. Membranes were subsequently incubated with *Klebsiella* species antibody (73/28) (ThermoFisher Scientific, Waltham, MA) at a 1:2000 dilution for 30 min, followed by washing three times for five minutes each with 0.5% PBS-Tween 20. Membranes were then incubated with goat anti-mouse antibody conjugated to alkaline phosphatase (1:2000) (Southern Biotech, Birmingham, AL) for 30 min and washed three times for five minutes with 0.5% PBS-Tween 20. Membranes were developed with 1-Step NBT/BCIP (ThermoFisher Scientific, Waltham, MA) for 15-30 min, rinsed with distilled water, and air-dried. Colony immunoblots were analyzed under a dissecting microscope.

LAMP

LAMP primers developed in this study are shown in Table 2. Individual LAMP reactions were performed in triplicate and contained 5 µl sample, 5 µl primer master mix [F3 (0.2 µM), B3 (0.2 µM), FIP (1.6 µM), BIP (1.6 µM), Loop (0.8 µM), Loop Probe (0.08 µM), and Quencher probe (0.16 µM)] and 15 µl ISO-001nd Isothermal Mastermix (OptiGene, West Sussex, UK). Hcp1 and UGH primers were multiplexed to produce a reaction to detect both *Klebsiella variicola* and *K. oxytoca*, requiring 0.4 µM Quencher probe. Negative controls used 5 µl ddH₂O and/or pathogen-free plant samples. LAMP reactions were run and analyzed using the iQ5 Multicolor Real-Time PCR Detection System (Bio-Rad, Hercules, CA) and a hand-held real-time assessment device (SMART-DART) (DiaGenetix Inc., Honolulu,

Oligonucleotide Primer Sequence (5'-3')		Source/Reference
LAMP		
Hcp1-F3	CCCCATACCTTTACAAGGCC	This Study
Hcp1-B3	CGTGGATCCTCCAGGTGATT	This Study
Hcp1-FIP	GCTGAGTTTCACTCTTGAATACCAGCATTATGAACTTCGAACGT	This Study
Hcp1-BIP	ACCGGGGTTAACTGTGGTGTACTGCAGGCTAACGCTTTCAAC	This Study
Hcp1-Loop	TGCAAACTTTCCTCGAATGACA	This Study
Hcp1-Loop Probe	/56-FAM/ ACGCTGAGGACCCGGATGCGAATGCGGATGCGGATGCCGATGCAAACTTTCCTCGAATGACA	This Study
UGH-F3	TGCCAGCAGACATTGACG	This Study
UGH-B3	CCGACCACTACGAACGGT	This Study
UGH-FIP	CGATTGCGATCTGCGGCCTGCGCCAGTTTCTGGTAACCG	This Study
UGH-BIP	TCTGCAGCCAGCAGCAGATCGCCTGCGAGTATTTCTTCCG	This Study
UGH-Loop	GCGCATAAGCACTTTCCGG	This Study
UGH-Loop Probe	/56-FAM/ ACGCTGAGGACCCGGATGCGAATGCGGATGCGGATGCCGATTTTGCGCATAAGCACTTTCCGG	This Study
Quencher probe	TCGGCATCCGCATCCGCATTCGCATCCGGGTCCTCAGCGT/3BHQ_1/	Kubota et al., [11]
PCR		
gyrA -A	CGCGTACTATACGCCATGAACGTA	Brisse and Verhoef, [42]
gyrA -C	ACCGTTGATCACTTCGGTCAGG	Brisse and Verhoef, [42]
DGGE		
341F	CCTACGGGAGGCAGCAG	Muyzer et al., [25]
341F(-GC)	(CGCCCGCCGCGCCCCGCGCCCGTCCCGCCGCCCCCGCCCG-)CCTACGGGAGGCAGCAG	Muyzer et al., [25]; Meyer and Kuever, 2008
907RA	CCGTCAATTCATTTGAGTTT	Muyzer et al., [26]; Ishii and Fukui, [27]

*
/56-FAM/: 5' 6- carboxyfluorescein; /3BHQ_1/: 3' Black Hole Quencher 1

Table 2: Oligonucleotide primers used in this study*.

HI), under the following conditions: 65°C for 30 min, with fluorescence readings being taken at 30-s or 1-min intervals. One drop of mineral oil was added to the tops of each sample when using the SMART-DART to prevent evaporation.

Sensitivity assay

Bacterial DNA was isolated using the Wizard Genome DNA Purification Kit (Promega, Madison, WI). DNA was quantified using the NanoDrop 2000 spectrophotometer (Thermo Scientific, Waltham, MA) and diluted to a starting concentration of 1 ng/μl, with subsequent 10-fold serial dilutions made to reach a final concentration of 1 fg/μl. LAMP reactions were performed in triplicate, using ddH$_2$O as the negative control.

Results

DGGE

DGGE analyses were performed to determine its effectiveness in identifying Ralstonia and Klebsiella from inoculated ironwood. Samples were analyzed alongside positive controls (Figure 1). Positive controls consisting of purified DNA samples of Ralstonia, Klebsiella, and mixes of the two former (Figure 1, Lanes 1-3), show a characteristic Ralstonia and Klebsiella band (Figure 1, Lane 1 and Lane 2, respectively), with both bands appearing in the mixed sample (Figure 1, Lane 3). A mixed sample containing DNA from purified Ralstonia and Klebsiella, as well as DNA from Enterobacter aerogenes, E. cloacae and Escherichia coli, which are closely related to Klebsiella and found associated with ironwood, produced the characteristic Ralstonia and Klebsiella bands, however additional bands reflecting all of the additional bacteria contained within the sample were not present (Figure 1, Lane 4). Plants samples from ironwood inoculated with Ralstonia, Klebsiella, and both Ralstonia and Klebsiella produced the appropriate characteristic Ralstonia and Klebsiella bands (Figure 1, Lanes 5-7), with the Klebsiella band being faint in the plant sample with mixed infection (Figure 1, Lane 7). All characteristic bands were excised, re-amplified and sequenced. BLAST analysis of returned sequence data revealed that the characteristic Ralstonia and Klebsiella bands were indicative of each respective bacterium, thus demonstrating the utility of DGGE, and its potential ability to be used without sequencing by including appropriate controls. However, in the mixed sample, which contained other enteric bacteria (Figure 1, Lane 4), sequence data could not be obtained for the Klebsiella band. A strong band was produced for Ralstonia solanacearum, a weak band for Klebsiella, and no distinct bands for all other enteric bacteria. Similarities between bands of the enteric bacteria and Klebsiella could explain the lack of additional bands. Additionally, the Klebsiella band in the mixed plant sample was very faint (Lane 7). Inefficient recovery of DNA from the excised band is responsible for the inability to obtain sequence data.

Colony-blot assay

A Klebsiella-specific antibody was tested for specificity using pure bacterial suspensions in a colony-blot immunoassay format (Figure 2a). Positive reactions were obtained with all three Klebsiella spp. tested and negative reactions with R. solanacearum, P. fluorescens and the water control. However, Enterobacter aerogenes, E. coli, and Agrobacterium tumefasciens gave cross-reactions with the Klebsiella antibody assay. The colony-blot assay detected Klebsiella in spiked plant samples (Figure 2b). Positive controls made with pure bacterial suspensions of Klebsiella were positive, while negative controls consisting of Ralstonia, only macerated plant tissue, and water, were all negative. Plant samples containing Klebsiella variicola and K. oxytoca showed

positive reactions, while plant samples containing Ralstonia did not, thus demonstrating potential for use in detecting Klebsiella from plant tissue samples.

LAMP

Two sets of LAMP primers were designed to specifically detect K. oxytoca and K. variicola. Both LAMP reactions were highly specific, only detecting their respective Klebsiella species and producing no reaction with any of the other bacteria tested in this study (Table 1), as well as negative controls. Positive LAMP reactions were recorded visually via turbidity, using the specified primers and a LAMP reaction master mix (dNTP mix (1.2 mM), 10X ThermoPol Reaction Buffer (New England Biolabs, Ipswich, MA) (2 mM), betaine (1 M), MgSO$_4$ (4 mM), Bst DNA polymerase (New England Biolabs, Ipswich, MA) (8 U) and ddH$_2$O) that was described previously [32,33]. Under the current protocol, positive reactions were visible in real-time and commonly occurred within 10-20 minutes.

The sensitivity of the Hcp1-LAMP reaction was determined and DNA in the range of 5 ng to 5 pg showed positive reactions in approximately 10 minutes, while DNA in the range of 500 fg to 5 fg showed positive reactions in approximately 15 minutes. However, at the 5 fg range of DNA, positive reactions were not always obtained due to the low amount of DNA in the starting sample. The sensitivity of the UGH-LAMP reaction was less than that of the Hcp1-LAMP, with a detection limit of ~5 pg of DNA.

LAMP was performed on plant sample extracts following a 12-h and 24-h enrichment phase on solid media. All samples from known inoculated plants were positive by LAMP, while negative control plants were negative by LAMP. Non-Klebsiella bacteria that were

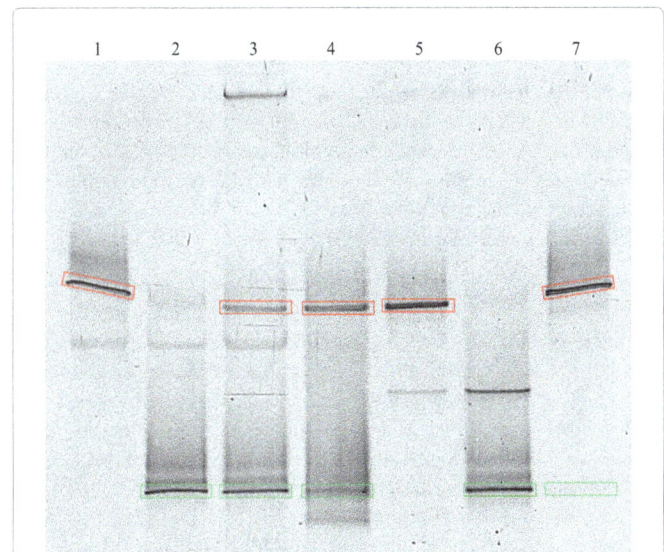

Figure 1: DGGE analysis using a linear denaturing-agent gradient of 35-65%. DNA from samples were amplified using DGGE-specific primers designed against the 16 rRNA gene. Lanes 1 and 2: positive controls using purified Ralstonia and Klebsiella DNA, respectively. Lane 3: additional positive control consisting of mixed purified Ralstonia and Klebsiella DNA. Lane 4: mixed sample consisting of mixed purified Ralstonia and Klebsiella DNA, as well as DNA from Enterobacter aerogenes, E. cloacae and Escherichia coli, which are closely related to Klebsiella and found associated with ironwood. Lane 5: Ralstonia-inoculated plant sample. Lane 6: Klebsiella-inoculated plant sample. Lane 7: plant sample inoculated with Ralstonia and Klebsiella. Red boxes indicate the characteristic Ralstonia band. Green boxes indicate the characteristic Klebsiella band. Characteristic bands were excised and sequenced from all lanes.

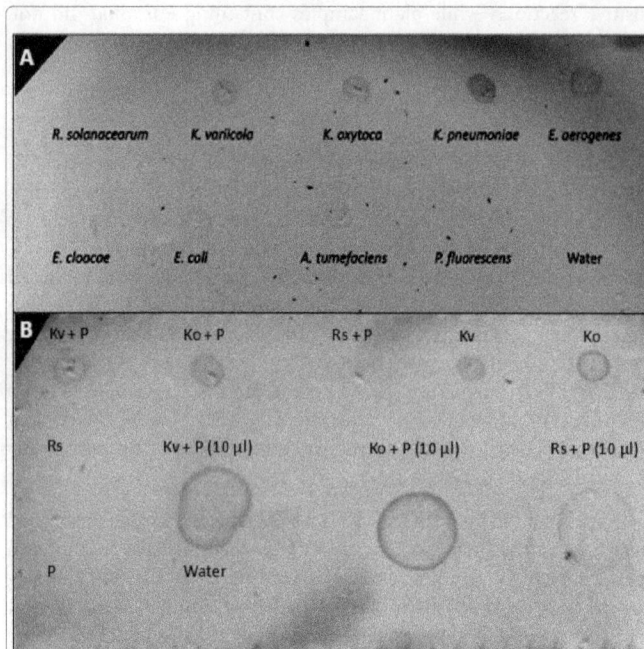

Figure 2: Colony-blot immunoassay using a *Klebsiella*-specific antibody. Nitrocellulose membranes were spotted with (A) 1 μl bacterial suspension (ODA600=0.1) or (B) spike plant samples and allowed to dry. A colony-blot immunoassay was performed on membranes using a *Klebsiella*-specific antibody. Secondary antibody was a goat anti-mouse antibody conjugated to alkaline phosphatase. Membranes were developed with 1-Step NBT/BCIP, which yields a purple color as a positive reaction. A faint spot is visible for the 10 μl *Ralstonia*-spiked plant sample; however that was attributed to background staining of the membrane by plant tissue itself, as seen by the faint spot for the non-spiked plant sample. Kv: *Klebsiella variicola*; Ko: *Klebsiella oxytoca*; Rs: *Ralstonia solanacearum*; P: plant extract.

present in the plant samples were tested separately and were negative with LAMP. In order to make a single test for both *K. variicola* and *K. oxytoca*, the LAMP primers were multiplexed into a single reaction. The multiplexed test was able to detect both *Klebsiella* spp.; however the sensitivity of the combined assay was compromised, as known to occur during multiplexing [34,35].

LAMP was further tested directly on plant sample extracts from two co-inoculated plants. Extracts were streaked onto plates and the first two sections of Plant 1 showed growth of *Klebsiella* and *Ralstonia*, while only *Ralstonia* grew from sections 3-5. In Plant 2, *Klebsiella* grew only from the first five sections. Initial testing using turbidity revealed that plant inhibitors did affect the LAMP reaction to a certain degree (data not shown). For real-time LAMP detection, extracts were processed, using various methods, and tests revealed no significant difference in results between direct boiling, 1:1 sample and TE buffer with boil, Chelex extraction, or GES buffer, using buffers 1 and 2 and sample at a 0.5:0.5:1 ratio. All *Klebsiella*-negative samples, including negative and *Ralstonia* controls, were all negative by LAMP. The LAMP reaction did not react with boiled samples from Plant 2 at sections 1, 3 and 4, TE-buffered samples from Plant 2 at section 2, and GES-buffered samples from Plant 1 at section 2. All other samples were positive by LAMP, including all Chelex-extracted samples. Included in the LAMP test was one *Klebsiella*-inoculated plant sample that had DNA extracted using the PowerPlant Pro DNA isolation kit, which is explicitly designed to extract DNA from plants and remove plant inhibitors. This sample was positive by LAMP, and reactions were the most robust; however, this kit is high in cost and therefore not practical for field use/studies. The

SMART-DART hand-held real-time fluorescence monitoring device performed equally to the in-lab real-time machine.

Discussion

The genus *Klebsiella* is comprised of opportunistic pathogens that are frequently isolated from humans and animals [36]. In humans, *Klebsiella pneumoniae* and *K. oxytoca* are particularly involved with nosocomial infections [36,37], such as septicemia, pneumonia and urinary tract infection. In animals, *Klebsiella* spp. are mostly associated with sepsis, infections of the urinary and respiratory tracts, and mastitis, with *Klebsiella pneumoniae* and *K. oxytoca* being isolated frequently from domestic livestock [38-40]. These disease syndromes can cause serious economic consequences in some cattle herds [41]. The ecological habitats of *Klebsiella* include surface water, sewage, soils and plants, as well as mucosal surfaces of mammals [42]. Recently, a new species of *Klebsiella* was proposed, *K. variicola*, having clinical and plant-associated isolates [43].

In Guam, the association found between ironwood decline and *Ralstonia* led to the discovery of an association between *Klebsiella* and the same disease [11,12]. An association was also found between wetwood symptoms observed in infected trees and *Klebsiella* spp., which were determined to be *Klebsiella oxytoca* and *K. variicola* [13,14]. The current study, which focused on evaluation and design of diagnostic tools, will enable a more extensive survey of Guam ironwood trees to further investigate the association between *Klebsiella*- and *Ralstonia*-infected ironwood trees. DGGE, an immunoassay and a molecular-based assay (LAMP) were all tested for their ability to identify *Klebsiella* in ironwood.

DGGE is a method used to discern bacterial populations in mixed environmental samples. However, the complexity of DGGE makes it less desirable as a routine test for field studies. Additionally, results generally require DNA sequencing of bands following gel separation, which is not readily available to researchers in the field. Here, we attempted to circumvent the need for sequencing. Knowing what specific bacteria to look for and by running appropriate controls, we thought characteristic bands could be found that would be representative of *Ralstonia* and *Klebsiella*. While we were able to demonstrate this, the inability of DGGE to separate bands from closely related *Enterobacter* species contained in a mixed sample showed the possibility for false positive results. This fact, in combination with the laboriousness of DGGE, resulted in not recommending it as a possible test to survey Guam ironwood for *Klebsiella* infection.

Immunostrips (Agdia, Inc.) are available for detection of *Ralstonia*, but there are none for detection of *Klebsiella*. We explored the possibility of making an immunoassay that would be later commissioned into immunostrips for rapid detection of *Klebsiella*. We decided to test a commercial *Klebsiella* antibody for its potential as a rapid detection system, using a simple colony-blot test. As with most immunoassays, we observed cross-reactions with closely related bacteria. The *Klebsiella* antibody was made using *Klebsiella aerogenes*. *Klebsiella* species are constantly being reclassified into other genera and/or species. Interestingly enough, a very strong cross reaction was observed with *Enterobacter aerogenes*, even stronger than the interaction with *Klebsiella oxytoca* and *K. variicola*. Since this *Klebsiella* antibody reacted with a closely related *Enterobacter* sp. associated with ironwoods, this immunoassay is not recommended for routine field testing of *Klebsiella*-infected ironwood in Guam.

The potential for false positive results using the immunoassay and the inability to clearly separate bands using DGGE led to the design of a specific molecular-based assay using loop-mediated amplification technology. The LAMP assay proved to be the most useful in this study. It was able to specifically detect *Klebsiella*, having no positive reactions with non-*Klebsiella* bacteria, and the two primer sets allowed differential detection of *Klebsiella oxytoca* and *K. variicola*. LAMP primers in this study were designed to amplify a hemolysin-coregultaed protein (Hcp1) family type VI secretion system (T6SS) effector (*K. variicola* specific) and an unsaturated glucuronyl hydrolase (UGH) (*K. oxytoca* specific). These regions were found by whole-genome comparisons. BLAST searches of these genes resulted in matches only to each designated *Klebsiella* species, demonstrating that these primer sets should be highly specific for each *Klebsiella* species to which they were designed. LAMP tests using non-target bacteria confirmed the specificity of these genes as targets. Testing a broader panel of non-target bacteria would improve specificity claims; however, in silico data from BLAST searches combined with test results provided in this study, together, suggest a high specificity.

As with any molecular-based assay, plant inhibitors are always a concern. Once an effective assay is developed, sample processing remains the underlying factor. The LAMP assays presented here were affected by plant inhibitors to an extent. Plant samples produced some false negatives, but there were no false positive results and Chelex extraction was able to remove enough inhibitors to allow proper detection directly from woody tissues. Ultimately, to avoid any problems with plant inhibitors, it is recommended to simply enrich samples on solid media [17,44,45]. *Klebsiella* grows extremely fast and a simple 12-h incubation on media will eliminate all false results.

LAMP was chosen because it is the most widely researched isothermal nucleic acid method, which allows greater trouble-shooting during development [46]. LAMP is comparable to equivalent PCR-, immunoassay- or culture-based detection methods [47], with detection limits as low as five copies [48]. Moreover, its isothermal nature makes it suitable for field diagnostics.

Molecular methods such as PCR have been evaluated for rapid identification of *Klebsiella* in human clinical specimens [37,49]. However, *Klebsiella* species are difficult to identify and are often misclassified in clinical microbiology laboratories [16,50,51]. Phenotypic distinction between *K. pneumoniae* and *K. oxytoca* isolates based on existing biochemical tests are time consuming, laborious and not very reliable [52]. Current procedures, such as BIOLOG and API systems, often fail to differentiate between species of *Klebsiella* [50,53]. Differentiation of *Klebsiella* is a key component of clinical responses to infection. In Japan, in 2013, a patient with sepsis caused by *Klebsiella variicola* died due to misidentification as *Klebsiella pneumoniae* by a commonly used automated identification system [54]. The need for diagnostic tests that can properly identify *Klebsiella* at the species level has been nationally recognized, as the National Institute of Health and the National Institute of Allergy and Infectious Disease recently put out a request for proposal for an R01 grant that included *Klebsiella* as a target organism (Funding Opportunity Announcement Number RFA-AI-14-019).

The LAMP assays developed for *Klebsiella* have potential as clinical diagnostic tests because LAMP is less sensitive than PCR to inhibitory substances present in biological samples [55] such as serum [56], CSF [57], swabs [58], and heat-treated blood [59], saving time and money required for sample processing steps [60]. As a result, this technology has been used widely for molecular detection of several microorganisms, including clinical and plant-associated bacterial

[32,61-73], fungal [74-76], viral [47,77-84], and parasitic [57,59,85-88] pathogens, and is therefore suitable as a rapid field test.

Due to the isothermal nature of LAMP, it offers an easy and rapid assay for clinical medicine [89]. The LAMP reactions presented in this study not only specifically detect *Klebsiella*, but also distinguish *K. oxytoca* and *K. variicola*. Moreover, the target regions that these LAMP primers were designed to could become the new focus for developing *Klebsiella*-detection systems, having application to numerous molecular detection platforms. In conjunction with portable real-time fluorescence monitoring devices, such as the SMART-DART, these LAMP reactions could provide beneficial point-of-care diagnostics for clinical, veterinary and agricultural *Klebsiella* spp.

Acknowledgements

This work was supported in part by USDA National Institute for Food and Agriculture, Project HAW00987-H, administered by the College of Tropical Agriculture and Human Resources, University of Hawai'i at Mānoa. This work was also supported in part by WPDN-201303063-01 and by projects administered by the University of Guam: US Forest Service project 12-DG-11052021-236, USDA NIFA, RREA project 231395 and USDA NIFA, McIntire-Stennis project 1005476. The authors thank Daniel Jenkins and Ryo Kubota of DiaGenetix Inc. (Honolulu, HI) for providing the Isothermal Mastermix and SMART-DART.

References

1. Morton JF (1980) The Australian pine or beefwood (Casuarina equisetifolia L.), an invasive "weed" tree in Florida. Proc Fla State Hort Soc 93: 87-95.

2. Schlub KA (2010) Investigating the ironwood tree (Casuarina equisetifolia) decline in Guam using applied multinomial modeling: Louisiana State University.

3. Mersha Z, Schlub RL, Spaine PO, Smith JA, Nelson SC (2010) Visual and quantitative characterization of ironwood tree (Casuarina equisetifolia) decline on Guam. Phytopathology 100: S82.

4. Mersha Z, Schlub RL, Moore L (2009) The state of ironwood (Casuarina equisetifolia subsp. equisetifolia) decline of the Pacific island of Guam. Phytopathology 99: S85.

5. Schlub RL, Schlub KA, Alvarez AM, Aime CM, Cannon PG, et al. (2012) Integrated perspective on tree decline of ironwood (Casuarina equisetifolia) on Guam. Proceedings of the 60th Annual Western International Forest Disease Work Conference: October 8-12, 2012 2012; Lake Tahoe, CA: 51-60.

6. Schlub RL, Kubota R, Alvarez AM (2013) Casuarina equisetifolia decline in Guam linked to colonization of woody tissues by bacteria. Phytopathology 103: S2.128.

7. Schlub RL, Mendi R, Aiseam C, Mendi R, Davis J, et al. (2012) Survey of wood decay fungi on Casuarina equisetifolia (ironwood) on the islands of Guam and Saipan. Phytopathology 102: S416.

8. Orian G (1961) Diseases of Filao (Casuarina equisetifolia) forest in Mauritius. Revue agricole et sucrière de l'île Maurice 40: 17-45.

9. Ali MIM, Anuratha CS, Sharma JK (1991) Bacterial wilt of Casuarina equisetifolia in India. Eur J Forest Pathol 21: 234-238.

10. Schlub RL, Moore A, Marx BD, Schlub KA, Kennaway L, et al. (2011) Decline of Casuarina equisetifolia (ironwood) trees on Guam: symptomology and explanatory variables. Phytopathology 101: S216.

11. Kubota R, Schell M, Peckham G, Rue J, Alvarez AM, et al. (2011) In silico genomic subtraction guides development of highly accurate, DNA-based diagnostics for Ralstonia solanacearum race 3 biovar 2 and blood disease bacterium. J Gen Plant Pathol 77: 182-193.

12. Kubota R, Vine BG, Alvarez AM, Jenkins DM (2008) Detection of Ralstonia solanacearum by loop-mediated isothermal amplification. Phytopathology 98: 1045-1051.

13. Ayin CM, Schlub RL, Alvarez AM (2013) Identification of bacteria associaed with decline of ironwood trees (Casuarina equisetifolia) in Guam. Phytopathology 103: S2, 10.

14. Ayin CM, Schlub RL, Yasuhara-Bell J, Alvarez AM (2015) Identification and characterization of bacteria associated with decline of ironwood (Casuarina equisetifolia) in Guam. Australas Plant Pathol 44: 225-234.

15. Schneider KL, Marrero G, Alvarez AM, Presting GG (2011) Classification of plant associated bacteria using RIF, a computationally derived DNA marker. PLoS One 6: e18496.

16. Alves MS, Dias RC, de Castro AC, Riley LW, Moreira BM (2006) Identification of clinical isolates of indole-positive and indole-negative Klebsiella spp. J Clin Microbiol 44: 3640-3646.

17. Schaad NW, Jones JB, Chun W (2001) Laboratory Guide for Identification of Plant Pathogenic Bacteria, APS Press, St. Paul, Minnesota, USA.

18. Murdoch CW, Campana RJ (1983) Bacterial species associated with wetwood of elm. Phytopathology 73: 1270-1273.

19. Tiedemann G, Bauch J, Bock E (1977) Occurrence and significance of bacteria in living trees of Populus nigra L. Eur J Forest Pathol 7: 364-374.

20. Bellstedt DU, Pirie MD, Visser JC, de Villiers MJ, Gehrke B (2010) A rapid and inexpensive method for the direct PCR amplification of DNA from plants. Am J Bot 97: e65-68.

21. De Boer SH, Ward LJ, Li X, Chittaranjan S (1995) Attenuation of PCR inhibition in the presence of plant compounds by addition of BLOTTO. Nucleic Acids Res 23: 2567-2568.

22. Llop P, Caruso P, Cubero J, Morente C, López MM (1999) A simple extraction procedure for efficient routine detection of pathogenic bacteria in plant material by polymerase chain reaction. J Microbiol Methods 37: 23-31.

23. Singh RP, Nie X, Singh M, Coffin R, Duplessis P (2002) Sodium sulphite inhibition of potato and cherry polyphenolics in nucleic acid extraction for virus detection by RT-PCR. J Virol Methods 99: 123-131.

24. Samarakoon T, Wang SY, Alford MH (2013) Enhancing PCR amplification of DNA from recalcitrant plant specimens using a trehalose-based additive. Appl Plant Sci 1.

25. Muyzer G, de Waal EC, Uitterlinden AG (1993) Profiling of complex microbial populations by denaturing gradient gel electrophoresis analysis of polymerase chain reaction-amplified genes coding for 16S rRNA. Appl Environ Microbiol 59: 695-700.

26. Muyzer G, Teske A, Wirsen CO, Jannasch HW (1995) Phylogenetic relationships of Thiomicrospira species and their identification in deep-sea hydrothermal vent samples by denaturing gradient gel electrophoresis of 16S rDNA fragments. Arch Microbiol 164: 165-172.

27. Ishii K, Fukui M (2001) Optimization of annealing temperature to reduce bias caused by a primer mismatch in multitemplate PCR. Appl Environ Microbiol 67: 3753-3755.

28. Díez B, Pedrós-Alió C, Marsh TL, Massana R (2001) Application of denaturing gradient gel electrophoresis (DGGE) to study the diversity of marine picoeukaryotic assemblages and comparison of DGGE with other molecular techniques. Appl Environ Microbiol 67: 2942-2951.

29. Altschul SF, Gish W, Miller W, Myers EW, Lipman DJ (1990) Basic local alignment search tool. J Mol Biol 215: 403-410.

30. Alvarez AM, Kaneshiro WS, Vine BG (2005) Diversity of Clavibacter michiganensis subsp. michiganensis populations in tomato seed: What is the significance? Acta Hort 695: 205-213.

31. Kaneshiro WS: Detection and characterization of virulent, hypovirulent, and nonvirulent Clavibacter michiganensis subsp. michiganensis. Honolulu: University of Hawai`i at Manoa; 2003.

32. Kubota R, Vine BG, Alvarez AM, Jenkins DM (2008) Detection of Ralstonia solanacearum by loop-mediated isothermal amplification. Phytopathology 98: 1045-1051.

33. Yasuhara-Bell J, Kubota R, Jenkins DM, Alvarez AM (2013) Loop-mediated amplification of the Clavibacter michiganensis subsp. michiganensis micA gene is highly specific. Phytopathology 103: 1220-1226.

34. Kanagawa T (2003) Bias and artifacts in multitemplate polymerase chain reactions (PCR). J Biosci Bioeng 96: 317-323.

35. Polz MF, Cavanaugh CM (1998) Bias in template-to-product ratios in multitemplate PCR. Appl Environ Microbiol 64: 3724-3730.

36. Podschun R, Ullmann U (1998) Klebsiella spp. as nosocomial pathogens: epidemiology, taxonomy, typing methods, and pathogenicity factors. Clin Microbiol Rev 11: 589-603.

37. Neuberger A, Oren I, Sprecher H (2008) Clinical impact of a PCR assay for rapid identification of Klebsiella pneumoniae in blood cultures. J Clin Microbiol 46: 377-379.

38. Brisse S, Duijkeren Ev (2005) Identification and antimicrobial susceptibility of 100 Klebsiella animal clinical isolates. Vet Microbiol 105: 307-312.

39. Kikuchi N, Kagota C, Nomura T, Hiramune T, Takahashi T, et al. (1995) Plasmid profiles of Klebsiella pneumoniae isolated from bovine mastitis. Vet Microbiol 47: 9-15.

40. Wilson DJ, Gonzalez RN, Das HH (1997) Bovine mastitis pathogens in New York and Pennsylvania: prevalence and effects on somatic cell count and milk production. J Dairy Sci 80: 2592-2598.

41. Waller KP, Unnerstad H (2004) Klebsiella mastitis: a potential threat to milk production. Svensk Veterinärtidning 56: 11-17.

42. Brisse S, Verhoef J (2001) Phylogenetic diversity of Klebsiella pneumoniae and Klebsiella oxytoca clinical isolates revealed by randomly amplified polymorphic DNA, gyrA and parC genes sequencing and automated ribotyping. Int J Syst Evol Microbiol 51: 915-924.

43. Rosenblueth M, Martínez L, Silva J, Martínez-Romero E (2004) Klebsiella variicola, a novel species with clinical and plant associated isolates. Sysetm Appl Microbiol 27: 27-35.

44. Brisse S, Grimont F, Grimont PAD: The genus Klebsiella. In: Dworkin M, Falkow S, Rosenberg E, Schleifer K-H, Stackebrandt E (eds) The Prokaryotes. vol. 6. New York: Springer; 2006: 159-196.

45. Gephardt P, Murray RGE, Costilow RN, Nester EW, Wood WA, et al. 1981 Manual of Methods for General Bacteriology, ASM Press, Washington D.C.

46. Craw P, Balachandran W (2012) Isothermal nucleic acid amplification technologies for point-of-care diagnostics: a critical review. Lab Chip 12: 2469-2486.

47. Fujino M, Yoshida N, Yamaguchi S, Hosaka N, Ota Y, et al. (2005) A simple method for the detection of measles virus genome by loop-mediated isothermal amplification (LAMP). J Med Virol 76: 406-413.

48. Iwamoto T, Sonobe T, Hayashi K (2003) Loop-mediated isothermal amplification for direct detection of Mycobacterium tuberculosis complex, M. avium, and M. intracellulare in sputum samples. J Clin Microbiol 41: 2616-2622.

49. Kovtunovych G, Lytvynenko T, Negrutska V, Lar O, Brisse S, et al. (2003) Identification of Klebsiella oxytoca using a specific PCR assay targeting the polygalacturonase pehX gene. Res Microbiol 154: 587-592.

50. Monnet D, Freney J (1994) Method for differentiating Klebsiella planticola and Klebsiella terrigena from other Klebsiella species. J Clin Microbiol 32: 1121-1122.

51. Hansen DS, Aucken HM, Abiola T, Podschun R (2004) Recommended test panel for differentiation of Klebsiella species on the basis of a trilateral interlaboratory evaluation of 18 biochemical tests. J Clin Microbiol 42: 3665-3669.

52. Chander Y, Ramakrishnan MA, Jindal N, Hansen K, Goyal SM (2011) Differentiation of Klebsiella pneumoniae and K. oxytoca by multiplex polymerase chain reaction. Intern J Appl Res Vet Med 9: 138-142.

53. Monnet D, Freney J, Brun Y, Boeufgras JM, Fleurette J (1991) Difficulties in identifying Klebsiella strains of clinical origin. Zentralbl Bakteriol 274: 456-464.

54. Seki M, Gotoh K, Nakamura S, Akeda Y, Yoshii T, et al. (2013) Fatal sepsis caused by an unusual Klebsiella species that was misidentified by an automated identification system. J Med Microbiol 62: 801-803.

55. Kaneko H, Kawana T, Fukushima E, Suzutani T (2007) Tolerance of loop-mediated isothermal amplification to a culture medium and biological substances. J Biochem Biophys Methods 70: 499-501.

56. Ihira M, Sugiyama H, Enomoto Y, Higashimoto Y, Sugata K, et al. (2010) Direct detection of human herpesvirus 6 DNA in serum by variant specific loop-mediated isothermal amplification in hematopoietic stem cell transplant recipients. J Virol Methods 167: 103-106.

57. Njiru ZK, Mikosza AS, Matovu E, Enyaru JC, Ouma JO, et al. (2008) African trypanosomiasis: sensitive and rapid detection of the sub-genus Trypanozoon by loop-mediated isothermal amplification (LAMP) of parasite DNA. Int J Parasitol 38: 589-599.

58. Enomoto Y, Yoshikawa T, Ihira M, Akimoto S, Miyake F, et al. (2005) Rapid diagnosis of herpes simplex virus infection by a loop-mediated isothermal amplification method. J Clin Microbiol 43: 951-955.

59. Poon LLM, Wong BWY, Ma EHT, Chan KH, Chow LMC, et al. (2006) Sensitive and inexpensive molecular test for falciparum malaria: detecting Plasmodium falciparum DNA directly from heat-treated blood by loop-mediated isothermal amplification. Clin Chem 52: 303-306.

60. Mori Y, Notomi T (2009) Loop-mediated isothermal amplification (LAMP): a rapid, accurate, and cost-effective diagnostic method for infectious diseases. J Infect Chemother 15: 62-69.

61. Enosawa M, Kageyama S, Sawai K, Watanabe K, Notomi T, et al. (2003) Use of loop-mediated isothermal amplification of the IS900 sequence for rapid detection of cultured Mycobacterium avium subsp. paratuberculosis. J Clin Microbiol 41: 4359-4365.

62. Hanaki K, Sekiguchi J, Shimada K, Sato A, Watari H, et al. (2011) Loop-mediated isothermal amplification assays for identification of antiseptic- and methicillin-resistant Staphylococcus aureus. J Microbiol Methods 84: 251-254.

63. Harper SJ, Ward LI, Clover GR (2010) Development of LAMP and real-time PCR methods for the rapid detection of Xylella fastidiosa for quarantine and field applications. Phytopathology 100: 1282-1288.

64. Koide Y, Maeda H, Yamabe K, Naruishi K, Yamamoto T, et al. (2010) Rapid detection of mecA and spa by the loop-mediated isothermal amplification (LAMP) method. Lett Appl Microbiol 50: 386-392.

65. Kubota R, LaBarre P, Singleton J, Beddoe A, Weigl BH, et al. (2011) Non-Instrumented Nucleic Acid Amplification (NINA) for Rapid Detection of Ralstonia solanacearum Race 3 Biovar 2. Biol Eng Trans 4: 69-80.

66. Lalande V, Barrault L, Wadel S, Eckert C, Petit JC, et al. (2011) Evaluation of a loop-mediated isothermal amplification assay for diagnosis of Clostridium difficile infections. J Clin Microbiol 49: 2714-2716.

67. Lin GZ, Zheng FY, Zhou JZ, Gong XW, Wang GH, et al. (2011) Loop-mediated isothermal amplification assay targeting the omp25 gene for rapid detection of Brucella spp. Mol Cell Probes 25: 126-129.

68. McKenna JP, Fairley DJ, Shields MD, Cosby SL, Wyatt DE, et al. (2011) Development and clinical validation of a loop-mediated isothermal amplification method for the rapid detection of Neisseria meningitidis. Diagn Microbiol Infect Dis 69: 137-144.

69. Misawa Y, Yoshida A, Saito R, Yoshida H, Okuzumi K, et al. (2007) Application of loop-mediated isothermal amplification technique to rapid and direct detection of methicillin-resistant Staphylococcus aureus (MRSA) in blood cultures. J Infect Chemother 13: 134-140.

70. Neonakis IK, Spandidos DA, Petinaki E (2011) Use of loop-mediated isothermal amplification of DNA for the rapid detection of Mycobacterium tuberculosis in clinical specimens. Eur J Clin Microbiol Infect Dis 30: 937-942.

71. Temple TN, Johnson KB (2011) Evaluation of loop-mediated isothermal amplification for rapid detection of Erwinia amylovora on pear and apple fruit flowers. Plant Dis 95: 423-430.

72. Temple TN, Stockwell VO, Johnson KB (2007) Development of a rapid detection method for Erwinia amylovora by loop-mediated isothermal amplification (LAMP). XI International Workshop on Fire Blight 739: 497-503.

73. Yamazaki W, Seto K, Taguchi M, Ishibashi M, Inoue K (2008) Sensitive and rapid detection of cholera toxin-producing Vibrio cholerae using a loop-mediated isothermal amplification. BMC Microbiol 8: 94.

74. Lucas S, da Luz Martins M, Flores O, Meyer W, Spencer-Martins I, et al. (2010) Differentiation of Cryptococcus neoformans varieties and Cryptococcus gattii using CAP59-based loop-mediated isothermal DNA amplification. Clin Microbiol Infect 16: 711-714.

75. Niessen L, Vogel RF (2010) Detection of Fusarium graminearum DNA using a loop-mediated isothermal amplification (LAMP) assay. Int J Food Microbiol 140: 183-191.

76. Sun J, Najafzadeh MJ, Vicente V, Xi L, de Hoog GS (2010) Rapid detection of pathogenic fungi using loop-mediated isothermal amplification, exemplified by Fonsecaea agents of chromoblastomycosis. J Microbiol Methods 80: 19-24.

77. Curtis KA, Rudolph DL, Owen SM (2008) Rapid detection of HIV-1 by reverse-transcription, loop-mediated isothermal amplification (RT-LAMP). J Virol Methods 151: 264-270.

78. Curtis KA, Rudolph DL, Owen SM (2009) Sequence-specific detection method for reverse transcription, loop-mediated isothermal amplification of HIV-1. J Med Virol 81: 966-972.

79. Dinh DT, Le MT, Vuong CD, Hasebe F, Morita K (2011) An Updated Loop-Mediated Isothermal Amplification Method for Rapid Diagnosis of H5N1 Avian Influenza Viruses. Trop Med Health 39: 3-7.

80. Hatano B, Goto M, Fukumoto H, Obara T, Maki T, et al. (2011) Mobile and accurate detection system for infection by the 2009 pandemic influenza A (H1N1) virus with a pocket-warmer reverse-transcriptase loop-mediated isothermal amplification. J Med Virol 83: 568-573.

81. Imai M, Ninomiya A, Minekawa H, Notomi T, Ishizaki T, et al. (2006) Development of H5-RT-LAMP (loop-mediated isothermal amplification) system for rapid diagnosis of H5 avian influenza virus infection. Vaccine 24: 6679-6682.

82. Kurosaki Y, Grolla A, Fukuma A, Feldmann H, Yasuda J (2010) Development and evaluation of a simple assay for Marburg virus detection using a reverse transcription-loop-mediated isothermal amplification method. J Clin Microbiol 48: 2330-2336.

83. Okafuji T, Yoshida N, Fujino M, Motegi Y, Ihara T, et al. (2005) Rapid diagnostic method for detection of mumps virus genome by loop-mediated isothermal amplification. J Clin Microbiol 43: 1625-1631.

84. Parida M, Posadas G, Inoue S, Hasebe F, Morita K (2004) Real-time reverse transcription loop-mediated isothermal amplification for rapid detection of West Nile virus. J Clin Microbiol 42: 257-263.

85. Bakheit MA, Torra D, Palomino LA, Thekisoe OM, Mbati PA, et al. (2008) Sensitive and specific detection of Cryptosporidium species in PCR-negative samples by loop-mediated isothermal DNA amplification and confirmation of generated LAMP products by sequencing. Vet Parasitol 158: 11-22.

86. Lau YL, Meganathan P, Sonaimuthu P, Thiruvengadam G, Nissapatorn V, et al. (2010) Specific, sensitive, and rapid diagnosis of active toxoplasmosis by a loop-mediated isothermal amplification method using blood samples from patients. J Clin Microbiol 48: 3698-3702.

87. Matovu E, Kuepfer I, Boobo A, Kibona S, Burri C (2010) Comparative detection of trypanosomal DNA by loop-mediated isothermal amplification and PCR from flinders technology associates cards spotted with patient blood. J Clin Microbiol 48: 2087-2090.

88. Nkouawa A, Sako Y, Li T, Chen X, Wandra T, et al. (2010) Evaluation of a loop-mediated isothermal amplification method using fecal specimens for differential detection of Taenia species from humans. J Clin Microbiol 48: 3350-3352.

89. Nagamine K, Watanabe K, Ohtsuka K, Hase T, Notomi T (2001) Loop-mediated isothermal amplification reaction using a nondenatured template. Clin Chem 47: 1742-1743.

25

The Effect of Blue-light-emitting Diodes on Antioxidant Properties and Resistance to *Botrytis cinerea* in Tomato

Kangmin Kim[1#], Hee-Sun Kook[1#], Ye-Jin Jang[1], Wang-Hyu Lee[2], Seralathan Kamala-Kannan[1], Jong-Chan Chae[1*] and Kui-Jae Lee[1*]

[1]Division of Biotechnology, Advanced Institute of Environment and Bioscience, Chonbuk National University, Iksan 570-752, Korea
[2]Department of Agricultural Biology, College of Agriculture and Life Sciences, Chonbuk National University, Jeonju 561-756, Korea
#These authors contributed equally to this work

Abstract

In higher plants, blue-light is mainly perceived by cryptochromes and phototropins, which subsequently orchestrates phototropism, chloroplast relocation, stomatal opening, rapid inhibition of hypocotyl elongation and leaf expansion. Blue-light signaling is also known to mediate the plant responses to biotic stresses, but relevant mechanisms are largely unknown. Here, we demonstrated that blue LED (Light Emitting Diode)-driven inhibition of gray mold disease was highly correlated with the increases in cellular protectants like proline, antioxidants and ROS (Reactive Oxygen Species) scavenger activities. After twenty one days of exposure to various wavelengths of LED lights, blue-LED treated tomato displayed significant increases in proline accumulation in the leaves and stems, whereas red- and green-LED treated tomato exhibited the lower proline contents. Similarly, the blue-LED treatment increased the amount of polyphenolic compounds in tomatoes, compared to other wavelength of LED lights. The activities of various ROS (Reactive Oxygen Species) scavenging enzymes were also slightly increased under the blue-LED lighted conditions. Finally, blue-LED significantly suppressed symptom development of tomato infected by gray mold. Combined results suggest that blue LED light inhibits the development of gray mold disease, which can be mechanistically explained by the enhanced proline accumulation and antioxidative processes at least in partial.

Keywords: Tomato; LED (Light Emitting Diode); Antioxidative defense system; *Botrytis cinerea*, Blue light

Introduction

Tomato (*Lycopersicon esculentum* Mill.) is the one of major greenhouse vegetable crops throughout the world. In terms of nutritional value, tomato is an excellent source of vitamin A and C, carotenoids, α-tocopherol, as well as phenolic compounds as antioxidants [1]. However, the quantity or quality levels of such phytochemicals vary considerably depending on genotype, and/or environmental conditions [2]. In greenhouse environment, phytopathogenic attack leads to retarded growth, damages to cell viability and eventually reduction of plant productivity [3,4]. In particular, the gray mold is the most widespread fungal disease in plants and caused by *Botrytis cinerea*, a necrotrophic fungal pathogen attacking fruits, vegetables and flowers of horticultural crops [3]. This fungus has great adaptability under broad environmental conditions, and is well known to rapidly develop fungicide resistance lines [5]. In horticultural aspects, optimization of cultivation conditions, such as temperature, humidity and light regimes might mitigate disease *via* modulating metabolite levels, and/or cellular compositions.

Light critically regulates the plant growth by regulating the various morphological and physiological changes of grown plants [6,7]. Recently, light-emitting diode (LED) as artificial light source for plant growing in controlled-environment have a variety of advantages, such as small volume, durability, longevity and selectable narrow-waveband emissions [8]. A few studies using this technology have been carried out on the effect of the light spectral quality on the plant growth and morphogenesis, as well as physiological responses by photooxidative changes [7,9,10]. It has been reported that red light, mainly perceived by *Phytochromes (Pyr)*, is important for shoot growth including stem elongation on strawberries [11]. Wang et al. [12] found that disease resistance to *Sphaerotheca fuliginea* in cucumber plants was induced by red light. On the other hand, blue lights are perceived by majorly two different receptor family, cryptochromes and phototropins. The activation of blue light signaling modulates the various physiological and developmental processes, such as phototropism, chloroplast relocation, stomatal opening, rapid inhibition of hypocotyl elongation and leaf expansion [13]. It is known that blue light also regulates the responses against biotic environmental stresses. *Arabidopsis CRY1* (*Cryptochrome1*) positively regulated resistance to *Pseudomonas syringae*, potentially *via* effector-triggered R protein-mediated local resistance [14]. In addition, stability of some R proteins is modulated by blue light receptors [15,16]. Nevertheless, the mechanistic basis underlying such blue light-driven protection is largely unknown.

In general, the oxidative burst involving generation of reactive oxygen species (ROS) is the earliest cellular responses, following recognition of a variety of bacterial and fungal pathogens [17]. The enhanced production of ROS is pre-requisite for hypersensitive response (HR) related to programmed cell death in systemic acquired resistance (SAR), but also damage to the major cellular components, such as DNA, lipids and proteins [18]. In a systemic tissues, in order to minimize oxidative stress by excess ROS, plants have developed detoxifying mechanisms consisting of antioxidants and some ROS scavenging enzymes, such as peroxidase, superoxide dismutase (SOD), catalase (CAT), ascorbate peroxidase (APX) [18]. Besides acting as an osmoprotectant, proline

*Corresponding authors: Kui-Jae Lee, Division of Biotechnology, Advanced Institute of Environment and Bioscience, Chonbuk National University, Iksan 570-752, Republic of Korea, Korea
E-mail: leekj@jbnu.ac.kr

Jong-Chan Chae, Division of Biotechnology, Advanced Institute of Environment and Bioscience, Chonbuk National University, Iksan 570-752, Korea, E-mail: chae@jbnu.ac.kr

also plays antioxidative roles by bringing concentrations of ROS within compatible ranges under the stressed conditions [19]. On the other hand, certain wavelengths of lights have been reported to increase antioxidative actions against abiotic challenges. In broad bean leaves, for instance, significantly increase of catalase activity under red light contributed to scavenging of hydrogen peroxide generated by *Botrytis cinerea* infection [20]. In addition, enhanced activities of CAT, as well as increased contents of total polyphenol and proline were observed in flax resistant to powdery mildew [21]. Regardless of these reports, the effects of light quality on the antioxidative status of plants are still largely open questions.

Here, we examined the roles of white-, blue-, red- and green-LED into the growth of tomato and analyzed the contents of proline, total phenol and the activities of antioxidant enzymes. Moreover, to shed light on the correlation between increased antioxdative capacity driven by LED lights and resistance to pathogenic attacks, we monitored the disease development of tomato under blue-LED lights.

Materials and Methods

Plant growth conditions

Tomato (cv. Toy-mini tomato) seeds were surface-sterilized in 70% ethanol for 5 min and in 2% sodium hypochlorite for 20 min, followed by several rounds of washing with sterile water. Seeds were transferred to two sheets of sterile filter paper moistened with deionized water, and then germinated at 25°C under dark condition for three days. The uniformly germinated seedlings were transplanted into plastic culture tray (25×50×5 cm, the outer size of tray) containing peat-vermiculite media (Uddeumi, Sunghwa Co., Korea), and were grown for two weeks. For nutrition supply, the half-strength Hoagland solution [22] was irrigated. Growth chamber was set as the photoperiod of 18 hr and 22 ± 1°C under a relative humidity 60-65% with a photon flux density of 150 μmol/m²s. The pH of nutrient solution was maintained at 5.8. At two weeks after transplanting, the tomato seedlings were planted into plastic pots (10×9 cm²), at a density of one plant per pot and grown in phytotron chamber (130×60×180 cm, Woniltech, Ltd., Korea), for twenty one days under different light-emitting diode (LED) conditions. After morphological measurements, the leaves and stems tissues of tomato were ground to fine powder in liquid nitrogen for biochemical analyses.

Light treatment conditions

Each lighting treatment was conducted in separately controlled chambers (ODTech, Ltd., Korea), to be free from spectral interference among treatments. The LED array chambers were programmed to provide an 18 h light/6 h dark photoperiod at photosynthetic photon flux (PPF) maintained of approximately 150 μmol/m²s. All tomato plants were grown under four different light sources with broad-spectrum-white LED (BSWL, 420-680 nm) as a control, blue LED (460 nm), red LED (635 nm) and green LED (520 nm). Light quality and quantity were estimated using a Testo545 light meter (Testo, Germany).

Determination of proline content

Determination of free proline content was performed as previously described [23]. Tomato leaf and stem samples (0.5 g) were homogenized in 3% (w/v) sulfosalicylic acid and filtered through filer paper. Filtrate (2 mL) was reacted with acid ninhydrin (2 mL) and 30% glacial acetic acid (2 mL), and then heated at 100°C for 1 h. The reaction was extracted with 4 mL toluene for 30 min at room temperature; and the absorbance of the toluene fraction aspired from the liquid phase was measured at 520 nm. The proline concentration was determined based on a standard curve drawn with pure proline and expressed as μmol proline g^{-1} FW.

Determination of total phenolic compounds

The amount of total phenolics was determined using the Folin-Ciocalteu method [24]. Tomato leaf and stem samples (0.5 g) from each LED treatment were stirred slightly in 10 mL of 80% aqueous methanol. The suspension was sonicated for 5 min and collected by centrifugation. Samples (500 μL) were reacted with Folin-Ciocalteu's reagent (2.5 mL) and 7.5% sodium carbonate (2 mL) at room temperature for 30 min. The absorbance of the reaction product was measured at 765 nm. The total phenolic concentration was determined using gallic acid as a standard, and expressed as gallic acid equivalents in milligrams per gram of dry matter.

Antioxidant enzyme analysis

The activities of SOD, CAT, APX and GR were determined spectrophotometrically. SOD activity was assayed at 560 nm by determining the inhibition rate of nitroblue tetrazolium reduction, with xanthine oxidase as a hydrogen peroxide generating agent [25,26]. CAT activity was assayed at 240 nm by measuring the conversion rate of hydrogen peroxide to water and oxygen molecules [27]. APX activity was determined at 290 nm following the oxidation of ascorbate to dehydroascorbate, as described by Nakano and Asada [28]. GR activity was determined at 340 nm by measuring the reduction kinetics of oxidized glutathione [29].

Detached leaf assay

Botrytis cinerea (No. 40574) was obtained from the Korean Agricultural Culture Collection (KACC; Suwon, Korea). *B. cinerea* were incubated on potato dextrose agar (PDA) medium (MB Cell, Los Angeles, USA) containing 4% potato starch, 20% dextrose and 15% agar at 24°C in the dark. After two weeks, spores on the medium were suspended with sterilized distilled water (DW). Spore concentrations were adjusted to the approximately $5.7×10^5$ mL^{-1} using a hemocytometer. If not otherwise stated, the spore suspension was used at the same concentration throughout the experiments. Detached leaves from 4-week-old tomato plants were washed with DW and placed on wet filter paper in Petri dishes. The leaves were inoculated with 10 μL drops of the *B. cinerea* spore suspension ($5.7×10^5$ mL^{-1}), and then kept under blue or broad-spectrum-white LED light conditions for 10 days. Disease severity on leaves infected with *B. cinerea* was visibly assessed on a scale of 0 ('no symptoms') to 4 ('51 to 100% symptoms'), according to the method of Rajkumar et al. [30].

Statistical analysis

Data were analyzed by a general linear model and multiple comparisons among the treatments were conducted by Tukey's honestly significant difference (HSD), using the statistical analysis program, Statistix (Statistix 9 Analytical Software, USA). The significance of differences among samples was determined at 95% confidence level.

Results and Discussion

In many plant species, proline is a major organic osmolyte that maintains osmotic balances, induces expression of stress responsive genes and functions to stabilize sub-cellular structures, scavenges free radicals and buffers cellular redox potential under stress conditions [31]. To explore the effect of different wavelength of light on the accumulation of proline, we quantitated the amount of proline in leaves and stem of tomato grown under LED light having different wavelengths. There was

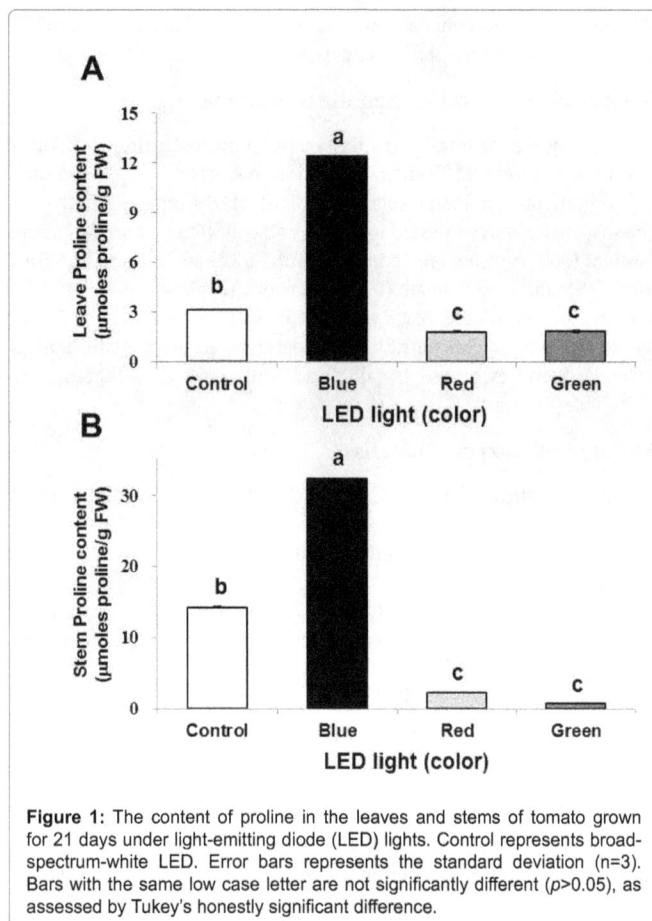

Figure 1: The content of proline in the leaves and stems of tomato grown for 21 days under light-emitting diode (LED) lights. Control represents broad-spectrum-white LED. Error bars represents the standard deviation (n=3). Bars with the same low case letter are not significantly different (p>0.05), as assessed by Tukey's honestly significant difference.

a considerable difference in the content of proline of tomato seedlings lightened with different LED sources. Compared to broad-spectrum-white LED (BSWL), when tomato seedlings were treated with blue-LED, proline contents were increased in leaves and stems by about 296% and 127%, respectively. Each differences were statistically significant based on Tukey's HSD test with α=0.05 (p<0.05). In contrast, red and green LED lights significantly decreased the amount of accumulated proline (p<0.05, Figure 1), compared to BSWL emitted conditions.

We also investigated the antioxidant capacity of tomato leaves and stems cultivated under different LEDs. In plants, antioxidant defense systems include various antioxidants, such as carotenoids, tocopherol, flavonoids, ascorbate and phenolic compounds, which play important roles in protection from photooxidative damage [11,21]. To learn how different wavelengths of light modulate antioxidation capacity in partial, we measured the contents of phenolic compounds from tomato grown under different colored LED lights. When blue LED was engaged, the content of total phenolic compounds both in leaves (1.3 fold) and stems (1.2 fold) was significantly increased (Tukey's test with α=0.05, p<0.05), compared to BSWL conditions (Figure 2). On the other hand, under red and green light conditions, the content of total phenolic compounds of leaves showed a similar levels (Tukey's test with α=0.05) to BSWL conditions. However, in stems, red and green LED significantly decreased the contents of phenolic compounds by 49% and 37%, respectively, compared to BSWL LED. Luthria et al. [32] reported that the quantity and composition of phenolic compounds in plants bearing edible fruits is significantly influenced by the quality of light. Furthermore, Johkan et al. [33] reported that the content of phenolic

compounds was increased in red leaf lettuce, as a result of supplemental blue light radiation [33]. Our results are in agreement with data published by Johkan et al. [33], in which the content of polyphenols and antioxidant activity were shown to be greatly increased in lettuce seedlings treated with blue LED light. Taken together, our results clearly indicated that the contents of proline and total phenolic compounds in leaves and stems of tomato plants was considerably influenced by the spectral quality of LEDs. Especially, blue-LED dramatically increased the contents of proline and phenolic compounds in vegetative tissues in plants.

We also assessed the activity of antioxidant enzymes (e.g. SOD, CAT, APX, and GR) of tomato grown under each LED lights (Figure 3). In our study, no significant differences (Tukey's HSD test with α=0.05) between the BSWL, blue, red and green light treatments were found for SOD activity in tomato leaves. However, in the case of stem, SOD activity was significantly increased by 29% under blue-LED treatment, while SOD activities under the red- and green-LED treated stems were decreased by 16% and 35%, respectively, compared to BSWL LED treatment as a control (Tukey's HSD test with α=0.05, Figure 3A and 3B). Under the red- and green-LED treatment, however, CAT activity in leaves and stems were noticeably decreased by about 28% to 18% and 63% to 39%, respectively, compared to BSWL LED light (Tukey's HSD test with α=0.05, Figure 3C and 3D). In contrast, CAT activity in blue LED-treated tomato was increased about 15% in leaves compared to that in BSWL LED light treatment (Tukey's HSD test with α=0.05). Being similar to our results, Schmidt et al. [34] reported that the activation of catalase enzyme in winter rye leaves was more enhances by blue light

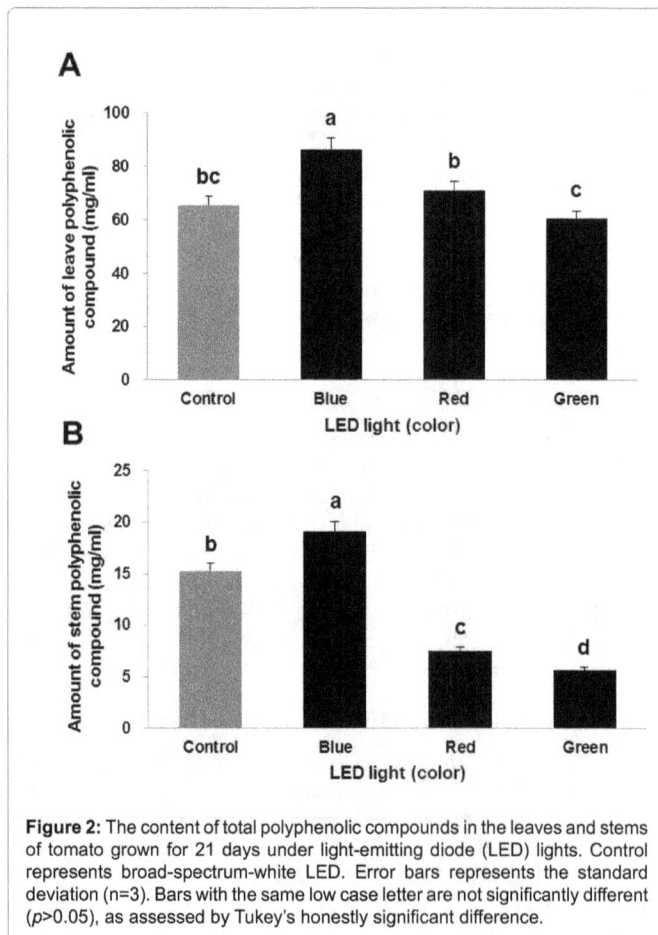

Figure 2: The content of total polyphenolic compounds in the leaves and stems of tomato grown for 21 days under light-emitting diode (LED) lights. Control represents broad-spectrum-white LED. Error bars represents the standard deviation (n=3). Bars with the same low case letter are not significantly different (p>0.05), as assessed by Tukey's honestly significant difference.

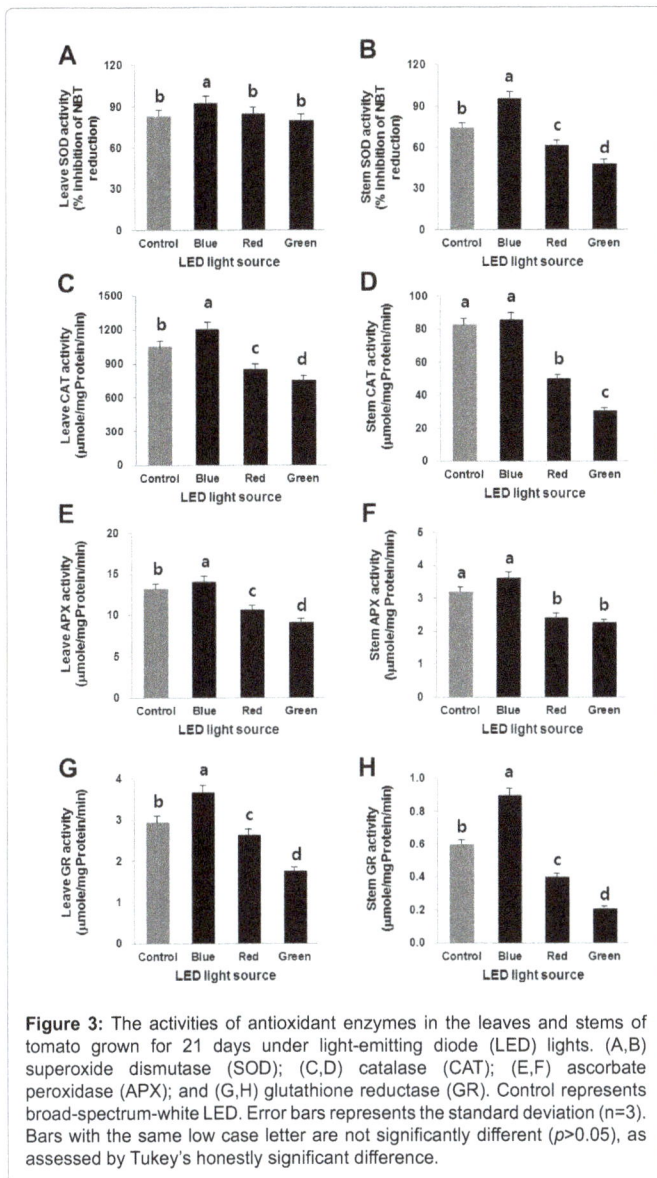

Figure 3: The activities of antioxidant enzymes in the leaves and stems of tomato grown for 21 days under light-emitting diode (LED) lights. (A,B) superoxide dismutase (SOD); (C,D) catalase (CAT); (E,F) ascorbate peroxidase (APX); and (G,H) glutathione reductase (GR). Control represents broad-spectrum-white LED. Error bars represents the standard deviation (n=3). Bars with the same low case letter are not significantly different ($p>0.05$), as assessed by Tukey's honestly significant difference.

3.33, respectively. Such differences are statistically significant based on (Tukey's HSD test with $\alpha=0.05$, Figures 4A and 4B). In tomato leaves, Kuzniak and Sklodowska [36] found that activity increases of peroxisomal antioxidant enzymes, such as superoxide dismutase (SOD), catalase (CAT) and glutathione peroxidase (GSH-Px) can contribute to the inhibition of pathogen-induced leaf senescence by *Botrytis cinerea* infection. Khanam et al. [37] reported that enhanced catalase activity under red light treatment contributes to the inhibition of lesion formation and fungal development on broad bean leaves infected with *Botrytis cinerea*. In addition, Grote and Claussen [38] reported that the proline content in tomato leaves is increased by pathogen attack such as *phytophthora nicotianae*, as well as light intensity. Thus, numerous studies have suggested that physiological resistance of plants to environmental stresses, including pathogen attack, is closely connected with specific light treatments, as well as effective antioxidative mechanisms. Similarly, our results suggest that blue-LED light suppress the development of gray mold, and/or the propagation of *B. cinerea* in tomato potentially *via* enhanced accumulation of proline and antioxidative responses.

In conclusion, current study suggests that blue-LED is highly efficient to protect crop plants from pathogenic attacks, at least where artificial lights are applied as main light sources. In mechanistic level, such advantages of blue-LED are ascribed into the increased production

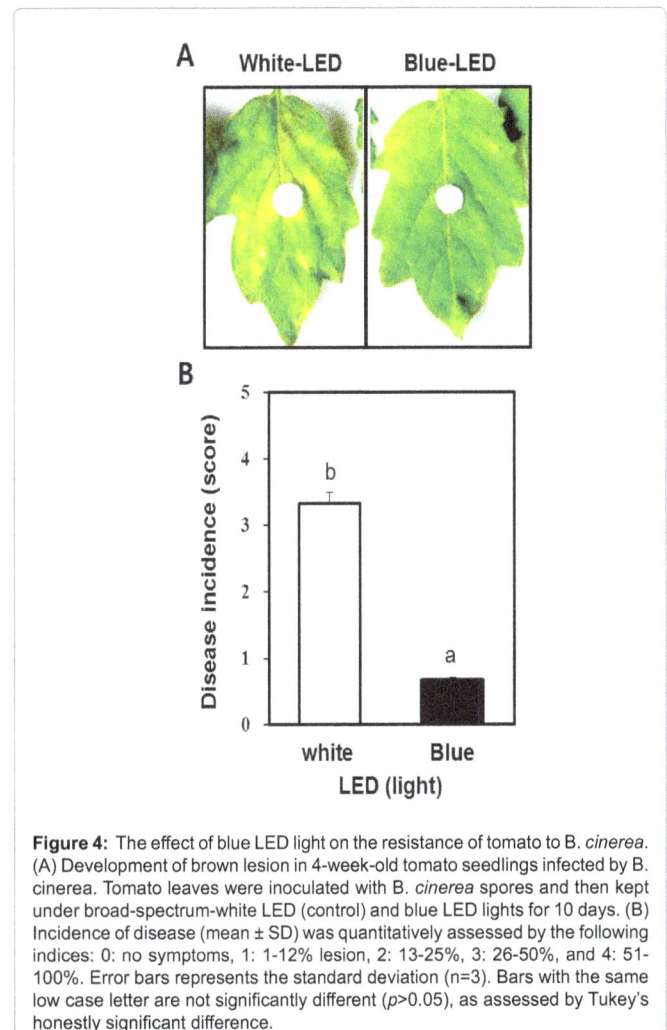

Figure 4: The effect of blue LED light on the resistance of tomato to B. *cinerea*. (A) Development of brown lesion in 4-week-old tomato seedlings infected by B. cinerea. Tomato leaves were inoculated with B. *cinerea* spores and then kept under broad-spectrum-white LED (control) and blue LED lights for 10 days. (B) Incidence of disease (mean ± SD) was quantitatively assessed by the following indices: 0: no symptoms, 1: 1-12% lesion, 2: 13-25%, 3: 26-50%, and 4: 51-100%. Error bars represents the standard deviation (n=3). Bars with the same low case letter are not significantly different ($p>0.05$), as assessed by Tukey's honestly significant difference.

treatment than that in red or far-red light treatment. APX activity in blue-LED light treatment was also increased in leaves and stems by 7% and 13%, respectively, compared to that in BSWL LED light treatment (Tukey's HSD test with $\alpha=0.05$, Figures 3E and 3F). In addition, in the case of GR activity of leaves and stems, the blue-LED treatment increased 1.4-fold and 2.1-folds (in leaves), 2.2-fold and 4.2-fold (in stems), compared to BSWL LED control (Tukey's HSD test with $\alpha=0.05$, Figure 3G and 3H). In the leaves of tall fescue, Xu et al. [35] found that the activities of catalase (CAT), ascorbate peroxidase (APX) and glutathione reductase (GR) are increased by light treatment. Combined together, results of the present study indicated that blue-containing LED radiation had positive effects on the action of antioxidant defense mechanisms in tomato seedlings.

Inferred from the outstanding effects of blue-LED on proline contents and antioxidation capacities, we examined whether blue-LED light treatment increased defense ability of tomato to gray mold disease caused by *B. cinerea*. Under the blue-LED light- and BSWL-treated tomato leaves, disease incidence of gray mold was 0.67 and

of osmoprotectants and antioxidants, including ROS scavenging enzymes. Nevertheless, it must be prompted which signaling pathway are activated by blue light to gain more insight into the light wavelength-dependent developmental modifications in plants.

Acknowledgment

This research was supported by the National Research Foundation of Korea (NRF) grant funded by the government (MEST) (No.2011-0020202), and a grant (PJ007408) from the Rural Development Administration (RDA).

References

1. Fanasca S, Colla G, Maiani G, Venneria E, Rouphael Y, et al. (2006) Changes in antioxidant content of tomato fruits in response to cultivar and nutrient solution composition. J Agric Food Chem 54: 4319-4325.

2. Gahler S, Otto K, Bohm V (2003) Alterations of vitamin C, total phenolics, and antioxidant capacity as affected by processing tomatoes to different products. J Agric Food Chem 51: 7962-7968.

3. Williamson B, Tudzynski B, Tudzynski P, Van Kan JAL (2007) Botrytis cinerea, the cause of grey mould disease. Mol Plant Pathol 8: 561-580.

4. Robert-Seilaniantz A, Navarro L, Bari R, Jones JD (2007) Pathological hormone imbalances. Curr Opin Plant Biol 10: 372-379.

5. Kretschmer M, Hahn M (2008) Fungicide resistance and genetic diversity of Botrytis cinerea isolates from a vineyard in Germany. J Plant Dis Protect 115: 214-219.

6. Drozdova IS, Bondar VV, Bukhov NG, Kotov AA, Kotova LM, et al. (2001) Effects of light spectral quality on morphogenesis and source-sink relations in radish plants. Russ J Plant Physl 48: 415-420.

7. Li Q, Kubota C (2009) Effects of supplemental light quality on growth and phytochemicals of baby leaf lettuce. Environ Exp Bot 67: 59-64.

8. Massa GD, Kim HH, Wheeler RM, Mitchell CA (2008) Plant productivity in response to LED lighting. HortScience 43: 1951-1956.

9. Yorio NC, Goins GD, Kagie HR, Wheeler RM, Sager JC (2001) Improving spinach, radish, and lettuce growth under red light-emitting diodes (LEDs) with blue light supplementation. Hort Sci 36: 380-383.

10. Urbonaviciute A, Samuoliene G, Brazaityte A, Duchovskis P, Ruzgas V, et al. (2009) The effect of variety and lighting quality on wheatgrass antioxidant properties. Zemdirbyste 96: 119-128.

11. Samuoliene G, Brazaityte A, Urbonaviciute A, Sabajeviene G, Duchovskis P (2010) The effect of red and blue light component on the growth and development of frigo strawberries. Zemdirbyste 97: 99-104.

12. Wang H, Jiang YP, Yu HJ, Xia XJ, Shi K, et al. (2010) Light quality affects incidence of powdery mildew, expression of defence-related genes and associated metabolism in cucumber plants. Eur J Plant Pathol 127: 125-135.

13. Shimazaki KI, Doi M, Assmann SM, Kinoshita T (2007) Light regulation of stomatal movement. Annu Rev Plant Biol 58: 219-247.

14. Xu Y, Sun X, Jin J, Zhou H (2010) Protective effect of nitric oxide on light-induced oxidative damage in leaves of tall fescue. J Plant Physiol 167: 512-518.

15. Jeong RD, Chandra-Shekara AC, Barman SR, Navarre D, Klessig DF, et al. (2010) Cryptochrome 1 and phototropin 2 regulate resistance protein-mediated viral defense by negatively regulating an E3 ubiquitin ligase. Proc Natl Acad Sci U S A 107: 13583-13543.

16. Jeong RD, Chandra-Shekara AC, Barman SR, Navarre D, Klessig DF, et al. (2010) Cryptochrome 1 and phototropin 2 regulate resistance protein-mediated viral defense by negatively regulating an E3 ubiquitin ligase. Proc Natl Acad Sci U S A 107: 13583-13543.

17. Torres MA, Jones JDG, Dangl JL (2006) Reactive oxygen species signaling in response to pathogens. Plant Physiol 141: 373-378.

18. Sharma P, Jha AB, Dubey RS, Pessarakli M (2012) Reactive oxygen species, oxidative damage, and antioxidative defense mechanism in plants under stressful conditions. J Bot 2012: 1-26.

19. Hayat S, Hayat Q, Alyemeni MN, Wani AS, Pichtel J, et al. (2012) Role of proline under changing environments: A review. Plant Signal Behav 7: 1456-1466.

20. Islam SZ, Rahman MZ, Khanam NN, Ueno M, Kihara J, et al. (2011) Disease suppression by light-enhanced antioxidant system in broad bean. Curr Top Plant Biol 12: 55-61.

21. Ashry NA, Mohamed HI (2011) Impact of secondary metabolites and related enzymes in flax resistance and or susceptibility to powdery mildew. Afr J Biotechnol 7: 78-85.

22. Hoagland DR, Arnon DI (1950) The water culture method for growing plants without soil. California Agricultural Experiment Station Circular, 347, University of California, Berkley, CA, USA.

23. Bates LE, Waldren RP, Teare ID (1973) Rapid determination of free proline for water stress studies. Plant Soil 39: 205-207.

24. Singleton VL, Orthofer R, Lamuela-Raventos RM (1999) Analysis of total phenols and other oxidation substrates and antioxidants by means of folin-ciocalteu reagent. Meth Enzymol 299: 152-178.

25. Obeley LW, Spitz DR (1984) Assay of superoxide dismutase activity in tumor tissue. Meth Enzymol 105: 457-467.

26. Kang JH (2004) Modification of Cu, Zn-superoxide dismutase by oxidized catecholamines. J Biochem Mol Biol 37: 325-329.

27. Beers RF, Sizer IW (1952) A spectrophotometric method for measuring the breakdown of hydrogen peroxide by catalase. J Biol Chem 195: 133-140.

28. Nakano Y, Asada K (1981) Hydrogen peroxide is scavenged by ascorbate specific peroxidase in spinach chloroplasts. Plant Cell Physiol 22: 867-880.

29. O'kane D, Gill V, Boyd P, Burdon RH (1996) Chilling, oxidative stress and antioxidant responses in Arabidopsis thaliana callus. Planta 198: 371-377.

30. Rajkumar M, Lee WH, Lee KJ (2005) Screening of bacterial antagonists for biological control of phytophthora blight of pepper. J Basic Microb 45: 55-63.

31. Phang JM, Liu W, Zabirnyk O (2010) Proline metabolism and microenvironmental stress. Annu Rev Nutr 30: 441-463.

32. Luthria DL, Mukhopadhyay S, Krizek DT (2006) Content of total phenolics and phenolic acids in tomato (Lycopersicon esculentum Mill.) fruits as influenced by cultivar and solar UV radiation. J Food Compos Analysis 19: 771-777.

33. Johkan M, Shoji K, Goto F, Hashida S, Yoshihara T (2010) Blue light-emitting diode light irradiation of seedlings improves seedling quality and growth after transplanting in red leaf lettuce. Hort Sci 45: 1809-1814.

34. Schmidt M, Grief J, Feierabend J (2006) Mode of translational activation of the catalase (cat1) mRNA of rye leaves (Secale cereale L.) and its control through blue light and reactive oxygen. Planta 223: 835-846.

35. Xu Y, Sun X, Jin J, Zhou H (2010) Protective effect of nitric oxide on light-induced oxidative damage in leaves of tall fescue. J Plant Physiol 167: 512-518.

36. Kuzniak E, Sklodowska M (2005) Fungal pathogen-induced changes in the antioxidant systems of leaf peroxisomes from infected tomato plants. Planta 222: 192-200.

37. Khanam NN, Ueno M, Kihara J, Honda YM, Arase S (2005) Suppression of red light-induced resistance in broad beans to Botrytis cinerea by salicylic acid. Physiol Mol Plant P 66: 20-29.

38. Grote D, Claussen W (2001) Severity of root rot on tomato plants caused by Phytophthora nicotianae under nutrient- and light-stress conditions. Plant Pathol 50: 702-707.

26

Plants as Antiviral Agents

Hoda MA Waziri*

Plant Virus and Phytoplasma Research Department, Plant Pathology Research Institute, Agricultural Research Center, Cairo, Egypt

Abstract

Viruses are microorganisms that infect all kinds of living organisms including plants, and cause remarkable lose in crop production. Although pesticides showed that they can protect plants from pest infections, there are no effective substances that can be used as potent virucides. Therefore, there is a continuous demand to produce chemicals in order to stop and cure viral infections in plants. However, toxicity and carcinogenicity issues were always attributed to chemical pesticides. Screening of natural products shined in the dark to find new safe virucides. The philosophy of selecting those plants is oriented towards plants that can protect themselves from viral infections. Plants have been reported as virus inhibitors and are able to prevent infection of viruses by inducing systemic resistance in non-infested parts of the plants, such as *Boerhaavia diffusa, Clerodendrum aculeatum*. Other plants defend themselves against virus infections; these plants contain ribosome inactivating proteins [RIPs], such *as Phytolacca Americana, Mirabilis jalapa, Dianthus caryophullus*. The methods used for extraction, separation, identification of those antiviral compounds are documented and discussed in this review.

Keywords: Plant virus; Antiviral activities; Antiviral evaluation; Systemic resistance inducers; Ribosome-inactivating proteins

Introduction

Viruses are microorganisms which are able to infect all kinds of living organisms and the most important hosts are humans, animals, plants, bacteria and fungi. However few viruses are able to infect more than one organism [1]. An enormous numbers of plant viruses have been identified, and almost reached one thousand [2]. Plant viruses are extended through higher plants which are considered to be natural hosts [3]. Mostly all kinds of cultivated and non-cultivated plants could be infected with viruses, but each virus has a different host range [4]. *Tobacco Mosaic Virus* [TMV] for example is capable of infecting over 1000 species in 85 different plant families [4], on the other hand, some viruses are unable to infect but only few species in the grass family [3], for instance, *Citrus Tristeza Virus* [CTV] is able to infect few species in Citrus genus [4]. Virus infection can be transmitted from an infected plant to any healthy plants all over the field, and it is impossible to cure or to make a plant free virus once it has been became infected. Therefore, perennial crop virus infection could be problematic. In fact, the infected plants must be eliminated and a new one replanted which will take a long time before we can get any economic return [3]. Crop losses due to virus diseases all over the world are estimated by 60 billion US dollars per year [3]. Losses in tomato could reach up to 100% by *Tomato Yellow Leaf Curl Virus* [TYLCV] in many countries around the world [5], also the loss caused by TMV and *Cucumber Mosaic Virus* [CMV], reached a billion dollar [3]. Africa suffers from losses which are evaluated by around 50,000 tons of cocoa beans each year caused by *Cacao swollen shoot virus. Rice tungro virus* in the Southeast Asia causes losses with an estimate of $1.5 billion dollars annually, the economic losses estimated by 1 billion dollar for *Tomato spotted wilt virus* which is able to infect different crops beside tomato, peanut and tobacco [1].

Materials and Methods

Classical chemicals used as virucides

Different strategies were used to control viral diseases which couldn't have been effective in decreasing or avoiding the virus infection, despite controlling fungi or bacteria were very effective; especially by chemical means [2]. In fungal diseases using fungicides is a very important way to protect crops from infections and to decrease fungal diseases. However, for virus diseases there is no such direct way available to control it so far [6]. Researches for virucide seem to be far behind when compared to similar areas using chemical compounds such as herbicide, fungicide and insecticide. Antiviral substances are strongly in demand to control virus diseases, but it has been documented that the agriculture field lacks antiviral chemicals [3]. Several chemicals have been found to be able to control virus replication and suppress virus disease symptoms [7], such as "benlate" and "bavistin", but unfortunately these chemicals failed to have any effect on the quantity of the virus in the leaves. Another problem is that many of these chemicals have negative properties such as, phytotoxic effects, probably bad effect on humans, animals, and the environment, so none of these compounds are used in applied fields for controlling plant viruses. These disadvantages make the regulations for the registration of any new chemical virucides very restricted and increases firmness of the regulations in many countries. The near future probably will not see any considerable progress of chemical virucides [8].

A combination of heat treatment or meristem tip culture and chemical treatments in a few cases may have been useful [6]. In apple shoot cultures ribavirin [virazole], an analog of guanosine has been used to eliminate *Apple Chlorotic Leaf Spot Virus* [ACLSV] in combination with tissue culture [9,10], used it in Cymbidium cultures to eradicate *Odontoglossum Ringspot Virus* [ORSV]. Potato viruses have been eliminated in potato meristem culture by using 2,4-dioxohexahydro-l,3,5-triazine [11], and potato stem cuttings [12]. Several compounds have been tested as antiviral against viruses infecting plants. These chemicals were produced from plants, other organisms and also synthetic organic chemicals, all of them nearly if applied to the

*Corresponding author: Hoda Waziri MA, Plant Virus & Phytoplasma Research Department, Plant Pathology Research Institute, Agricultural Research Center, Cairo, Egypt, E-mail: h_waziri@yahoo.com

leaves before inoculation, or shortly after inoculation affect virus infectivity. For instance, a component produced from *Phytophthora megaspermaf s. P. glycinea* called "glucan" has an inhibiting effect on a number of viruses but the mode of action remains unknown [13]. Several scientists reviewed work in this field as [14-16]. Many synthetic analogs for these bases including nucleic acids [purine and pyrimidine] have been studied very well and the research continues for compounds of this type [17]. Virazole was found to have a wide-spectrum effect on the virus system of experimental animals [18].

The antiviral activity of Virazole has been studied in different plant virus systems. It suppresses and slows down the systemic infection with *Tomato Spotted Wilt Virus* [TSWV] and also tobacco plants if pretreated with it. It also decreases the concentration of CMV and *Alfalfa Mosaic Virus* [AMV] in tissue culture plant. However, in meristem tip cultures, virus free plants were obtained with or without virazol being added to the medium. Murphy et al. reported that Plant Growth Promoting Rhizoctinia [PGPR] reduced the severity of the infection of *Tomato Mottle Virus* [ToMoV] which infects the tomato plants. Hence, PGPR was suggested to be used in the management of ToMoV on tomato plants within the integrated program. Also, it was observed that, when a high dose of nitrogen fertilizers were used, suppression of virus disease symptoms occurred but without an effect on the virus concentration [19]. Significant efforts have been made to find components that can be used as inhibitors of virus infection and replication that works as fungicides and give direct protection to the crops against fungi, but such substances are not available yet [20].

It is very important to evaluate the new antiviral chemicals. They need to be effective only on the virus metabolism, and stop virus replication, but have no effect on plant metabolism which is not easy indeed [3].

Plant as a source of virucides

The Ancient Egyptians and Chinese used plants for treating a lot of important health issues, and different diseases, also for preparing hundreds of medicinal products; this is shown in their records [21]. Eighty percent of the world's population use medicinal plants products for fighting different pathogens. Many records are published about medicinal plants used in folk medicine for their high level of effectiveness as antiviral agent. Some of them have been approved ability to treat viral infections in animals and people [22,23]. In 1925 Duggar and Armstrong found the first plant inhibitors, they discovered that TMV replication was inhabited by the extracts from different plants [24]. During the following 25 years, a lot of progress was achieved in the discovery of virus inhibitors [25,26]. The knowledge was very limited in the field of basic science in the 1950s when they started to search for antiviral agent, although luck interfered [27].

In Europe after the Second World War in 1952 the need for antiviral agents encouraged research in this field [28]. Boots drug company [Nottingham, England] screened less than three hundred different plants in search for an antiviral against influenza. These efforts were the first practical trial in the discovery of antiviral agents [29]. In 1964 a breakthrough occurred in this field. A plant survey was done and succeeded in characterizing antiviral substances as antibacterial, antifungal, and antiviral actions [30]. A great deal of knowledge became available about virus replication, accumulated through the previous thirty years which facilitated finding and recognizing more antiviral agents with high efficacy against virus diseases [31].

Antiviral agents must also be safe for the environment and field applications; therefore, scientists responsible for plant protection are very active and are always working hard trying to find antivirals with such characteristics [3].

Mechanisms of antiviral activities

Plant viruses depend on plant cell for replication [3]. The relationship between the virus replication and the host cell is very intimate; this is the first problem in the progress and thus needs a very effective antiviral. The second problem is the late identification of viral diseases when the treatment is not highly effective [32]. Antiviral's main target is viral nucleic acid, there are two kinds of viral nucleic acids which are Deoxyribonucleic Acid [DNA], and Ribonucleic Acid [RNA], containing the code responsible for viral replication in host cell and the spread of the virus and viral proteins, but there is not enough information about the stage of penetration at the virus propagation [33]. A good antiviral agent must stop the multiplication in the infected cell should not have any toxic effect on normal cells [32]. Antiviral chemical agents can stop the spread of the virus, control the visible symptoms and induce nature resistance by the host [3].

Methods for investigating antiviral agent

Screening and extraction for plants: It is amazing that antiviral products could be found in many different natural resources, for instance fungi, marine fauna and flora bacteria, and plants, however, the screening is usually focus on plants which are known to have medicinal background [34]. Screening for natural products can be explored by two methods the first by collecting plants randomly and the second by searching for plants that are common and known to have medicinal effects. In a study to evaluate both ways, it was found that, the second way resulted production of higher number of plants with active substances than the first way [35].

Mirabilis jalapa extracts from different parts of the plant [root, leaves and stem], induced inhibition of plant virus activity [36]. The root extracts of *B. diffusa* induced the highest antiviral activity when compared with the other *B. diffusa* parts extract and it was found to have antiviral effects on different hosts [37]. An extract of *Potentilla arguta* root, and the branch tip extract of *Sambucus racemosa* were used. Both were proven to have inhibitory effect [28]. Pokeweed [*Phytolacca americana*] contains 3 types of antiviral proteins PAP well defined as [PAP-I, PAP-II, and PAP-III] it is obtained from leaves but at different times and different season. Spring leaves, early summer leaves, and late summer leaves, sequentially [38].

Plants produce 2 types of natural metabolic products, where the first one is termed "primary metabolites" which is very important since it is very essential for plant growth and development, the second termed "secondary metabolites" it is extremely varied in structure, produced in an enormous variety and plays an important role in the relations between plants and their biotic and abiotic resistance [21]. Many antiviral natural substances have been obtained during the previous fifty years. Some have been used as a crude extract that may contain a number of non-active substances, while others have been used as purified or partially purified, thus the chemicals included are not identified, nor the mechanism of these compounds [20]. The kind of solvent that is used in plant extraction depends on the kind of the chemicals needed to be obtained, the purpose of use, the conditions of extraction and the storage medium [21].

Separation and identification of the compounds: Separation for plant extract materials is important, because the extracted sap has a mixture of different components and structures; usually Chromatography is used [21].

The determination of the detector depends on the structure of the component needed to be analyzed and the available analytical system. The same structure can be analyzed by more than one system [39]. The High Performance Liquid Chromatography System [HPLC] are almost available everywhere. UV or Diode Array Light [DAD] detectors quantify the analyzed component. Another common detector is Fluorescence [21]. In the last few years, extraction methods for phytochemedicinal products preparation and isolation, have significantly improved due to increased concern in traditional medicine [40]. There are numbers of Books and literature available in the market, for example to start with: Plant Drug Analysis: A Thin Layer Chromatography Atlas [41], and Laboratory Handbook for the Fractionation of Natural Extracts [42]. They all describe standardized extraction procedures for preparing antiviral component from medicinal plants [43].

The evaluation methods for antiviral agents: The evaluation for antiviral chemicals mostly has been done depending on two viruses which are TMV and PVX [44]. For field application, using the antiviral chemical substances as a spray on the foliar parts is the most common way in application, because this method is easy, simple and inexpensive equipments are not needed [20,44]. Granular method is convenient, promising, and most importantly it is safe on the environment, because the substance kept at the same spot of application [20]. Seed treatment is ignored and is not being used anymore as one of the application methods [45]. Substances used as antivirals need substantial methods for evaluation for testing the infected tissues several times after being treated [46]. Development of the local lesion assay makes evaluation and quantification for virus replication and inhibition easier and more reliable [25,26]. Local Lesion assay is used for viruses which have hosts reacting with the infection and this result in a local lesion. Local lesion method is a reliable assay to evaluate most of the compounds as an antiviral [20]. The number of local lesions is an indication of the effectiveness of the antiviral. A fewer number of local lesions means more antiviral effect [47]. The inhibition effect using antiviral proteins was evaluated in percentage using the following formula:

Inhibition rate=[C-T] × 100/C

C is the local lesion number mean in the plants used as a control, while T is the local lesion number mean in the treated plants [48]. Serological methods such as ELISA for example are available and faster [20]. Infected plants after using the tested antiviral substances were examined by ELISA double antibody sandwich [49], to estimate virus inhibition due to a correlation with the virus concentration, then Real time Polymerase Chain Reaction [PCR] was used for those plants that reacted negatively with ELISA test for more sensitive evaluation of the virus concentration [50].

Resistance acquired by antiviral plants

Natural resistance: Some plants have natural resistance and are able to defend themselves by preventing the virus from replication inside the plant cells [4]. Hundreds of cells are invaded by the virus after infection, but the infection is localized in the area of the leaf and does not spread to the other cells, hence it must be a kind of natural resistance that stops the virus replication [46].

Systemic resistance inducers: Some plants have virus inhibitors these are reported to prevent infection of viruses and to induce systemic resistance in non-treated parts of the plants [51-53]. For instance glycoproteins obtained from the root of *Boerhaavia diffusa* substance incites the antiviral system in the treated plants. The glycoprotein inhibits the infection of the virus by blocking virus replication [51,52] and prompts the plant system to produce new proteins by activating

the defense mechanism in susceptible hosts [54]. Host plants sprayed with *Boerhaavia diffusa* glycoprotein were found to have high antiviral activity in the sap extracted from the leaves [54-56], however in the sap extracted from the control plants [non treated], such activity was not detected . The explanation is that sap from treated leaves contains some antiviral agent [AVA] protein, which is not found in the control plants [55]. In a number of susceptible hosts *B. diffusa* glycoprotein was found to be successful in inciting high systemic resistance by stimulating the immunity system.

Whereas, the resistance induced by *Clerodendrum aculeatum* is systemic. Leaf extract from *Clerodendrum aculeatum* when applied as a spray it could prevent infection of some viruses transmitted by white fly and mechanically in a number of local lesion hosts and a systemic hosts [57,58].

Ribosome-inactivating proteins [RIPs]: Several plants have the ability to be protected against virus infection, since they possess effective inhibiting substances, a few of them are purified and identified. These plants have ribosome inactivating proteins [RIPs] which are considered to be as defense related protein [59]. Pokeweed Antiviral Proteins [PAPs] from *Phytolacca Americana* [60], Mirabilis Antiviral Protein [MAP] from *Mirabilis jalapa* [61], and Carnation Antiviral Protein [dianthins] from *Dianthus caryophullus* [58]. *Phytolacca* is a genus that has several Ribosome-Inactivating Proteins [RIPs]. In 1925, PAP was discovered in the plant of pokeweed [24]. PAP has a very high ability to inhibit protein assembling, and can stop the transmission of a number of plant viruses [62]. PAP antiviral activity has several features which make it very powerful against all types of viruses and this ability as an antiviral is quite strong even at minimal concentrations.

Mirabilis jalapa extract from different parts of the plant [root, leaves, and stem], induces an inhibiting effect for the plant virus activity. The leaf extract of *Mirabilis jalapa*, when sprayed 24-hours before the virus inoculation, stops the symptoms of virus disease on a few systemic hosts. The infectivity test indicated that, 50-60% decrease in virus content was noticed in the treated plants. *Mirabilis jalapa* extract was able to control a number of viruses, which are spread by insects in systemic hosts. It was able to stop the increase of aphids and whiteflies population. The mechanical transmission of *Turnip Mosaic Virus* [TMV] and *Potato Virus Y* [PVY] were inhibited by Mirabilis Antiviral Protein [MAP] [36]. The sap extracted from carnation leaves induces inhibition of virus infection [63-65] *Dianthus caryophyllus* has two types of protein inhibitors, which are Dianthin 30 and, Dianthin 32 they are isolated from the leaves [66]. The proteins of carnation also induce systemic resistance [67].

Novel antiviral agents: Chemicals are considered very important, and should play a significant job with highly effectiveness and minimum side effects. Production of "intelligent" and environment-friendly plant antiviral agents will give a significant protection for plants against viruses, crop yields will increase and the quality will improve without any side effects on users, customer needs, and the environment [3]. Ten years of research for making an innovation by discovering a novel green antiviral plant agents has led to successfully finding several new compounds with significant bioactivity as antivirus. Some novel structures with high plant antiviral bioactivity, such as α- aminophosphonate derivative Dufulin [Bingduxing], chiral cyanoacrylates, and GU188 were discovered. After successfully passing systemic R&D work, a brand new compound was commercialized as an antiviral. It is able to prevent the losses caused by virus infection, it called Duflin. It is a new antiviral agent for plants. Duflin mode of action is activating the plant immune system, by decreasing the virus ability for infection and causing an aggregation to the virus particles. The need for green technology laid the way for developing safe and systemic new antiviral agent for plants [3].

Nanotechnology and nanoscience caused a boom in research and applications recently. Plant extracts or plant biomasses and microorganisms are being used in an eco-friendly way as biological organisms for production of nanoparticles, and as an alternative to chemical and physical methods [68]. A number of plant extracts or plant biomasses are used for extracellular biosynthesis of silver and gold nanoparticles, and they have been proved to be successful. Using plant extracts to synthesize metal nanoparticles is an important branch of biosynthesis of nanoparticles [69]. *C. colocynthis* which is widely grown in Egypt, Sudan and some African countries where it is used in folk medicine [70,71] is also used. The aqueous extracts of this plant from its fruits, seeds, leaves and roots are being used to synthesize silver nanoparticles in a simple and not expensive method [69].

Natural sources are still in focus, and remain an important source for more substances which have an antiviral effect. It is vital to keep searching for antiviral agents that are valuable and with new structures [28]. Advances in plant virus elimination could be possible, by the new compound effective as an antiviral in human or animal medicine, keeping in mind that their effectiveness on plant viruses may be different [50].

Conclusion

Nature is still fertile and rich with an enormous number of different kinds of plants possessing an antiviral effect. They are highly variable; each regional area has its own special kind of plants which are not available elsewhere around the world. While screening for plant antivirals it's better to focus on plants known for their medicinal effect in folk medicine, to maximize the valuable results, saves time and effort. Also screening should expand on-to food crops, which has been proven not to have any toxic effect. Further studies need to be done on the plant extract to separate the different compounds determine the effective substances using standard methodology. Chemical virucide is not a magic solution for the plant virus diseases, but it's important and a vital link in the control system chain of the control system, it may even be expected to have the upper hand in the control system. Environment friendly measures and procedures should be kept in mind while working on the research and the production of chemical virucides. On the other hand, to increase the number of the plant antiviral agents used as a virucide for plants, we can apply the different plant antiviral agents that have been proven successful as antiviral compounds for human beings and animals, especially those which have a broad- spectrum antiviral activity. This is to save time and to guarantee that they successfully meet the food and drug regulations. At the moment nanotechnology is focused on the field of pharmaceuticals and medicine, but in the next decade it will be very important in the field of agriculture for the search and development of better, safer and more effective antimicrobial and antiviral agents.

Acknowledgments

This work was sponsored by grant-ship of the Scientific and Projects section [#AD1401/01] at the Bibliotheca Alexandria. The author would like to thank Miss Enas Mostafa, Mr. Nader Zakhary volunteers at the Bibliotheca Alexandria for their efforts to write and edit this review.

References

1. Gergerich RC, Dolja VV (2006) Introduction to plant viruses, the invisible foe. The Plant Health Instructor, DOI: 10.1094/PHI-I-2006-0414-01.

2. Hadidi A, Khetarpal RK, Koganezawa H (1998) Plant virus disease control. The American Phytopathological Society, St. Paul, USA: APS press.

3. Song B (2010) Environment-friendly antiviral agents for plants. Philadelphia, USA: Springer Science + Business Media LLC.

4. Wassenegger M, Pelissier T (1998) A model for rna-mediated gene silencing in higher plants. Plant Molecular Biology 37 349-62.

5. Pico B, Diez MJ, Nuez F (1996) viral diseases causing the greatest economic losses to the tomato crop. Ii. The tomato yellow leaf curl virus — a review. ScientiaHorticulturae 67: 151-96.

6. Matthews R (2001) Plant virology. New York, USA: Academic Press.

7. Cassells AC (1983) Chemical control of virus diseases of plants. Prog Med Chem 20: 119-155.

8. Stace-Smith R (1990) Tissue culture in plant viruses. Boca Raton, USA: CRC Press.

9. Hansen AJ, Lane WD (1985) Elimination of apple chlorotic leaf spot virus from apple shoots cultures by ribavirin. Plant Disease 69: 134-135.

10. Toussaint A, Kummert J, Maroquin C, Lebrun A, Roggemans J (1993) Use of virazole R to eradicate odontoglossum ringspot virus from in vitro cultures of cymbidium sw. Plant Cell, Tissue and Organ Culture 32: 303-309.

11. Borissensko S, Schuster G, Schmygla W (1985) Obtaining a high percentage of explants with negative serological reactions against viruses by combining potato meristem culture with phytoviral chemotherapy. Phytopathology Z. 114: 185-188.

12. Bittner H, Schenk G, Schuster G (1987) Chemotherapeutical elimination of potato virus x from potato stem cuttings. Journal of Phytopathology 120: 90-92.

13. Kopp M, Rouster J, Fritig B, Darvill A, Albersheim P (1989) Host-Pathogen Interactions : XXXII. A Fungal Glucan Preparation Protects Nicotianae against Infection by Viruses. Plant Physiol 90: 208-216.

14. Tomlinson JA (1981) Chemotherapy of plant viruses and virus diseases. In: Harris KM, Maramorosch M (eds.) Pathogens, vectors and plant diseases: Approaches to control. New York, USA: Academic Press.

15. White RF, Antoniw JF (1983) Direct control of diseases. Crop pro. 2,259-271.

16. Verma H, Baranwal V, Srivastava S (1998) Antiviral substances of plant origin. In: Hadidi A, Khetarpal RK, Koganezawa H (eds.) Plant Virus Disease Control. 539 Minnesota, USA: APS press 154-162.

17. Dawson W, Boyd C (1987) Modifications of nucleic acid precursors that inhibit plant virus multiplication. Phytopathology 77 477-80.

18. Sidwell RW, Huffman JH, Khare GP, Allen LB, Witkowski JT, et al. (1972) Broad-spectrum antiviral activity of Virazole: 1-beta-D-ribofuranosyl-1,2,4-triazole-3-carboxamide. Science 177: 705-706.

19. Murphy JF, Zehnder GW, Schuster DJ, Sikora EJ, Polston JE, et al. (2000) Plant growth-promoting rhizobacterial mediated protection in tomato against tomato mottle virus. Plant Disease 84: 779-784.

20. Hansen AJ, Stace-Smith R (1989) Antiviral chemicals for plant disease control. Critical Reviews in Plant Sciences 8 45-88.

21. Osbourn AE, Lanzotti V (2009) Plant-derived natural products: Synthesis, function, and application. Philadelphia, USA: Springer Science + Business Media LLC.

22. Thyagarajan SP, Thiruneelakantan K, Subramanian S, Sundaravelu T (1982) In vitro inactivation of HBsAg by Eclipta alba Hassk and Phyllanthus niruri Linn. Indian J Med Res 76 Suppl: 124-130.

23. Venkateswaran PS, Millman I, Blumberg BS (1987) Effects of an extract from Phyllanthus niruri on hepatitis B and woodchuck hepatitis viruses: in vitro and in vivo studies. Proc Natl Acad Sci U S A 84: 274-278.

24. Duggar BM, Armstrong JK (1925) The effect of treating the virus of tobacco mosaic with the juices of various plants. Ann Missouri Botan Garden 12: 259-366.

25. Matthews RE (1954) Effects of some purine analogues on tobacco mosaic virus. J Gen Microbiol 10: 521-532.

26. Francki RI, Matthews RE (1962) Some effects of 2-thiouracil on the multiplication of turnip yello mosaic virus. Virology 17: 367-380.

27. Kinchingto D, Kangro DH, Jeffries DJ (1995) Design and testing of antiviral

compounds. In: Desselberger U (eds.) Medical Virology: A practical approach, New York, USA: Oxford University Press, 147-71.

28. Jassim SA, Naji MA (2003) Novel antiviral agents: a medicinal plant perspective. J Appl Microbiol 95: 412-427.

29. Chantrill BH, Coulthard CE, Dickinson L, Inkley GW, Morris W, et al. (1952) The action of plant extracts on a bacteriophage of Pseudomonas pyocyanea and on influenza A virus. J Gen Microbiol 6: 74-84.

30. Kucera LS, Herrmann EJ (1966) Antiviral agents. Annual report in medicinal Chemistry 1 129-35.

31. Abdel-Haq N, Chearskul P, Al-Tatari H, Asmar B (2006) New antiviral agents. Indian J Pediatr 73: 313-321.

32. Duggar BM, Armstrong JK (1925) The effect of treating the virus of tobacco mosaic with the juices of various plants. Annals of the Missouri Botanical Garden 12: 359-66.

33. Abonyi D, Abonyi M, Esimone C, Ibezim E (2009) Plants as sources of antiviral agents. African Journal of Biotechnology 8: 3989-3994.

34. Cos P, Vanden B, Tde B, Vlietinck A (2003) Plant substances as antiviral agents: An update (1997-2001). Current Organic Chemistry 7: 1163-1180.

35. Williams JE (2001) Review of antiviral and immunomodulating properties of plants of the Peruvian rainforest with a particular emphasis on Una de Gato and Sangre de Grado. Altern Med Rev 6: 567-579.

36. Verma H, Kumar V (1980) Prevention of plant virus diseases by mirabilis jalapa leaf extract. New Botanist 7: 87-91.

37. Verma H, Awasthi L (1979) Antiviral activity of Boerhavia diffusa root extract and the physical properties of the virus inhibitor. Canadian Journal of Botany 57: 926-32.

38. Rajamohan F, Venkatachalam TK, Irvin JD, Uckun FM (1999) Pokeweed antiviral protein isoforms PAP-I, PAP-II, and PAP-III depurinate RNA of human immunodeficiency virus (HIV)-1. Biochem Biophys Res Commun 260: 453-458.

39. Lesney MS (2004) HPLC takes the head in identifying food phytochemicals. Today's Chemist at Work 2: 32–36.

40. Ong ES (2004) Extraction methods and chemical standardization of botanicals and herbal preparations. J Chromatogr B Analyt Technol Biomed Life Sci 812: 23-33.

41. Wagner H (1996) Plant drug analysis: A thin layer chromatography atlas. Philadelphia, USA: Springer Science & Business Media.

42. Houghton P, Raman A (1998) A laboratory manual for the fractionation of natural extracts. London, UK: Chapman & Hall. Hudson JB (1990) Antiviral compounds from plants. Boca Raton, USA: CRC Press.

43. Mukhtar M, Arshad M, Ahmad M, Pomerantz RJ, Wigdahl B, et al. (2008) Antiviral potentials of medicinal plants. Virus Res 131: 111-120.

44. Dawson W, Schlegel D (1976) The sequence of inhibition of tobacco mosaic virus synthesis by actinomycin d, 2-thiouracil, and cycloheximide in a synchronous infection. Phytopathology 66: 177-81.

45. Nyland G (1973) Tetracycline therapy of pear decline and x-disease in peach and cherry. In: IX International Symposium on Fruit Tree Virus Diseases. Kent, UK: ISHS Acta Horticulturae 44.

46. Gera A, Spiegel S, Loebenstein G (1986) Production, preparation, and assay of an antiviral substance from plant cells. Methods Enzymol 119: 729-734.

47. Pardee K, Ellis P, Bouthillier M, Towers GH, French C (2004) Plant virus inhibitors from marine algae. Canadian Journal of Botany 82: 304-309.

48. Yang J, Jin GH, Wang R, Luo ZP, Yin QS (2012) Spinaciaoleracea proteins with antiviral activity against tobacco mosaic virus. African Journal of Biotechnology 11: 6802-8.

49. Clark MF, Adams AN (1977) Characteristics of the microplate method of enzyme-linked immunosorbent assay for the detection of plant viruses. J Gen Virol 34: 475-483.

50. Spak J, Holý A, Pavingerová D, Votruba I, Spaková V, et al. (2010) New in vitro method for evaluating antiviral activity of acyclic nucleoside phosphonates against plant viruses. Antiviral Res 88: 296-303.

51. Verma H, Awasthi L, Mukerjee K (1979) Prevention of virus infection and multiplication by extracts from medicinal plants. Journal of Phytopathology 96: 71-76.

52. Verma H, Awasthi L, Mukerjee K (1979) Induction of systemic resistance by antiviral plant extracts in non-hypersensitive hosts. Zeitschrift fur Pflanzenkrankheiten und Pflanzenschutz 8: 735-746.

53. Verma H, Baranwal VK, Srivastava S (1998) Breeding for resistance to plant viruses. In: Hadidi A, Khetarpal RK, Koganezawa (eds.) Plant Virus Disease Control, APS Press, Paul, MN, USA, pp.154-162.

54. Verma H, Awasthi L (1980) Occurrence of a highly antiviral agent in plants treated with Boerhaavia diffusa inhibitor. Canadian Journal of Botany 58: 2141-2144.

55. Verma HN, Mukerjee K, Awasthi LP (1980) Determination of molecular weight of a polypeptide inducing resistance against viruses. Naturwissenschaften 67: 364-365.

56. Singh A (2006) Comparative studies on control of certain ailments of plants, mice and cancer cell lines and in vitro stimulation of growth of plants and virus resistance using phytoproteins from Boerhaaviadiffusa and Clerodendrumaculeatum. Lucknow, India: LucknowUniversit, PhD thesis.

57. Verma H, Chowdhury B, Rastogi P (1984) Antiviral activity in leaf extracts of Different Clerodendrum Species Zeitschrift fur pflanzenkantheiten und flanzenschutz 91: 34-41.

58. Verma H, Varsha N, Baranwal VK (1995) Endogenousvirus inhibitors from plants, their physical and biological properties. In: Chessin M, DeBorde D, Zipf A (eds.) Antiviral proteins in higher plants. Boca Raton, USA: CRC Press.

59. Tumer NE, Hudak K, Di R, Coetzer C, Wang P, et al. (1999) Pokeweed antiviral protein and its applications. Curr Top Microbiol Immunol 240: 139-158.

60. Barbieri L, Aron GM, Irvin JD, Stirpe F (1982) Purification and partial characterization of another form of the antiviral protein from the seeds of Phytolacca americana L. (pokeweed). Biochem J 203: 55-59.

61. Kubo S, Ikeda T, Imaizumi S, Takanami Y, Mikami Y (1990) A potent plant virus inhibitor found in mirabilis jalapa l. Annals of the Phytopathological Society of Japan 56: 481-487.

62. Irvin JD (1995) Antiviral proteins from phytolacca. In: Chessin M, DeBorde D, Zipf A (eds.) Antiviral proteins in higher plants. Boca Raton, USA: CRC press 65-94.

63. Ragetli HW, Weintraub M (1962) Purification and characteristics of a virus inhibitor from Dianthus caryophyllus L. I. Purification and activity. Virology 18: 232-240.

64. Ragetli HW, Weintraub M (1962) Purification and characteristics of a virus inhibitor from Dianthus caryophyllus L. II. Characterization and mode of action. Virology 18: 241-248.

65. Van kammen A, Noordam D, Thung TH (1961) The mechanism of inhibition of infection with tobacco mosaic virus by an inhibitor from carnation sap. Virology 14: 100-108.

66. Stirpe F, Williams DG, Onyon LJ, Legg RF, Stevens WA (1981) Dianthins, ribosome-damaging proteins with anti-viral properties from Dianthus caryophyllus L. (carnation). Biochem J 195: 399-405.

67. Plobner L, Leiser R (1990) Induction of virus resistance by carnation proteins. In: Proceedings of the international congress on virology, Berlin, Germany 21-26.

68. Surana R, Aher A, Pal S, Deore U (2011) Evaluation of anthelmintic activity of Ixora coccinea. International Journal of Pharmacy & Life Sciences 2: 813-814.

69. Sadowski Z (2010) Biosynthesis and application of silver and gold nanoparticles. In: Perez DP (ed.) Nanotechnology and Nanomaterials. Rijeka, Croatia: InTech 257-76.

70. Kumar S, Kumar D, Manjusha, Saroha K, Singh N, et al. (2008) Antioxidant and free radical scavenging potential of Citrullus colocynthis (L.) Schrad. methanolic fruit extract. Acta Pharm 58: 215-220.

71. Gurudeeban S, Ramanathan T (2010) Antidiabetic effect of citrulluscolocynthis in alloxon-induced diabetic rats. Inventi Rapid: Ethnopharmacology 1: 112-5.

27

Molecular Determination and Characterization of Phytoplasma 16S rRNA Gene in Selected Wild Grasses from Western Kenya

Adam OJ[1*], Midega CAO[1], Runo S[2] and Khan ZR[1]

[1]International Centre of Insect Physiology and Ecology (icipe), Nairobi, Kenya

[2]Kenyatta University, Nairobi, Kenya

*Corresponding author: Adam OJ, 1International Centre of Insect Physiology and Ecology (icipe), P.O Box 30, Mbita 40305, Kenya
E-mail: okinyiadam@gmail.com

Abstract

Napier grass (*Pennisetum purpuruem*) production for zero grazing systems has been reduced to rates of up to 90% in many smallholder fields by the Napier stunt (Ns) disease caused by phytoplasma sub-group 16SrXI in western Kenya. It is hypothesized that several other wild grasses in Kenya could be infected by phytoplasmas that would otherwise pose a significant threat to Napier, other important feeds and food crops. This study therefore sought to detect and identify phytoplasma strains infecting wild grasses in western Kenya using 16S ribosomal RNA (ribonucleic acid) gene as well as identify wild grass species hosting phytoplasmas in 646 wild grass samples that were collected in October 2011 and January 2012 during a random crossectional survey conducted in Bungoma and Busia counties of western Kenya. DNA was extracted and nested polymerase reaction (nPCR) used to detect phytoplasmas. Two sub-groups of phytoplasmas were detected in eight grass species observed to grow near infected Napier fields. Only one of the two phytoplasmas reported was related to the Ns phytoplasma. There was a strong association between proportions of phytoplasma infection and the grass species collected (p = 0.001). *C. dactylon*, *D. scalarum*, *B. brizantha*, poverty grass and P. maximum had high proportions of infection and were abundantly distributed in western Kenya hence considered wild phytoplasma hosts. *E. indica* and *C. ciliaris* were scarcely distributed and had low infection rates. There was statistically significant difference in proportions of infection per location of survey (p = 0.001). Phytoplasma subgroups 16SrXI and 16SrXIV were the only phytoplasma genotypes distributed among wild grasses in western Kenya. Phytoplasma subgroup 16SrXIV predominantly infects only *C. dactylon* and *B. brizantha* grasses while phytoplasma subgroup 16SrXI is broad spectrum and infects a large number of wild grasses. In general, there is a diversity of wild grasses hosting phytoplasmas in western Kenya. These host grasses may be the reason for the high rates observed in the spread of Ns disease in western Kenya by acting as reservoirs for Ns phytoplasma.

Keywords: Phytoplasma; *Pennisetum purpureum* Characterization; Napier stunt disease; *Cynodon dactylon* ; *Brachiaria brizantha*; *Digitaria scalarum*

Introduction

Napier grass (*Pennisetum purpuruem*) is an indigenous tropical African clumping grass which grows up to 5 meters tall. It is mainly vegetatively propagated through cuttings of about 3 to 4 centimeters in length and clump splitting. It has been widely used as fodder crop and for environmental sustenance, by stabilizing soils as well as acting as windbreaks [1]. In Kenya, napier grass has been employed in a new 'Push-Pull' management strategy for maize stem borers [2].

Napier grass has been used by many farmers in Kenya as the major livestock feed. It is as well sold to generate additional revenue. The increased population results in land subdivision which decreases farm size, hence, resulting in the adoption of the zero grazing system by most farmers which uses large amounts of fodder such as napier and other wild grasses that are cut and carried home for stall feeding [3].

A disease that attacks and greatly reduces the productivity of napier grass has been identified particularly in regions of western Kenya. Napier stunt disease (Ns) is a newly identified disease caused by a phytoplasma that adversely affects napier production at a rate between 30% and 90% observed in many smallholder fields. The year 2004, the disease is estimated to have affected over 23,298 km^2 of napier grass

crop, an estimated 2 million households (about 30% of the population) in Western and Rift Valley provinces of Kenya [3].

Many grass diseases across the world have been attributed to phytoplasma infection. Four varieties of phytoplasmas were identified in seven species of grasses growing near sugarcane crops. These phytoplasmas were observed to be related to sugarcane white leaf phytoplasma that causes sugarcane disease in Asia [4]. Phytoplasma has also been reported to cause cynodon white leaf (CWL) disease in the Bermuda grass (*Cynodon dactylon*) [5], hyparrhenia white leaf disease (HWLD) in *Hyparrhenia rufa* grass [6], rice yellows dwarf disease (RYD) in rice [7] and *sorghum grassy* shoot (SGS) in sorghum crop plants (*Sorghum stipoideum*) [4]. This is an indication that several other wild grass species could be infected by specific phytoplasma strains, hence; act as phytoplasma reservoirs which pose a threat to important feeds and food crops as well as reduce the forage supply of such wild grass strains for dairy farming.

The elimination of alternative phytoplasma hosts around napier farms as well as bioengineering of phytoplasma resistant variety of napier grass would constitute components for the management of phytoplasma diseases. This study identified phytoplasma wild host range among wild grasses and there genotypic distribution in western

Kenya necessary for the establishment of viable management and prevention strategies for the spread of Ns disease in napier grass and other important fodder.

Materials and Methods

Sample collection

Both phytoplasma symptomatic and asymptomatic wild grasses near Ns affected Napier fields from Bungoma and Busia counties were collected in this study (Table 1). Approximately 16 fields from each county were chosen from different sub agro-ecological zones as replications. An itinerary for each area was set up. Fields in each area were chosen at random. In each field an average of 20 grass samples were obtained. The first samples were taken at the edge of every field which formed the base. Along transects placed 1-3m apart depending on the width of the field, one sample was collected per quadrat (1m x 1m) thrown 1-3m apart throughout the entire length of the transect. The numbers of plants were counted for the grasses. The collected samples were air-dried and transported to the international center for insect physiology and ecology (ICIPE-TOC) for further taxonomic identification, laboratory screening and phytoplasma diagnosis.

County	Location	Samples
Bungoma	Bungoma Township	22
	Mlaha 1	18
	Mlaha 2	20
	Kibabii 1	22
	Kibabii 2	20
	Bisunu	21
	Lwandanyi 1	20
	Lwandanyi 2	20
	Luya	20
	Milo	20
	Chetambe	20
	Kimilili	20
	Kibabii 3	19
	Kimaeti	20
	Kokare	20
	Katakwa	20
	Netima	20
Busia	Wakhungu-Odiado	21
	Bumala	20
	Bukhayo East	20
	Otimong'	21
	Nambale	21
	Elugulu	20
	Marachi	20
	Bukhayo West	20
	Bukhayo Central	20
	Bwamani	21
	Bulanda 1	20
	Bulanda 2	20
	Busia Township, ADC	20
	Aget	20
	Busia Township, BP	20
TOTAL		646

Table 1: Total number of grass samples collected for this study.

DNA extraction and PCR amplification

Total DNA was extracted from 300 mg of leaf tissues by CTAB (cetyl trimethyl-ammonium bromide) method [8] and modified as described by Khan et al. [9]. DNA pellets were suspended in 50 µL deionized distilled water and the DNA suspensions stored at - 20°C.

PCR assays to amplify the phytoplasma DNA were performed using universal primer pairs P1/P6 [10] and NapF/NapR. The reaction mixture in the initial PCR contained 1.0 µL of template DNA of each sample at 100 ng/µL, 1.0 µL of both P1 and P6 primers for each sample, 2.5 µL of dNTPs (deoxyribonucleotide triphosphates) for each sample, 0.25 µL of Taq DNA polymerase (GenScript) for each sample and 2.5 µL of 1x Taq Polymerase buffer (GenScript) for each sample used. The PCR reaction mixture was gently vortexed for 10 seconds to mix and 22.25 µL of the mixture added to PCR tubes containing 1.0 µL of each template. A 35 cycle PCR was conducted using P1/P6 primer pair in a PTC-100® Thermal cycler (MJ Research, Incorporated, Lincoln Street, Massachusetts, USA) as follows; denaturation of DNA at 94°C for 2 minutes for 1 cycle, annealing of the primers at 52°C for 2 minutes for the first reaction and 72°C for 3 minutes for the subsequent reactions and elongation reaction at 72°C for 10 minutes for 1 cycle [11-13]. The second amplification of the primary PCR products of the 16S rDNA fragment was carried out using a reaction mixture containing NapF/NapR primer pair (Inqaba BiotecTM). The second round reaction was performed using 0.5 µL of the first PCR amplicons. From each of the second PCR amplicons, 6.0 µL of the DNA was mixed with 4.0 µL of 6X loading dye (SIGMA-ALDRICH®) prior to loading into the gel wells. Electrophoresis was carried out at 70 volts for 30 minutes on 1% (w/v) agarose gel containing 0.3 µg/ mL ethidium bromide in 1x TAE (22.5mM Tris-acetate 1mM EDTA; pH 8.0) buffer. The gels were observed under UV transilluminator at 312nm wavelength to visualize the bands.

DNA Purification and Sequencing of polymerase chain reaction (PCR) products

The expected ≤ 800 bp nested polymerase chain reaction (nPCR) products obtained from the positive grass samples were purified on GenScript Quick Clean II PCR Extraction kit (GenScript® Centennial Ave, Piscataway Township, NJ 08854, USA) as per the manufacturer's protocol and directly sequenced. A total of 81 DNA amplicons that tested positive for phytoplasma were run on a gel at 1 µL (Figure 1) out of which 33 representative samples (Table 2) were submitted for sequencing. To avoid redundancy, representative samples were selected based on the species of the grasses as well as the location the samples were obtained from. Sequencing was carried out in both directions (Forward and Reverse) using BigDye Terminator Cycle Sequencing in a DNA automated sequencer (SegoliLab, International Livestock Research Institute - ILRI, Nairobi Kenya).

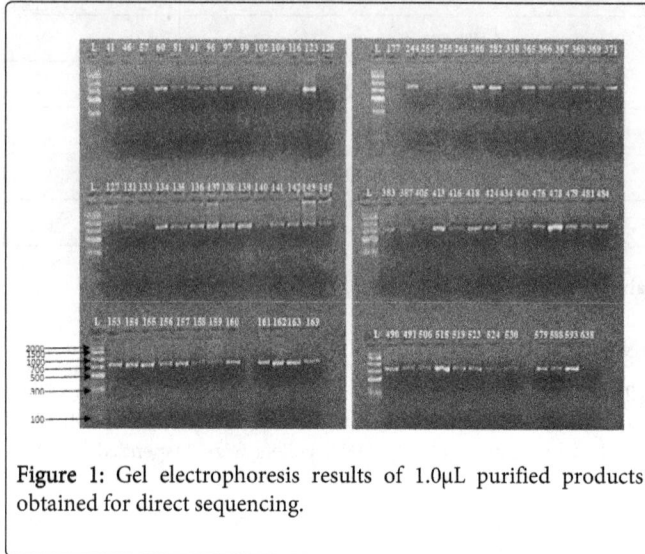

Figure 1: Gel electrophoresis results of 1.0μL purified products obtained for direct sequencing.

Results

Sequence homology and phylogenetic relationships

The partial 16S rRNA genome sequences were assembled and edited using BioEdit sequence alignment editor [14]; gaps and ambiguities were eliminated from the final sequences. Partial full-length 16S rRNA gene sequences were converted to MEGA files for phylogenetic analysis by DNA neighbor-joining method using MEGA version 5.05 software [11] and the phylogenetic tree constructed with 1,000 bootstraps replications.

The 16S rRNA gene sequences of 33 phytoplasmas isolated from the wild grasses in this study were compiled in FASTA format and compared with each other and with 16 other reference phytoplasmas from NCBI Genbank database (appendix).

All phytoplasma sequences analyzed in this study aligned themselves in two discrete clades when compared to each other as depicted by phylogenetic tree (Figure 2). This study did not identify any novel phytoplasma strain (16S rRNA group/ subgroup) from all the sequences characterized since the divergence of all the phytoplasma sequences retrieved in this study (Table 2) was below the recommended threshold of 97.5% sequence similarity (divergence of less than 2.5%) used in defining a novel phytoplasma species falling within the provisional status '*Candidatus*' as per the International Research Program on Comparative Mycoplasmology [15]. This study, however, did not employ the use of 16S-23S rDNA spacer region in characterizing the phytoplasmas detected.

Sample No.	Host Grass	County	Location	16S rRNA Group	Partial sequence
46	*C. dactylon*	Bungoma	Mlaha2	16SrXIV	16S rRNA gene
96	*P. maximum*	Bungoma	Kibabii2	16SrXI	16S rRNA gene
97	*B. brizantha*	Bungoma	Kibabii2	16SrXIV	16S rRNA gene
102	*B. brizantha*	Bungoma	Bisunu	16SrXIV	16S rRNA gene
123	*C. dactylon*	Bungoma	Bisunu	16SrXIV	16S rRNA gene
134	*C. dactylon*	Bungoma	Lwandanyi1	16SrXIV	16S rRNA gene
136	*C. dactylon*	Bungoma	Lwandanyi1	16SrXIV	16S rRNA gene
138	*C. dactylon*	Bungoma	Lwandanyi1	16SrXIV	16S rRNA gene
139	*C. dactylon*	Bungoma	Lwandanyi2	16SrXIV	16S rRNA gene
155	*C. dactylon*	Bungoma	Lwandanyi2	16SrXIV	16S rRNA gene
157	*C. dactylon*	Bungoma	Lwandanyi2	16SrXIV	16S rRNA gene
169	*B. brizantha*	Bungoma	Luhya	16SrXIV	16S rRNA gene
266	Other	Bungoma	Kimaeti	16SrXI	16S rRNA gene
282	*B. brizantha*	Bungoma	Kimaeti	16SrXI	16S rRNA gene
365	*B. brizantha*	Busia	Bumala	16SrXI	16S rRNA gene
366	Other	Busia	Bumala	16SrXI	16S rRNA gene
413	*D. scalarum*	Busia	Otimong'	16SrXI	16S rRNA gene
416	*C. dactylon*	Busia	Otimong'	16SrXIV	16S rRNA gene

418	Poverty grass	Busia	Otimong'	16SrXI	16S rRNA gene
478	*C. dactylon*	Busia	Marachi	16SrXI	16S rRNA gene
479	*B. brizantha*	Busia	Marachi	16SrXI	16S rRNA gene
484	Poverty grass	Busia	Marachi	16SrXI	16S rRNA gene
490	*D. scalarum*	Busia	Bukhayo west	16SrXI	16S rRNA gene
491	*C. dactylon*	Busia	Bukhayo west	16SrXI	16S rRNA gene
515	*D. scalarum*	Busia	Bukhayo central	16SrXI	16S rRNA gene
519	*P. maximum*	Busia	Bukhayo central	16SrXI	16S rRNA gene
523	*D. scalarum*	Busia	Bukhayo central	16SrXI	16S rRNA gene
524	*B. brizantha*	Busia	Bukhayo central	16SrXI	16S rRNA gene
530	*D. scalarum*	Busia	Bwamani	16SrXI	16S rRNA gene
579	Poverty grass	Busia	Bulanda2	16SrXI	16S rRNA gene
588	*E. indica*	Busia	Busia Township, ADC	16SrXI	16S rRNA gene
593	*D. scalarum*	Busia	Busia Township, ADC	16SrXI	16S rRNA gene
642	*P. maximum*	Busia	Busia Township, BP	16SrXI	16S rRNA gene

Table 2: Phytoplasma isolates, location of collection, host plant, and associated 16Sr groups retrieved in this study.

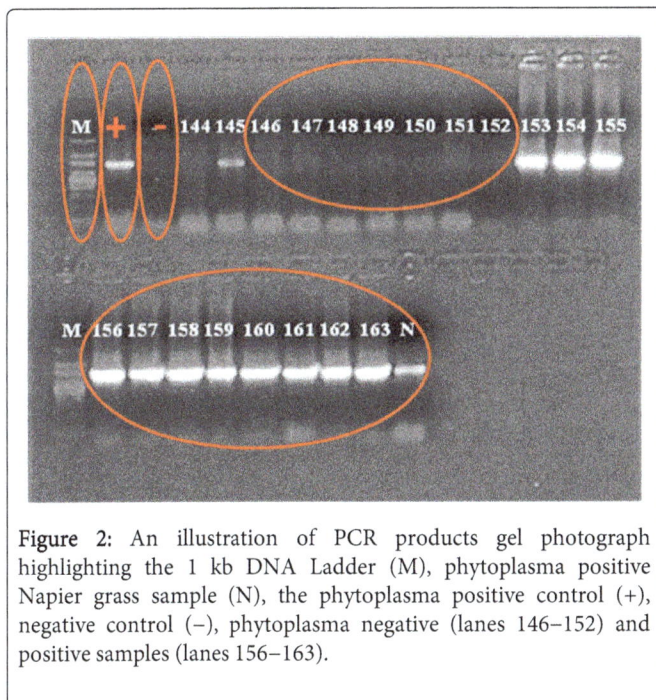

Figure 2: An illustration of PCR products gel photograph highlighting the 1 kb DNA Ladder (M), phytoplasma positive Napier grass sample (N), the phytoplasma positive control (+), negative control (−), phytoplasma negative (lanes 146–152) and positive samples (lanes 156–163).

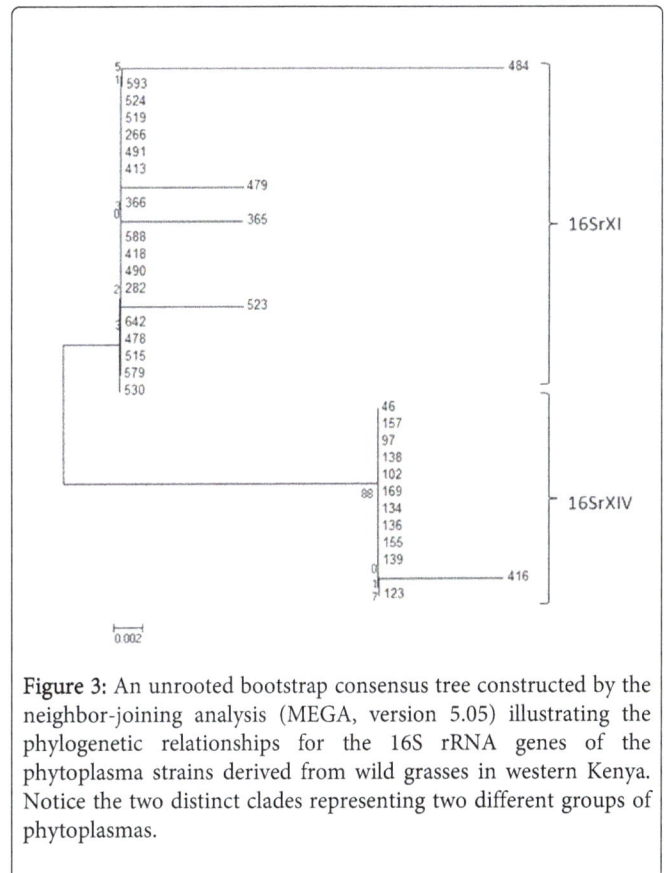

Figure 3: An unrooted bootstrap consensus tree constructed by the neighbor-joining analysis (MEGA, version 5.05) illustrating the phylogenetic relationships for the 16S rRNA genes of the phytoplasma strains derived from wild grasses in western Kenya. Notice the two distinct clades representing two different groups of phytoplasmas.

Since this is a less significant taxonomic tool as compared to the 16S rDNA sequence [12]. It is recommended by IRPCM that phytoplasmas which differ with less than 2.5% of 16S rDNA nucleotide positions should be regarded as putative species when characterization is supported by data based upon molecular markers such as plant host range, insect vector transmission and serological studies rather than on 16S rDNA sequence [12].

The NapF/NapR reactions from samples 46(*C. dactylon*), 97(*B. brizantha*), 102(*B. brizantha*), 123(*C. dactylon*), 134(*C. dactylon*), 136(*C. dactylon*), 138(*C. dactylon*), 139(*C. dactylon*), 155(*C. dactylon*), 157(*C. dactylon*), 169(*B. brizantha*), 416(*C. dactylon*) yielded the expected 800 bp amplicons. Multiple sequence alignment via MEGA software version 5.05 showed 99% identity with each other (Figure 2). BLAST (basic local alignment search tool) search program (www.ncbi.nih.gov/BLAST) showed that the above sequences were 99% similar to '*Ca. Phytoplasma cynodontis*' (accession no. EU999999.1) and '*Ca. Phytoplasma cynodontis*' (accession no. FJ348654.1). There was also 98% similarity with Bermuda grass white leaf isolates BGWL 1SL and PG as well as other several phytoplasma strains in group 16SrXIV from the NCBI database (Figure 3). BGWL disease was first reported in Bermuda grass (*C. dactylon*) in Kenya in 2010 [13].

On the other hand, multiple sequence alignment using MEGA software version 5.05 of the partial 16S rRNA gene sequences for the samples 479(*B. brizantha*), 524(*B. brizantha*), 366(Other), 418(Poverty grass), 365(*B. brizantha*), 478(*C. dactylon*), 579(Poverty grass), 490(*D. scalarum*), 523(*D. scalarum*), 491(*C. dactylon*), 515(*D. scalarum*), 413(*D. scalarum*), 530(*D. scalarum*), 642(*P. maximum*), 484(Poverty grass), 266(Other), 519(*P. maximum*), 282(*B. brizantha*), 588(*E. indica*), 96(*P. maximum*) and 593(*D. scalarum*) showed 99% identity with each other (Figure 3).

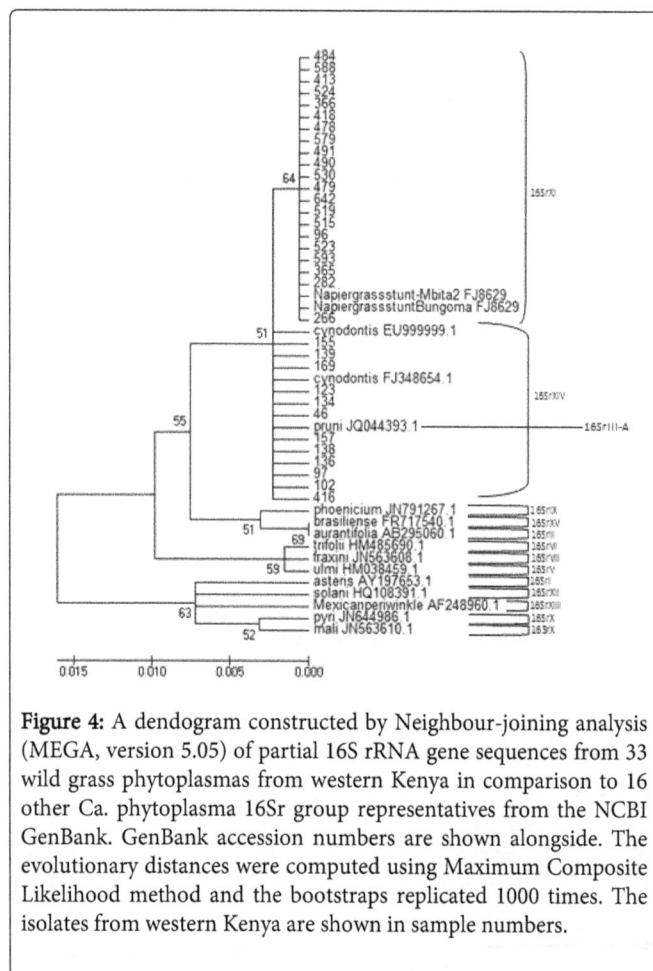

Figure 4: A dendogram constructed by Neighbour-joining analysis (MEGA, version 5.05) of partial 16S rRNA gene sequences from 33 wild grass phytoplasmas from western Kenya in comparison to 16 other Ca. phytoplasma 16Sr group representatives from the NCBI GenBank. GenBank accession numbers are shown alongside. The evolutionary distances were computed using Maximum Composite Likelihood method and the bootstraps replicated 1000 times. The isolates from western Kenya are shown in sample numbers.

A BLAST search carried out on the above sequences revealed that there was 98-100% sequence similarity with the Ns (Napier grass stunt) phytoplasma isolate Mbita 2 (accession no. FJ862999.2) and

Napier grass stunt phytoplasma isolate Bungoma (accession no. FJ862998.2) from Kenya, all of which belong to the phytoplasma group '*Ca. Phytoplasma oryzae*' (group 16SrXI), as depicted in the phylogenetic tree (Figure 4).

Association between phytoplasma infection and grass species

From the 81 phytoplasma infections registered in this study (Table 3), *C. dactylon* had the highest proportion of total infections at 38%, followed by *D. scalarum* at 17.3%, *B. brizantha* had 16%, poverty grass and *P. maximum* had 7.4% and 4.9% respectively while *E. indica* and *C. ciliaris* had the least proportions of phytoplasma infection at 2.5% and 1.2% respectively

Grass species	PCR status		Proportion of infection	Total
	0	1		
Brachiaria brizantha	71(0.8452)	13(0.1548)	13(16.0000)	84
Cenchrus ciliaris	0	1(1.0000)	1(1.2000)	1
Cymbopogon nardus	2(1.0000)	0	0	2
Cynodon dactylon	55(0.6395)	31(0.3605)	31(38.3000)	86
Digitaria scalarum	286(0.9533)	14(0.0467)	14(17.3000)	300
Echinichloa pyramidalis	2(1.0000)	0	0	2
Eleusine indica	6(0.7500)	2(0.2500)	2(2.5000)	8
Eragrostis curvula	4(1.0000)	0	0	4
Hyparrhenia pilgerama	6(1.0000)	0	0	6
Other	65(0.8784)	9(0.1216)	9(11.1000)	74
Panicum maximum	28(0.8750)	4(0.1250)	4(4.9000)	32
Pennisetum polystachion	5(1.0000)	0	0	5
Pennisetum purpureum	1(1.0000)	0	0	1
Poverty grass	24(0.8000)	6(0.2000)	6(7.4000)	30
R. cochinchinensis	1(1.0000)	0	0	1
Setaria incrassata	2(0.6667)	1(0.3333)	1(1.2000)	3
Sorghum versicolor	2(1.0000)	0	0	2
Sporobolus pyramidalis	4(1.0000)	0	0	4
Themeda triada	1(1.0000)	0	0	1
Total	565	81	81(100)	646
Chi square test	75.787(a)			
df	18			
Likelihood Ratio	68.054			
P Value (≤0.05)	0.0001			

Table 3: Total grass species, their phytoplasma statuses and the proportions of infection.

Other grasses that were not identified constituted 11.1%. The proportions of phytoplasma infection per grass species were compared using two-sided Chi-Square tests at 95% confidence interval as summarized in the Table 2 From the test carried out, there was a strong association between proportions of phytoplasma infections and grass species (p = 0.0001).

Association between 16S rRNA sub-group and grass species

Of all the grass samples that had positive phytoplasma infections, 33 were chosen for sequencing and phylogenetic analyses (Table 2). Two wild grass species that registered positive phytoplasma infections; *B. brizantha* and *C. dactylon* were infected by both phytoplasma subgroup 16SrXIV and 16SrXI. *B. brizantha* had 16SrXI: 16SrXIV infection ratio of 4:3 while *C. dactylon* had 16SrXI: 16SrXIV infection ratio of 2:9 (Figure 5). The remaining grass species positive for phytoplasma were entirely infected by phytoplasma subgroup 16SrXI (Table 4).

Grass species	16SrXI	16SrXIV	Not done	Total
B. brizantha	4(57.14%)	3(42.86%)	77	84
C. ciliaris	0	0	1	1
C. nardus	0	0	2	2
C. dactylon	2(18.18%)	9(81.81%)	75	86
D. scalarum	6(100%)	0	294	300
E. pyramidalis	0	0	2	2
E. indica	1(100%)	0	7	8
E. curvula	0	0	4	4
H. pilgerama	0	0	6	6
Other	2(100%)	0	72	74
P. maximum	3(100%)	0	29	32
P. polystachion	0	0	5	5
P. purpureum	0	0	1	1
Poverty grass	3(100%)	0	27	30
R. cochinchinensis	0	0	1	1
S. incrassata	0	0	3	3
S. versicolor	0	0	2	2
S. pyramidalis	0	0	4	4
T. triada	0	0	1	1
Total	21	12	613	646

Table 4: A table of grass species collected and their associated 16S rRNA sub-group.

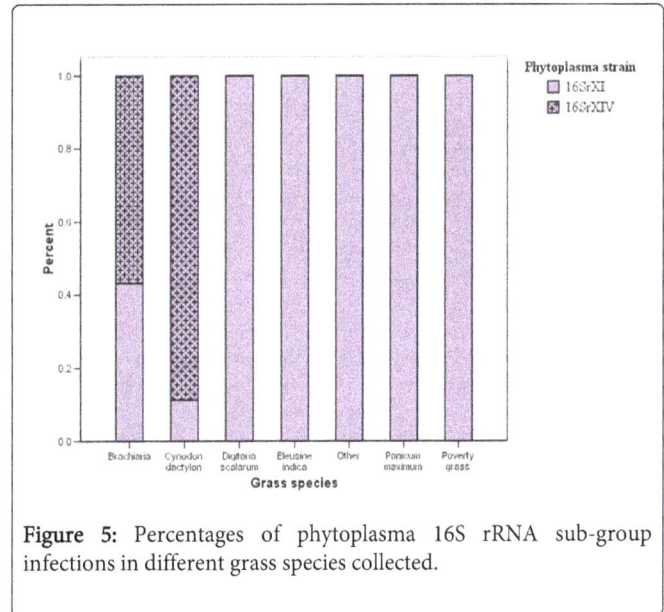

Figure 5: Percentages of phytoplasma 16S rRNA sub-group infections in different grass species collected.

Discussion and Conclusions

This study found out that there was a strong association between proportions of phytoplasma infection and the grass species collected. *C. dactylon*, *B. brizantha*, *D. scalarum*, *P. maximum* and poverty grass generally act as wild phytoplasma hosts and are abundantly distributed in western Kenya. *E. indica* and *C. ciliaris* are scarcely distributed in western Kenya even though they play host to phytoplasma. There were substantial differences in proportions of phytoplasma infection per location of survey. There seems to be a trend in phytoplasma genotypic distribution in this study. The observed ecological isolation could be as a result of exclusive association of particular phytoplasmas with particular grass plant and/or insect host range in particular geographical regions. Gundersen et al observed that, two or more phytoplasma strains could exhibit specificity for preferred host plant in specific locations that may, to a large extent, reflect transmitting insect (vector) feeding behavior (Gundersen et al, 1996). This natural phytoplasmal ecological diversity may be exploited in the investigation of the epidemiology of phytoplasma-related diseases, hence the prevention of the spread of phytoplasma diseases. This was, however not verified as insect vectors were not collected and determined for correlation analysis in this study.

Phytoplasma subgroups 16SrXI and 16SrXIV were the only phytoplasma genotypes distributed among wild grasses in western Kenya. '*Ca. Phytoplasma cynodontis*' (subgroup 16SrXIV): the causative agent of Bermuda grass white leaf disease (BGWLD) predominantly infects only *Cynodon dactylon* and *Brachiaria brizantha* wild grass types. This concurs with the findings made by Marcone et al where an association was made between BGWLD in *C. dactylon* (Bermuda grass) and '*Ca phytoplasma cynodontis*', as well as *Brachiaria* white leaf disease [12]. Marcone et al demonstrated that phytoplasmas associated with Brachiaria white leaf disease and carpet grass white leaf showed 16S rDNA sequences identical or nearly similar to those of Bermuda grass white leaf in *C. datylon*. On the other hand '*Ca Phytoplasma oryzae*' (subgroup 16SrXI): the causative agent of Ns disease exhibited a broad pathogenic potential in this study and infects a large number of wild grasses, most importantly; *P. maximum*, *D. scalarum*, poverty grass and *B. brizantha*.

Isolate	Acronyms	Phytoplasma species	16S rRNA Group-subgroup	Host species	Location	NCBI Accession No.	Literature
Aster yellows	MIAY	Ca. P asteris	16SrI	Cannabis sativa L	India	EU439257.1	[16]
Napier grass stunt	NSD	Ca. P oryzae	16SrXI	P. purpuruem	Kenya, Mbita	FJ862999.2	[17]
Napier grass stunt	NSD	Ca. P oryzae	16SrXI	P. purpuruem	Kenya, Bungoma	FJ862997.2	[17]
Bermuda grass white leaf	BGWL	Ca. P cynodontis	16SrXIV	Cynodon dactylon	China	EU999999.1	[18]
Rice yellow Dwarf	RYD	Ca. P oryzae	16SrXI	Oryza sativa	Vietnam	JF927999.1	[19]
Peanut witches'-broom	PWB	Ca. P aurantifolia	16SrII	Citrus araurantifolia	Oman, Rumis	AB295060.1	[20]
X-disease	PX11Ct1	Ca. P pruni	16SrIII-A	stone fruits, Prunus	U.S.A/ canada	JQ044393.1	[21]
Stolbur	STOL	Ca. P solani	16SrXII-A	Solanum tuberosum	Romania/ Russia	HQ108391.1	[22]
Elm yellows	EY20_SRB	Ca. P ulmi	16SrV-A	Ulmus spp	Serbia	HM038459.1	[24]
Clover proliferation	CP	Ca. P trifolii	16SrVI	Calotropis gigantean	India: Gorakhpur	HM485690.1	[26]
Ash yellows	AY	Ca. P fraxini	16SrVIIA	Graminella nigrifrons	Canada	JN563608.1	[24]
Pigeonpea witches'-broom	PPWB	Ca. P phoenicium	16SrIX	Blueberry	U.S.A	JN791267.1	[25]
Apple proliferation	AP	Ca. P mali	Gn-16SrXA	Graminella nigrifrons	Canada	JN563610.1	[23]
Apple proliferation	AP	Ca. P pyri	16SrX	Cacopsylla pyri	Portugal	JN644986.1	[26]
Mexican periwinkle viresc	MPWV	Unidentified	16SrXIII-A	Catharanthus roseus	U.S.A	AF248960.1	[27]
Bermuda grass white leaf	BGWL	Ca. P cynodontis	16SrXIV	Dicanthium annulatum	India	FJ348654.1	[28]
Hibiscus witches'-broom	HWB	Ca. P brasiliense	16SrXV	Prunus persica	Azerbaijan	FR717540.1	[29]

Appendix: Acronyms and GenBank accession numbers of phytoplasma 16S rDNA sequences used for phylogenetic analysis.

References

1. Jones P, Devonshire BJ, Holman TJ, Ajanga S (2004) Napier grass stunt: a new disease associated with a 16SrXI Group phytoplasma in Kenya. Insect Science and its Application, 21: 375-380.
2. Khan ZR, Pickett JA, Wadhams LJ, Muyekho F (2001) Habitat management strategies for the control of cereal stem borers and striga in maize in Kenya. Insect Science and its Application 21: 375-380.
3. Mulaa MA, Muyekho F, Ajanga S, Jones P, Boa E (2004) Napier stunt disease vector identification and containment of the disease on farm. Kenya Agricultural Research Institute (KARI).
4. Blanche KR, Tran-Nguyen LT, Gibb KS (2003) Detection, identification and significance of phytoplasmas in grasses in northern Australia. Plant Pathology 52: 505-512.
5. Salehi M, Izadpanah K, Siampour M, Taghizadeh M (2009) Molecular characterization and transmission of bermuda grass white leaf phytoplasma in Iran. Journal of Plant Pathology, 91: 655-661.
6. Obura E, Masiga D, Midega CAO, Otim M, Wachira F, et al. (2011) Hyparrhenia grass white leaf disease, associated with 16SrXI phytoplasma. New Disease Reports.
7. Lee LM, Pastore M, Vibio M, Danielli A, Attathom S, et al. (1997) Detection and Characterization of a Phytoplasma associated with annual Blue Grass (Poa annua) white leaf disease in Southern Italy. European Journal of Plant Pathology 103: 251-254.
8. Doyle JJ, Doyle JL (1990) Isolation of plant DNA from fresh tissue. Focus 12: 13-15.
9. Khan S, Qureshi MI, Kamaluddin Alam T, Abdin MZ (2006) Protocol for isolation of genomic DNA from dry and fresh roots of medicinal plants suitable for RAPD and restriction digestion. African Journal of Biotechnology 6: 175-178.
10. Deng S, Hiruki C (1991) Amplification of 16S rRNA genes from culturable and non-culturable mollicutes. Journal of microbiological Methods 14: 53-61.
11. Tamura K, Dudley J, Nei M, Kumar S (2007) MEGA4: Molecular Evolutionary Genetics Analysis (MEGA) software version 4.0. Molecular Biology and Evolution 24: 1596-1599.

12. Marcone C, Schneider B, Seemuller E (2004) 'Candidatus phytoplasma cynodontis', the phytoplasma associated with Bermuda grass white leaf disease. International journal of Systematic and Evolutionary Microbiology 54: 1077-1082.

13. Obura E, Masiga D, Midega CAO, Wachira F, Pickett JA, et al. (2010) First report of a phytoplasma associated with bermuda grass white leaf disease in Kenya. New Disease Reports 21, 23.

14. Hall TA (1999) BioEdit: a user-friendly biological sequence alignment editor and analysis program for Windows 95/98/NT. Nucl. Acids. Symp Ser 41: 95-98.

15. IRPCM (International Research Program on Comparative Mycoplasmology) Phytoplasma/ Spiroplasma Working Team– Phytoplasma Taxonomy Group. (2004). 'Candidatus Phytoplasma', a taxon for the wall-less, non-helical prokaryotes that colonize plant phloem and insects. International Journal of Systematic and Evolutionary Microbiology, 54: 1243-125.

16. Raj SK, Snehi SK, Khan MS, Kumar S (2008) 'Candidatus Phytoplasma asteris' (group 16SrI) associated with a witches'-broom disease of Cannabis sativa in India. Plant pathology 57: 1173.

17. Obura E, Midega CA, Masiga D, Pickett JA, Hassan M, et al. (2009) Recilia banda Kramer (Hemiptera: Cicadellidae), a vector of Napier stunt phytoplasma in Kenya. Naturwissenschaften 96: 1169-1176.

18. Li Z, Wu K, Zheng X, Zhang C, Wu Y, et al. (2008) First report of Bermuda grass white leaf phytoplasma infecting Cynodon dactylon in China. Unpublished.

19. Trinh HX, Ngo BG, Mai QV, Nguyen TD, Dang TV, et al. (2011) Diversity of phytoplasmas in sugarcane and rice in Vietnam. Unpublished.

20. Natsuaki T, Al-Zadjali AD (2007) Detection, identification and molecular characterization of Candidatus Phytoplasma aurantifolia isolates associated with lime plants (Citrus araurantifolia) and other citrus species. Unpublished.

21. Davies RE (2011) 'Candidatus Phytoplasma pruni', a novel taxon associated with X-disease of stone fruits, Prunus spp.: multilocus characterization based on 16S rRNA, secY, and ribosomal protein genes. Unpublished.

22. Ember I, Acs Z, Munyaneza JE, Crosslin JM, Kolber M (2011) Survey and molecular detection of phytoplasmas associated with potato in Romania and southern Russia. European Journal of Plant Pathology 130: 367-377.

23. Arocha-Rosete Y (2011) Identification of Graminella nigrifrons as a potential vector for phytoplasmas identified in Canada. Unpublished.

24. Priya M, Chaturvedi Y, Rao GP, Raj SK (2010) First report of phytoplasma 'Candidatus Phytoplasma trifolii' (16SrVI) group associated with leaf yellows of Calotropis gigantea in India. New Disease Reports 22: 29.

25. Lee IM, Bottner-Parker KD, Zhao Y, Bertaccini A, Davis RE (2012) Differentiation and classification of phytoplasmas in the pigeon pea witches'-broom group (16SrIX): an update based on multiple gene sequence analysis. Int J Syst Evol Microbiol. In press.

26. Sousa E, Marques A, Cardoso F (2011) The first report of Candidatus phytoplasma pyri in the pear-leafhopper Psylla pyri in Portugal. Unpublished.

27. Dally EL, Bottner KD, Davis RE (2000) Revised Subgroup Classification of Group V Phytoplasmas and Placement of Flavescence Doree-Associated Phytoplasmas in Two Distinct Subgroups. Unpublished.

28. Rao GP, Mall S, Marcone C (2008) First report of Bermuda grass white leaf phytoplasma infecting Dicanthium annulatum grass in India. Unpublished.

29. Balakishiyeva G, Qurbanov M, Mammadov A, Bayramov S, Foissac X (2010) Identification of 'Ca. P. brasiliense' from a yellowing peach tree in Azerbaijan and development of a specific detection test. Unpublished.

28

The Potential of Cell-free Cultures of *Rhizobium leguminosarum, Azotobacter chroococcum* and Compost Tea as Biocontrol Agents for Faba Bean Broomrape (*Orobanche crenata* Forsk.)

Yasser El-Halmouch[1,4]*, Ahlam Mehesen[2] and Abd El-Raheem Ramadan El-Shanshoury[3,4]

[1]*Botany Department, Faculty of Science, Damanhour University, Damanhour 22511, Egypt*
[2]*Soils, Water and Environment Research Institute, Agriculture Research Center, Giza 12619, Egypt*
[3]*Botany Department, Faculty of Science, Tanta University, Tanta 31527, Egypt*
[4]*Biotechnology Department, Faculty of Science, Taif University, Taif 21974, Kingdom of Saudi Arabia*

Abstract

In the present study, cell-free cultures of four isolates of *Rhizobium leguminosarum*, an isolate of *Azotobacter chroococcum* and compost tea were investigated for their biocontrol potential against the root parasitic weed *Orobanche crenata*. Individual cell-free cultures of *Azotobacter chroococcum* or *Rhizobium* sp., dual and mixture of cell-free cultures of *Rhizobium* spp. or compost tea were applied to infested pots in greenhouse conditions. The treatments showed variable effects on many developmental parameters of both faba bean and broomrape. Significant decrease in the number of broomrape attachments, dry weight of the attached tubercles on faba bean roots and the reduction in percentage of broomrape seed germination were recorded. Compost tea, individual and mixture of *R. leguminosarum* isolates were more reducing on broomrape germination and growth than *A. chroococcum* alone did; being the former treatment is the best. The reduction in broomrape incidence by compost tea was due to certain phenotypic mechanisms, which acted alone or in combination. These mechanisms included negative effect of natural stimulant broomrape on seed germination, prevention of radical penetration inside the host roots, parasite yield reduction, and thus increasing the growth and vitality of faba bean. *In vitro* experiment indicated that seed germination percentage of broomrape was also negatively affected by the combination of root-exudates and compost tea. Radical apexes of the germinated seeds were distorted. These distortions may prevent the radicals to follow up the infestation. In conclusion, the study presents the potential of *R. leguminosarum* isolates and compost tea in biocontrol of broomrape. More investigations should be carried out with viable bacterial cells on the parasite plant before use in sustainable agricultural systems.

Keywords: Broomrape; Biocontrol; Faba bean; Root exudates; *R. leguminosarum*; *A. chroococcum*; Compost tea

Introduction

Phytoparasitic weeds are known as destructive parasites on many agricultural crops in the Mediterranean region, Eastern Europe and North Africa [1,2]. *Orobanche crenata* is the most dangerous and the most widespread *Orobanche* species in the Mediterranean region and Western Asia. It is a major constraint for faba beans, field peas, lentils, vetches and various forage legumes [3]. Unfortunately, several strategies have been employed to control broomrape with little success [4,5]. The potential of natural enemies as biological control agents has received more attention in recent years. The main impact of the biocontrol agent is the reduction of the seed germination [6,7], germ tube-host attachment or seed production, resulting in the prevention of supplementary infestation and seed dissemination, and leading to a reduction of the seed bank in the soil [8]. Germination of *Orobanche* spp. is stimulated by root exudates from crop hosts, but in the absence of a host, seeds can remain viable for 10 years or more [9], making it difficult for any crop rotation to be efficient. Toxins from soil-borne pathogens that inhibit *Orobanche* seed germination could prevent attachment of the parasite. The most common soil-borne pathogens isolated from diseased *Orobanche* plants belong to the genus *Fusarium* [10].

Information concerning rhizobacteria antagonistic to *Orobanche* is very few. Zermane et al. [11] identified some *Pseudomonas* and *Ralstonia* strains as natural antagonists of *Orobanche*. Mabrouk et al. [12] showed that symbiosis with some non-pathogenic *Rhizobium leguminosarum* strains could induce both better development and lower susceptibility in pea to *O. crenata*. Some rhizobacteria referred to as PGPR (plant growth promoting rhizobacteria) or PHPR (plant health-promoting rhizobacteria) have the ability to improve plant growth, and/or root health [13]. Induced resistance in the nodulated pea was characterized by low activity of the root exudates in triggering *Orobanche* seed germination, and by the induction of necrosis of most of the *Orobanche* seedlings before and after attachment to host roots [12].

Organic soil amendments, especially composts, can provide a rich source of plant disease suppressive microorganisms and large populations and high diversity of microorganisms with biological control potential [14]. Many microfloral species can be released from the compost, such as *Bacillus* spp., *Enterobacter* spp., *Flavobacterium balustinum* and *Pseudomonas* spp., and fungi such as *Penicillium* spp., *Gliocladium virens* and several *Trichoderma* spp. that act as biocontrol

*Corresponding author: Yasser El-Halmouch, Biotechnology Department, Faculty of Science, Taif University, Taif 21974, P.O. Box 888, Kingdom of Saudi Arabia
E-mail: halmouch@yahoo.com

agents [15]. These beneficial microorganisms can provide nutrients that stimulate the proliferation of antagonistic bacteria and fungi in the rhizosphere [16,17]. Due to the scarce of information about the control of *Orobanche* by free living and symbiotic nitrogen fixing bacteria, this study aimed to evaluate the potential of individual and mixture of *R. leguminosarum* isolates, *A. chroococcum* and compost tea to control faba bean infection by *O. crenata*.

Materials and Methods

Source of bacterial isolates and herbicide solutions

All *Rhizobium leguminosarum* isolates (302, 312, 317 and 313), *Azotobacter chroococcum* and compost tea were obtained from the Micrbiology Department, National Research Centre, Sakha, Egypt.

Microbial isolates were cultured in Erlenmeyer flasks (250 mL), each containing 150 mL of selective liquid culture medium; Yeast Manitol extract for *R. leguminosarum* isolates, which consists of the following components in g/L: Yeast extract 1, Mannitol 10, Dipotassium phosphate 0.5, Magnesium sulphate 0.2, Sodium chloride 0.1, Calcium carbonate 1. Basic nitrogen Burk's medium was also used for *Azotobacter* that consists of the following components in g/L: Magnesium sulphate 0.2, Dipotassium phosphate 0.8, Monopotassium phosphate 0.2, Calcium sulphate 0.13, Ferric chloride 0.001, Sodium molybdate 0.001 and glucose 20 as carbon source. Flasks were inoculated with the growth appeared on one slant, previously grown at 25°C for 24-48 hrs. Inoculated flasks were incubated at 25 ± 2°C in a rotary shaker at 160 rpm for 24 hrs. Six ml were transferred to 150 ml liquid culture and incubated at 25 ± 2°C for 3 days at 160 rpm on a rotary shaker. Bacterial growth was separated by centrifugation at 4000×g for 20 min, and then the crude cell-free fermentation culture was used as a soil drench throughout the growing season in the green house.

Pot experiment

Pot experiments were performed in 2010 and 2011 in greenhouse of National Research Centre, Sakha, Egypt. *O. crenata* seeds (4 mg kg^{-1} of soil; about 1300 seeds) were mixed with a 1:1:1 peat–sand–clay mixture in a 3 L pot. Five infested pots were prepared for each microbial treatment, watered and protected from the light for 1 week at 25°C to obtain preconditioned broomrape seeds; then three seeds of *V. faba* Misr1 and *V. faba* Giza 843 cultivars were sown separately into each pot. Each pot recieved 250 ml of crude microbial cell-free culture. Two weeks after faba bean emergence, seedlings were thinned to one per pot. Broomrape-free pots were included as control. The two cultivars were grown under greenhouse conditions at 25°C with 300 µmol m^{-2}s^{-1} PAR and a 16 h photoperiod. Twelve weeks after sowing, faba bean roots and broomrape attachments were uprooted and observed under a binocular microscope. Broomrape attachments on host roots were counted and the dry weight/plant of the entire parasite was recorded. In addition, some other parameters like dry weight of broomrape, number and weight of of bacterial nodules, total nitrogen and chlorophyll content of faba bean plants (*V. faba* Misr1 and *V. faba* Giza 843) were recorded.

Biochemical analysis

Total nitrogen content: 0.3 gm of finely ground oven dried plant samples were digested by 2 ml of concentrated sulfuric acid (H$_2$SO$_4$), then potassium sulphate, 8 g; copper sulphate, 1 g and mercuric oxide, 1 g were added as catalyst. Digestion was done using Micro-Kjeldahl technique with continuous heating until the mixture turned colorless. The final volume was made up to 25 ml with distilled water in a volumetric flask. 5 ml of this volume was analyzed by steam distillation

in the presence of NaOH (40%) and ammonia were in boric acid (10%), plus three drops of indicator containing; 6 ml methyl red (0.16% in 95% ethyl alcohol); 12 ml bromocresol green (0.04% in water) and 6 ml of 95% alcohol. Titration was made with 0.01 M HCl, until the indicator turns from green to pink. Then, total nitrogen was calculated as mg in the sample [18].

Chlorophyll content: Extracts of plant material were obtained from green leaves of faba bean by direct immersion of faba bean sample (500 mg) in N,N-dimethylformamide (DMF), as recommended by Moran and Porath [19]. Immersed sample was mixed in a Sorvall Omni-Mix for 3 min and then centrifuged at 27×10^3 Xg for 10 min. The extract stored in the dark for one to two days at 4°C, prior to spectroscopic examination. Spectrophotometric measurements were made by means of Varian Techtron model 635 UV-VIS scanning spectrophotometer, calibrated at 703 nm, using the 0.2 nm band width measuring beam and 1 ml cuvette having a path length of 1 cm [20].

Root exudates collection and *in vitro* germination of *O. crenata*

Two seeds of each faba bean cultivar were sown and grown in a pot filled with sand. Root exudates were collected using the double pot technique, as described by Parker et al. [21]. An aliquot of 100 ml of root exudates was collected from each pot and sterilized by filtration through 0.22 µm bacterial filters before storage for several weeks at -20°C. Pre-conditioned surface-sterilized broomrape seeds were transferred to small Petri dishes (3.5 cm), containing mixture of 250 µl root exudates plus 250 µl of microbial culture-filtrate or compost tea filtrate, and then incubated at 25°C for one week. Viability and germination tests were performed according to Linke [22]. Germination was determined after observation of radical emergence under a binocular microscope.

Statistical analysis

All results were subjected to one-way ANOVA and the means were compared according to the Student–Newman–Keuls (SNK) multiple range test (P ≤ 0.05).

Results

In the two pot experiments, many parameters were used to evaluate the resistance of faba bean cultivars to *O. crenata*; two specific for the parasite (number of broomrape attachments and dry weight), and six specific for the host (faba bean dry weight, bacterial nodules number and dry weight, total nitrogen and chlorophyll a and b content). Although none of the tested microbial treatment was absolutely inhibitory to the parasite, remarkable variations were observed in faba bean response to broomrape, according to the treatment.

First pot experiment

Effect of the microbial treatment on parasite: In the first pot experiment done in 2010, significant variations among the treatment were obtained in broomrape number (Figure 1). Resistant cultivar Misr 1 always exhibited lower number of broomrape tubercles than the susceptible cultivar Giza 843. Microbial treatment and compost tea reduced broomrape attachments fixed in both faba bean cultivars.

In Misr 1, a significant reduction in total number of broomrape tubercles was recorded by all microbial treatments and compost tea, except *Azotobacter chroococcum*. About 86.5% of reduction in broomrape attachments was observed using compost tea as a biocontrol agent (Figure 1). Among the bacterial treatments, the mixture of *R. leguminosarum* 313 and 317 isolates exhibited the inhibitoriest effect

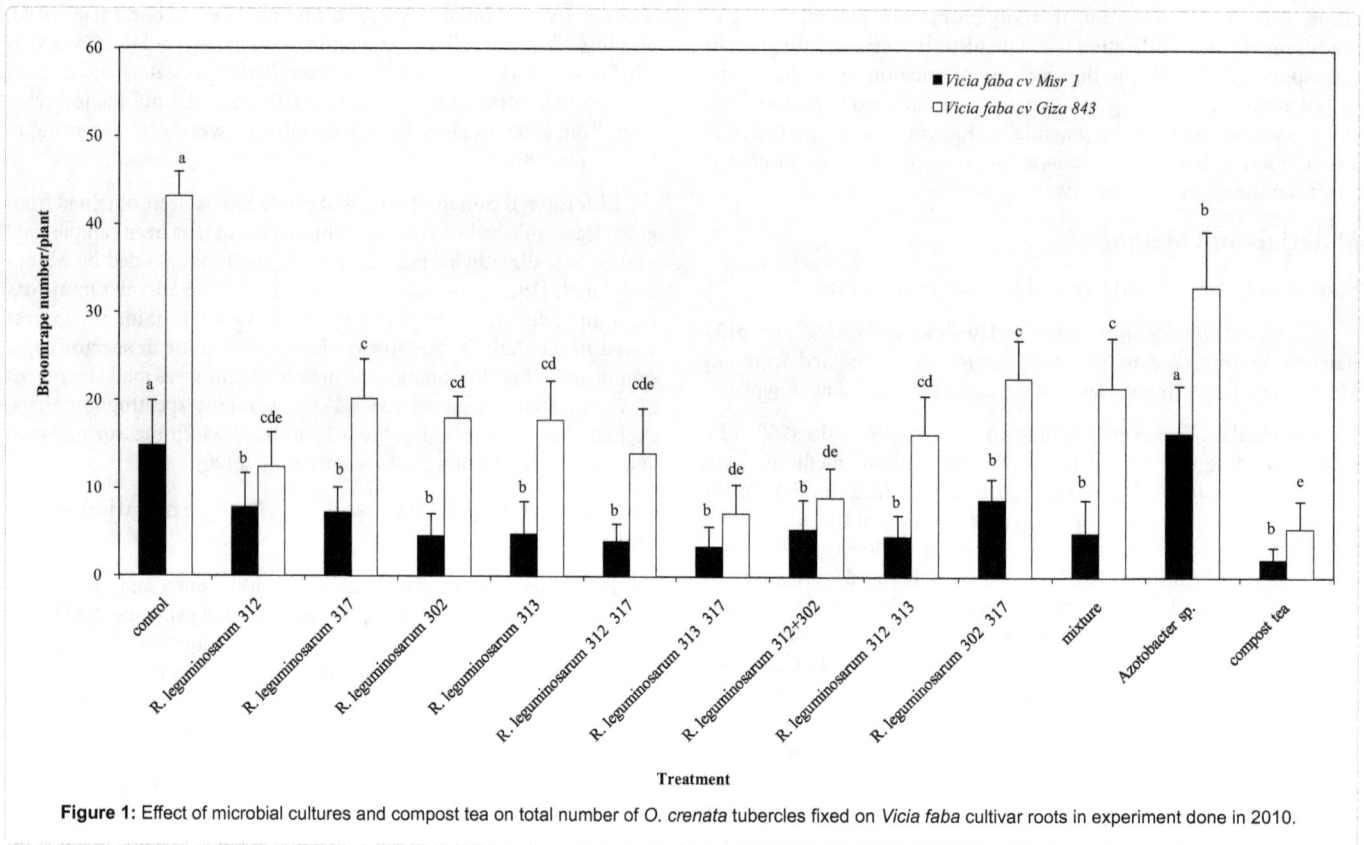

Figure 1: Effect of microbial cultures and compost tea on total number of *O. crenata* tubercles fixed on *Vicia faba* cultivar roots in experiment done in 2010.

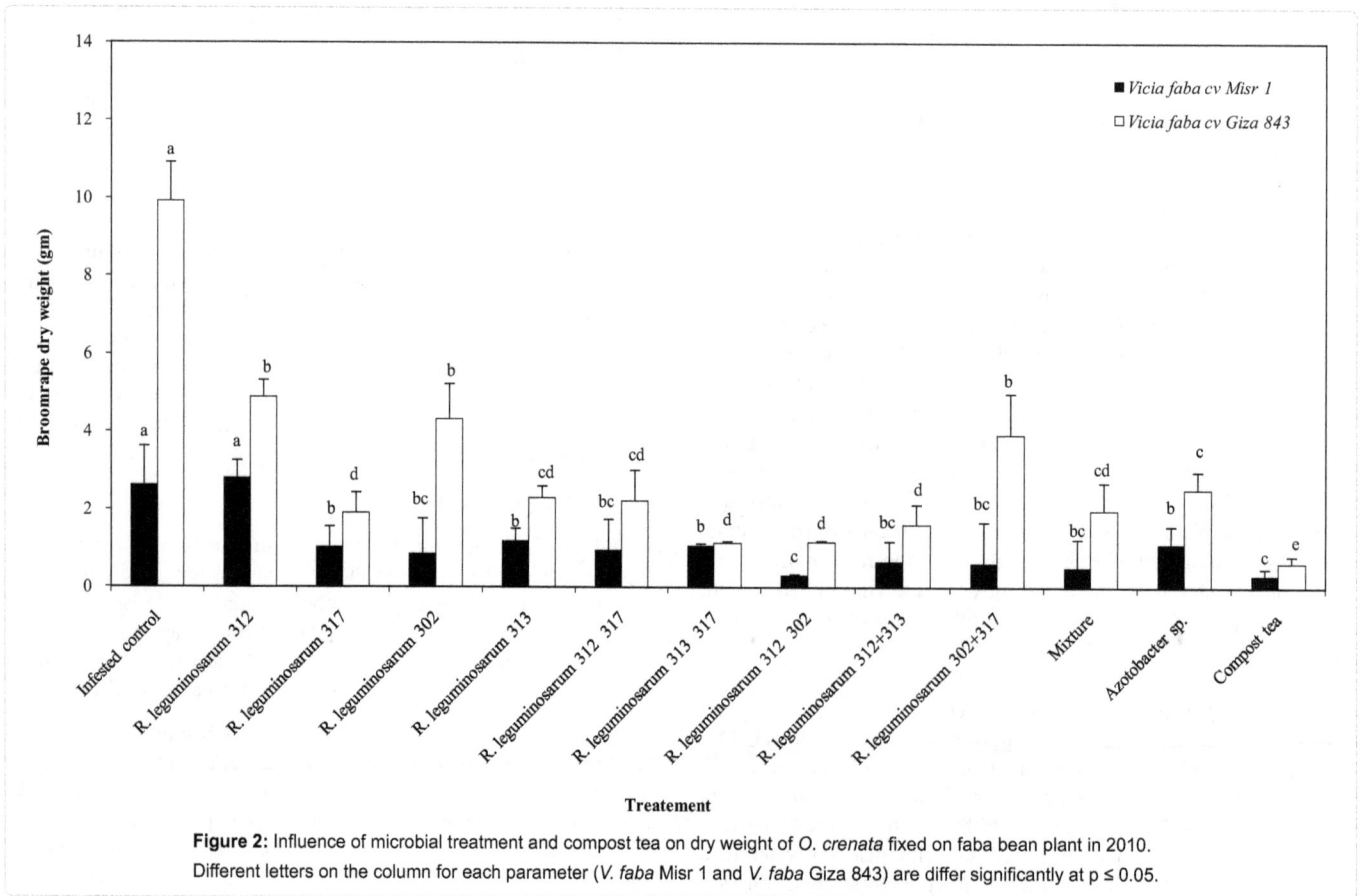

Figure 2: Influence of microbial treatment and compost tea on dry weight of *O. crenata* fixed on faba bean plant in 2010.
Different letters on the column for each parameter (*V. faba* Misr 1 and *V. faba* Giza 843) are differ significantly at p ≤ 0.05.

of broomrape attacked Misr 1. However, more than 77% reduction in broomrape tubercles was recorded. Broomrape tubercles fixed on Giza 843 varied significantly and ranged between 5.6-33 per faba bean plant. Compost tea produced about 87% reduction in number of broomrape tubercles and considered the most effective biocontrol agent.

Dry weight of broomrape treated with the microbial strains and compost tea was significantly lower than those of non-treated control (Figure 2). Most of the mixed microbial treatments were more effective than individual treatments. The treatments with *R. leguminosarum* 312+302 and compost tea were inhibitorier to broomrape growth on Misr 1 roots than individual treatment. Compost tea was the best inhibitory, that caused 88% and 94% dry weight reduction in broomrape fixed on Misr 1 and Giza 843, respectively (Figure 2).

Effect of the microbial treatment on hosts: All faba bean growth and metabolic parameters varied significantly by microbial and compost tea treatments. In *V. faba* Misr 1, utilization of compost tea as a biocontrol agent significantly increased the shoot dry weight, total nitrogen and chlorophyll b content by 97%, 179% and 94%, respectively, compared to the infested control (Table 1). The mixture of *R. leguminosarum* isolates increased nodule number and their dry weight by 5.3 and 3.3 times, respectively, over the infested control. Chlorophyll a was increased two times by dual culture filtrates of isolates 312 and 317, compared to the infested control.

All the tested growth parameters of *V. faba* Giza 843 varied significantly in all microbial and compost tea treatments, in comparison with the infested control. An increase in shoot dry weight, total nitrogen and chlorophyll b content was recorded by 3, 2 and 1.6 times, over the infested control, respectively, in case of compost tea treatment. The mixture of all *R. leguminosarum* isolates enhanced both nodule number and dry weight by eight and two folds, respectively.

Second pot experiment

Effect of the microbial treatment on parasite: In the second experiment done in 2011, expression of resistance in faba bean by number of broomrape tubercle showed a considerable variation among microbial and compost tea treatments for the two cultivars (Misr1 and Giza 843), that ranged between 2.6 to 18.2 for Misr 1 and from 15.4 to 48.8 for Giza 843.

In Misr 1, dual inhibitory effect of isolates 312+317 and 312+313, and compost tea significantly varied compared to other treatments on broomrape tubercles number (Figure 3). The highest percentage of reduction was obtained by the treatment with *R. leguminosarum* 312+313 and compost tea (80% and 61% respectively). No reduction was recorded by the treatment with *R. leguminosarum* 313+317, *R. leguminosarum* 302+317 and Azotobacter. In contrast, these treatments induced the broomrape attachment. However, broomrape tubercles fixed on Giza 843 roots varied significantly according to the treatment (Figure 3). The dual treatment with 312 and 302 isolates induced broomrape attachment, compared with the control. *R. leguminosarum* 312, *R. leguminosarum* 317, dual isolates of 302 and 317 and compost tea were the most inhibitory for broomrape attachment. They gave about 51%, 52% and 60% reduction, respectively. Cultivar resistance to *Orobanche* based on broomrape dry weight varied significantly among the treatments. About 93% and 96% reduction in broomrape dry weight were obtained by compost tea in Misr 1 and Giza 843, respectively (Figure 4).

Effect of the microbial treatment on hosts: All faba bean growth parameters (shoot dry weight, nodule number, nodule dry weight, total nitrogen and chlorophyll a contents) for the two cultivars (Misr 1 and Giza 843) varied significantly by the treatments, except chlorophyll b content in *V. faba* Giza 843. In *V. faba* Misr 1, a highly significant reduction in all parameters was obtained in infested non-treated control, compared with non-infested non-treated control. In contrast, significant increases in shoot dry weight, nitrogen content, chlorophyll a and chlorophyll b contents, were recorded by 81.2%, 165%, 94.1% and 86.4%. respectively, using compost tea (Table 2). Moreover, nodule number and dry weight were significantly higher due to the treatment with the mixture of all *R. leguminosarum* isolates than the infested control (57.57 and 460 gm/plant, respectively). They increased by 4.49 and 3.41 times over than those of infested non-treated control. More than 100% of chlorophyll b content was induced in Misr 1 using *R. leguminosarum* 312+317.

For cultivar Giza 843, compost tea increased shoot dry weight and nitrogen content by 279% and 143%, respectively, compared with the infested control. High increases of nodule number, nodule dry weight, nitrogen and chlorophyll a contents were observed due to treatment

Treatments	Shoot dry weight (gm/plant)		Nodule number/ plant		Nodule dry weight (mg/plant)		Total nitrogen content (mg/plant)		Chlorophyll a content (mg/cm²)		Chlorophyll a content (mg/cm²)	
	V. faba Misr1	*V. faba* Giza 843	*V. faba* Misr1	*V. faba* Giza 843	*V. faba* Misr1	*V. faba* Giza 843	*V. faba* Misr1	*V. faba* Giza 843	*V. faba* Misr1	*V. faba* Giza 843	*V. faba* Misr1	*V. faba* Giza 843
Non-infested control	6.51e	8.28e	18.03g	17.51e	165d	138g	51.58h	56.88h	0.31e	0.27d	0.27b	0.17d
Infested control	5.89e	4.36f	9.82h	5.31f	118e	238ef	36.98g	47.74j	0.22f	0.18g	0.17e	0.15d
R. leguminosarum 312	6.74e	10.57d	27.52e	26.33d	255c	273e	54.52e	76.72e	0.44b	0.21f	0.29ab	0.16d
R. leguminosarum 317	9.12d	9.45d	31.06de	25.51d	245c	225f	71.45d	72.23g	0.38bc	0.24e	0.23c	0.16d
R. leguminosarum 302	7.20e	9.67d	32.82cd	26.02d	333b	248ef	84.89b	91.15b	0.41b	0.23ef	0.29ab	0.16d
R. leguminosarum 313	8.37d	10.79d	29.53de	25.05d	338b	235ef	70.19d	65.07i	0.31e	0.22ef	0.22d	0.17d
R. leguminosarum 312+317	10.15b	11.62c	35.05c	32.01bc	333b	418b	88.92b	71.69g	0.46a	0.29c	0.29ab	0.20bc
R. leguminosarum 313+317	9.14d	12.33b	41.51b	30.04c	353b	375c	86.18b	71.13g	0.35cde	0.27d	0.27b	0.18cd
R. leguminosarum 312+302	8.65d	12.68b	42.04b	34.04b	323b	360c	86.19b	74.08f	0.33de	0.29c	0.21cd	0.21b
R. leguminosarum 312+313	8.77d	12.40b	41.51b	33.83b	338b	320d	87.39b	77.99e	0.35cde	0.30b	0.21cd	0.25a
R. leguminosarum 302+317	8.84d	12.43b	43.06b	31.54bc	320b	320d	75.75c	80.69d	0.34cde	0.26d	0.20d	0.23ab
Mixture	10.53b	13.41b	52.09a	43.57a	390a	468a	91.55b	83.56c	0.41b	0.35a	0.26b	0.24a
Azotobacter chroococcum	9.66c	12.65b	23.11f	19.51e	225c	248f	70.05d	70.72g	0.36bcd	0.28cd	0.22cd	0.24a
Compost tea	11.58a	14.44a	28.51d	25.04d	328b	315d	103.13a	96.59a	0.41b	0.24e	0.33a	0.24a

Different letters on the column for each parameter (V. faba Misr 1 & V. faba Giza 843) are differ significantly at p≤ 0.05.

Table 1: Influence of different microbial treatments and compost tea on shoot dry weight, nodule number, nodule dry weight, total nitrogen content and chlorophyll a and chlorophyll b contents in faba bean in experiment done in 2010.

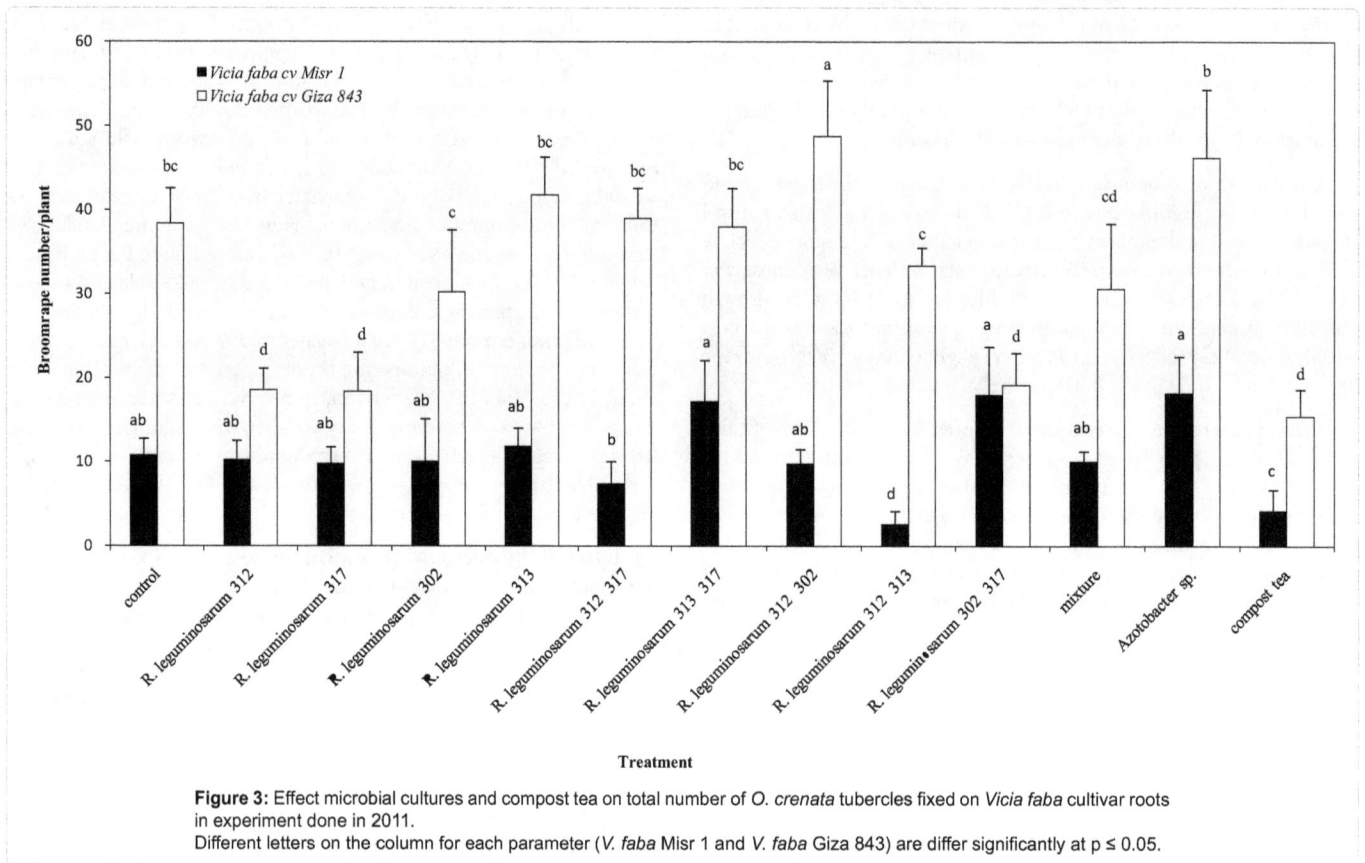

Figure 3: Effect microbial cultures and compost tea on total number of *O. crenata* tubercles fixed on *Vicia faba* cultivar roots in experiment done in 2011.
Different letters on the column for each parameter (*V. faba* Misr 1 and *V. faba* Giza 843) are differ significantly at p ≤ 0.05.

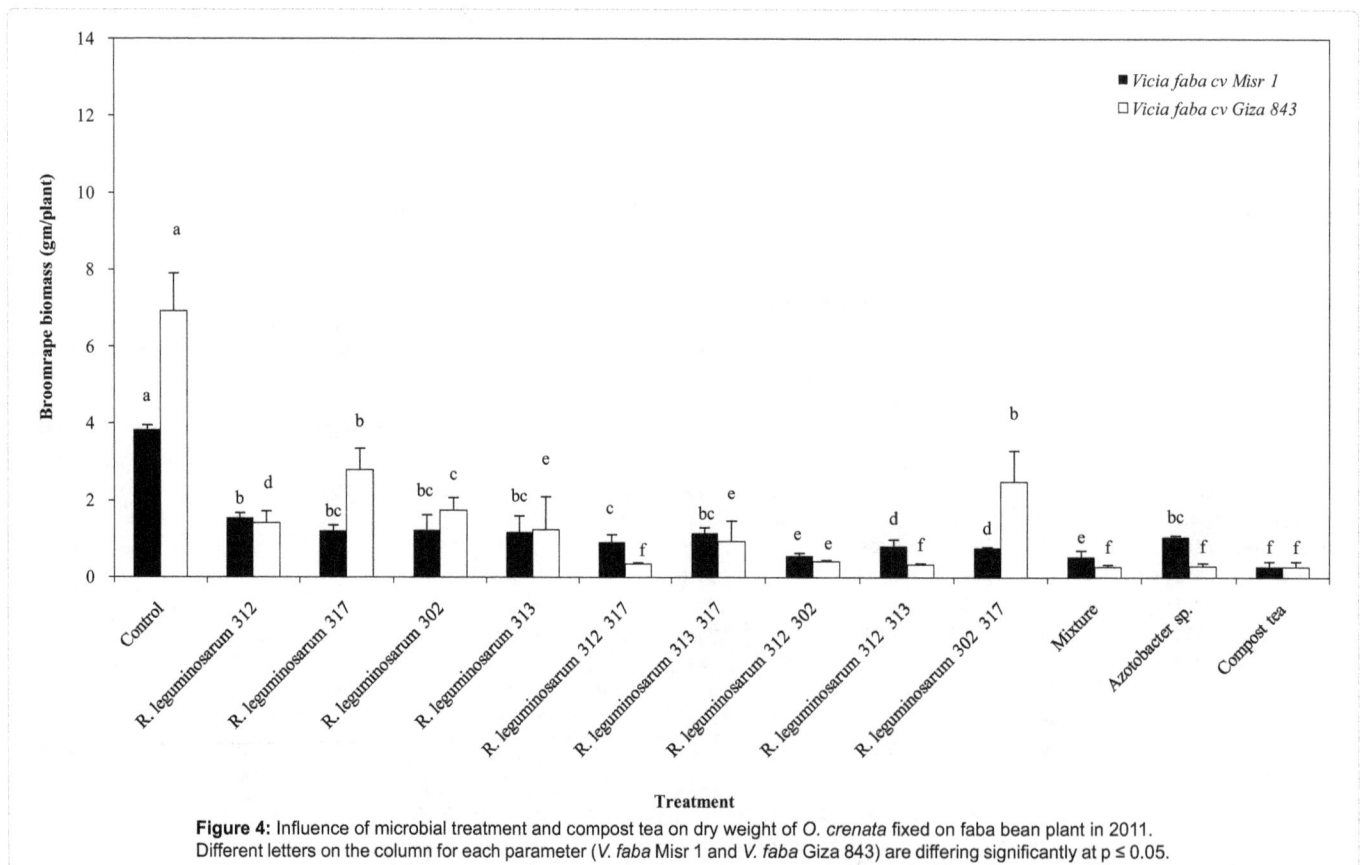

Figure 4: Influence of microbial treatment and compost tea on dry weight of *O. crenata* fixed on faba bean plant in 2011.
Different letters on the column for each parameter (*V. faba* Misr 1 and *V. faba* Giza 843) are differing significantly at p ≤ 0.05.

Treatments	Shoot dry weight (gm/plant)		Nodule number/ plant		Nodule dry weight (mg/plant)		Total nitrogen content (mg/plant)		Chlorophyll a content (mg/cm²)		Chlorophyll b content (mg/cm²)	
	V. faba Misr1	V. faba Giza 843	V. faba Misr1	V. faba Giza 843	V. faba Misr1	V. faba Giza 843	V. faba Misr1	V. faba Giza 843	V. faba Misr1	V. faba Giza 843	V. faba Misr1	V. faba Giza 843
Non-infested control	9.13^{d}	8.14^{g}	19.33^{e}	17.83^{d}	213^{d}	150^{g}	47.81^{h}	51.61^{g}	0.26^{b}	0.28^{b}	0.31^{c}	0.17^{a}
Infested control	7.27^{e}	4.04^{h}	12.82^{f}	8.25^{e}	135^{e}	243^{ef}	38.41^{i}	40.82^{h}	0.17^{e}	0.19^{c}	0.22^{d}	0.16^{a}
R. leguminosarum 312	8.23^{d}	9.38^{f}	27.31^{d}	27.52^{c}	258^{cd}	255^{ef}	61.30^{g}	73.23^{f}	0.29^{ab}	0.23^{b}	0.44^{ab}	0.17^{a}
R. leguminosarum 317	8.45^{e}	11.12^{d}	32.33^{c}	27.33^{c}	268^{c}	225^{f}	82.33^{de}	76.84^{e}	0.23^{c}	0.26^{b}	0.38^{ab}	0.18^{a}
R. leguminosarum 302	7.21^{e}	10.12^{d}	36.11^{c}	27.03^{c}	250^{cd}	235^{f}	85.61^{f}	72.52^{f}	0.35^{ab}	0.23^{b}	0.43^{ab}	0.19^{a}
R. leguminosarum 313	7.43^{e}	11.23^{d}	32.51^{c}	25.52^{c}	253^{cd}	275^{ef}	70.80^{f}	76.42^{e}	0.22^{cd}	0.24^{b}	0.13^{c}	0.18^{a}
R. leguminosarum 312+317	10.33^{c}	11.93^{c}	26.51^{e}	35.53^{b}	340^{b}	393^{b}	88.91^{c}	86.11^{b}	0.29^{ab}	0.32^{b}	0.46^{a}	0.21^{a}
R. leguminosarum 313+317	9.35^{c}	10.73^{d}	42.52^{b}	34.04^{b}	390^{b}	338^{cd}	86.81^{d}	81.11^{d}	0.27^{b}	0.33^{b}	0.35^{b}	0.21^{a}
R. leguminosarum 312+302	8.33^{d}	10.81^{d}	43.81^{b}	36.06^{b}	363^{b}	382^{bc}	84.72^{de}	82.51^{c}	0.21^{cd}	0.31^{b}	0.33^{c}	0.22^{a}
R. leguminosarum 312+313	8.89^{d}	9.61^{ef}	46.33^{b}	34.30^{b}	360^{b}	320^{d}	85.42^{de}	83.23^{c}	0.21^{cd}	0.29^{b}	0.35^{b}	0.24^{a}
R. leguminosarum 302+317	9.03^{d}	11.52^{d}	42.11^{b}	33.16^{b}	343^{b}	335^{cd}	77.81^{e}	89.63^{b}	0.23^{d}	0.27^{b}	0.34^{b}	0.24^{a}
Mixture	10.5^{b}	13.21^{b}	57.57^{a}	47.55^{a}	460^{a}	443^{a}	93.52^{b}	96.93^{a}	0.26^{b}	0.35^{a}	0.41^{ab}	0.24^{a}
Azotobacter chroococcum	9.66^{d}	10.33^{d}	22.82^{e}	19.35^{d}	248^{cd}	245^{ef}	71.23^{f}	75.13^{e}	0.22^{cd}	0.28^{b}	0.36^{b}	0.24^{a}
Compost tea	13.17^{a}	15.23^{a}	34.33^{c}	23.53^{c}	280^{c}	298^{de}	101.71^{a}	99.53^{a}	0.33^{a}	0.27^{b}	0.41^{ab}	0.24^{a}

Different letters on the column for each parameter (*V. faba* Misr 1 and *V. faba* Giza 843) are differ significantly at $p \le 0.05$.

Table 2: Influence of different microbial treatments and compost tea on shoot dry weight, nodule number, nodule dry weight, total nitrogen content, chlorophyll a and chlorophyll b contents in faba bean in experiment done in 2011.

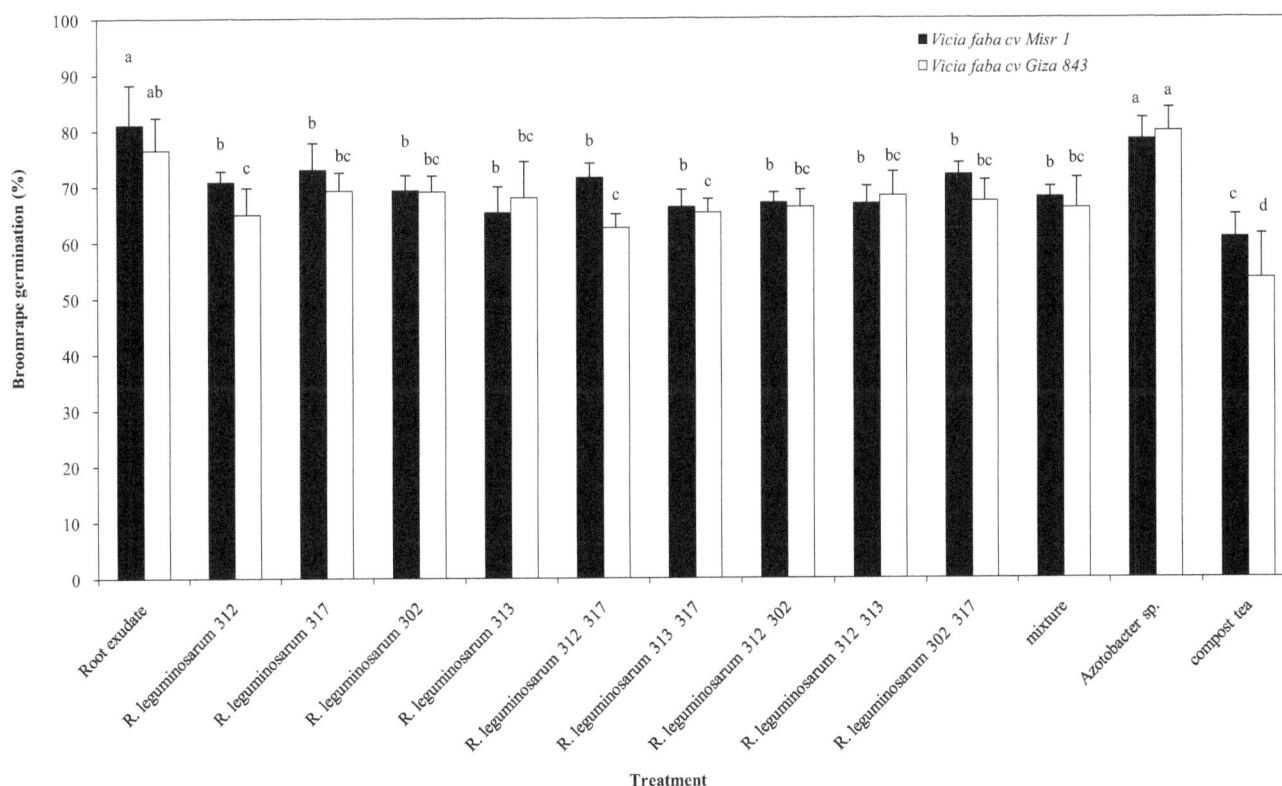

Figure 5: Influence of microbial culture filterate and compost tea on broomrape germination in presence of Faba bean root exudates. Different letters on the column for each parameter (*V. faba* Misr 1 & *V. faba* Giza 843) are differ significantly at $p \le 0.05$.

with the mixture of *R. leguminosarum* isolates, compared with the infested control. The highest chlorophyll a content (0.46 mg/cm²) was obtained using *R. leguminosarum* 312+317. On the other hand, chlorophyll b content varied insignificantly among all treatments and the controls. Germination of *O. crenata* seeds was strongly stimulated by faba bean root exudates. About 81% and 76.5% were obtained by root exudates of Misr 1 and Giza 843, respectively (Figure 5).

In vitro assay of microbial isolates on the broomrape germination: To test the inhibitory effects of different treatments on broomrape germination, broomrape seeds were mixed separately with root exudates (1:1) and their effects were analyzed. Broomrape seed germination percentage reached more than 75% by the two tested root exudates. All microbial culture filtrates significantly decreased broomrape germination, except those of *A. chroococcum*, compared to

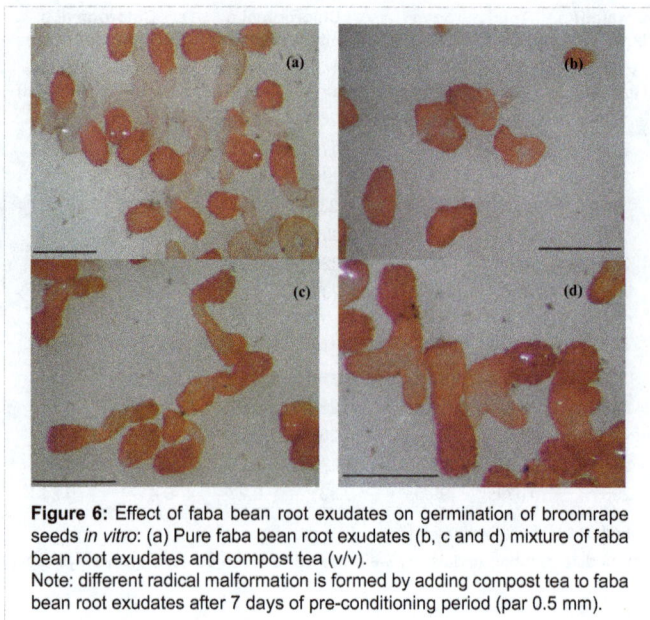

Figure 6: Effect of faba bean root exudates on germination of broomrape seeds *in vitro*: (a) Pure faba bean root exudates (b, c and d) mixture of faba bean root exudates and compost tea (v/v).
Note: different radical malformation is formed by adding compost tea to faba bean root exudates after 7 days of pre-conditioning period (par 0.5 mm).

the treatment with root exudates of Misr 1 alone (as a control). The lowest percentage of germination was obtained by compost tea.

In the presence of Giza 843 exudates, relatively lower percentage of germination of the parasite was observed for all microbial culture filtrates, except those of *Azotobacter* sp. Culture filtrate of *Azotobacter* sp. was slightly effective in triggering broomrape germination. Significantly, the lowest percentages of broomrape germination (60.5%) were observed for compost tea.

On the other hand, microscopic examination of germinated broomrape seeds treated by the compost monitored certain malformations in the form of germ tubes (Figure 6).

Discussion

Non-pathogenic rhizobacteria can induce a systemic resistance in plants against many pathogens like fungi, bacteria and viruses. The present study confirm that using rhizobacteria filtrates induced resistance of faba bean against root-parasitic weed, as shown by the significant decrease in the number of broomrape attachments, dry weight of the attached tubercles on faba bean roots and the reduction in percentage of broomrape seed germination. *R. leguminosarum* play a very important role in agriculture by inducing the number and activity of nitrogen-fixing nodules on the roots of legumes. This symbiosis can reduce the requirements for added nitrogenous fertilizer during the growth of leguminous crops [23,24]. This may give a power for faba bean to de tolerate the infectivity of broomrape. The defense elicited by the tested isolates of *R. leguminosarum* and compost tea is strongly efficient. Our results are in agreement with Mabrouk et al. [12], who showed that symbiosis with some non-pathogenic *R. leguminosarum* strains could induce in pea both better development and lower susceptibility to *O. crenata*. Pseudomonas spp. is typical rhizosphere bacteria with high root-colonizing abilities [25]. These species have been identified as the major group of rhizobacteria with potential as effective candidates for biological control of weeds [26]. In this connection, Zermane et al. [27] showed high biocontrol activity against both *O. crenata* and *O. foetida* species and positively influenced faba bean growth by five isolates of *P. fluorescens*. The nitrogen-fixing bacterium *Azospirillum brasilense*

has been reported to inhibit germination and radical growth of *O. aegyptiaca* [28] and *Striga hermonthica* [29].

The statistical difference in the total number of *O. crenata* tubercles recorded in 2010 (Figure 1) and 2011 (Figure 3) may be due to some of individual genetic variations in the host, which lead to variation in many metabolic processes. This assumption was confirmed by the studies of Appels and Dvorak [30] and Saghai-Maroof et al. [31], who found clear variations in the ribosomal classes among individuals in wheat and barley, respectively.

Orobanche requires the presence of host root-exuded germination stimulants in order to germinate and orientate its radical to the host root [32,33]. In this study, minimization the impact of stimulant producing plants (faba been) by microbial culture filtrate or compost tea solution could be a suitable option for reducing *O. crenata* infection. In this context, our investigation showed a reduction in broomrape seed germination by *R. leguminosarum* isolates. These results are in agreement with Mabrouk et al. [34], who stated that the germination of *O. crenata* seeds *in vitro* decreased significantly after inoculation with P.SOM and P.1236 Rhizobium strains. Similarly, bacteria isolates obtained from the soil reduced *Striga* seeds germination significantly [29,35]. However, there are other reports of effective inhibition of broomrape germination by microorganisms without active destruction of the seed. For example, purified toxins produced by Fusarium species significantly reduced *O. ramosa* germination [36,37]. Combination of *P. fluorescens* and *P. putida* recovered from suppressive soils in sorghum fields in Nigeria inhibited the seed germination of *Striga hermonthica* [38]. Composts can play a promising role in the biological control of plant diseases. Generally, they provide the slow release of adequate quantities of bio-available energy sustaining the introduced biocontrol agents. Composts affect the release of nutrients to plants directly through their nutrients content or indirectly by their effect on the cation-exchange capacity [39]. Composts may offer further advantages over pesticides and herbicides, in addition to their nutrient contents. They introduce many biological control agents (BCAs) to the soil and establish themselves in the ecosystem.

There are many reports that demonstrate the ability of compost tea to suppress a wide range of both air- and soil-borne plant pathogens [40,41]. Direct application of compost tea on tomato plants significantly reduced disease symptoms caused by three tomato pathogens: *Alternaria alternata*, *Botrytis cinerea* and *Pyrenochaeta lycopersici* [42]. Bharathi et al. [43] reported that PGPR-microorganisms involving in compost tea, can induce the systemic resistance in the plants against different pathogens. In the present study, in addition to the reduction in broomrape seed germination by compost tea, abnormal germ tubes were formed in the presence of natural stimulants. In accordance with our finding, Miche´ et al. [29] reported that abnormal germ tubes of *Striga* were formed by some bacterial isolates obtained from sorghum soil fields in the presence of the synthetic germination stimulant (GR24). Our work suggested that the reduction in broomrape incidence by compost tea may be due to certain phenotypic mechanisms which acted alone or in combination, including negative effect of natural stimulant broomrape seed germination, radical deformation which hinders the ability to penetrate faba bean root. All these deleterious effects on *O. crenata* increased growth and vitality of faba bean. These are in agreement with other workers who stated that the compost tea enhance crop fertility by introducing microorganisms that might aid in soil nutrient retention and extraction, and by adding soluble nutrients, further adding to their potential value as a part of an integrated crop management plan [40,44-46]. In conclusion, the present study has shown the potential use of the mixture of *R. leguminosarum* isolates and compost tea to suppress *O. crenata* in area cultivated with *V. faba* Further detailed investigations are needed to better understand the mechanisms of microbial induction to the resistance in faba bean plants to broomrape.

Acknowledgements

The authors thank Dr. W. Al Rodeny for providing faba bean seeds and T. El-Sakhawy for greenhouse culture management.

References

1. Parker C, Riches C (1993) Parasitic weeds of the world: Biology and control. CAB International, Wallingford, UK.

2. Sauerborn J (1991) The economic importance of the *phytoparasites Orobanche* and *Striga*. In: Ransom JK (Eds.), Proceedings of the 5th International Symposium of Parasitic Weeds. Nairobi, Kenya 137-143.

3. Rubiales D, Ptrez-de-Luque A, Cubero JI, Sillero JC (2003) Crenate broomrape (*Orobanche crenata*) infection in field pea cultivars. Crop Prot 22: 865-872.

4. Joel DM, Hershenhorn Y, Eizenberg H, Aly R, Ejeta G, et al. (2007) Biology and management of weedy root parasites. In: Janick J (Ed.), Horticultural Reviews, John Wiley & Sons, Hoboken, NJ, USA 33: 267-349.

5. Pérez-de-Luque A, Fondevilla S, Pérez-Vich B, Aly R, Thoiron S, et al. (2009) Understanding *Orobanche* and *Phelipanche*–host plant interaction and developing resistance. Weed Res 49: 8-22.

6. El-Halmouch Y, Benharrat H, Thalouarn P (2006) Effect of root exudates from different tomato genotypes on broomrape (*O. aegyptiaca*) seed germination and tubercle development. Crop Prot 25: 501-507.

7. Müller-stöver D, Kohlschmid E, Sauerborn J (2009) A novel strain of *Fusarium oxysporum* from Germany and its potential for biocontrol of *Orobanche ramosa*. Weed Res 49: 175-182.

8. Thomas H, Heller A, Sauerborn J, Müller-Stöver D (1999) *Fusarium oxysporum* f. sp. orthoceras, a potential mycoherbicide, parasitizes seeds of *Orobanche cumana* (sun-flower broomrape): A cytological study. Ann Bot 83: 453-458.

9. Cubero JI, Moreno MT (1979) Agronomical control and sources of resistance in *Vicia faba* to *O. crenata*. In: Bond DA, Scarascia-Mugnozza GT, Poulsen MH (Eds.), Some current research on *Vicia faba* in Western Europe, Commission of the European Communities, Brussels, Belgium 41-80.

10. Amsellem Z, Barghouthi S, Cohen B, Goldwasser Y, Gressel J, et al. (2001) Recent advances in the biocontrol of *Orobanche* (broomrape) species. Biocontrol 46: 211-228.

11. Zermane N, Kroschel J, Souissi T (2004) Options for biological control of the parasitic weed Orobanche in North Africa. In: Peters KJ, Kirschke D, Manig W (Eds.), Rural poverty reduction through research for development and transformation: International research on food security, natural resource management and rural development–Book of abstracts, Humboldt-University, Berlin, Germany.

12. Mabrouk Y, Zourgui L, Sifi B, Delavault P, Simier P, et al. (2007). Some compatible *Rhizobium leguminosarum* strains in peas decrease infections when parasitized by *Orobanche crenata*. Weed Res 47: 44-53.

13. Sikora RA (1992) Management of the antagonistic potential in agricultural ecosystems for the biological control of plant parasitic nematodes. Ann Rev Phytopathol 30: 245-270.

14. Alfano G, Lustrato G, Lima G, Vitullo D, Ranalli G (2011) Characterization of composted olive mill wastes to predict potential plant disease suppressiveness. Biol Control 58: 199-207.

15. Litterick AM, Harrier LA, Wallace P, Watson CA, Wood M (2004) The role of uncomposted materials, composts, manures and compost extracts in reducing pest and disease incidence and severity in sustainable temperate agricultural and horticultural crop production–A review. Crit Rev Plant Sci 23: 453-479.

16. Green SJ, Inbar E, Michel FC, Hadar Y, Minz D (2006) Succession of bacterial communities during early plant development: Transition from seed to root and effect of compost amendment. Appl Environmen Microbiol 72: 3975-3983.

17. Noble R, Coventry E (2005) Suppression of soil-borne plant diseases with composts: A review. Biocontrol Sci Technol 15: 3-20.

18. Nelson DW, Sommers LE (1973) Determination of total Nitrogen in plant J Agron 65: 109-112.

19. Moran R, Porath D (1980) Chlorophyll determination in intact tissues using N,N-dimethylformamide. Plant Physiol 65: 478-479.

20. Moran R (1982) Formulae for determination of chlorophyllous pigments extracted with N, N-dimethylformamide. Plant Physiol 69: 1376-1381.

21. Parker C, Hitchcock AM, Ramaiah KV (1977) The germination of Striga species by crop root exudates: Techniques for selecting resistant crop cultivars. In: Proceedings Sixth Asian-Pacific Weed Science Society Conference, Jakarta, Indonesia 67-74.

22. Linke KH (2001) Seed features, germination and seed bank. In: Kroschel J (Ed), A technical manual for parasitic weed research and extension, Kluwer Academic Publishers, Dordrecht, The Netherlands 35-40.

23. Allen EN, Allen EK (1981) The leguminosae, a source book of characteristics. Uses and nodulation, WISK, University of Wisconsin, USA.

24. Zakhia F, De Lajudie P (2001) Taxonomy of rhizobia. Agronomie 21: 569-576.

25. Seenivasan N, Lakshmanan PL (2003) Biocontrol potential of native isolates of *Pseudomonas fluorescens* against rice root nematode. J Ecobiol 15: 69-72.

26. Kremer RJ, Kennedy AC (1996) *Rhizobacteria* as biocontrol agents of weeds. Weed Technol 10: 601-609.

27. Zermane N, Souissi T, Kroschel J, Sikora R (2007) Biocontrol of broomrape (*Orobanche crenata* Forsk.) and *Orobanche foetida* Poir. by *Pseudomonas fluorescens* isolate Bf7-9 from the faba bean rhizosphere. Biocontrol Sci Technol 17: 483-497.

28. Dadon T, Bar Nun N, Mayer AM (2004) A factor from *Azospirillum brasilense* inhibits germination and radicle growth of *Orobanche aegyptiaca*. Israeli J Plant Sci 52: 83-86.

29. Miché L, Bouillant ML, Sallé G, Bally R (2000) Physiological and cytological studies on the inhibition of Stiga seed germination by the plant growth-promoting bacterium *Azospirillum brasilence*. Eur J Plant Pathol 106: 347-351.

30. Appels R, Dvorak J (1982) The wheat ribosomal DNA spacer region: Its structure and variation in populations and among species. Theor Appl Genet 63: 337-348.

31. Saghai-Maroof MA, Soliman KM, Jorgensen RA, Allard RW (1984) Ribosomal DNA spacer-length polymorphisms in barley: Mendelian inheritance, chromosomal location, and population dynamics. Proc Natl Acad Sci 81: 8014-8018.

32. Fernández-Aparicio M, Flores F, Rubiales D (2009) Recognition of root exudates by seeds of broomrape (*Orobanche* and *Phelipanche*) species. Ann Bot 103: 423-431.

33. Whitney PJ, Carstein C (1981) Chemotropic response of broomrape radicals to host roots exudates. Ann Bot 48: 919-921.

34. Mabrouk Y, Zourgui L, Sifi B, Belhadj O (2007) The potential of *Rhizobium* strains for biological control of *Orobanche crenata*. Biologia 62: 139-143.

35. Bouillant ML, Miche L, Ouedraogo O, Alexandre G, Jacoud C, et al. (1997) Inhibition of *Striga* seed germination associated with sorghum growth promotion by soil bacteria. C R Acad Sci III-vie 320: 159-162.

36. Andolfi A, Boari A, Evidente A, Vurro M (2005) Metabolites inhibiting germination of *Orobanche ramose* seeds produced by *Myrothecium verrucaria* and *Fusarium compactum*. J Agric Food Chem 53: 1598-1603.

37. Zonno MC, Vurro M (2002) Inhibition of germination of *Orobanche ramosa* seeds by *Fusarium toxins*. Phytoparasitica 30: 519-524.

38. Ahonsi MO, Berner DK, Emechebe AM, Lagoke ST (2002) Selection of rhizobacterial strains for suppression of germination of *Striga hermonthica* (Del.) Benth. seeds. Biol Control 24: 143-152.

39. Hoitink HJ, Boehm MJ (1999) Biocontrol within the context of soil microbial communities: A substrate-dependent phenomenon. Ann Rev Phytopathol 37: 427-446.

40. Scheuerell SJ, Mahaffee WF (2002) Compost tea: Principles and prospects for plant disease control. Compost Sci Utilization 10: 313-338.

41. Scheuerell SJ, Mahaffee WF (2004) Compost tea as a container medium drench for suppressing seedling damping-off caused by *Pythium ultimum*. Phytopathology 94: 1156-1163.

42. Pane C, Celano G, Villecco D, Zaccardelli M (2012) Control of *Botrytis cinerea*, *Alternaria alternata* and *Pyrenochaeta lycopersici* on tomato with whey compost-tea applications. Crop Prot 38: 80-86.

43. Bharathi R, Vivekananthan R, Harish S, Ramanathan A, Samiyappan R (2004) Rhizobacteria-based bio-formulations for the management of fruit rot infection in chilliest. Crop Prot 23: 835-843.

44. Gea FJ, Santos M, Diánez F, Tello JC, Navarro MJ (2012) Effect of spent mushroom compost tea on mycelial growth and yield of button mushroom (*Agaricus bisporus*). World J Microbiol Biotechnol 28: 2765-2769.

45. Ingham ER (2005) Compost tea: Promises and practicalities. The IPM Practitioner. The newsletter of integrated pest management 27: 1-5.

46. Kannangara T, Forge T, Dang B (2006) Effects of aeration, molasses, kelp, compost type, and carrot juice on the growth of *Escherichia coli* in compost teas. Compost Sci Utilization 14: 40-47.

29

Population Genetic Structure among Iranian Isolates of *Fusarium verticillioides*

Hassan Momeni[1]* and Fahimeh Nazari[2]

[1]Department of Plant Pathology, Iranian Research Institute of Plant Protection (IRIPP), Tehran, Iran
[2]Department of Plant Pathology, University of Tarbiat Modarres, Tehran, Iran

Abstract

The genetic structure among Iranian populations of *Fusarium verticillioides* from the main corn growing areas of five provinces including Ardabil, Fars, Mazandaran, Khorasan, and Khuzestan were evaluated using VCG, RAPD and rep-PCR. Sixty-one isolates of *F. verticillioides* were placed in 14 Vegetative Compatibility Groups and 19 haplotypes. VCG3 with 14 members (23% of all isolates) was the most frequent VCG. RAPD-PCR and rep-PCR generated multiple distinct products demonstrated considerable variability among the isolates of different VCGs. Haplotype 1(HP1) had the highest frequency (0.57) in the population and was present in isolates from the majority of the locations in this study. All molecular phenotypes were distributed randomly across the various locations. Although there are some consistent between geographical origin of the isolates and their genetic similarity but VCG groups were distributed among different geographical locations and there was no correlation between geographical distribution and VCG groups. Gene diversity was 0.2909 and populations of *F. verticillioides* were placed in five distinct groups based on geographical origin. The highest genetic distance observed between Fars and Khuzestan (0.1801) and the smallest genetic distance was obtained between Ardabil and Khuzestan (0.0589). Analysis of molecular variance (AMOVA) showed a significant difference among populations of *F. verticillioides*. According to our results PhiPT was equal to 0.176 (p<= 0.001) and 82% of genetic variance occurred within populations and only 18% was found among populations. Moghan in Ardabil province is the main site for seed producing in Iran and the seeds that are produced there are distributed all over the country. In this study isolates of Moghan were located besides the isolates from other regions in different clusters. So it is presumed that the infected seeds from Moghan can be a major source for the spreading of the disease through all corn growing areas in Iran.

Keywords: *Fusarium verticillioides*; Corn ear rot; Rep-PCR; RAPD; Genetic diversity

Introduction

Fusarium ear rot is the most common fungal disease of corn ears that is caused by several species of Fusarium. Symptoms of the disease are a white to pink- or salmon-colored mold, beginning anywhere on the ear or scattered throughout. *Gibberella moniliformis* Wineland [anamorph *Fusarium verticillioides* (Sacc.) Nirenberg] is genetically the most intensively studied species in Fusarium section Liseola. [1]. Although yield usually is not much affected, kernel infection by Fusarium is of concern because of the loss of grain and seed quality and the potential occurrence of fumonisins and other mycotoxins [2]. *F. verticillioides* is the major species that causes ear rot on corn in Iran and is the most commonly reported fungal species associated with maize plants (*Zea mays* L.). During recent years, the disease is so severe that in some fields the entire crop has to be discarded. Host range and plant-fungus interactions are of significant interest in terms of understanding the distribution, biology, and population dynamics of this mycotoxigenic fungus [3]. The fungus can be found in plants or residues in maize fields in the United States at some time during the growing season [4].

Infection of developing corn kernels may occur through the silks, through holes and fissures in the pericarp or at points where the pericarp is torn by the emerging seedling and as a result of systemic infection of the corn plant by *F. verticillioides* [5]. *F. verticillioides* produces abundant, mostly single-celled microconidia in long chains [6].

Plant pathologist should study the population genetics of plant-pathogenic fungi, because pathogens evolve. Pathogen populations must constantly adapt to changes in their environment to survive [7]. Defining the genetic structure of populations is a logical first step in studies of fungal population genetics because the genetic structure of a population reflects its evolutionary history and its potential to evolve [7]. Knowledge of the genetic structure of pathogens is useful for developing control strategies, as the amount of genetic variation present within a population indicates how rapidly a pathogen can evolve. This information may eventually be used to predict how long control measures such as fungicides and resistant cultivars are likely to be effective [8].

Lineages that are capable of fusing (anastomosis) and forming stable and functional heterokaryons are known as sexually or vegetatively compatible, the former being frequently described as members of the same group of vegetative compatibility or vegetative compatibility group [9].

The genetic of vegetative compatibility in the entire Fusarium genus is modeled on the basic results obtained primarily with *F. verticillioides* [10]. Although Vegetative compatibility groups (VCG) is relatively a simple way to distinguish between strains that are morphologically identical [11], but as a tool for population genetic analysis in *F. verticillioides* it has not proven particularly useful, as most of the strains in a population are in different VCGs and thus the information obtained primarily is that no two strains are identical, i.e., clones in these populations are rare [12]. According to Danielsen et

*Corresponding author: Momeni H, Department of Plant Pathology, Iranian Research Institute of Plant Protection (IRIPP), Tehran, Iran
E-mail: hmomeni5@gmail.com

al. [13] Vegetative compatibility identified 34 vegetative compatibility groups (VCGs), of which 29 had one member and 5 had two members. Their results demonstrated that natural populations of *F. verticillioides* in Costa Rica consist of genetically diverse that represent a potential risk for disease development in corn crops.

Genetic diversity among *F. verticillioides* isolates was analyzed using VCG and RAPD [14]. According to their results RAPD could differentiate VCGs except in two cases. Genetic diversity among Iranian isolates of *F. verticillioides* was analyzed using VCGs [15]. Their results demonstrated that natural populations of *F. verticillioides* in Iran are genetically highly divergent and include isolates representing a potential risk for disease development. Genetic diversity of *F. verticillioides* was investigated by Mohammadi et al. [16] using VCGs in Iran among isolates that were recovered from Seed samples had been collected from the major producing area in Khuzestan and Ardabil provinces. Specific relation was not observed between VCGs and geographic origin of the isolates in their study and genetic diversity among population of *F. verticillioides* was very high. Isolates of *F. verticillioides* were recovered from diseased sugarcanes in Iran [17]. According to their results Forty-eight VCGs of *F. verticillioides* were isolated and none of the VCGs was common.

A simple procedure that can be used to detect infection by *F. verticillioides* from infected plant tissues has been developed [18]. A Polymerase Chain Reaction–Based assays was used for species-specific detection of Fusarium [14,19]. This technique has been successfully used to assess genetic variability within many plant pathogenic fungi, including Fusarium section *Liseola* [4,20-22]. There are two detailed genetic maps of Fusarium species available, one for *F. verticillioides* [1] and the other for *F. graminearum* [23].

Genetic diversity among 41 isolate of *F. verticillioides* collected from rice in Iran was determined using vegetative compatibility groups and RAPD. High level of genetic diversity was observed among *F. verticillioides* isolates [24].

Edel et al. [25] used ERIC and REP primers as molecular methods along with RFLP and PCR-amplified IGS for characterization of *Fusarium oxysporum* strains. Good correlation was found between the groupings obtained by the three methods. According to their results discrimination of closely related strains within IGS genotypes could be achieved by ERIC- or REP-PCR fingerprinting, which is the most efficient procedure in terms of simplicity and rapidity.

Karimi Dehkordi et al. [26] used rep-PCR to determine genetic diversity of 55 isolates of *F. verticillioides* from infected ears and stems of *Zea maize* and *Oryzae sativa* from different corn and rice producing areas of Iran. Their results suggested that *F. verticillioides* isolates from rice and corn are genetically different and that rep-PCR is a convenient and rapid method for analysis of genetic diversity and strain differentiation in *F. verticillioides*.

McDonald et al. [27] investigated the potential of repetitive-sequence-based polymerase chain reaction (rep-PCR) fingerprinting of fungal genomic DNA as a rapid and simple alternative to random amplified polymorphic DNA (RAPD) analysis in the study of phylogenetic relationships, and also as a diagnostic method among some species of Tilletia.

Jedryczka et al. [28] used REP, ERIC and BOX primers for rep-PCR genomic fingerprinting to assess the ability of rep -PCR genomic fingerprinting methods to characterize a collection of 90 isolates of *Leptosphaeria maculans* from Poland.

The present survey was undertaken in order to obtain a current picture of genetic structure among Iranian populations of *F. verticillioides* collected from the main corn producing areas. A second analysis was conducted to determine how many molecular phenotypes (haplotypes) were present in the population.

Materials and Methods

Sample collection and fungal isolation

A survey of corn fields was conducted in 2012-2013. Ears with symptoms resembling fusarium ear rot were collected from fields (each field ≥ 1 ha) in the 5 main corn producing provinces including 14 locations (Figure 1). A total of 5 infected ears were collected per field.

Seed samples were surface disinfected in 5.25% solution of Naocl for 1 min, rinsed three times in sterilized water and air dried on sterile paper towel. Sterilized corn seeds were cut in half and plated cut side down onto Nash and Snyder selective medium [29,30] that allows formation of easily recognizable colonies [31]. Fusarium colonies were transferred to Carnation-Leaf Agar (CLA), Potato Dextrose Agar (PDA) and Spezieller Nahrstoffarmer Agar (SNA), and were incubated for 7 days at 25°C [32]. Sixty-one Single spore isolates of *F. verticillioides* were identified based on the morphological criteria of Leslie and Summerell [10].

Recovered isolates were grouped into 5 populations based on geographical distances. Each population represented one province.

Pathogenicity test

Pathogenicity test was carried out according to Danielsen et al. [13] using SC301 corn seeds. The inoculums of isolates were prepared by transferring plugs of PDA containing 7-day old *F. verticillioides* to vials containing autoclaved toothpicks and PDB medium. After 7 week, 3 plants (stalks) per isolates were inoculated by insertion infected toothpicks about 5 cm above soil level and the inoculated site were sealed with parafilm. The control was the stalks that were inoculated with sterile toothpicks. The length of the necrotic region at the insertion point was measured and compared with the control.

Vegetative compatibility groups

VCGs assignment was based on complementation of nitrate non-utilizing (*nit*) mutants. Heterokaryon formation was demonstrated by pairing mutants that were unable to reduce nitrate [33]. Generation of mutants from 61 *F. verticillioides* isolates carried out according to Correll et al. [10] on PDA, malt agar (MA), corn meal agar (CMA) and minimal medium (MM) amended with 3 to 4.5 % potassium Cholorate [34]. Hyphal tips from Fast-growing, Chlorate-resistant sectors were transferred to minimal medium. Colonies with an expanding thin mycelium were considered *nit* mutants. All *nit* mutants showed wild-type growth on PDA.

The *nit* mutants were assigned phenotypically as *nit*1, *nit*3 or NitM based on differential growth on Minimal Medium (MM) amended with NaNo3, NaNo2, hypoxanthine, ammonium tartrate and Uric acid as sole nitrogen sources [35].

Complementation tests were carried out between NitM (or *nit*3) and *nit*1 mutants on minimal medium [36]. After 10 days in 25°C, Pairs of isolates that exhibited robust growth at the line of contact between the two colonies were determined as vegetatively compatible, otherwise were grouped in different VCGs [34,37]. All pairing were performed twice.

Figure 1: Provinces and locations where the isolates of *Fusarium verticillioides* were collected in Iran. 1-Mashhad 2-Neyshabur 3-Chenaran 4-Dashtenaz 5-Gharakheil 6-Moghan 7-Shushtar 8-Shush 9-Dezful 10-Andimeshk 11-Marvdasht 12-Zarghan 13-Fasa 14-Darab.

Genomic DNA extraction

Single spore colonies of all isolates were established and grown on PDA Medium. An inoculum disk was taken from each colony and used to inoculate 50 ml of liquid PDB (Potato Dextrose Broth) medium [38]. The cultures were incubated for 7 days, at 25°C, after which the mycelium was harvested, washed and used for extraction.

Total genomic DNA was isolated from mycelium by a microextraction protocol according to Möller et al. [39]. DNA was quantified by comparison with known amounts of genomic DNA on a 1.5% agarose. Appropriate dilution of the samples ensured a DNA sample of 10 ng genomic DNA for PCR reactions.

RAPD–PCR

DNA from individual single spore colonies was taken and used for each PCR reaction. Three previously identified primers by the name OPR11, OPR14, OPR16 [14] and four new primers including UBC682, UBC648, UBC199 and UBC196 (metabion international AG, Martinsried/Deutschland) were selected for PCR reactions. These primers were initially tested on three isolates and reactions repeated two times to insure of production of reproducible bands.

Reactions were performed with a BioRad thermocycler (Icycler model) in a 25 µl total volume containing 50 ng genomic DNA, 1.25 X PCR buffer (Fermentaz, Germany), 0.2 mM of each of the four dNTPs, 2.5 mM MgCl2, 12.5 pmol of each primers and 1 U *Taq* DNA Polymerase (Fermentaz, Germany). Reaction conditions consisted of an initial denaturation step at 95°C for 2 min, followed by 35 cycles at 94°C for 1 min, 36°C for 1 min and 72°C for 2 min, and then a final extension step at 72°C for 7 min. The fragment analysis was performed on 2% agarose gels in 1XTBE buffer.

Rep-PCR reactions

Rep-PCR reactions was carried out using BOX (5′- CTA CGG CAA GGC GAC GCT GAC G-3′), ERIC (ERIC1: 5′-ATG TAA GCT CCT GGG GAT TCA C-3′ ERIC2: 5′-AAG TAA GTG ACT GGG GTG AGC G-3′) and REP (REP1: 5′-IIII CGI CGI CAT CIG GC-3′ REP2: 5′-ICG ICI TAT CIG GCC TAC-3′) primers [40].

Reaction conditions was the same as RAPD-PCR but PCR components was different including 30 ng genomic DNA, 1X PCR buffer (Fermentaz, Germany), 0.1 mM of each of the four dNTPs, 1.5 mM MgCl2, 5 pmol of each forward and reverse primers, and 1 U *Taq* DNA Polymerase (Fermentaz, Germany) in a 25 µl total volume.

Data analysis

Haplotype: Determination of molecular phenotype (Haplotype) was carried out according to the DNA banding patterns of all seven primers that were used in RAPD based on Kolmer et al. [41]. So each isolates of *F. verticillioides* was given a seven-digit number that shows the haplotype of that isolates.

Genetic diversity: Sixty-one isolates of *F. verticillioides* from different geographical regions along with an isolate of *Fusarium proliferatum* were analyzed for genetic diversity based on pooled data that was obtained from RAPD and rep-PCR reactions. Genetic diversity within each population and for the entire isolates (Ht) computed by the program POPGENE Version 1.31. [42]. Genetic diversity was calculated as $H = (1 - \sum pi^2)$, where i is the frequency of allele i at the locus [43]. Gene flow that is shown with *Nm* (Number of migrants) is calculated as $Nm = 0.5(1 - G_{ST})/G_{ST}$, and G_{ST} is the Coefficient of gene differentiation [43]. Genetic distances were calculated between different populations based on Nei [43,44] and a dendrogram was generated using the unweighted pair group method with arithmetic averaging (UPGMA). The consensus tree was displayed using TREEVIEW v. 1.6.6 [45]. Shannon's information index (I) as a measure of gene diversity was estimated as well.

Analysis of molecular variance (AMOVA): Distribution of genetic variation and genetic structure was evaluated using analysis of molecular variance (AMOVA) that computed by GenAlEx6 [46]. We used AMOVA to estimate the partitioning of the total genetic diversity among and within the 5 studied populations including Fars, Khuzestan, Ardabil, Mazandaran and Khorasan. PhiPT is a measure facilitating AMOVA is calculated with the formulae of *PhiPT = AP / (WP + AP)*, where AP is estimated variance Among Populations and WP is estimated variance Within Populations. Genetic Distance (GD) matrix was obtained between all isolates of *F. verticillioides*.

Cluster analysis: Cluster analysis was computed by the help of NTSYSpc-2.02e. The input file was an Excel with binary data including 1 for the presence and 0 for the absence of each amplified band. The SimQual program was used to calculate the Dice similarity coefficients [47]. The resulting similarity matrix was used for unweighted pair group method with arithmetic averages (UPGMA) based dendrogram [48] using the sequential agglomerative hierarchical nested cluster analysis (SAHN) module of NTSYSpc. An isolate of *F. proliferatum* was used as outlier.

Results

Identification of VCGs

Based on mycological characteristics, 61 isolates were identified as *F. verticillioides* [10]. Pathogenecity test demonstrated that all isolates are pathogenic and inoculated corn stalks showed discoloration. Nitrate non-utilizing (*nit*) mutants were recovered from all 61 isolates of *F. verticillioides* and used in complementation tests and each isolate was assigned to a unique VCG group (Table 1).

Ninty-four percent of the sectors recovered were unable to utilize nitrate as the sole nitrogen source. A total number of 434 mutants of *F. verticillioides* were obtained with 49% of *nit*1, 29% of *nit*3 and 22% of NitM. Isolates of *F. verticillioides* were grouped into 14 VCGs based on complementation tests between NitM of one isolates with *nit1* or *nit3* of other isolates. NitM of isolates Fv1, Fv8, Fv14, Fv28, Fv31, Fv39, Fv41, Fv44, Fv47, Fv49, Fv53 and Fv58 were assigned as testers for 12 VCG groups and Two VCG groups including VCG13 and VCG14 that have only one member, no tester were considered. VCG3 with 14 members (23% of all isolates) was the largest and the most frequent VCG.

Determination of molecular phenotype based on RAPD data

According to the patterns of all seven random decamer primers (Table 2), 19 molecular phenotypes (haplotypes) were determined among 61 isolates of *F. verticillioides* (Table 1).

Haplotype 1(HP1) had the highest frequency (0.57) in the population and was present in isolates from the majority of the locations in this study. Fourteen molecular phenotypes occurred only once in the population. All molecular phenotypes were distributed randomly across the various locations. We didn't found a clear consistent between haplotype of the isolates and their VCGs.

VCG group	Haplotype group	Haplotype (primer OPR11-OPR14-OPR16-UBC196-UBC199-UBC648-UBC682)	Locations	Province (population)	Isolates
VCG1	HP1	1111111	Fasa	Fars	Fv1
VCG3	HP1	1111111	Fasa	Fars	Fv18
VCG3	HP1	1111111	Fasa	Fars	Fv19
VCG3	HP1	1111111	Fasa	Fars	Fv20
VCG3	HP1	1111111	Darab	Fars	Fv14
VCG3	HP1	1111111	Darab	Fars	Fv15
VCG3	HP1	1111111	Darab	Fars	Fv16
VCG3	HP1	1111111	Darab	Fars	Fv17
VCG8	HP14	1112212	Darab	Fars	Fv46
VCG3	HP1	1111111	Marvdasht	Fars	Fv21
VCG3	HP1	1111111	Marvdasht	Fars	Fv22
VCG3	HP1	1111111	Zarghan	Fars	Fv23
VCG3	HP1	1111111	Zarghan	Fars	Fv24
VCG3	HP1	1111111	Zarghan	Fars	Fv25
VCG1	HP1	1111111	Dashte naz	Mazandaran	Fv2
VCG3	HP1	1111111	Dashte naz	Mazandaran	Fv26
VCG3	HP1	1111111	Dashte naz	Mazandaran	Fv27
VCG12	HP19	1121111	Dashte naz	Mazandaran	Fv59
VCG1	HP2	1114131	Gharakheil	Mazandaran	Fv3
VCG1	HP1	1111111	Gharakheil	Mazandaran	Fv4
VCG9	HP1	1111111	Gharakheil	Mazandaran	Fv47
VCG9	HP15	1111112	Gharakheil	Mazandaran	Fv48
VCG1	HP1	1111111	Neyshabur	Khorasan Razavi	Fv5
VCG2	HP2	1114131	Neyshabur	Khorasan Razavi	Fv8
VCG5	HP12	1111211	Mashhad	Khorasan Razavi	Fv35
VCG5	HP12	1111211	Mashhad	Khorasan Razavi	Fv36
VCG5	HP12	1111211	Chenaran	Khorasan Razavi	Fv37
VCG5	HP1	1111111	Chenaran	Khorasan Razavi	Fv38
VCG1	HP1	1111111	Moghan	Ardabil	Fv6
VCG2	HP4	2111411	Moghan	Ardabil	Fv9
VCG5	HP1	1111111	Moghan	Ardabil	Fv31
VCG5	HP9	1141141	Moghan	Ardabil	Fv32
VCG1	HP3	1124131	Moghan	Ardabil	Fv7
VCG2	HP5	2111111	Moghan	Ardabil	Fv10
VCG2	HP1	1111111	Moghan	Ardabil	Fv11
VCG2	HP6	1421111	Moghan	Ardabil	Fv12
VCG2	HP1	1111111	Moghan	Ardabil	Fv13
VCG7	HP1	1111111	Moghan	Ardabil	Fv41
VCG7	HP1	1111111	Moghan	Ardabil	Fv42
VCG7	HP1	1111111	Moghan	Ardabil	Fv43
VCG5	HP10	1112331	Moghan	Ardabil	Fv33
VCG5	HP11	1111131	Moghan	Ardabil	Fv34
VCG11	HP13	1111121	Moghan	Ardabil	Fv52
VCG11	HP1	1111111	Moghan	Ardabil	Fv53
VCG11	HP1	1111111	Moghan	Ardabil	Fv54
VCG11	HP15	1111112	Moghan	Ardabil	Fv55
VCG11	HP18	1111113	Moghan	Ardabil	Fv56
VCG11	HP1	1111111	Moghan	Ardabil	Fv57
VCG4	HP1	1111111	Dezful	Khuzestan	Fv28
VCG6	HP1	1111111	Dezful	Khuzestan	Fv39
VCG6	HP1	1111111	Dezful	Khuzestan	Fv40
VCG10	HP13	1111121	Dezful	Khuzestan	Fv49
VCG4	HP7	1111141	Andimeshk	Khuzestan	Fv29

VCG4	HP8	1111411	Andimeshk	Khuzestan	Fv30
VCG13	HP1	1111111	Andimeshk	Khuzestan	Fv60
VCG14	HP13	1111121	Andimeshk	Khuzestan	Fv61
VCG8	HP1	1111111	Shush	Khuzestan	Fv44
VCG10	HP16	1112231	Shush	Khuzestan	Fv50
VCG10	HP17	1112111	Shush	Khuzestan	Fv51
VCG8	HP13	1111121	Shushtar	Khuzestan	Fv45
VCG12	HP13	1111121	Shushtar	Khuzestan	Fv58

Table 1: Populations, Locations, Haplotypes and VCG groups of the isolates of *Fusarium verticillioides* that were used in this study.

Size range of scorable bands	Polymorphic bands	G+C (%)	5'-Sequence-3'	Primer
300-2500	16	60	5'-CTG CGA CGG T-3'	UBC682
150-2500	19	70	5'-GCA CGC GAG A-3'	UBC648
200-1800	14	80	5'-GCT CCC CCA C -3'	UBC199
200-2000	17	80	5'-CTC CTC CCC C -3'	UBC196
300-3000	14	60	5'- GTA GCC GTC T -3'	OPR11
300-2500	16	60	5'- CAG GAT TCC C -3'	OPR14
150-2000	20	70	5'- CTC TGC GCG T -3'	OPR16

Table 2: Characteristics of RAPD primers that were used in PCR reactions with the isolates of *Fusarium verticillioides* in this study.

Population genetic diversity and differentiation

The evaluation was carried out based on pooled data derived from rep-PCR and RAPD. Nei's analysis [44] of gene diversity estimated gene flow (*Nm*) 1.7624 and *Gst* equal to 0.2210. Genetic diversity for the entire collection (Ht) was 0.2909. A dendrogram based on Nei's [44] genetic distance with the method UPGMA and modified from NEIGHBOR procedure of PHYLIP Version 3.5 was obtained and populations were placed in 5 distinct groups based on geographical origin (Figure 2).

The highest genetic distance observed between Fars and Khuzestan (0.1801) and the smallest genetic distance was obtained between Ardabil and Khuzestan (0.0589) (Table 3).

Results from analysis of molecular variance (AMOVA) showed a significant difference among 5 populations of *F. verticillioides* (Table 4). According to our results PhiPT was equal to 0.176 (p<= 0.001) and 82% of genetic variance occurred within populations and only 18% was found among populations.

Cluster analysis

When subjected to NTSYS-pc version 2/02e with UPGMA cluster analysis and Dice coefficient four clusters (I, II, III and IV) were found on the level of genetic similarity of 41% (Figure 3).

As expected the isolate of *F. proliferatum* was placed separately in the dendrogram. Subdividing of cluster I lead to 6 further groups (A to F). All of the isolates in group A have been originated from the Province Fars only. With one exception all isolates in group B are from Ardabil province. Isolates in group C are from Mazandaran (4 isolates) and Khorasan Razavi (3 isolates) and one isolates has been originated from Khuzestan. Except the isolate Fv20 from Fars other three isolates in group D are from Ardabil. Group E has only one member from Ardabil. Isolates in group F are from Ardabil and Khuzestan that shows the highest genetic identity in Figure 2. Cluster II has isolates from three provinces including Fars, Khuzestan and Ardabil. Isolates in Cluster III compromised isolates from Khorasan and Mazandaran, two populations that shows more genetic identity comparing with other populations. All isolates in Cluster IV are from Ardabil.

We didn't found a clear separation of Haplotype and VCG distribution among populations that evaluated in this study although there are some consistent between geographical origin of the isolates and their genetic similarity. Isolates Fv1 and Fv14 showed the highest similarity (87%) both of them from Fars province but from different location, Fasa and Darab.

Discussion

The results presented here demonstrated that the selected *F. verticillioides* populations from Iran consist of highly genetically diverse isolates indicated by the high level of VCG and Haplotype polymorphism. These results are in agreement with results from other studies on *F. verticillioides*, demonstrating that this fungus is genotypically highly diverse [13,15,36,49].

This may important because pathogen populations with high

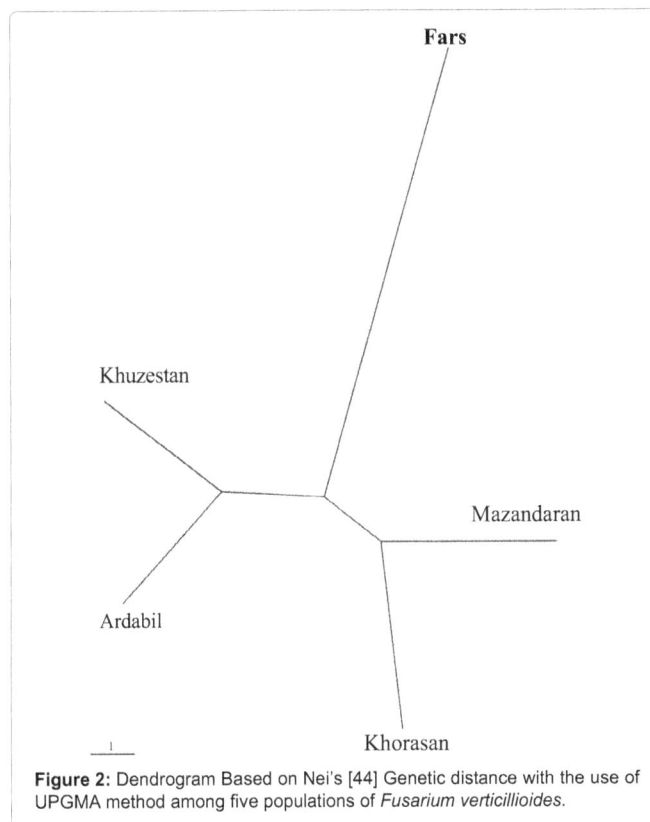

Figure 2: Dendrogram Based on Nei's [44] Genetic distance with the use of UPGMA method among five populations of *Fusarium verticillioides*.

POP ID	Fars	Mazandaran	Khorasan	Ardabil	Khuzestan
Fars	****	0.8783	0.8526	0.9141	0.8352
Mazandaran	0.1298	****	0.9305	0.9251	0.8982
Khorasan	0.1595	0.0721	****	0.9179	0.8769
Ardabil	0.0898	0.0778	0.0857	****	0.9428
Khuzestan	0.1801	0.1074	0.1314	0.0589	****

Table 3: Nei's genetic identity (above diagonal) and genetic distance (below diagonal) between five populations of *Fusarium verticillioides* in this study.

Source	df	SS	MS	Est. Var.	Variation (%)
Among Pops	4	355.704	88.926	5.423	18%
Within Pops	56	1425.148	25.449	25.449	82%
Total	60	1780.852		30.872	100%

*P value=0.001 and number of permutations is 999.

Table 4: AMOVA results for five populations of *Fusarium verticillioides*.

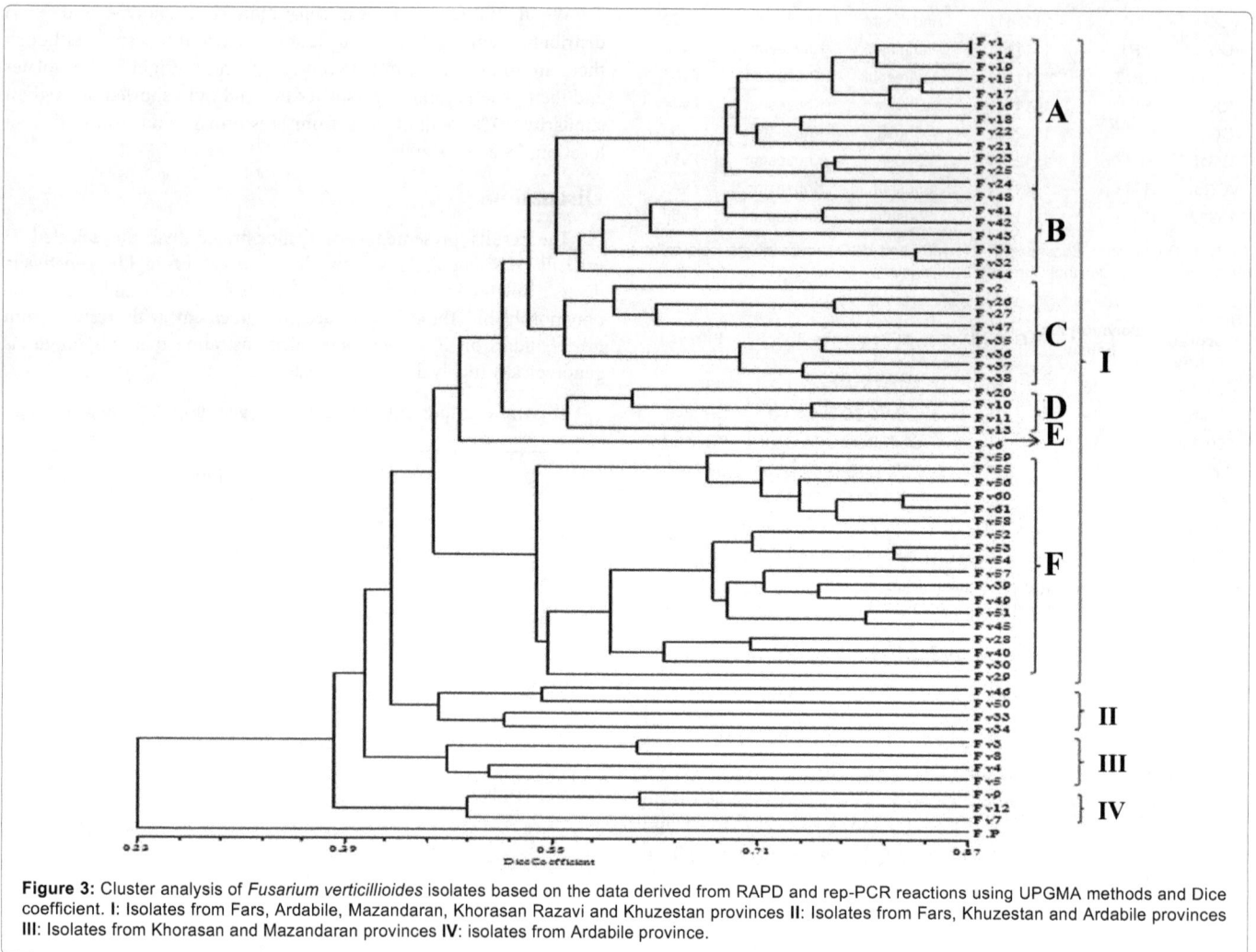

Figure 3: Cluster analysis of *Fusarium verticillioides* isolates based on the data derived from RAPD and rep-PCR reactions using UPGMA methods and Dice coefficient. **I:** Isolates from Fars, Ardabile, Mazandaran, Khorasan Razavi and Khuzestan provinces **II:** Isolates from Fars, Khuzestan and Ardabile provinces **III:** Isolates from Khorasan and Mazandaran provinces **IV:** isolates from Ardabile province.

genetic variation are potentially capable of rapidly evolving responses to changing environmental conditions [50]. A possible explanation for the high levels of genetic diversity found among isolates of *F. verticillioides* could be natural chance mutations, combined with the fact that the fungus can produce abundant numbers of spores in a relatively short period of time. However high levels of genetic variation are usually due to recombination, which occurs sexually through mating or asexually through the parasexual cycle [14]. Since parasexuality is dependent on the formation of a heterokaryon, parasexual recombination occurs only between members of the same VCG [9]. The high degree of VCG polymorphism in this study supports the assumption that the genotypic diversity of *F. verticillioides* is primarily caused by recombination during the sexual state. In spite of sexual state that maintains high level of diversity in the population, asexual reproduction decrease diversity because of selection and genetic drift.

Farrokhi-Nejad and Leslie [51] also found high degree of diversity among isolates of *F. verticillioides* in USA. In a similar investigation extensive variation was detected among isolates of *F. verticillioides* by Huang et al. [14]. Comparing with their results we saw some consistence between geographical separation and genetic clustering. Although this correlation was not perfect and several exceptions were observed. Isolates of Moghan in Ardabil province were scattered among isolates of other populations indicating that the initial inoculums of *F. verticillioides* may be distributed through country with

seeds that are mainly produced in Moghan. Seed-borne inoculums might be important for long distance gene flow, but compared to the large fungal population in the soil, its effect on population diversity is probably small [14]. Moghan in Ardabil province is the main site for seed producing in Iran and the seeds that are produced there are distributed over the country. So infected seeds in Moghan are very important for the spreading of fungus through the maize producing areas. In this study isolates of Moghan are seen besides other isolates in different clusters (Figure 3). Spores of this pathogen are distributed by wind between corn fields [52] but long distance distribution is made up mainly by seed infection.

VCGs assignment based on complementation of nitrate non-utilizing (*nit*) mutants results in grouping isolates into 14 VCGs in this study accounting for a genetic diversity (number of VCGs/number of isolates) of 0.23 that was lower comparing with the diversity that was obtained with some other studies [14,51,53]. Most of the isolates in this study have been collected from Moghan where is the hot spot of the disease and is the main corn seed production site in Iran. This can cause low number of VCG groups and haplotypes. Long distance gene flow with seedborne inoculums from Moghan to other provinces/ populations can have an effect on low diversity comparing with other studies. Isolates of Ardabil have been distributed through 11 of 19 haplotypes that were achieved in this study. The frequency of *nit1* mutants was higher than the frequency of the other types of *nit* mutants that is in agreement with the results of Bowden and Leslie [54] and Puhalla [33]. Sectoring frequency of *F. verticilioides* has been shown to be heritable and to vary among isolates [34]. The wide range of sectoring frequency in plant pathogenic fungi on different concentrations of chlorate has also been suggested as a selective advantage for rapid adaptation to environmental stresses such as fungicides and host resistance [35].

VCG groups were distributed among different geographical locations and there was no correlation between geographical distribution, VCG groups and genetic similarity of the isolates. Of 14 VCG groups 6 contain isolates from more than one population (province).

Unlike the VCG assays that are based on only one marker, RAPD uses a numerous markers. Of 19 haplotypes that were obtained among 61 isolates of *F. verticilioides* (0.31 haplotype per isolate) in our study 15 of them belong to only one VCG indicating that RAPD shows more diversity than VCG assignment.

In consistence with our results, Zamani et al. [55] and Bahmani et al. [24] didn't find a close relationship between VCGs and RAPDs in Iran, but unlike our results Huang et al. [14] found a clear relationship between VCGs and RAPDs data and they found 0.66 haplotype per isolate which is more than the diversity we obtained in this study.

According to Bodker et al. [56], in a population with high VCG diversity, it usually is not possible to correlate VCG to another trait. In contrast, in populations those consist of only a few VCGs, these VCGs can sometimes be correlated to other trait.

Evidence for the usefulness of the RAPD technique for evaluation genetic diversity among isolates of *F. verticillioides* was provided by many researchers [14,57]. Use of rep-PCR with three primers including BOX, ERIC and REP as a molecular marker beside RAPD, increased the reproducibility, specificity and credibility of the results.

Cluster analysis of combined data derived from rep-PCR and RAPD resolved the isolates into four main clusters and six more groups within the first cluster and that all of them were distant from *F. proliferatum* which is included as outlier. Cluster analysis also showed that the maximum similarity among isolates of *F. verticillioides* was approximately 87% and that no identical isolates were detected, indicating that every isolate was a unique genotype. The highest genetic distance observed between Fars and Khuzestan (0.1801) and the smallest genetic distance was obtained between Ardabil and Khuzestan (0.0589) (Table 3).

There are some consistent between geographical origin of the isolates and their genetic similarity, although we didn't found any correlation between haplotypes and geographical distribution of isolates. Furthermore there is no clear clustering of isolates according to VCG groups.

Some of previous studies on *F. verticillioides* have shown no geographic clustering of isolates [14,16,17]. Their results indicated that the various geographical populations are not genetically isolated, which may be due to dissemination of propagules by biotic and abiotic factors. We consider the distribution of corn seeds and vegetative material over a wide geographical area to be another cause of genetic variation of *F. verticillioides* in Iran.

Nineteen haplotypes were identified among the 61 isolates tested. Many researchers have used random amplified polymorphic DNA as a marker for assessing genetic diversity in *F. verticillioides* and some other fungal species [58,59]. Khalil et al. used RAPD analysis

combination with pathogenicity assays to study the taxonomy of five *Fusarium* species [38]. Kini et al. used RAPD for determination of genetic variation among *F. verticillioides* isolates that were collected from seeds of different host species [57]. MacDonald and Chapman [21] divided isolates of *F. verticillioides* from Kenya into two subgroups based on RAPDs, with some correlation to the tissue-origin of isolates. Results of this study indicated that isolates of *F. verticillioides* in Iran are genetically diverse populations. One of the strategies for the control of the disease can be limitations of seed movement from the main producing seed production site in Moghan, Ardabil to other corn producing areas and also seed treatment with suitable fungicides.

Acknowledgement

Financial support by Iranian Research Institute of Plant Protection (IRIPP) (to H. M.) is gratefully acknowledged.

References

1. Jurgenson JE, Zeller KA, Leslie JF (2002) Expanded genetic map of *Gibberella moniliformis* (*Fusarium verticillioides*). Appl Environ Microbiol 68: 1972-1979.

2. Munkvold GP, Desjardins AE (1997) Fumonisins in maize: Can we reduce their occurrence? Plant Dis 81: 556-565.

3. Glenn AE, Hinton DM, Yates IE, Bacon CW (2001) Detoxification of corn antimicrobial compounds as the basis for isolating *Fusarium verticillioides* and some other *Fusarium* species from corn. Appl Environ Microbiol 67: 2973-2981.

4. Amoah BK, Rezanoor HN, Nicholson P, MacDonald MV (1995) Variation in the *fusarium* section Liseola: Pathogenecity and genetic studies of *Fusarium moniliforme* Sheldon from different hosts in Ghana. Plant Pathol 44: 563-720.

5. Glenn AE, Meredith FI, Morrison WH 3rd, Bacon CW (2003) Identification of intermediate and branch metabolites resulting from biotransformation of 2-benzoxazolinone by *Fusarium verticillioides*. Appl Environ Microbiol 69: 3165-3169.

6. Glenn AE, Richardson EA, Bacon CW (2004) Genetic and morphological characterization of a *Fusarium verticillioides* conidiation mutant. Mycologia 96: 968-980.

7. McDonald BA (1997) The population genetics of fungi: tools and techniques. Phytopathology 87: 448-453.

8. McDonald BA, Linde C (2002) Pathogen population genetics, evolutionary potential, and durable resistance. Annu Rev Phytopathol 40: 349-379.

9. Leslie JF (1993) Fungal vegetative compatibility. Annu Rev Phytopathol 31: 127-150.

10. Leslie JF, Summerell BA (2006) Fusarium laboratory workshops-A recent history. Mycotoxin Res 22: 73-74.

11. Kedera CJ, Leslie JF, Claflin LE (1994) Genetic diversity of Fusarium section Liseola (*Gibberella fujikuroi*) in individual maize stalks. Phytopathology 84: 603-607.

12. Sidhu GS (1986) Genetics of *Gibberella Fujikuroi*. VIII. Vegetative compatibility groups. Can J Bot 64: 117-121.

13. Danielsen S, Meyer UM, Jensen FD (1998) Genetic characteristics of *Fusarium verticillioides* isolates from maize in Costa Rica. Plant Pathol 47: 615-622.

14. Huang R, Galperin M, Levy Y, Perl-Treves R (1997) Genetic diversity of *Fusarium moniliforme* detected by vegetative compatibility groups and random amplified polymorphic DNA markers. Plant Pathol 46: 871-881.

15. Gohari MA, Javan-Nikkhah M, Hedjaroude GA, Abbasi M, Rahjoo V, et al. (2008) Genetic diversity of *Fusarium verticillioides* isolates from maize in Iran based on vegetative compatibility grouping. J Plant Pathol 90: 113-116.

16. Mohammadi A, Mofrad NN (2011) Investigation on genetic diversity of *Fusarium verticillioides* isolated from corn using vegetative compatibility groups and relation of VCGs to the pathogenicity. J Agr Tech 7:143-148.

17. Mohammadi A, Farrokhi Nejad R, Mofrad NN (2012) *Fusarium verticillioides* from Sugarcane, Vegetative Compatibility Groups and Pathogenicity. Plant Prot Sci 48: 80-84.

18. Murillo I, Cavallarin L, San-Segundo B (1998) The development of a rapid PCR assay for detection of *Fusarium moniliforme*. Euro J Plant Pathol 104: 301-311.

19. Schilling AG, Möller EM, Geiger HH (1996) Polymerase chain reaction-based assays for species-specific detection of *Fusarium culmorum*, *F. graminearum*, and *F. avenaceum*. Phytopathology 86: 515-522.

20. Amoah BK, MacDonald MV, Rezanoor HN, Nicholson P (1996) The use of the random amplified polymorphic DNA technique to identify mating groups in the *Fusarium* section Liseola. Plant Pathol 45: 115-250.

21. MacDonald MV, Chapman R (1997) The incidence of *Fusarium moniliforme* on maize from central America, Africa and Asia during 1992-1995. Plant Pathol 46: 112-125.

22. Voigt K, Schleier S, Brückner B (1995) Genetic variability in Gibberella Fujikuroi and some related species of the genus *Fusarium* based on random amplification of polymorphic DNA (RAPD). Curr Genet 27: 528-535.

23. Jurgenson JE, Bowden RL, Zeller KA, Leslie JF, Alexander NJ, et al. (2002) A genetic map of Gibberella zeae (*Fusarium graminearum*). Genetics 160: 1451-1460.

24. Bahmani Z, Nejad FR, Nourollahi K, Fayazi F, Mahinpo V (2012) Investigation of *Fusarium verticillioides* on the Basis of RAPD analysis, and Vegetative Compatibility in Iran. Plant Pathol Microbiol 3:147.

25. Edel V, Steinberg C, Avelange I, Laguerre G, Alabouvette C (1995) Comparison of three molecular methods for the characterization of *Fusarium oxysporum* strains. Phytopathology 85: 579-585.

26. Dehkordi KM, Javan-Nikkhah M, Morid B, Rahjoo V, Hajmansoor S (2013) Analysis of the association between *Fusarium verticillioides* strains isolated from rice and corn in Iran by molecular methods. Eur J Exp Biol 3: 90-96.

27. McDonald JG, Wong E, White GP (2000) Differentiation of *Tilletia* species by rep-PCR genomic fingerprinting. Plant Dis 84:1121-1125.

28. Jedryczka M, Rouxel T, Balesdent MH (1999) Rep-PCR based genomic fingerprinting of isolates of *Leptosphaeria maculans* from Poland. Eur J Plant Pathol 105: 813-823.

29. Nash SM, Snyder WC (1962) Quantitative estimations by plate counts of propagules of the bean root rots *Fusarium* in field soils. Phytopathology 52: 567-572.

30. Nelson PE, Toussoun TA, Marasas WFO (1983) *Fusarium* species. An illustrated manual for identification. Pennsylvania State University Press, University Park, PA.

31. Chen Y, Wang JX, Zhou MG, Chen CJ, Yuan, SK (2007) Vegetative Compatibility of *Fusarium graminearum* Isolates and genetic study on their carbendazim-resistance recombination in China. Phytopathology 97: 1584-1589.

32. Gerlach W, Nirenberg H (1982) The genus *Fusarium*: a pictorial atlas. Mitteilungen aus der Biologischen Bundesanstalt für Land-und Forstwirtschaft. Berlin-Dahlem, Germany.

33. Puhalla JE (1985) Classification of strains of *Fusarium oxysporum* on the basis of vegetative compatibility. Can J Bot 63:179-183.

34. Klittich C, Leslie JF (1988) Nitrate reduction mutants of *fusarium moniliforme* (*gibberella fujikuroi*). Genetics 118: 417-423.

35. Pasquali M, Dematheis F, Gilardi G, Gullino ML, Garibaldi A (2005) Vegetative Compatibility Groups of *Fusarium oxysporum* f. sp. lactucae from Lettuce. Plant Dis 89: 237-240.

36. Correll JC, Klittich CJR, Lesli JF (1987) Nitrate nonutilizing mutants of *Fusarium oxysporum* and their use in vegetative compatibility tests. Phytopathology 77: 1640-1646.

37. Aqeel AM, Pasche JS, Gudmestad NC (2008) Variability in morphology and aggressiveness among North American vegetative compatibility groups of *Colletotrichum coccodes*. Phytopathology 98: 901-909.

38. Khalil MS, Abdel-sattar MA, Aly IN, Abd-Elsalam K, Verret JA (2003) Genetic affinities of *Fusarium* spp. and their correlation with origin and pathogenicity. Afr J Biotechnol 2: 109-113.

39. Möller EM, Bahnweg G, Sandermann H, Geiger HH (1992) A simple and efficient protocol for isolation of high molecular weight DNA from filamentous fungi, fruit bodies, and infected plant tissues. Nucleic Acids Res 20: 6115-6116.

40. Versalovic J, Schneider M, De Bruijn FJ, Lupski JR (1994) Genomic fingerprinting of bacteria using repetitive sequence-based polymerase chain reaction. Meth Mol Cell Biol 5: 20-25.

41. Kolmer JA, Liu JQ, Sies M (1995) Virulence and molecular polymorphism in *Puccinia recondita* f.sp. tritici in Canada. Phytopathology 85: 276-285.

42. Yeh FC, Yang RC, Boyle T (1999) POPGENE version 1.31.Microsoft Window-based Freeware for Population Genetic Analysis. Quick User Guide. A joint Project Development by University of Alberta and the Centre for International Forestry Research.

43. Nei M (1973) Analysis of gene diversity in subdivided populations. Proc Natl Acad Sci USA pp: 3321-3323.

44. Nei M (1978) Estimation of average heterozygosity and genetic distance from a small number of individuals. Genetics 89: 583-590.

45. Page RD (1996) TreeView: an application to display phylogenetic trees on personal computers. Comput Appl Biosci 12: 357-358.

46. Peakall R, Smouse PE (2012) GenAlEx 6.5: genetic analysis in Excel. Population genetic software for teaching and research--an update. Bioinformatics 28: 2537-2539.

47. Dice LR (1945) Measures of the amount of ecologic association between species. Ecology 26: 297-302.

48. Sneath PHA, Sokal RR (1973) Numerical taxonomy: the principles and practice of numerical classification. San Francisco, CA: Freeman Press.

49. Chulze SN, Ramirez ML, Torres A, Leslie JF (2000) Genetic variation in *Fusarium* section Liseola from no-till maize in Argentina. Appl Environ Microbiol 66: 5312-5315.

50. McDonald BA, Miles J, Nelson LR, Pettway RE (1994) Genetic variability in nuclear DNA in field population of *Stagonospora nodorum*. Phytopathology 84: 250-255.

51. Farrokhi-Nejad R, Leslie JF (1990) Vegetative compatibility group diversity with populations of *Fusarium moniliforme* isolated from corn seed. (Abstr.) Phytopathology 80: 1043.

52. Ooka JJ, Kommedahl T (1977) Wind and rain dispersal of *Fusarium moniliforme* in corn fields. Phytopathology 67: 1023-1026.

53. Campbell CL, Leslie JF, Farrokhi-Nejad R (1992) Genetic diversity of *Fusarium moniliforme* in seed from two maize cultivars. Phytopathology 82: 1082.

54. Bowden RL, Leslie JF (1992) Nitrate-nonutilizing mutants of Gibberella zeae (*Fusarium graminearum*) and their use in determining vegetative compatibility. Exp Mycol 16: 308-315.

55. Zamani MR, Motallebi M, Rostamian A (2004) Characterization of Iranian isolates of *Fusarium oxysporum* on the basis of RAPD analysis, virulence and vegetative compatibility. J Phytopath 152: 449-453.

56. Bodker L, Lewis BG, Coddington A (1993) The occurrence of a new genetic variant of *Fusarium oxysporum* f.sp. pisi. Plant Pathol 42: 833-838.

57. Kini KR, Leth V, Mathur SB (2002) Genetic variation in *Fusarium moniliforme* isolated from seeds of different host species from Burkina Faso based on Random Amplified Polymorphic DNA analysis. J Phytopath 150: 209.

58. Meijer G, Megnegneau B, Linders EGA (1994) Variability for isozyme, vegetative compatibility and RAPD markers in natural populations of phomopsis subordinaria. Mycol Res 98: 267-276.

59. Wang PH, Lo HS, Yeh Y (2001) Identification of F. o. *cucumerinum* and F. o. luffae by RAPD-generated DNA probes. Lett Appl Microbiol 33: 397-401.

30

Morphological, Molecular Identification and SSR Marker Analysis of a Potential Strain of *Trichoderma/Hypocrea* for Production of a Bioformulation

Mohammad Shahid*, Mukesh Srivastava, Antima Sharma, Vipul Kumar, Sonika Pandey and Anuradha Singh

Biocontrol Laboratory, Department of Plant Pathology, Chandra Shekhar Azad University of Agriculture and Technology, Kanpur, 208002, India

Abstract

Seven different strains of *Trichoderma* are isolated from wilt infected leguminous crops of an Indian state and tested for their antagonistic activity against *Fusarium* (soil borne pathogen) which is expressed as a zone of inhibition in the culture plates. The seven strains are identified as *Trichoderma viride*, *T. harzianum*, *T. asperellum*, *T. koningii*, *T. atroviride*, *T. longibrachiatum*, and *T. virens*. Upon successful identification, morphological description and sequencing of the isolated strains with the help of universal ITS primers, the sequences are submitted to NCBI and allotted with the accession numbers JX119211, KC800922, KC800921, KC800924, KC008065, JX978542 and KC800923, respectively. Genetic variability studies reveal that a percentage of polymorphism in SSRs is obtained within the seven strains of *Trichoderma* species which is comparatively higher (>77%) than with RAPD primers (~50%). This study aims at selecting the best strain of *Trichoderma* species (*Trichoderma viride 01PP*) and then preparing a simple bioformulation that is cheap, easy to apply and readily accessible to the farmers. Shelf life of the prepared bioformulation is even checked for 180 days and it is concluded that the number of propagules start declining from 30th day onwards when the bioformulation is prepared in talc as a carrier material.

Keywords: Antagonism; Biocontrol agent; *Trichoderma*; Shelf life; Polymorphism; Genetic variability

Abbreviations: PDA: Potato Dextrose Agar; PDB: Potato Dextrose Broth; EDTA: Ethylene Diamine Tetra-Acetic Acid; w/v: weight/volume; CTAB: Cetyl Trimethyl Ammonium Bromide; PCR: Polymerase Chain Reaction; SSR: Simple Sequence Repeats; TAE: Tris base, Acetic acid and EDTA; CMC: Carboxy Methyl Cellulose; CSAU: Chandra Shekhar Azad Agriculture University; BCA: Biological Control Agent; RFLP: Restriction Fragment Length Polymorphism; ITCC: Indian Type Culture Collection; NCBI: National Centre for Biotechnology Information; ITS: Internal Transcribed Spacer; LDPE: Low Density Poly Ethylene

Introduction

The genus *Trichoderma* has its own significance in the agricultural industry due to its varied activities ranging from being a valuable antagonist against the soil-borne pathogens to acting as a provider of nutrition to the soil as well. Several scientists have worked on how this genus acts as a potential biocontrol agent against a range of pathogenic fungi. Harman et al. [1] have even reported *Trichoderma* as opportunistic, avirulent plant symbionts. They have explained the features of *Trichoderma* as to how it colonizes the roots that eventually proves beneficial to the soil in terms of nutrition and plant growth increasing crop productivity simultaneously.

The biocontrol activity of *Trichoderma* is of immense importance not only to agriculture and its crops but also the environment as it does not accumulate in the food chain and thus does no harm to the plants, animals and humans [2]. The genes and gene products involved in the biocontrol mechanism of *Trichoderma* provide a vast array of research to the scientists in Biotechnology and Bioinformatics as well.

The infrageneric classification by Bisset [3] shows significant morphological similarities between *Trichoderma* and *Hypocrea* and have defined genus *Trichoderma* to include the anamorphs of *Hypocrea*.

The morphology of *Trichoderma* spp. is very interesting to study as there are a finite number of morphological descriptors to study and disseminate the genus and its features [4,5]. It is believed that the identification of any microorganism becomes quite easy by a careful morphological observation; hence, a detailed morphological description of some of the commercially important strains of *Trichoderma* has been carried out in this study. Samuels [6] described the systematics, the sexual stage and the ecology of *Trichoderma* and mentioned in his study that the morphology of *Trichoderma* is not only limited to a few characters but many species may be included in this genus due to their geographical distribution.

Druzhinina and Kubicek [7] studied and brought forth the species concepts and biodiversity in *Trichoderma* and *Hypocrea* by aggregating the morphological, physiological and genetic studies and presented an update on the taxonomy and phylogeny of a number of taxa. This helped us in understanding that the identification of *Trichoderma* only on the basis of morphology is not of high precision. Thus, molecular identification and characterization comes under investigation that would help in evaluating the genetic diversity between the species.

Kumar et al. [8], Shahid [9] and Sagar et al. [10] focused on the molecular identification and analysis of the genetic variability of a specific strain of *Trichoderma* based on antagonistic and RAPD analysis in some leguminous crops (Pigeonpea, Chickpea and Lentil) produced in Uttar Pradesh (India). RAPD analysis with a set of 20 OPA primers was carried out on 5 isolates of the same strain (*Trichoderma longibrachiatum*) collected from different soil samples of Pigeonpea. This resulted in a significant amount of genetic variability where more

*Corresponding author: Mohammad Shahid, Biocontrol Laboratory, Department of Plant Pathology, CSA University of Agriculture and Technology, Kanpur, India, E-mail: shahid.biotech@rediffmail.com

than 50% of the amplified fragments in each case were polymorphic. Thus, it was concluded that there was good genetic variability among the isolates under study.

Based on the genetic variability studies done earlier, the study now focuses upon developing a strain-specific molecular marker solely for the identification of Trichoderma species. rRNA based analysis is thought to be the best one to explore the microbial diversity and identify new strains [11].

The study also includes the behavior of these BCAs against fungal wilt pathogens affecting leguminous crops. Fusarium wilt causes huge loss to the leguminous crops in India every year ranging from 15 to 20% thereby reducing the production of important legumes. Various management strategies such as use of resistant cultivars are been undertaken to prevent the crops and soils as well from the wilt caused by Fusarium as it may last for several years. Thus, it becomes necessary to derive a cheap and better way to fight against the pathogen and increase the crop production. Bioformulation containing Trichoderma has emerged as an effective alternative to this problem and thus has been disseminated in this report. But, before preparing a bioformulation with Trichoderma, the effect of media, temperature and pH on the growth and sporulation of Trichoderma species should be known [12,13]. Trichoderma species, when grown either in PDA or PDB within a pH range of 7-7.5 and at an optimum temperature range of 25-30°C gives the best growth and sporulation rates both.

Talc-based bioformulation of Trichoderma [14] has proven beneficial to the wilt infected leguminous crops but an important aspect to be taken into prior consideration is the shelf life of spores that are present in talc. Various methods and measures are still to be taken that can result in the longevity, competitiveness and survival of Trichoderma on fields.

Materials and Methodology

Isolation and selection of strains

Trichoderma strains were isolated from the soil of pulse fields of various districts of Uttar Pradesh (India) and were tested against phythopathogenesis. The most promising isolates were selected for biochemical, molecular and disease suppressiveness tests. Initially, a total of seven strains were identified and were selected for further study. Based on the descriptions of Bissett [3], we classified these fungi as: Trichoderma anamorph and Hypocrea teleomorph. The isolates were screened for antagonistic activity towards the major soil borne fungi such as Fusarium solani, Rhizoctonia solani, Pythium ultimum, Macrophomina phaseolina, Sclerotinia sclerotiorum, Phytophthora, Fusarium oxysporum and Sclerotium cepivorum that were previously isolated and identified in the Biocontrol Laboratory, Department of Plant Pathology, CSAUAT, Kanpur, India.

in vitro bioassay

in vitro bioassay was conducted between the Trichoderma isolates and the phytopathogenic fungi in petridishes containing PDA. Isolates which showed a marked effect towards pathogens were selected and used for further study. Each Trichoderma isolate was separately inoculated into 100 ml Potato Dextrose Broth and incubated at 20°C for 10 days. After incubation, the cultures were filtered through 0.22 mm Millipore filters and the aliquots (2 ml) of these filtrates were placed in sterile petridishes and 25 ml of 1/4 strength PDA at 45°C was added. Once the agar solidified, mycelial discs of the pathogens (7 mm in diameter) obtained from actively growing colonies were placed gently on the centre of the agar plates. The petridishes were incubated

at 20°C for 6 days. There were three replicates for each experiment and the growth reduction of the pathogens was recorded.

Morphological descriptors such as colony morphology, colony color, colony edge and others of each strain were studied.

SSR analysis

DNA was extracted using CTAB method from all seven isolates and quantified using agarose gel electrophoresis. A total of 20 SSR primers i.e., SSR 1-20 were selected. PCR was programmed with an initial denaturing for 4 minutes at 94°C; followed by 35 cycles of denaturation for 1 minute at 94°C; annealing at 36°C for 1 minute; extension for 90 seconds at 70°C, and a final extension for 7 minutes at 72°C in a Primus 96 advanced gradient Thermocycler. PCR product (20 µl) was mixed with loading buffer (8 µl) containing 0.25% Bromophenol Blue, 40% (w/v) sucrose in water and then loaded in 2% agarose gel with 0.1% ethidium bromide for examination by horizontal electrophoresis.

Electrophoresis

The amplification products were analyzed by electrophoresis according to Sambrook and Russell [15] in 2% agarose in TAE buffer (for a litre of 50X TAE Stock solution, we used: 242 g Tris Base, 57.1 ml Glacial Acetic Acid and 100 ml 0.5 M EDTA), stained with 0.2 µg/ml ethidium bromide. Nucleic acid bands were photographed and detected by BioRad Gel Doc system.

Preparation of bioformulation

Talc powder was evaluated as carrier material to produce bioformulation of Trichoderma sp. The carrier was dried under sun, powdered (sieve pore, 1mm) and sterilized at 1.05 kg/cm² pressure for 30 min. The substrate was mixed with 7 days old culture of respective Trichoderma spp. which were previously grown on potato dextrose agar in 2:1 (solid culture) w/v and CMC 5 gm/kg was added as adjuvant. Fifty grams of such mixture was then filled in polypropylene bags (25x30 cm) tied and stored at 25 ± 2°C. Observations on colony forming units (cfu) of Trichoderma spp. was recorded initially and at monthly interval up to 6 months for shelf life study.

Seed treatment

Required quantity of fungicide (Vitavax @ 2 gm/kg seed), insecticide (Chloropyriphos 20 EC @ 8ml/kg seed), biocontrol agent (Trichoderma viride @ 4 gm/kg seed) and biofertilizer Rhizobium culture @ 1 packet/ acre or 30 gm/kg seeds) along with different combination with 100 seeds of lentil and Chickpea taken from the healthy fields and 100 seeds of lentil and chickpea taken from the infected fields were used for studies.

Results

Isolation and bioassay

A total of seven isolates of Trichoderma species were isolated from the soil of pulse fields of various districts of Uttar Pradesh, India. These include Trichoderma viride, T. harzianum, T. asperellum, T. koningii, T. atroviride, T. longibrachiatum and T. virens.

All tested strains in genus Trichoderma had high or moderate antagonistic activity towards pathogens expressed as a zone of inhibition and fungal growth reduction by using culture filtrate. Among all isolated strains, T. harzianum and T. viride were found to be the most effective species against all pathogens.

Trichoderma strains that were isolated and taken into consideration in this study have been validated and submitted to the Indian Type

Strain No.	Name of Bioagent	ITCC Accession No	GenBank Accession No.	Strain code	Source
T1	T. viride	8315	JX119211	01PP	Hardoi (U.P., India)
T2	T. harzianum	6796	KC800922	Th azad	CSA, Kanpur (U.P., India)
T3	T. asperellum	8940	KC800921	T$_{asp}$/CSAU	CSA, Kanpur (U.P., India)
T4	T. koningii	5201	KC800924	T$_k$ (CSAU)	CSA, Kanpur (U.P., India)
T5	T. atroviride	7445	KC 008065	71L	Hardoi (U.P., India)
T6	T. longibrachiatum	7437	JX978542	21PP	Kaushambi (U.P., India)
T7	T. virens	4177	KC800923	T$_{vi}$ (CSAU)	CSA, Kanpur (U.P., India)

Table 1: Details of Trichoderma strains.

Name of Strains	Colony Growth rate (cm/day)	Colony color	Reverse color	Colony edge	Mycelial form	Mycelial color	Conidiation	Conidiophore branching	Conidia wall	Conidial color	Chlamydospores
T. viride	8-9 in 3 days	Dirty green	Dark greenish	Smooth	Floccose to Arachnoid	Watery white	Ring like zones	Ball like structure	Rough	Green	Not observed
T. harzianum	8-9 in 3 days	Dark green	Colorless	Wavy	Floccose to Arachnoid	Watery white	Ring like zones	Highly branched, regular	Smooth	Dark Green	Not observed
T. asperellum	5-6 in 3 days	Snow white green	Orange	Smooth	Floccose	Watery White	Ring like zones	Branched, regular	Smooth	Green	Not observed
T. koningii	7-8 in 3 days	Dirty green	Yellowish	Smooth	Floccose to Arachnoid	Watery white	Ring like zones	Highly branched, regular	Rough	Grayish Green	Not observed
T. atroviride	5-6.5 in 3 days	Light dark effuse	Colorless	Effuse	Floccose to Arachnoid	Watery white	Irregular	Irregular	Rough	Yellowish Green	Not observed
T. longibrachiatum	8-9 in 4 days	White to green	Colorless	Effuse	Floccose to Arachnoid	Watery white	Circular zones	Rarely re-branched	Smooth	Green	Not observed
T. virens	8-9 in 3 days	Snow white	Colorless	Smooth	Floccose to Arachnoid	Watery White	Flat	Highly branched, regular	Smooth	Dirty Green	Not observed

Table 2: Morphological descriptors used for the characterization of native isolates of Trichoderma spp.

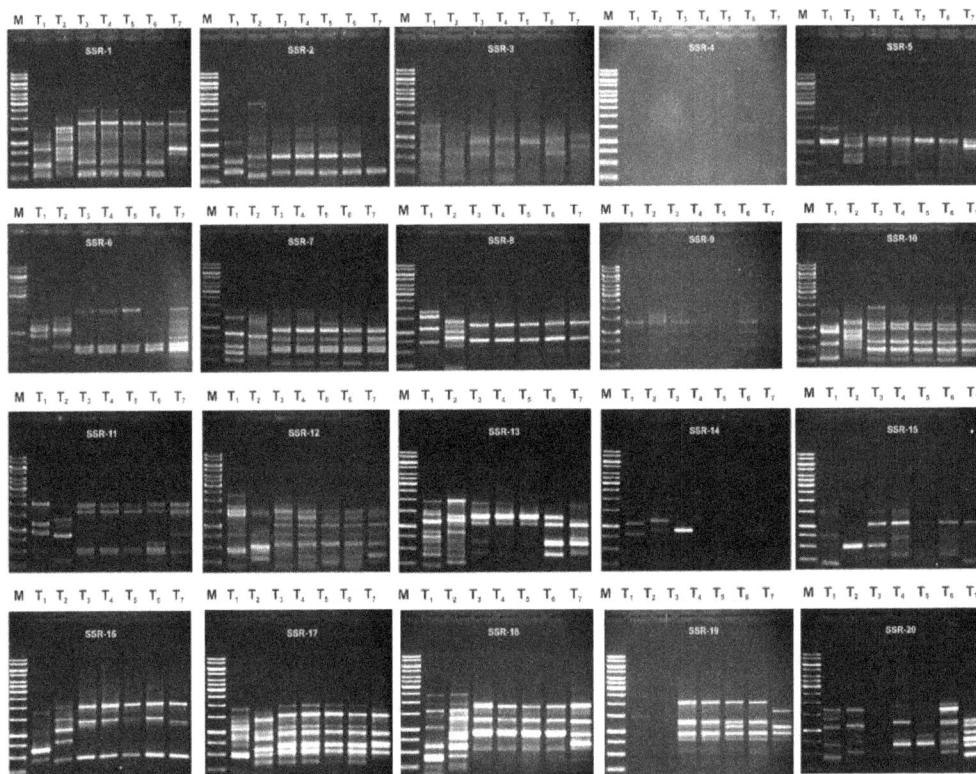

Figure 1: Molecular Analysis of Trichoderma spp. using SSR Primers.

Name of SSR Primer	Primer Sequence (5' – 3')	Amplified product	Total no. of bands	No. of Polymorphic bands	No. of Monomorphic bands
SSR 1.	F: GAAACAACACCGAAATACAC R: CAAGTCAGATGAAGTTTG	YES	6	2	4
SSR 2.	F: GACTCATACTTTGTTCTTAGCAG R: GAACGGAGCGGTCACATTAG	NO	-	-	-
SSR 3.	F: CAAGCTGACGCCTATGAAGA R: CTTTCACTCACTCAACTCTC	YES	4	3	1
SSR 4.	F: CATGGTGGAATAGTGATGGC R: CTCCATACACCACTCATTCAC	YES	4	3	1
SSR 5.	F: CCAAATACTGCAACACACCG R: GTTCCCATCAAGGCAGAAGG	NO	-	-	-
SSR 6.	F: CCATGCATACGTGACTGC R: GTTGACTGTTGGTGTAAGTG	YES	6	3	3
SSR 7.	F: GTTATCTTCCAGCGTC R: GATATACAATCAGAGATG	NO	-	-	-
SSR 8.	F: GGGAATTTGTGGAGGGAAG R: CCTCAGAATGTCCCTGTC	YES	3	1	2
SSR 9.	F: GCGGCGAGCAAATAAAT R: GGAGAATAAGAGTGAAATG	NO	-	-	-
SSR 10.	F: CCGTAAGAATAGGTGTC R: GGAAAATAGGGTGGAAAG	YES	7	2	5
SSR 11.	F: GAACTCAGTTTCTCATTG R: GAACATATCCAATTATCATC	YES	10	10	0
SSR 12.	F: GTATGTGCTTGTATGCTTC R: GAACGGAGCGGTCACATTA	NO	-	-	-
SSR 13.	F: CCACGTATGTGACTGTATG R: GAAAGAGAGGCTGAAACTTG	YES	12	11	1
SSR 14.	F: GGTAGGTGAGATAGTTG R: GGAGCAAGAAGAAGCAG	YES	11	11	0
SSR 15.	F: GGAATTTATCACACTATCTC R: GACTCCCAACTTGTATG	YES	7	7	0
SSR 16.	F: GTACATTGAACAGCATCATC R: CAATAGGGCATGAAAGGAG	NO	-	-	-
SSR 17.	F: CACATATGAAGATTGGTCAC R: CATTTATGTCTCACACACAC	NO	-	-	-
SSR 18.	F: GTGTGTACCTAAAGCCTTG R: GTAAGTTGATCAAACGCCC	YES	5	3	2
SSR 19.	F: GTGTGCATGGTGTGTG R: CCATCCCCCTCTATC	NO	-	-	-
SSR 20.	F: CACGACTATCCCACTTG R: CTTACTTTCTTAGTGCTATTAC	YES	9	9	0
GRAND TOTAL			84	65	19

Table 3: SSR Amplification and their corresponding PCR products for bioagent *Trichoderma spp*

Culture Collection (ITCC) and GenBank (NCBI) database where particular accession numbers have been allotted to the specific strain of each species (Table 1).

Morphological description

Morphological study of the *Trichoderma* strains has been done and the characteristics include various parameters such as colony growth rate, colony color, colony edge, mycelial form, growth pattern and speed along with morphology of conidia and phialids, conidia color, shape and size etc. were studied for the identification of each strain of the genus *Trichoderma* (Table 2).

Molecular characterization of *Trichoderma spp.* using SSR markers

A set of 20 SSR primers were used in this study (Table 3). The preliminary studies indicate that the *Trichoderma* spp. isolates under study had very good diversity and there is a strong possibility to get the isolate-specific primers that will be utilized for identification of the particular *Trichoderma* isolates with a good biological potential from the field isolates without undergoing the cumbersome bioassay. All

reproducible polymorphic bands were scored and analyzed following UPGMA cluster analysis protocol and computed *in silico* into similarly matrix using NTSYSpc.

The size of the fragments (molecular weight in base pairs) was estimated by using 1 kb ladder marker which was run along with the amplified products. In the gel, "1" indicates the presence of a band whereas "0" indicates the absence of any band. Out of the seven strains of *Trichoderma* spp. tested, the percentage of polymorphism in SSRs obtained was more than 77%. This shows that there is a complete variability within the strains of *Trichoderma* spp. being isolated from different fields of the Indian State (Uttar Pradesh). This would enable us to develop a potential strain possessing competitive ability, growth promoting characters and inducing resistance in plants (Figure 1).

In the dendrogram shown in Figure 2, all the 7 isolates of *Trichoderma spp.* were distinctly divided into two major clusters **A** and B at 20 units. Isolate $T_k/CSAU$ and *01PP* spanned the extremes of the entire dendrogram. Genetic dissimilarity ranged from a lowest value of 0.143 to a highest value of 0.857 (between $T_{asp}/CSAU$ and T_k (*CSAU*). Isolates T_k (*CSAU*), *71L*, *21PP*, and T_{vi} (*CSAU*) were assigned to cluster

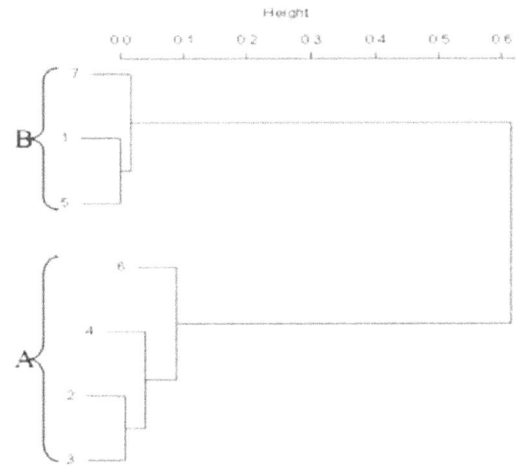

Figure 2: Dendrogram for *Trichoderma spp.* isolates as revealed by SSR markers; where: 1: *Th azad*, 2: *71L*, 3: *21PP*, 4: T_k *(CSAU)*, 5: T_{asp}/*CSAU*, 6: T_{vi}*(CSAU)*, 7: *01PP*.

Strain Name	Locus	Definition	Primer used	Sequence bp
T. viride 01PP	JX119211	*Hypocrea rufa* isolate 01PP-8315/11 18S ribosomal RNA gene, partial sequence; Internal transcribed spacer 1, 5.8S ribosomal RNA gene, and internal transcribed spacer 2, complete sequence; and 28 S ribosomal RNA gene, partial sequence.	ITS-1tccgtaggtgaacctgcgg ITS-2 tcctccgcttattgatatgc	1173bp
T. harzianum Th azad	KC800922	*Trichoderma harzianum* isolate Th-azad/CSAU 6796 18S ribosomal RNA gene, partial sequence; internal transcribed spacer 1, 5.8S ribosomal RNA gene, and internal transcribed spacer 4, partial sequence.	ITS-1agagtttgatcctggctcag ITS-4ggttaccttgttacgactt	546bp
T. asperellum T_{asp}/*CSAU*	KC800921	*Trichoderma asperellum* isolate T_{asp}(CSAU)-8940 18s ribosomal RNA gene, partial sequence; internal transcribed spacer 1, 5.8s ribosomal RNA gene, and internal transcribed spacer 4, partial sequence.	ITS-1tccgtaggtgaacctgcgg ITS-4 tcctccgcttattgatatgc	641 bp
T. koningii T_k *(CSAU)*	KC800924	*Trichoderma koningii* isolate T_k *(CSAU)* 5201 18S ribosomal RNA gene, partial sequence; internal transcribed spacer 1, 5.8S ribosomal RNA gene, and internal transcribed spacer 4, partial sequence.	ITS-1 tctgtaggtgaacctgcgg ITS-4 ggaagtaaaagtcgtaacaagg	206 bp
T. atroviride 71L	KC008065	*Trichoderma atroviride* strain TAU8 18S ribosomal RNA, partial sequence; internal transcribed spacer 1, 5.8S ribosomal RNA, and internal transcribed spacer 2 complete sequence; and 28S ribosomal RNA partial sequence.	ITS-1tcctccgcttattgatatgc ITS-2 ggaagtaaaagtcgtaacaagg	627bp
T longibrachiatum 21PP	JX978542	*Trichoderma longibrachiatum* strain 21PP 18S ribosomal RNA gene, partial sequence; internal transcribed spacer 1, 5.8S ribosomal RNA gene, and internal transcribed spacer 2, complete sequence; and 28S ribosomal RNA gene, partial sequence	ITS-1tcctccgcttattgatatgc ITS-2 ggaagtaaaagtcgtaacaagg	664 bp
T. virens T_{vi} *(CSAU)*	KC800923	*Trichoderma virens* isolate T_{vi} (CSAU)-417718S ribosomal RNA gene, partial sequence; internal transcribed spacer 1, 5.8S ribosomal RNA gene, and internal transcribed spacer 4, partial sequence.	ITS-1 tcctccgcttattgatatgc ITS-4 ggaagtaaaagtcgtaacaagg	635 bp

Table 4: Submission of the gene sequences at NCBI Database.

'A'. Genetic dissimilarity among the entries in this cluster ranged from a lowest of 14.3% (between *21PP* and T_{vi} *(CSAU)* to a highest of 35.7% (between T_{vi} *(CSAU)* and T_k *(CSAU)*. The other cluster 'B' comprised of three isolates namely *Th azad*, T_{asp} *(CSAU)* and *01PP* were grouped together. The genetic dissimilarity in this group ranged from 33.3% between T_{asp} *(CSAU)* and *01PP* to 75% between *Th azad* and *01PP*.

The molecular identification and characterization of *Trichoderma* isolates was conducted with the help of universal ITS primers also. Four ITS primer sequences were used for the identification of all seven isolates of *Trichoderma* and were then submitted to NCBI database. The details of the strains submitted can be seen in Table 4.

Bioformulation and its validation under *in vitro* conditions

Talc-based bioformulation of *Trichoderma* is prepared as it is relatively cheap and easily accessible to farmers for use on fields. It can be stored in plastic bags for long as it has been observed that storing the talc-based bioformulation in plastic bags increases the shelf-life of *Trichoderma* preserving its bioefficiency simultaneously.

The shelf life of all the seven isolates was also ascertained at ambient environment prevailing during a period of 6 months on the basis of spore load per gram. The talc based powder of the bioagent was prepared (*Trichoderma viride 01PP* (spore+mycelium) 1.0% w/w+Talc 98.5% w/w+0.5% carboxyl methyl cellulose) and used for shelf life, bioefficacy etc. studies. The talc based bioformulation was stored in LDPE pouches. The powder was dull white in color, pH 7.0, moisture 8% and cfu of 29.7×10^6. It was found that the bioformulation has good shelf life up to six months and then the spores started declining.

Shelf life of *Trichoderma* in talc as a carrier material was determined

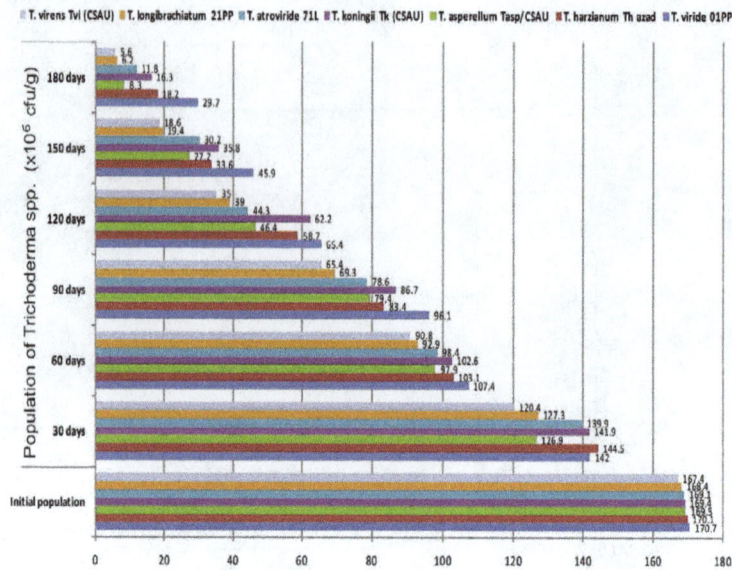

Figure 3: Effect of Talc as a carrier on the population of *Trichoderma spp.*

Sl. No.	Treatment	Germination %	Root length (cm)	Shoot length (cm)	Seedling length (cm)	Dry weight	Vigour index-I	Vigour index-II
1.	*Trichoderma viride*	86.67	7.58	11.70	19.28	0.14	1671.00	12.13
2.	*Vitavax*	85.00	7.31	11.26	18.57	0.13	1578.45	11.05
3.	*Chlorpyriphos*	81.00	7.27	8.87	16.14	0.11	1307.34	8.91
4.	*Rhizobium*	81.33	6.45	9.69	16.14	0.11	1312.67	8.95
5.	*Trichoderma viride+Vitavax*	90.00	7.91	13.08	20.99	0.14	1889.10	12.60
6.	*Trichoderma viride+Chlorpyriphos*	77.67	5.90	10.41	16.31	0.11	1266.80	8.54
7.	*Trichoderma viride+Rhizobium*	84.33	6.59	9.35	15.94	0.11	1344.22	9.28
8.	*Vitavax+Chlorpyriphos*	67.67	5.51	9.35	14.86	0.12	1005.58	8.12
9.	*Vitavax+Rhizobium*	73.67	4.41	8.34	12.75	0.11	939.29	8.10
10.	*Chlorpyriphos+Rhizobium*	84.67	3.95	7.16	11.11	0.11	940.68	9.31
11.	*Trichoderma viride+ Vitavax+Chlorpyriphos*	77.33	6.57	10.85	17.42	0.12	1347.09	9.28
12.	*Trichoderma viride+ Vitavax+Rhizobium*	78.33	5.84	8.92	14.76	0.12	1156.15	9.40
13.	*Trichoderma viride+ Chlorpyriphos+ Rhizobium*	80.00	5.57	9.74	15.31	0.11	1224.80	8.80
14.	*Vitavax+Chlorpyriphos+Rhizobium*	79.33	5.70	8.83	14.53	0.10	1152.66	7.93
15.	*Vitavax+Chlorpyriphos+Vitavax+Rhizobium*	74.00	5.45	9.05	14.5	0.10	1073.00	7.40
16.	Control	66.33	3.87	7.10	10.97	0.10	727.64	6.63
	CD=5%	5.69	1.22	0.63	1.72	0.02	321.60	1.81
	S.D.	2.79	0.60	0.31	0.86	0.01	157.88	2.03

Table 5: Evaluation with special reference to the use of pesticides in seed treatment in combination with bioagents.

at a time interval of 30 days that further indicated that the number of propagules started declining from 30th day onwards. Talc-based bioformulation was found to be the best material to retain maximum number of viable propagules i.e., 29.7×10^6 cfu/g at 180 days of storage. It has also been found that the isolates can retain their viability up to 120 days in all the cases (Figure 3).

Under natural conditions, application of talc-based solid formulation of *Trichoderma* in soil provides protection against wilt disease in leguminous crops. Higher reduction in wilt was obtained in lentil and pigeon pea crops. As compared with the control and other strains, application of *Trichoderma viride 01PP* was more effective in reducing the wilt disease caused by *Fusarium* in Pigeonpea. *Trichoderma* species can act as biocontrol agents through different synergistic mechanisms. However, it is difficult to predict the degree of synergism and the behavior of a BCA in natural system. Considering

that the environmental conditions are important, the right selection of BCAs, which begins with a safe characterization of biocontrol strains in the new taxonomic schemes of *Trichoderma*, is equally important since the exact identification of strains at the species level is the first step in utilizing the full potential of fungi in specific applications. *Trichoderma* species play an important role in controlling fungal plant pathogens, especially soil borne fungal pathogens. Strains of *Trichoderma* can produce extracellular enzymes and antifungal antibiotics, they may also be competitors to fungal pathogens, promote plant growth, and induce resistance in plants.

The different pre-sowing seed treatments when taken from healthy fields showed different response for all seven seed quality attributes i.e. germination, root length, shoot length, seedling length, dry weight, vigour index I and vigour index II. The data revealed that *Vitavax* followed by treatment with *Trichoderma viride* were found superior

Morphological, Molecular Identification and SSR Marker Analysis of a Potential Strain...

199

(Table 5). Among all the treatments, control exhibited the poorest performance in most of the characters under study.

The commercial use of *Trichoderma* BCAs must be preceded by precise identification, adequate formulation, and studies about the synergistic effects of their mechanisms of biocontrol. *T. viride 01PP* have been reported as the most important BCAs against plant pathogenic fungi. The strain distribution in several genotypes could also support the idea of developing antifungal formulations in which different *Trichoderma* BCAs could be combined. The use of *Trichoderma*-based products is not only safe for the farmers and consumers but it also proves friendly to the environment.

Discussion

The morphological characters of the fungus under study agree very closely with the description given by Vasudeva and Srinivasan [16] and Booth [17]. Cornea et al. [18] found that the molecular analysis using ITS-RFLP and PCR with specific primers allow the confirmation of previous taxonomic determination of *Trichoderma harzianum* and *T. viride*. However, an increased intra-specific molecular polymorphism was observed using several arbitrary primers (RAPD) analysis. Shahid et al. [19] and Singh et al. [20] also reported that germination and seedling length along with seedling dry weight are important attributes, which determine the quality of seed of any seed lot. Besides these quality seed parameters seed vigour index also plays very crucial role in predicting the fate of any seed lot under biotic and abiotic stress conditions.

Conclusion

It is concluded from this study that *Trichoderma* has been successfully isolated, identified, characterized and used as an effective biocontrol agent against wilt caused by other pathogenic fungi. The seven strains of *Trichoderma* have been isolated from wilt infected leguminous crops and tested in the laboratory for the identification of pathogens infecting the crops.

The strains have been examined morphologically and at molecular level as well. Specific markers have been defined that could quickly identify specific strains and amplify them. The genetic variability among the strains is also studied with the help of a set of SSR markers. The effect of enzyme activities during interaction with the pathogen is also counted and the data reveals the best carbon source for the enzyme for its induction.

In the end, a talc based bioformulation is prepared that showed beneficial effects when applied on wilt infected crops on pulse fields.

Acknowledgement

The authors are grateful to the financial support granted by the Indian Council of Agricultural Research under the Niche Area of Excellence on "Exploration and Exploitation of Trichoderma as an antagonist against soil borne pathogen" running in Biocontrol Laboratory, Department of Plant Pathology, Chandra Shekhar Azad University of Agriculture and Technology, Kanpur-208002, Uttar Pradesh, India.

References

1. Harman GE, Howell CR, Viterbo A, Chet I, Lorito M (2004) *Trichoderma* species—opportunistic, avirulent plant symbionts. Nat Rev Microbiol 2: 43-56.

2. Monte E, Llobell A (2003) *Trichoderma* in organic agriculture. V Congreso Mundial del Aguacate 725-733.

3. Bissett J (1991) A revision of the genus *Trichoderma* . II. Infrageneric classification. Canadian Journal of Botany 69: 2357-2372.

4. Gams W, Bissett J (1998) Morphology and identification of *Trichoderma*. In: Kubicek C and Harman G (Eds). *Trichoderma* and *Gliocladium*. Taylor & Francis, London, UK.

5. Gams W, Meyer W (1998) What exactly is *Trichoderma harzianum*? Mycologia 90: 904-915.

6. Samuels GJ (2006) *Trichoderma*: systematics, the sexual state, and ecology. Phytopathology 96: 195-206.

7. Druzhinina I, Kubicek CP (2005) Species concepts and biodiversity in *Trichoderma and Hypocrea*: from aggregate species to species clusters? J Zhejiang Univ Sci B 6: 100-112.

8. Kumar V, Shahid M, Singh A, Srivastava M, Biswas SK (2011) RAPD Analysis of *Trichoderma longibrachiatum* isolated from Pigeonpea Fields of Uttar Pradesh. Indian J Agric Biochem 24: 80-82.

9. Shahid M (2012) Evaluation of Antagonistic activity and Shelf life study of *Trichoderma viride* (01 PP-8315/11). Advances in Life Sciences 1: 138-140.

10. Sagar MSI, Meah MB, Rahman MM, Ghose AK (2011) Determination of genetic variations among different Trichoderma isolates using RAPD marker in Bangladesh. J Bangladesh Agril Univ 9: 9-20.

11. Shahid M (2013) Sequencing of 28SrNA Gene for Identification of *Trichoderma longibrachiatum* 28 CP/ 74444 Species in Soil Sample. International Journal of Biotechnology for Wellness Industries 2: 84-90.

12. Singh A, Shahid M, Pandey NK, Kumar S, Srivastava M, Biswas SK (2011) Influence of temperature, pH and media for growth and sporulation of *Trichoderma atroviride* and its Shelf life study in different carrier based formulation. J Pl Dis Sci 6: 32-34.

13. Shahid M, Singh A, Srivastava M, Sachan CP, Biswas SK (2011) Effect of seed treatment on Germination and Vigour in Chickpea. Trend in Biosciences 4: 205-207.

14. Shahid M (2012) Molecular characterization and variability of *Trichoderma longibrachiatum* based on antagonistic and RAPD analysis in legume crops of Uttar Pradesh. J Botan Soc Bengal 66: 105-110.

15. Sambrook J, Russell DW (2001) Agarose Gel Electrophoresis. CSH Protocols.

16. Vasudeva RS, Srinivasan KV (1952) Studies on the wilt disease of lentil *(Lens esculenta Moench)*, Indian Phytopath 5: 23-32.

17. Booth C (1971) The genus *Fusarium*. CMI, Kew, Surrey, England.

18. Cornea CP, Pop A, Matei S, Ciuca M, Voaides C, et al. (2009) Antifungal action of new *Trichoderma* species Romanian isolates on different plant pathogens. Biotechnol and Biotechnol 23: 766-770.

19. Shahid M, Singh A, Srivastava M, Mishra RP, Biswas SK (2011) Effect of temperature, pH and media for growth and sporulation of *Trichoderma longibrachiatum* and self life study in carrier based formulations. Ann Pl Protec Sci 19: 147-149.

20. Singh A, Shahid M, Sachan CP, Srivastava M, Biswas SK (2013) Effect of seed treatment on Germination and Vigour in Lentil. J Pl Dis Sci 8: 124-125.

31

Silent Mutation: Characterization of its Potential as a Mechanism for Sterol 14α-Demethylase Resistance in *Cercospora beticola* Field Isolates from the United States

James O Obuya[1]*, Anthony Ananga[1] and Gary D. Franc[2]

[1]*Center for Viticulture and Small Fruit Research, Florida A&M University, 6505 Mahan Drive, Tallahassee, FL 32317, USA*
[2]*Plant Sciences Department-3354, University of Wyoming, 1000 E. University Ave, Laramie, WY 82071, USA*

Abstract

Sterol demethylation inhibitors (DMIs) are considered among the most effective fungicides used to control Cercospora leaf spot (CLS), caused by *Cercospora beticola* Sacc., in sugar beet. Resistance to DMI fungicides has been reported in the *C. beticola* population from the United States, but the molecular mechanism is not known. It is considered that genetic changes in the *C. beticola* 14α-demethylase (*CbCyp51*) gene may be contributing to DMI resistance. The study investigated a silent mutation (GAG to GAA) at codon 170 as a potential mechanism for *C. beticola* DMI resistance. The *CbCyp51* gene was obtained from DMI-sensitive and -resistant isolates, cloned into a plasmid vector, transformed in an isogenic yeast R-1, and tested for DMI sensitivity. Transformed yeast showed low ED_{50} values (0.02 - 0.09 µg ml^{-1}) as compared to high ED_{50} values from *C. beticola* DMI-resistant isolates (21 - 65 µg ml^{-1}). The finding did not support our hypothesis that a silent mutation in the *CbCyp51* gene may be associated with *C. beticola* DMI resistance. Furthermore, genetic analysis of the *CbCyp51* gene found no mutation in 2 *C. beticola* DMI-resistant isolates from the Central High Plains. Further studies will be required to investigate additional mechanisms which have been associated with DMI resistance in fungi. Thus, we could not develop a molecular-based assay for the rapid detection of *C. beticola* DMI resistance, because no mutation was found in the *CbCyp51* gene. Currently, fungicide sensitivity assay could be the best method screen for *C. beticola* DMI resistance.

Keywords: Cercospora leaf spot; Silent mutation; DMIs; Sugar beet; *Beta vulgaris*

Introduction

Cercospora leaf spot (CLS) of sugar beet (*Beta vulgaris* L.), caused by *Cercospora beticola* Sacc., is the most destructive foliar disease worldwide [1,2]. Severe disease can cause a reduction in extractable sucrose, root yield, and increased concentration of impurities and leading to higher processing losses [3,4]. CLS can be controlled using integrated pest management (IPM) strategies which include planting resistant varieties, crop rotation, deep tillage, control of alternate host plants, or application of foliar fungicides [5-9]. The disease can be controlled by limiting canopy development through the regulation of water use in irrigated areas and nitrogen fertilization [6,9].

Sterol demethylation inhibitors (DMIs) are most widely used for CLS control in sugar beet [9-11]. DMIs protect plants against foliar diseases such as that caused by *C. beticola* [10,11], and members of this group have a medium- risk of resistance due to their single mode of action [12]. Resistance to DMIs has been associated with polygenic-controlled mechanism in which quantitative or additive interaction of several mutant genes can lead to a gradual shift in response to fungicides [9,11,12]. DMI resistance has been reported in fungi, and has been associated with multiple mechanisms [13-27]. DMI resistance in *Zymoseptoria tritici* (formerly known as *Mycosphaerella graminicola*) is due to single nucleotide polymorphic sites (SNPs) on the *Cyp51* gene which led to alterations of the protein [27,28]. In *Penicillium digitatum*, a 126 base pair (bp) insert found on the *Cyp51* gene was associated with DMI resistance [16]. Genetic analysis of the *P. digitatum Cyp51* gene found five tandem repeats on the 126 bp insert. Furthermore, the insert was present in the regulatory region and acted as a transcriptional enhancer, leading to overexpression of the *PdCyp51* gene [16]. DMI resistance has been associated with either uptake or efflux of DMIs which involved an adenosine triphosphate binding cassette (ABC) protein [14,15,29-31].

In *C. beticola*, DMI resistance was reported in field isolates from Greece in which four SNPs were found on the *CbCyp51* gene [32]. Two SNPs led to predicted amino acid changes at positions 297 (E297K) and 330 (I330T) and the mutations were present in *C. beticola* isolates that showed a moderate DMI resistance (MR) phenotype. The third SNP was predicted to lead to an amino acid substitution at position 384 (P384S) and was present in *C. beticola* isolate with a high DMI resistance (HR). The fourth SNP led to predicted synonymous mutation ('silent mutation') at codon 169 and was present on the *CbCyp51* gene in a few Greek *C. beticola* DMI-resistant isolates that showed an overexpression of the *Cyp51* gene. The study hypothesized that the silent mutation may likely be associated with overexpression of the *CbCyp51* gene leading to *C. beticola* DMI resistance [32]. Furthermore, it was indicated that a high ED_{50} value for tetraconazole (>1.0 µg ml^{-1}) was likely associated with overexpression of the *CbCyp51* gene, and was reported in *C. beticola* DMI-resistant isolates from the Red River Valley (RRV) region in the United States [33].

The research objective was to investigate a hypothesis that a silent mutation, on the *CbCyp51* gene, was potentially correlated with *C. beticola* DMI resistance. First, we analyzed the genetic changes on

---
*Corresponding author: James O Obuya, Center for Viticulture and Small Fruit Research, Florida A&M University, 6505 Mahan Drive, Tallahassee, FL 32317, USA
E-mail: james.obuya@famu.edu

the *CbCyp51* gene for DMI-resistant and -sensitive isolates. Second, we investigated whether a silent mutation on the *CbCyp51* gene had a potential role in conferring *C. beticola* DMI resistance. This was achieved by transforming the *CbCyp51* gene from DMI-resistant and -sensitive isolates in an isogenic yeast R-1 (ΔPdr-5), and determined ED_{50} values (effective dose leading to growth inhibition of 50%) for DMI fungicides used for disease control in sugar beet. Transformed yeast R-1 was chosen because it was the fastest method to test the hypothesis as compared to using the *C. beticola,* which required a complex transformation protocol.

Materials and Methods

Fungal strains

We analyzed 8 *C. beticola* isolates which included 4 DMI-resistant isolates (RR-08-553, RR-08-760, RR-08-762, and RR-08-940), a DMI-sensitive isolate RR-08-418 from the Red River Valley (RRV) region of the United States; and 2 DMI-resistant isolates (UW11-60 and UW11-81) from the 2010 CLS survey in the Central High Plains (Colorado, Nebraska, Montana and Wyoming) of the United States. One Greek DMI-resistant isolate GR-10-292 [32] was included in the analysis and compared with isolates from the United States. An isogenic yeast R-1 strain (*Ura3-* and *His1-*) was transformed with the *Cyp51* gene from either *C. beticola* DMI-sensitive or *C. beticola* DMI-resistant isolates. The yeast R-1 was a knockout strain lacking a multidrug ABC transporter-Pdr5 [34]. The yeast R-1 strain particularly lacked the protein required for an efflux mechanism which has been associated with a multidrug resistance in fungi [14,15,35].

Fungicides and growth media

DMIs used for fungicide sensitivity assays included tetraconazole (Sipcam Agro USA Inc., Roswell, GA), propiconazole (Syngenta Crop Protection, Greensboro, NC), and difenoconazole (Syngenta Crop Protection). Transformed and non-transformed yeast R-1 strains were cultured on modified selective dropout medium (SM) plates [36]. The modified SM medium included 68 g of yeast nitrogen base (BD Diagnostics, Sparks, MD) and 200 g of ammonium sulfate (Spectrum Quality Products Inc., Gardena, CA). The medium was supplemented with 0.44 µg ml⁻¹ of L-histidine (Sigma-Aldrich, St. Louis, MO) and amended with 300 µg ml⁻¹ of Geneticin G-418 (Teknova, Hollister, CA). However, control plates were supplemented with 0.44 µg ml⁻¹ of L-uracil (Sigma-Aldrich) because non-transformed yeast R-1 strains required L-uracil in order to grow on the modified SM medium [34].

DNA isolation

Mycelia on dry medium, preserved cryogenic vials (Corning Inc., NY) and frozen at -72°C were obtained using a sterile forcep (Fisher Scientific, Pittsburg, PA) to inoculate ~15 ml of nutrient broth (BD Diagnostics). Inoculated cultures were incubated at 27°C by shaking at 200 rpm on an orbital shaker (Lab-line orbital shaker, Romeoville, IL) for 7 days. Mycelia were harvested by straining the liquid medium through sterile cheesecloth, rinsed three times using sterile distilled water (5 ml), lyophilized for 48 h, and pulverized (~20 mg) under liquid nitrogen using a sterile mortar and pestle. We purified DNA using a plant mini kit (DNeasy®, Qiagen Inc., Valencia, CA) and measured absorbance at 260 nm using a spectrophotometer (Nanodrop®-1000, Wilmington, DE).

Sequencing and analysis of *Cyp51* gene from *C. beticola* isolates

Primer pairs (Table 1) were designed based on a consensus sequence

of the *Cyp51* gene from *C. beticola* isolates for PCR amplification. PCR reactions (25 µl) included DNA (2 µl), each of the primers (0.4 µM), dNTPs (200 µM each), 1X of a high fidelity buffer with 1.5 mM MgCl₂ (Phusion®: ThermoFisher Scientific, Lafayette, CO), 0.4 units µl⁻¹ of a high fidelity DNA polymerase (Phusion®: ThermoFisher Scientific), and a final volume was adjusted by the addition of PCR grade water. PCR run parameters included an initial denaturation at 98°C for 30 s, 25 cycles of denaturation at 98°C for 10 s, annealing at 60°C for 30 s, and extension at 72°C for 30 s with a final extension at 72°C for 10 min. PCR products were separated by gel electrophoresis on a 0.8% agarose gel and 0.5X of Tris borate-EDTA buffer (EMD Chemicals, Gibbstown, NJ), stained with ethidium bromide (IBI Scientific, Peosta, IA), and visualized under UV (570-640nm) illumination (UVP LLC, Upland, CA).

An aliquot (5 µl) was obtained from each PCR product, treated with ExoSAP-IT® reagent (2 µl) (USB Corporation, Cleveland, OH) to remove unincorporated primer pairs and non-specific fragments. The reaction was incubated at 37°C for 15 min and inactivated at 80°C for 15 min. Cleaned PCR products were sequenced at the Sequetech Corporation (Mountain View, CA) using a set of primers that included Cbdm-519F, CBdm906F, and CBdm2284R (Table 1). Three overlapping sequences of the *Cyp51* gene were obtained that included a partial fragment of the promoter region, a complete open reading frame (ORF), and a partial fragment of the 3' untranslated region (3' UTR).

Characterization of *Cyp51* gene from *C. beticola* isolates

Partial sequences were assembled using the web version of CAP3 program and were analyzed using the DNA Baser Software (HeracleSoftware, Germany). The assembled sequences were analyzed for fidelity in which the predicted nucleotides were compared to chromatograms from sequenced data. An alignment was obtained for assembled partial sequences and a sequence from a *C. beticola* DMI-sensitive isolate (GenBank accession # HM778021) and was performed using the *MEGA* ver. 5 software [37]. We determined predicted amino acid residues for the *Cyp51* protein for each *C. beticola* isolate, and compared those residues with predicted amino acid residues for *C. beticola* DMI-sensitive isolate (GenBank accession # ADW54535) using the *MEGA* ver. 5 software [37]. This was performed to identify any potential amino acid changes which could be present on the *C. beticola Cyp51* protein.

Plasmid construction and yeast transformation

Partial fragments of the *C. beticola Cyp51* gene were obtained using a pair of primers: Cbdm-519F and Cbdm2284R and a nested PCR was performed using a second pair of primers: CbdmBamHIF and CbdmNotIR (Table 1), to obtain PCR products with an open reading frame (ORF) of the *C. beticola Cyp51* gene (~1625 bp). Standard cloning was performed and was followed by sequencing to confirm

Primer	Sequences (5' to 3')	References
Cbdm-519F	gttgtatgccgctttggagt	This work
Cbdm208F	gcatcgacccgtacaagttc	This work
Cbdm906F	agaggtggcacacatgatga	This work
Cbdm2284F	ttgcttcaatactggatgctt	This work
CbdmBamHIF	cgatggatccgttgtatgccgctttg	This work
CbdmNotIR	tttgcggccgcagtgtgtccaagg	This work
CbdmCyp51F	tgccacgcgacgagacattcaagatgagc	This work
CbdmCyp51R	cagctcctttgctgaccagaccgtagc	This work

Table 1: Primers used for PCR amplification of *Cyp51* gene.

C. beticola isolate	Modified log-logistic model[1]					
	ED_{50}	b[2]	c[3]	d[4]	e[5]	f[6]
DMI-sensitive RR-08-418	0.96	0.81	2.34	7.86	0.005	55.53
	(1.2)[7]	(-0.2)	(-2)	(-2.1)	(-0.01)	(-53.8)
DMI-resistant RR-08-553	34.4	0.81	-4.91	15.15	10.62	24.07
	(-45.9)	(-0.7)	(-10.6)	(-1.5)	(-10.7)	(-26.8)
DMI-resistant RR-08-760	35.9	0.84	-4.28	13.67	11.19	22.49
	(-52.6)	(-0.9)	(-10.7)	(-1.6)	(-11.5)	(-26.8)
DMI-resistant RR-08-762	21	1.24	-1.55	14.14	10.77	16.22
	(-41.6)	(-2.7)	(-11.3)	(-1.6)	(-6.2)	(-21.1)
DMI-resistant RR-08-940	27.7	0.72	-5.04	13.67	4.24	40.68
	(-91.4)	(-1.3)	(-22.1)	(-1.5)	(-6.4)	(-76.3)
DMI-resistant GR-10-292	65.6	0.78	-5.39	9.43	9.49	37.82
	(-96.7)	(-0.6)	(-10.9)	(-1.5)	(-9)	(-26.1)

Table 2: Dose-response parameters for C. beticola isolates on potato dextrose agar amended with tetraconazole. Dose-response parameters were obtained using a four-parameter log-logistic model [45] and shown are model parameters and ED50 (medium effective dose reducing growth by 50%) for each C. beticola isolate on PDA amended with tetraconazole and compared with non-amended control.
[1] Modified log-logistic model (CRS.5) with fixed α = 0.25
[2] Steepness of the curve after the maximal hormetic effect
[3] Lower limit of the dose-response curve
[4] Upper limit of the dose-response curve
[5] Lower bound of the dose in which growth was reduced by 50%
[6] Denotes the theoretical upper bound of the hormetic effect
[7] Numbers in parentheses are standard errors

whether our insert which included the plasmid vector was present and determined its orientation [38]. Partial PCR products of the *Cyp51* gene were inserted between BamHI and NotI restriction sites of the pCM189-URA3 vector [39]. Plasmid constructs included a pCM189-URA3::mut*Cyp51* (*CbCyp51* from RRV DMI-resistant isolate), pCM189-URA3::mut*Cyp51* (*CbCyp51* from Greek DMI-resistant isolate), and pCM189-URA3::wt*Cyp51* (*CbCyp51* from DMI-sensitive isolate). Each construct was transformed in *E. coli* competent cells using a standard heat-shock procedure as described in a QIAgene expression kit (Qiagen, Inc.). Transformed *E. coli* competent cells were cultured on a Lauria-Bertani (LB) agar plates, supplemented with 100 μg ml⁻¹ of filter-sterilized ampicillin sodium salt (Sigma-Aldrich, St. Louis, MO), and incubated at 30°C overnight. LB broth (5 μl) was inoculated using a single colony of *E. coli* competent cells and incubated at 30°C overnight by shaking at 200 rpm on an orbital shaker (Lab-line orbital shaker, Romeoville, IL). Plasmid was purified from overnight cultures using a Qiaprep® miniprep kit (Qiagen Inc.), and transformed in isogenic yeast R-1 strain [MATα PDR1-3pdr5::KANMX4 ura3 his1 yor1 pdr10 pdr11 ycf1 pdr3] [34] using a *S. cerevisiae* direct transformation kit (Wako Chemicals USA, Inc., Richmond, VA).

Detection of C. beticola Cyp51 messenger RNA

Total RNA was extracted from overnight cultures from either transformed or non-transformed yeast R-1 strain using a yeast RNA kit (E.Z.N.A®: Omega Bio-Tek Inc., Norcross, GA). Total RNA concentration was determined by measuring the absorbance at 280 nm using a spectrophotometer (NanoDrop-1000®: Wilmington, DE). Purified total RNA (5 μl) was separated by gel electrophoresis in 1%

agarose gel, 1X Tris-borate-EDTA buffer (EMD Chemicals) and visualized as described above. The total RNA was reverse-transcribed to cDNA using a qScript® cDNA synthesis kit (Quanta Biosciences Inc., Gathersburg, MD), which included random oligonucleotide primers. The cDNA was PCR-amplified using a pair of primers such as Cbdm*Cyp51*F and Cbdm*Cyp51*R (Table 1). These primer pairs spanned a region of ~425 base pairs that included targeted partial sequence of the *Cyp51* gene and a partial fragment of the plasmid vector. PCR products were cleaned using the ExoSAP-IT® reagent (2 μl) and sequenced at the University of Wyoming-Nucleic Acid Exploration Facility (UW-NAEF) in both directions using Cbdm*Cyp51*F and Cbdm*Cyp51*R (Table 1). Directional cloning of the *Cyp51* gene insert within the plasmid vector was confirmed by PCR amplification. The sequences were subjected to nucleotide-nucleotide BLAST search [40] and compared with other sequences in the NCBI/GenBank® database.

Determination of dose-response

To determine dose-response for *C. beticola* isolates, a fungicide sensitivity assay was performed on a potato dextrose agar (BD Diagnostics, Sparks, MD) using the modified procedure [41]. PDA was amended with each DMI fungicide prepared to a final concentration of 0, 0.001, 0.01, 0.1, 1, 10, and 100 μg ml⁻¹ of the active ingredient and inoculated with *C. beticola* isolates prepared from sugar beet leaf extract agar (SBLEA) [42] plates. We grew yeast R-1 strain transformed with the *C. beticola Cyp51* gene overnight cultures (5 ml) in a yeast extract-peptone-dextrose (YPD) broth (BD Diagnostics). The YPD broth [36] was amended with 300 μg ml⁻¹ of Geneticin G-418 as described above. The overnight cultures were incubated at 30°C by shaking on a 150 rpm orbital shaker (Lab-line orbital shaker) to obtain a cell density of ca. 1 x 10⁶ cells ml⁻¹, which is equivalent to ~0.1 optical density (OD) at 600 nm [43]. Sterile tubes with YPD (5 ml) were either amended with DMI fungicide or non-amended controls, and each was inoculated with 0.5 x 10⁵ ml⁻¹ cells [43]. Additionally, each culture (5 ml) was amended with a DMI fungicide prepared to a final concentration such as 0, 0.01, 0.02, 0.04, 0.08, 0.1, 0.12, 0.16, and 0.32 μg ml⁻¹. Each amended YPD broth (5 ml) was inoculated either with transformed or non-transformed yeast R-1, and incubated at 30°C by shaking at 150 rpm on an orbital shaker (Lab-line orbital shaker) for 24 h. We measured optical density (OD) at 600 nm for cell cultures (200 μl) in 96-well plate.

Statistical analysis

Data analysis was performed using the language of R statistical computing environment [44] and estimated dose-response using a Cedergreen-Ritz-Streibig modified model [45]. Our null hypothesis was that a silent mutation at codon 170 on the *Cyp51* gene had a potential role in *C. beticola* DMI resistance. Hormesis model (CRS.5) was used because growth stimulation was observed at a low dose of tetraconazole. The model was a modified non-linear regression based on five parameters using a dose-response curve (*drc*) statistical add-on package [46,47]. Below was the expression that defined the relationship between response (y) and dose (x)

$$y = c + \frac{d - c + f \exp\left(\dfrac{1}{x^{\alpha}}\right)}{1 + \exp\{b[\ln(x) - \ln(e)]\}} \quad \text{(CRS.5)}$$

Parameters from the CRS.5 model were estimated at fixed alpha value (α=0.25). The dose-response estimated '*y*' which was the maximal response at zero dose, '*c*' was an estimate of lower limit of the dose-response curve, '*b*' estimated the steepness of the curve after the maximal hormetic effect, and '*e*' had no straightforward interpretation in this model, but it estimated a lower bound on the ED_{50} value [45]. Whereas ('$d + f$') estimated upper bound on growth stimulation ('f'>0) whereby a larger value of 'f' was correlated with increased growth stimulation (hormesis) as long as its value was positive. For instance, statistical analysis for hormesis was the same as testing for 'f' = 0 based on fixed α value (0.25). ED_{50} value for each *C. beticola* isolate was estimated based on the modified log-logistic model [45,47] and results were classified as either low (<0.01 μg ml^{-1}), medium (0.01 to 1.0 μg ml^{-1}), or high (>1.0 μg ml^{-1}) [33]. Dose response-curves were fitted based on radial growth of *C. beticola* or optical density (OD) at 600 nm of yeast measurements.

Prediction of *C. beticola Cyp51* mRNA model structure

Putative secondary structures of the *Cyp51* mRNA of *C. beticola* were predicted based on the DNA sequences of a DMI-sensitive isolate CB6-80 (GenBank accession # HM778021) and a DMI-resistant isolate RR-08-940 (GenBank accession # HM778022) using the *RNAfold* program of the Vienna RNA Websuite (http://rna.tbi.univie.ac.at/cgi-bin/RNAfold.cgi) [48]. Determination of secondary RNA structure was performed using the *RNAfold* and was based on a comparative sequence analysis [23]. The algorithm constrains prediction of minimum free energy (MFE), and provides the best thermodynamic folding by taking into consideration factors such as base-pairing and unpaired regions of the sequence [49]. Additionally, the algorithm computes MFE on the assumption of a 'nearest neighbor model' achieved through the application of empirical estimates of thermodynamic parameters on neighboring interactions and loop entropies, and used as a mechanism for scoring folding structures [23,49].

Results

Fungicide sensitivity

Growth inhibition was determined for each *C. beticola* isolate from the Red River Valley (RRV) and Central High Plains of the United States and compared with that for a DMI-resistant isolate from Greece. Radial growth was measured from PDA plates amended with DMIs, inoculated with each *C. beticola* isolate, and compared with corresponding measurements obtained from a non-amended PDA control (Figure 1).

Results revealed that *C. beticola* DMI-sensitive isolate RR-08-418 had low ED_{50} on tetraconazole of 0.96 μg ml^{-1} compared to DMI-resistant isolates with ED_{50} values included 21.0 and 65.6 μg ml^{-1} (Table 3). In addition, all *C. beticola* isolates showed hormesis [46,50] at low concentrations of tetraconazole (Figure 2). No ED_{50} value was obtained for propiconazole and difenoconazole, because these DMIs were too effective against the tested *C. beticola* isolates at doses below 0.01 μg ml^{-1}.

Dose-response curve for a DMI-sensitive isolate RR-08-418 was compared with that for five DMI-resistant isolate RR-08-553, DMI-resistant isolate RR-08-760, DMI-resistant isolate RR-08-762, DMI-resistant isolate RR-08-940, and DMI-resistant isolate GR-10-292. Points included are radial growth measurements at a given dose of tetraconazole. The lines were fitted using the dose-response curve [47].

Sequence analysis of *Cyp51* gene from *C. beticola* isolates

Approximately 2.4 kb of the *C. beticola Cyp51* gene was obtained that included a partial sequence of the promoter as well as an entire coding region for each isolate using a set of four primers (Table 1). Mutations were identified from 4 *C. beticola* field isolates when compared to a baseline sequence of a DMI-sensitive isolate. First, 2 RRV *C. beticola* isolates RR-08-553 and RR-08-940 contained an identical SNP (silent mutation) and was predicted not to lead an amino acid change at codon 170 (Figure 3). Second, we identified two SNPs (silent mutations), predicted not to lead an amino change at codon 355 for a DMI-resistant isolate RR-08-553 and codon 211 for a DMI-resistant isolate RR-08-760 (Table 3). Two non-identical SNPs were identified from the *C. beticola Cyp51* gene, in which one SNP was predicted to lead to amino acid substitution at position 12 (D12N) for a DMI-resistant isolate RR-08-762. A second SNP was predicted to lead to amino acid substitution at position 195 (P195A) for a DMI-resistant isolate RR-08-760. However, 2 *C. beticola* DMI-resistant isolates (UW11-60 and UW11-81) were identical to the baseline sequence of the *CbCyp51* gene obtained from a DMI-sensitive isolate RR-08-418 (Table 3).

Predicted *CbCyp51* Messenger RNA Structure of *C. beticola* Isolate

Predicted structure of the messenger RNA (mRNA) was determined using *RNAfold* based on the *C. beticola Cyp51* sequence of a DMI-sensitive isolate CB6-80 and a DMI-resistant isolate RR-08-940 (Figure 3). The putative mRNA structure for DMI-resistant isolate RR-08-940 with a silent mutation which did not lead to predicted amino acid substitution at position 170 had a slightly different folding pattern on the lower region as compared to that from DMI-sensitive isolate CB6-80 (Figure 4).

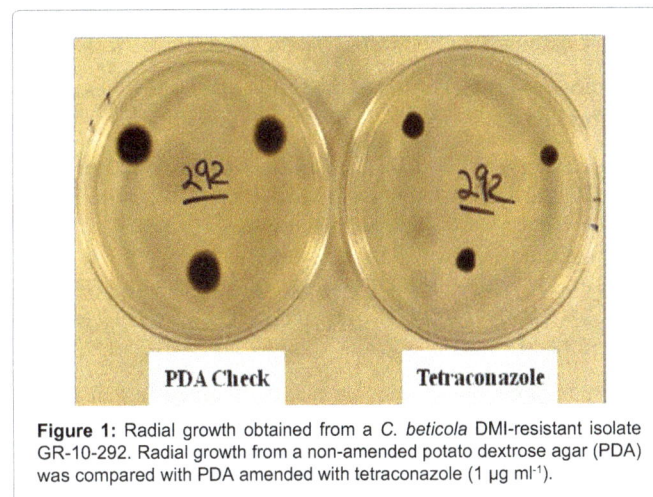

Figure 1: Radial growth obtained from a *C. beticola* DMI-resistant isolate GR-10-292. Radial growth from a non-amended potato dextrose agar (PDA) was compared with PDA amended with tetraconazole (1 μg ml^{-1}).

	C. beticola isolates	Origin of isolate	DMI Phenotype	Mutation(s)
1	RR-08-418	Red River Valley	Sensitive	baseline sequence
2	UW11-60	Central High Plains	Resistant	baseline sequence
3	UW11-81	Central High Plains	Resistant	baseline sequence
4	RR-08-553	Red River Valley	Resistant	GAG to GAA at 170 and AAG to AAA at 355 (silent mutations)
5	RR-08-760	Red River Valley	Resistant	CCT to GCT (P195A)
6	RR-08-762	Red River Valley	Resistant	GGC to GGG at 211 (silent mutation) and GAC to AAC (D12N)
7	RR-08-940	Red River Valley	Resistant	GAG to GAA at 170 (silent mutation)

Table 3: Single nucleotide polymorphic sites (SNPs) from partial sequences of the *Cyp51* gene for *C. beticola* isolates.

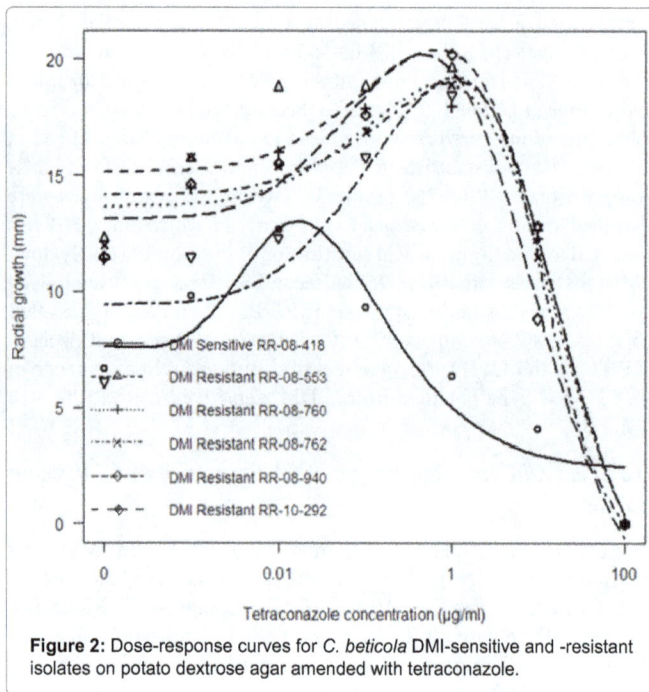

Figure 2: Dose-response curves for *C. beticola* DMI-sensitive and -resistant isolates on potato dextrose agar amended with tetraconazole.

The predicted mRNA folding structure from *C. beticola* DMI-resistant isolate RR-08-940 (GenBank accession # HM778022) showed a lower minimum free energy (mfe) of −565.92 kcal mol^{-1} as compared to a similar mfe of −567.82 kcal mol^{-1} from DMI-sensitive isolate CB6-80 (GenBank accession # HM778021).

Heterologous transcription of the *C. beticola Cyp51* messenger RNA in yeast

Results showed *C. beticola Cyp51* messenger RNA was produced in transformed yeast R-1 strain and was not found in control (Figure 5). PCR products were sequenced, aligned using ClustalW2 [22] and partial sequences showed a 100% identity (e-value = 0) to partial sequence of *C. beticola* DMI-sensitive isolate CB6-80 (GenBank accession # HM778021) or DMI-resistant isolate RR-08-940 (GenBank accession # HM778022).

Dose-response curves were estimated by measuring an optical density (OD) at 600 nm. Yeast R-1 strains included untransformed, transformed with plasmid, and transformed with *Cercospora beticola*: DMI-sensitive isolate RR-08-418 [Transformed Sensitive (418)], a DMI-resistant isolate RR-08-940 [Transformed Resistant (940)], and a DMI-resistant isolate GR-10-292 [Transformed Resistant (292)].

Non-transformed yeast R-1 strain, transformed with plasmid, and transformed with constructs with the *C. beticola Cyp51* gene from a DMI-sensitive isolate RR-08-418, a DMI-resistant RR-08-940, or DMI-resistant GR-10-292 showed medium ED$_{50}$ values of between 0.02 and 0.09 μg ml^{-1} in culture amended with tetraconazole (Table 4). Furthermore, all transformed yeast R-1 strains showed hormesis (growth stimulation) [46,50] at low concentrations of tetraconazole (Figure 6). Growth stimulation was significant especially at low tetraconazole dose (one-sided t-test, $p < 0.05$) for 3 yeast strains (Table 4). However, no response curve was obtained for propiconazole and difenoconazole because the two DMI fungicides were effective at all rates used.

Discussion

DMIs are used for CLS control in sugar beet [18] and resistance has been reported in *C. beticola* field isolates [11,18,33,51,52]. However, the mechanism of *C. beticola* DMI resistance is not known. DMI resistance

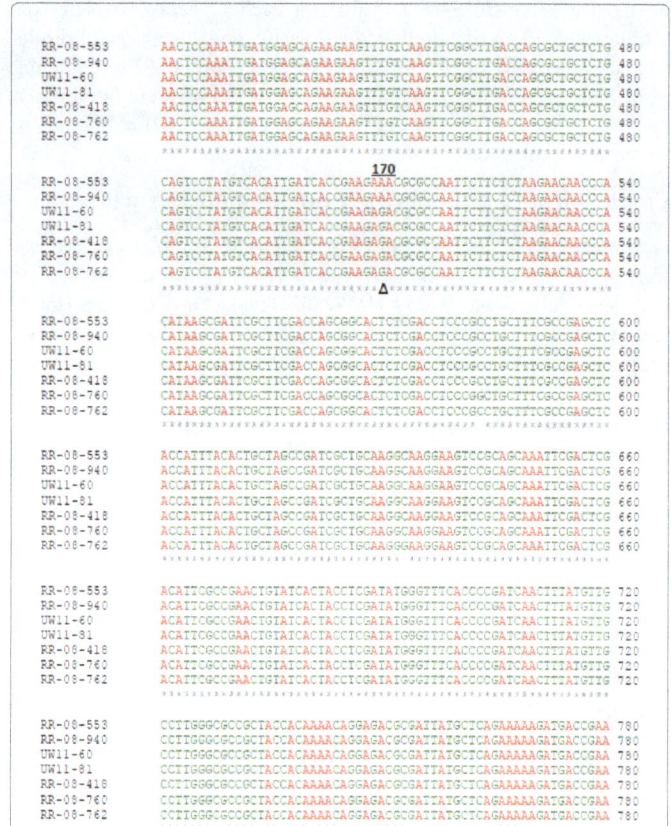

Figure 3: Alignment of partial sequences of the *Cercospora beticola Cyp51* gene. Shown are partial sequences from DMI-resistant (RR-08-553; RR-08-760; RR-08-762, RR-08-940, UW11-60; UW11-81); a point mutation (GAG to GAA) at codon 170 (Δ).

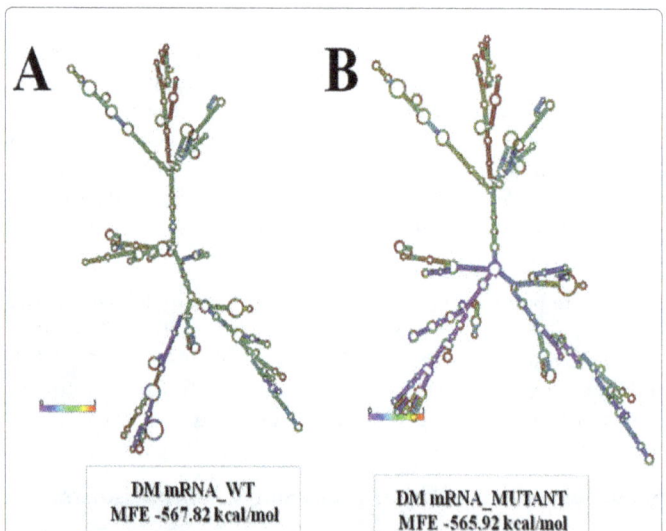

Figure 4: Putative *C. beticola Cyp51* messenger RNA structures. The putative mRNA structures were obtained for (**A**) DMI-sensitive isolate CB6-80 (WT) and (**B**) DMI-resistant isolate RR-08-940 (mutant).

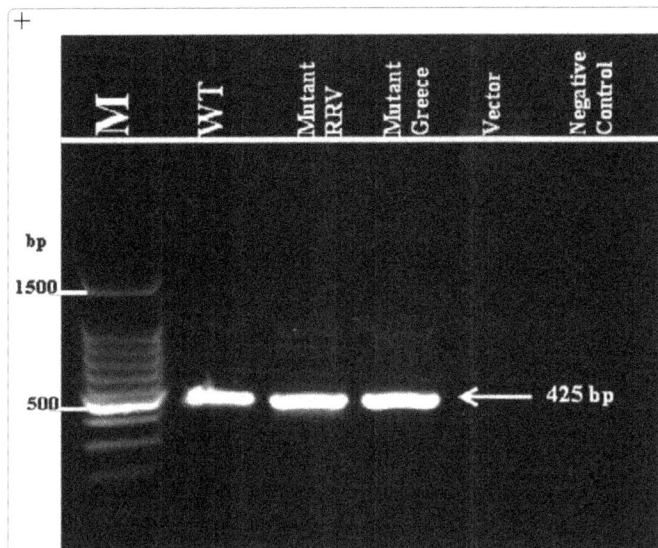

Figure 5: Transcripts from an isogenic yeast R-1 transformed with the *C. beticola Cyp51* gene. Shown are transcripts produced in yeast R-1 strain as well as PCR products (~425 bp each). Constructs included pCM189-URA3:*Cyp51* (RRV DMI-sensitive RR-08-418), pCM189-URA3:mut *Cyp51* (RRV DMI-resistant RR-08-940), and pCM189-URA3:mut*Cyp51* (Greek DMI-resistant GR-10-292). **M**: 100 bp DNA Marker (Promega).

has been associated with mutations predicted to lead to amino acid substitution on the *Cyp51* gene from other fungi [13-15,27,35]. Partial sequences of the *CbCyp51* gene were obtained from *C. beticola* isolates and analyzed for genetic changes. Genetic analysis found a few single nucleotide polymorphic sites (SNPs) within the coding region of the *CbCyp51* gene of five DMI-resistant isolates (Table 3). However, the genetic analysis found no single insertion sequence within the promoter region of *C beticola* DMI-resistant or -sensitive isolates as was reported for either *P. digitatum* DMI-resistant strains [16] or *Monilinia fruticola* DMI-resistant isolates [53]. Results indicated the SNPs identified from 5 DMI-resistant isolates were unlikely associated with *C. beticola* DMI resistance. This is because the genetic analysis found no mutations on the *CbCyp51* gene from 2 Central High Plains DMI-resistant isolates (Table 3). Similarly, a high degree of sequence variation was reported in the coding and the flanking regions for *C. beticola* isolates from the Red River Valley region in the United States, but none of those mutations could be associated with DMI-resistance because SNPs were present in isolates which showed both low and high ED_{50} values [33].

It was noted that among the genetic changes include a silent mutation in the *C. beticola* 14α-demethylase (*CbCyp51*) gene which may be associated with *C. beticola* DMI resistance. Our genetic analysis found a silent mutation (GAG to GAA) at codon 170 (Figure 3). An identical SNP was reported in a few DMI-resistant isolates from Greece [32] as well as two RRV DMI-resistant isolates. Hence, our study investigated whether the silent mutation at codon 170 had a potential role in conferring *C. beticola* DMI resistance. Putative mRNA structures of *C. beticola* DMI-sensitive isolate CB6-80 and *C. beticola* DMI-resistant isolate RR-08-940 were obtained using the *RNAfold*, (Figure 4). This was to determine whether the silent mutation at codon 170 may lead to changes in the *CbCyp51* mRNA structure which could be associated with *C. beticola* DMI resistance. Genetic mutations could potentially change mRNA structures leading to alteration in the amount of protein produced in cells. The *RNAfold* is considered to be a reliable program for finding and comparing hairpins free from pseudoknots (non-nested structural elements), in which, "the analysis

determines similarities between centroid and minimum free energy structure as well as using base-pair distance ensemble between two predicted structures" [48]. The *RNAfold* program computes the energy contributions of elementary substructures (which support parts of a structure) leading to prediction of secondary structure associated with the total minimum free energy for each mRNA structure and it is based on the nearest-neighbor thermodynamic method [23,49]. The silent mutation at codon 170 on the *Cyp51* messenger RNA for *C. beticola* DMI-resistant isolate RR-08-940 led to a putative mRNA structure with a total minimum free energy of –565.92 kcal mol^{-1} as compared to –567.82 kcal mol^{-1}, which was obtained for a *C. beticola* DMI-

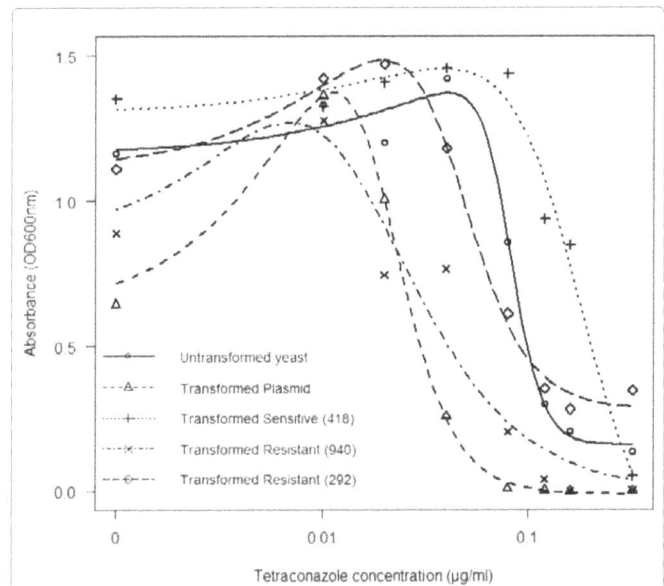

Figure 6: Dose-response curves for yeast R-1 strains in yeast extract-peptone-dextrose (YPD) culture treated with tetraconazole.

Yeast strain	Modified log-logistic model[1]					
	ED_{50}	b[2]	c[3]	d[4]	e[5]	f[6]
Untransformed yeast	0.04	5.76	0.15	1.17	0.08	2.07
	(0.004)[7]	(-1.8)	(-0.06)	(-0.06)	(-0.004)	(-0.96)
Transformed Plasmid	0.07	3.21	–0.01	0.64	0.02	20.66*
	(-0.007)	(-0.6)	(-0.04)	(-0.07)	(-0.002)	(4.06)
Transformed Sensitive (418)	0.05	2.51	–0.30	1.31	0.17	1.78
	(-0.007)	(-0.8)	(-0.36)	(-0.06)	(-0.029)	(-1.08)
Transformed Resistant (940)	0.02	1.65	–0.00	0.89	0.01	25.87*
	(-0.043)	(-0.2)	(-0.06)	(-0.07)	(0.006)	(-17)
Transformed Resistant (292)	0.09	2.89	0.28	1.12	0.04	6.93*
	(-0.005)	(-0.6)	(-0.06)	(-0.07)	(-0.005)	(-2.15)

Table 4: Dose-response parameters and ED50 (medium effective dose reducing growth by 50%) for non-transformed and transformed yeast R-1 strain in yeast extract-peptone-dextrose (YPD) broth amended with tetraconazole
The parameters were determined using a Cedergreen-Ritz-Streibig modified log-logistic model [45].
[1]Modified log-logistic model (CRS.5) with fixed α = 0.25 [45]
[2]Steepness of the curve after the maximal hormetic effect
[3]Lower limit of the dose-response curve
[4]Upper limit of the dose-response curve
[5]Lower bound on the dose at *d-c* reduced by 50%
[6]Denotes the theoretical upper bound of the hormetic effect
[7]Numbers in parentheses are standard errors
*Significantly different from zero (*p*<0.05)

sensitive isolate CB6-80 (Figure 4). Low folding free energy has been associated with low translation rates in addition to the occurrence of high transcript turnovers in *Saccharomyces cerevisiae* [54]. Our results found two different stable structures and were predicted to have similar free energies, although a minimum shift of –2.1 kcal mol^{-1} of predicted minimum free energy was observed between these two putative mRNA structures. The shift, however, was not significant to support our hypothesis that the silent mutation at codon 170 in the *CbCyp51* gene may likely be associated with *C. beticola* DMI resistance.

The genetic analysis of two RRV *C. beticola* DMI-resistant isolates found a few silent mutations on the coding region of the *Cyp51* gene (Table 3). This was determined using a partial fragment of the *CbCyp51* gene obtained from *C. beticola* DMI-sensitive RR-08-418 and DMI-resistant isolates (RR-08-940 and GR-10-292), cloned in a plasmid vector, and transformed into an isogenic yeast R-1 strain. Our results indicated that low concentrations of tetraconazole led to growth stimulation (hormetic response) of *C. beticola* isolates (Figure 2). Hormesis was first reported in yeast [55], and was later reported in fungi such as *Pythium aphanidermatum* mefonoxam-resistant strain, the causal organism of damping off disease in sugar beet, in which 10% growth stimulation was observed in culture amended with sub-lethal doses of mefonoxam [56]. A similar effect was reported for *Sclerotina homoeocarpa*, the causal organism of Dollar leaf spot in turf grass, in which growth stimulation was observed in culture amended with sub-lethal concentrations of trinexapac- ethyl [57]. However, we do not believe that hormesis may be associated with DMI resistance in *C. beticola* isolates because growth stimulation was mainly observed for low concentration of tetraconazole.

Results from heterologous transcription did not support our hypothesis that the silent mutation at codon 170 could be associated with *C. beticola* DMI resistance, because transformed yeast R-1 strains showed low ED$_{50}$ values between 0.01 and 0.17 µg ml^{-1} (Table 2 and Figure 6). However, we determined that the messenger RNA was produced (Figure 5), which proved that the insert was being transcribed in the yeast R-1 strain. A similar silent mutation was reported from a few *C. beticola* DMI-resistant isolates from the Red River Valley (RRV) region in the United States, but it was noted that the silent mutation was unlikely associated with *C. beticola* DMI resistance [33]. The silent mutation, however, was found from three variants of *C. beticola* isolates that showed either low or high EC$_{50}$ values, and it was suggested that the mutation was likely not associated with an overexpression in the *CbCyp51* gene [33]. The first incidence of DMI resistance was reported for two *C. beticola* isolates (UW11-60 and UW11-81) from the Central High Plains region, but the genetic analysis of the *CbCyp51* gene revealed that the isolates showed identical sequence to that of a DMI-sensitive isolate (RR-08-418). Hence, results did not support the hypothesis that the silent mutation at codon 170 was likely conferring *C. beticola* DMI resistance. Although propiconazole and difenoconazole were included in the study, no response curves were obtained because the fungicides were effective at low doses (<0.01 µg ml^{-1}) as compared to tetraconazole.

We could not develop a diagnostic assay for the rapid detection of *C. beticola* DMI resistance because the mechanism is not known. Hence, further studies will be required to investigate additional mechanisms which have been associated with DMI resistance in other fungi [14,15,35]. Currently, screening for *C. beticola* for DMI resistance could effectively be performed using fungicide sensitivity.

Acknowledgements

Funding support was received from the University of Wyoming Agricultural Experiment Station and the Western Sugar Cooperative. Dr. Gary Secor (North Dakota State University, Fargo, ND, USA) provided 5 RRV *C. beticola* isolates; Dr. Dimitri Georgakopoulos (Agricultural University, Athens, Greece) provided a Greek *C. beticola* isolate. Dr. John Golin (Catholic University of America, Washington, D.C., USA) provided an isogenic yeast R-1 strain and Dr. Peter Thorsness of the University of Wyoming, Laramie, WY, USA) provided a pCM189 cloning vector. We thank Dr. Andrew Kniss (Department of Plant Science, University of Wyoming, Laramie, WY, USA) for help with data analysis.

References

1. Jacobsen BJ, Franc GD (2009) Cercospora leaf spot. in: Harveson RM, Hanson LE, Hein GL (eds) Compendium of Beet Diseases and Pests, APS Press St. Paul, MN, 7-10.

2. Duffus JE, Ruppel EG (1993) Diseases. in: Cooke DA, Scott RK (eds)The Sugar Beet Crop: Science into Practice, Chapman & Hall, London, NY, 346-427.

3. Khan MFR, Smith LJ (2005) Evaluating fungicides for controlling Cercospora leaf spot on sugar beet. Crop Prot 24: 79-86.

4. Shane WW, Teng PS (1992) Impact of Cercospora leaf spot on root weight, sugar yield, and purity of *Beta vulgaris*. Plant Dis 76: 812-820.

5. Khan J, del Rio LE, Nelson R, Khan MFR (2007) Improving the Cercospora leaf spot management model for sugar beet in Minnesota and North Dakota. Plant Dis 91: 1105-1108.

6. Meriggi P, Rosso F, Ioannidis PM, J. AG (2000) Fungicide treatments against Cercospora leaf spot in sugarbeet (*Beta vulgaris* L.). in: Asher MJC, Holtschulte B, Richard Molard M, Rosso F, Steinracken G, Beckers R (eds) *Cercospora beticola* Saccâ-Biology, Agronomic Influence and Control Measures in Sugar Beet, Advances in Sugar Beet Research, Vol 2, IIBR, Brussels. 77-102.

7. Miller J, Rekoske M, Quinn A (1994) Genetic resistance, fungicide protection and variety approval policies for controlling yield losses from Cercospora leaf spot infection. J Sugar Beet Res 31: 7-12.

8. Smith GA, Martin SS (1978) Differential response of sugarbeet cultivars to Cercospora leaf spot disease. Crop Sci 18: 39-42.

9. Weiland J, Koch G (2004) Sugarbeet leaf spot disease (*Cercospora beticola* Sacc.)dagger. Mol Plant Pathol 5: 157-166.

10. Brown MC, Waller CD, Charlet C, Palmieri R (1986) The use of flutriafol based fungicides for the control of sugar beet diseases in Europe. in: 1986 British Crop Protection Conference-Pests and Diseases, Vol 3, British Crop Protection Council. 1055-1061.

11. Karaoglanidis GS, Loannidis PM, Thanassouloupoulos CC (2002) Changes in sensitivity of *Cercospora beticola* populations to sterol-demethylation-inhibiting fungicides during a four year period in Greece. Plant Pathol 51: 55-62.

12. Georgopoulos SG, Skylakakis G (1986) Genetic variability in the fungi and the problem of fungicide resistance. Crop Prot 5: 299-305.

13. Cools HJ, Fraaije BA (2008) Are azole fungicides losing ground against Septoria wheat disease? Resistance mechanisms in *Mycosphaerella graminicola*. Pest Manag Sci 64: 681-684.

14. Délye C, Bousset L, Corio-Costet MF (1998) PCR cloning and detection of point mutations in the eburicol 14alpha-demethylase (CYP51) gene from *Erysiphe graminis* f. sp. *hordei*, a "recalcitrant" fungus. Curr Genet 34: 399-403.

15. Délye C, Laigret F, Corio-Costet MF (1997) A mutation in the 14 alpha-demethylase gene of *Uncinula necator* that correlates with resistance to a sterol biosynthesis inhibitor. Appl Environ Microbiol 63: 2966-2970.

16. Hamamoto H, Hasegawa K, Nakaune R, Lee YJ, Makizumi Y, et al. (2000) Tandem repeat of a transcriptional enhancer upstream of the sterol 14alpha-demethylase gene (CYP51) in *Penicillium digitatum*. Appl Environ Microbiol 66: 3421-3426.

17. Joseph-Horne T, Hollomon DW (1997) Molecular mechanisms of azole resistance in fungi. FEMS Microbiol Lett 149: 141-149.

18. Karaoglanidis GS, Loannidis PM, Thanassouloupoulos CC (2000) Reduced sensitivity of *Cercospora beticola* isolates to sterol demethylation inhibiting fungicides. Plant Pathol 49: 567-572.

19. Kӧller W, Wilcox WF (1999) Evaluation of tactics for managing resistance of *Venturia inaequalis* to sterol demethylation inhibitors. Plant Dis 83: 857-863.

20. Leroux P, Walker AS (2011) Multiple mechanisms account for resistance

to sterol 14α -demethylation inhibitors in field isolates of *Mycosphaerella graminicola*. Pest Manag Sci 67: 44-59.

21. Ma Z, Michailides TJ (2005) Advances in understanding molecular mechanisms of fungicide resistance and molecular detection of resistant genotypes in phytopathogenic fungi. Crop Prot 24: 853-863.

22. Ma Z, Morgan DP, Felts D, Michailides TJ (2002) Sensitivity of *Botryosphaeria dothidea* from California pistachio to tebuconazole. Crop Prot 21: 829-835.

23. Mathews DH1, Sabina J, Zuker M, Turner DH (1999) Expanded sequence dependence of thermodynamic parameters improves prediction of RNA secondary structure. J Mol Biol 288: 911-940.

24. Romero RA, Sutton TB (1997) Sensitivity of *Mycosphaerella fijiensis*, causal agent of Black Sigatoka of banana, to propiconazole. Phytopathology 87: 96-100.

25. Schnabel G, Jones AL (2001) The 14alpha-Demethylasse(CYP51A1) gene is overexpressed in *Venturia inaequalis* strains resistant to myclobutanil. Phytopathology 91: 102-110.

26. Stanis VF, Jones AL (1985) Reduced sensitivity to sterol-inhibiting fungicides in field isolates of *Venturia inaequalis*. Phytopathology 75: 1098-1101.

27. Leroux P, Albertini C, Gautier A, Gredt M, Walker AS (2007) Mutations in the CYP51 gene correlated with changes in sensitivity to sterol 14 alpha-demethylation inhibitors in field isolates of *Mycosphaerella graminicola*. Pest Manag Sci 63: 688-698.

28. Cools HJ, Fraaije BA, Kim SH, Lucas JA (2006) Impact of changes in the target P450 CYP51 enzyme associated with altered triazole-sensitivity in fungal pathogens of cereal crops. Biochem Soc Trans 34: 1219-1222.

29. Decottignies A1, Goffeau A (1997) Complete inventory of the yeast ABC proteins. Nat Genet 15: 137-145.

30. Del Sorbo G, Andrade AC, Van Nistelrooy JG, Van Kan JA, Balzi E, et al. (1997) Multidrug resistance in *Aspergillus nidulans* involves novel ATP-binding cassette transporters. Mol Gen Genet 254: 417-426.

31. Tobin MB, Peery RB, Skatrud PL (1997) Genes encoding multiple drug resistance-like proteins in *Aspergillus fumigatus* and *Aspergillus flavus*. Gene 200: 11-23.

32. Nikou D, Malandrakis A, Konstantakaki M, Vontas J, Markoglou A, Ziogas B (2009) Molecular characterization and detection of overexpressed C-14 alpha-demethylase-based DMI resistance in *Cercospora beticola* field isolates. Pestic Biochem Physiol 95: 18-27.

33. Bolton MD, Birla K, Rivera-Varas V, Rudolph KD, Secor GA (2012) Characterization of CbCyp51 from field isolates of *Cercospora beticola*. Phytopathology 102: 298-305.

34. Sauna ZE1, Bohn SS, Rutledge R, Dougherty MP, Cronin S, et al. (2008) Mutations define cross-talk between the N-terminal nucleotide-binding domain and transmembrane helix-2 of the yeast multidrug transporter Pdr5: possible conservation of a signaling interface for coupling ATP hydrolysis to drug transport. J Biol Chem 283: 35010-35022.

35. de Waard MA, Andrade AC, Hayashi K, Schoonbeek HJ, Stergiopoulos I, et al. (2006) Impact of fungal drug transporters on fungicide sensitivity, multidrug resistance and virulence. Pest Manag Sci 62: 195-207.

36. Treco DA, Lundblad V (2001) Preparation of yeast media. in: Ausubel FM, Brent R, Kingston RE, Moore DD, Seidman JG, Smith JA, Struhl K (eds) Current protocols in molecular biology.

37. Tamura K, Peterson D, Peterson N, Stecher G, Nei M, et al. (2011) MEGA5: molecular evolutionary genetics analysis using maximum likelihood, evolutionary distance, and maximum parsimony methods. Mol Biol Evol 28: 2731-2739.

38. Sambrook J, Fritsch EF, Maniatis T (1989) Molecular cloning: a laboratory manual, Vol. 1. (4thedn). Cold Spring Harbor Laboratory Press, Cold Spring Harbor, NY.

39. Garí E, Piedrafita L, Aldea M, Herrero E (1997) A set of vectors with a tetracycline-regulatable promoter system for modulated gene expression in Saccharomyces cerevisiae. Yeast 13: 837-848.

40. Zhang Z, Schwartz S, Wagner L, Miller W (2000) A greedy algorithm for aligning DNA sequences. J Comput Biol 7: 203-214.

41. Briere SC, Franc GD, Kerr ED (2001) Fungicide sensitivity characteristics of *Cercospora beticola* isolates recovered from the High Plains of Colorado, Montana, Nebraska, and Wyoming. 1. Benzimidazole and triphenyltin hydroxide. J Sugar Beet Res 38: 111-120.

42. Calpouzos L, Stallknecht GF (1966) Photoperiodism by conidiophores of *Cercospora beticola*. Phytopathology 56: 702-704.

43. Gietz D, St Jean A, Woods RA, Schiestl RH (1992) Improved method for high efficiency transformation of intact yeast cells. Nucleic Acids Res 20: 1425.

44. Dean CB, Nielsen JD (2007) Generalized linear mixed models: a review and some extensions. Lifetime Data Anal 13: 497-512.

45. Cedergreen N, Ritz C, Streibig JC (2005) Improved empirical models describing hormesis. Environ Toxicol Chem 24: 3166-3172.

46. Knezevic SZ, Streibig JC, Ritz C (2007) Utilizing R software package for dose-response studies: the concept and data analysis. Weed Tech 21: 840-848.

47. Ritz C, Streibig JC (2005) Bioassay analysis using R. J Stat Softw 12: 1-22.

48. Gruber AR, Lorenz R, Bernhart SH, Neuböck R, Hofacker IL (2008) The Vienna RNA websuite. Nucleic Acids Res 36: W70-74.

49. Gardner PP, Giegerich R (2004) A comprehensive comparison of comparative RNA structure prediction approaches. BMC Bioinformatics 5: 140.

50. Schabenberger O, Tharp BE, Kells JJ, Penner D (1999) Statistical tests for hormesis and effective dosages in herbicide dose response. Agron J 91: 713-721.

51. Karaoglanidis GS, Thanassoulopoulos CC (2003) Cross-resistance patterns among sterol biosynthesis inhibiting fungicides (SBIs) in *Cercospora beticola*. Eur J Plant Pathol 109: 929-934.

52. Secor GA, Rivera VV, Khan MFR, Gudmestad NC (2010) Monitoring fungicide sensitivity of *Cercospora beticola* of sugar beet for disease management decisions. Plant Dis 94: 1272-1282.

53. Luo CX, Schnabel G (2008) The cytochrome P450 lanosterol 14alpha-demethylase gene is a demethylation inhibitor fungicide resistance determinant in *Monilinia fructicola* field isolates from Georgia. Appl Environ Microbiol 74: 359-366.

54. Ringnér M, Krogh M (2005) Folding free energies of 5'-UTRs impact post-transcriptional regulation on a genomic scale in yeast. PLoS Comput Biol 1: e72.

55. Stebbing AR (1982) Hormesis--the stimulation of growth by low levels of inhibitors. Sci Total Environ 22: 213-234.

56. Garzon CD, Molineros JE, YÃ¡nez JM, Flores FJ, del Mar JimÃ©nez-Gasco M, Moorman GW (2011) Sublethal doses of mefenoxam enhance Pythium damping-off of geranium. Plant Dis 95: 1233-1238.

57. Ok C, Popko JT, Campbell-Nelson K, Jung G (2010) In vitro assessment of *Sclerotinia homoeocarpa* resistance to fungicides and plant growth regulators. Phytopathology 100: S92-S92.

32

New Method for Isolation of Plant Probiotic Fluorescent Pseudomonad and Characterization for 2,4-Diacetylphluoroglucinol Production under Different Carbon Sources and Phosphate Levels

Sumant Chaubey, Malini Kotak and Archana G*

Department of Microbiology & Biotechnology Centre, Faculty of Science, The Maharaja Sayajirao University of Baroda, Vadodara-390 002, Gujarat, India

Abstract

Aims: Present work describes the new enrichment method for the isolation of effective root colonizing and rhizospheric competent strains of genus fluorescent Pseudomonad and study of metabolic regulation of 2,4-DAPG biosynthesis in them under carbon sources and Pi levels.

Methods and Results: Three rounds of plant assay was performed using root tip attached microorganism mixtures for the next round of root treatment followed by phonotypical separation of fluorescent colonies to isolate fluorescent pseudomonad strains from different crop and vegetables rhizospheres. Isolated strains were characterized for their Plant Growth Promoting Rhizobacteria (PGPR) traits *viz* phosphate solubilisation, production of siderophore, IAA, HCN, 1-aminocyclopropane-1-carboxylate /L-methionine utilization pathway and antifungal metabolites production. Isolated strains have shown high 2,4-diacetylphluoroglucinol production and strain G2 has shown 4.6 fold high production than *Pf* CHA0. Conclusions: Strain G1 and G8 supported 2,4-DAPG production under sucrose and found to be suitable biocontrol for sucrose rich rhizosphere. Strain G1 and G2 showed good 2,4-DAPG production at high Pi and will perform well in phosphate fertilizer supplemented soils.

Significance of Study: Identification of factors favorable for bio-control will facilitate the targeted application of specific strains to plant rhizosphere/soil type/fertilizer supplemented suitable to their biocontrol activity i.e. "prescription" controls.

Keywords: Fluorescent pseudomonad; PGPR traits; *PhlD*; 2,4-DAPG; Carbon sources; Pi

Introduction

Fluorescent pseudomonad represents a major group of the plant beneficial rhizobacteria present in various crop rhizospheres [1-3]. Fluorescent *Pseudomonas* sp. control plant diseases by antibiosis [4-6] competition for niches and nutrients i.e. effective root colonization [7]. Competitive root tip colonization by *Pseudomonas* strains can play an important role in the efficient control of soil borne crop diseases caused by fungi [8-10]. Three major types of molecules found to be involved in the antagonism towards soil borne fungal pathogens: siderophores, antibiotics and HCN [11-13]. Among the antibiotics the polyketide-2,4-diacetylphloroglucinol (2,4-DAPG) has received particular attention because of its broad-spectrum antifungal, antibacterial and antihelminthic activity [6,12,14,15]. *P. fluorescens* CHA0 isolated from a Swiss soil naturally suppressive to black root rot of tobacco caused by *Chalara elegans* (synanamorph *Thielaviopsis basicola,*). *Pf* CHA0 reduces the extent of disease caused by several root-pathogenic fungi such as *Thielaviopsis basicola, Gaeumannomyces graminis* var. *tritici* (Ggt), *Pythium ultimum, Rhizoctonia solani,* and *Fusarium oxysporum* [16]. In many of these studies, production of 2,4-DAPG has emerged to be a key factor in the biological control activity of *Pf* CHA0 [17]. *Pf* CHA0 has been used as a model organism to identify biosynthetic genes of HCN and 2,4-DAPG and to study their regulation [16,17].

Certain plant growth promoting rhizobacteria (PGPR) contain the enzyme ACC deaminase to lower endogenous levels of ethylene by hydrolyzing ACC into α-ketobutyrate and ammonia, which affects plant growth [18], while majority of soil microorganisms produce ethylene from methionine (L-MET) via the 2-keto-4-methylthiobutyric acid (KMBA) pathway. Methionine is deaminated to produce 2-keto-4-methylthiobutyric acid (KMBA), which is then oxidized to produce

ethylene by *Escherichia, Pseudomonas, Bacillus, Acinetobacter, Aeromonas, Rhizobium, and Corynebacterium* species [19]. Some bacterial strains have either ACC deaminase activity (*Pseuodomonas Putida* biotype A, A7), or the ability to produce ethylene from L-MET (*Acinetobacter calcoaceticus*, M9) or both (*Pseudomonas fluorescens,* AM3) [18].

Sugars constitute a major component in root exudates and a very labile source of carbon for microorganisms [20]. Jaeger et al. [21] has reported that sucrose availability was highest at the tip section of the grass root and decreased in progressively older sections. Effective root colonizer and plant growth promoting strains of fluorescent pseudomonas have isolated worldwide and it was found to be a time consuming isolation method. Method of isolation can be made easy by the enrichment of effective root colonizers and plant growth promoting fluorescent pseudomonad using specific enriching conditions. Kuiper et al. [22] described a method to select enhanced grass root tip colonizing bacteria. In this method a mixture of rhizosphere bacteria is applied on a sterile seedling. After plant growth in a gnotobiotic system

---

*Corresponding author: Archana G, Department of Microbiology and Biotechnology Centre, Faculty of Science, The Maharaja Sayajirao University of Baroda, Vadodara, Gujarat, India
E-mail: sumant.msu@gmail.com

those bacteria that have reached the root tip are isolated. These are subsequently used to inoculate a fresh sterile seedling, which again is allowed to grow [23]. After three of these enrichment cycles, excellent competitive root tip colonizers were obtained [15]. Kamilova et al. [15] used this method to select enhanced tomato root tip colonizers. To our knowledge, no procedures have been described, which facilitate the selection of effective root colonizing and plant beneficial fluorescent pseudomonad.

2,4-DAPG producing pseudomonad are commonly found in the rhizosphere of important crops such as cucumber, maize, pea, tobacco, tomato, and wheat and protect from severe phytopathogens [3,9,19,22,24-26]. In the fungus *Pythium ultimatum* var. *sporangiiferum*, 2,4-DAPG causes alterations of the plasma membrane, vacuolization, and the disintegration of cell contents [24], suggesting that it impedes the maintenance of membrane integrity. In bacteria, 2,4-DAPG may cause lysis by a novel antibiotic mechanism. For instance, methicillin-resistant *Staphylococcus aureus* will lyse within 2 h of exposure to 5 μM 2,4-DAPG[27], whereas *Vibrio parahaemolyticus* lyses more slowly and in response to higher 2,4-DAPG concentrations of 114 μM [28]. The biosynthetic locus for 2,4-DAPG includes *phlA*, *phlC*, *phlB*, and *phlD*, which are transcribed as an operon from a promoter upstream of *phlA* [29]. PhlD is responsible for the production of monoacetylphloroglucinol (MAPG), and PhlA, PhlC, and PhlB are necessary to convert MAPG to 2,4-DAPG. PhlD is especially interesting because of its homology to members of the highly conserved chalcone and stilbene synthase family of plant enzymes, which is suggestive of a common evolutionary origin [29]. Probes and primers specific for sequences in *phlD* have been used in combination with colony hybridization and polymerase chain reaction (PCR) to quantify population sizes of 2,4-DAPG producers in the rhizosphere environment [1,14,21,30].

Important obstacle to commercial application of efficient 2,4-DAPG producers is the inconsistency of their performance [31]. Although its ability to reduce the severity of diseases caused by soil borne fungal pathogens under laboratory conditions has been reported in several studies, inconsistent performance in commercial settings and field trials tends to be disappointing [25]. Understanding the sources of variability is key to overcoming this obstacle. Because a primary mechanism of disease suppression available to fluorescent pseudomonad is antibiosis [25], it is thought that variable performance might result from variation in production of antimicrobial compounds like 2,4-DAPG production. Variable performance might be because of variability in 2,4-DAPG production due to variations in environmental conditions, abiotic and biotic, that might confront bacterial metabolite production in the rhizosphere. Inorganic phosphate inhibited PHL production in different ARDRA groups of fluorescent pseudomonad to various degrees in the study by Duffy and Defago, 1999. PHL production by CHA0 was almost abolished by 10 mM phosphate, whereas 100 mM phosphate reduced production by Q2-87 by only 10-fold and no strain was insensitive to 100 mM phosphate [31]. Glucose but not glycerol enhanced 2,4-DAPG production in *P. fluorescens* Pf-5 and CHA0 [31], whereas in *P. fluorescens* F113 production of 2,4-DAPG and MAPG is stimulated by sucrose and Fe^{3+} ions but is poor in the presence of succinate [31]. Thus, much information about the mechanism and factors affecting the action of 2,4-DAPG are available, but a more insight is needed about the ecological interactions taking place in the soil and root environments, which might influence production of 2,4-DAPG [13,17]. This will help to customize the biocontrol strains for use in particular environments, we can understand how to prepare the inoculum for optimal performance, the environment can be modified

to be more favourable to strains, or strains could be constructed that are independent of environmental signals.

We have screened the enhanced root tip colonizers and plant growth promoting strain of fluorescent *Pseudomonas* and checked for their ability to control the disease caused by phytopathogen *Rhizoctonia bataticola*. In the present study, Pf CHA0 was used as a model organism. This study deals with isolation of rhizospheric fluorescent pseudomonad from various plants and ecologically diverse locations, characterization of their plant growth promoting traits and detail study on 2,4-Diacetyl phloroglucinol production in the isolates.

Materials and Methods

Isolation of efficient root colonizing and plant growth promoting fluorescent pseudomonad

Seed sterilization: Equal size *Vigna radiata* seeds were thoroughly washed with sterile distilled water. Seed were further treated with 1% $HgCl_2$ (For 2 minute) followed treatment of 70% ethanol (For 2 minute). Final wash of sterile distilled water to remove traces of $HgCl_2$ and seeds were transferred to sterile petri plates containing wet filter paper. Sterile seeds on soaked filter paper were incubated 30 ± 2°C and kept in dark. Seeds were allowed to germinated up to radicle size of 1 cm.

Enrichment method for isolation of fluorescent pseudomonad: Isolation of fluorescent pseudomonad involved the three successive round of plant inoculation. Root samples of were washed with distill water for two to three times to remove all the superficially attached bacteria. The suspensions of tightly attached bacteria were prepared in sterile 0.85% saline after vigorously vertex of root tip samples in 20 ml saline sample for nearly 30 minutes. In the method, a mixture of rhizospheric bacteria was applied on a germinated seedling and incubated for 1 hour and inoculated in Murashige and Skoog media containing 0.8% agar-agar as solidifying agent. The germinated *Vigna radiata* seedlings were allowed to grow at 30°C under maintained light-dark period. After 7 day, the plants growth was monitored in term of shoot and root weight. Plants showing the enhanced growth compare to uninoculated and/or *Pf* CHA0 were selected out for further study. The roots were washed twice with sterile distill water so that only those bacteria that have colonized efficiently remain attached. One cm of root tips were sliced from main and lateral roots to resuspended in 1.5 ml of 0.85% sterile saline and vortex vigorously for 30 minutes. Serial dilutions was performed and dilutions were spreaded evenly on the King's B Agar Medium (KMB). Plates were incubated at 30°C for 24 hours and were observed for the fluorescent colonies as fluorescent pseudomonad produces water soluble green fluorescence pigment when subjected to UV exposure. Screening of microorganisms was done on the basis of fluorescence of the colony. Whole zone/patches of fluorescent colonies was collected by wire loop and resuspended in 1.5 ml of 0.85% sterile saline and vortexes for making it uniform suspension. The suspensions of fluorescent colonies were used to inoculate sterile germinated seedlings and repeated for two more cycles of plant inoculation study. Fluorescent bacteria were purified by repeated streaking on KMB plates and were believed to be good plant growth promoting and root colonizing bacteria.

Identification bacterial cultures by biochemical methods: For the identification of fluorescent pseudomonad biochemical test like catalase test, oxidase Test, Hugh-Leifson's Oxidation-Fermentation test, Gram staining and arginine dihydrolase test was performed using protocol as described in Bergey's manual.

Identification bacterial cultures by molecular methods: Modified CTAB method was used for the extraction of genomic DNA. 1.5 ml of overnight grown cultures was centrifuged at 10,000 rpm for 5 minutes at 4°C. The supernatant was drained off and the pellet was resuspended in 200 μl of T. E. Pellet was vortexed vigorously to resuspend the pellet and then was kept at 60°C for 30 minute. 100 μl of 3M NaCl was added to it followed by 80 μl of 10% CTAB. It was mixed properly and then again kept at 60°C for 10 minutes. Equal volume of phenol-cholorform-isoamyl alcohol (25:24:1) was added and centrifuged at 10,000 rpm for 12 minutes at 4°C. Aqueous phase was collected and 2-3 volume of chilled 100% ethanol was added and kept for 1 hour. Further it was centrifuged at 10,000 for 10 minutes at 4°C and supernatant was drained off and the pellet was again washed with 70% ethanol. The ethanol was allowed to evaporate and dried DNA was resuspended in sterile distilled water.

Identification of fluorescent pseudomonad by molecular method involved the amplification of region including the 3' half of the 16S rDNA with the whole 16S-23S rRNA Internal Transcripted Spacer (ITS) sequence using specific primers [11]. Primer sequences ITS1F-5'-AAGTCGTA ACA AG GTAG-3' and ITS2R-5'-GACCATATATAACCCCAAG-3' was used to get amplicon size of 560 bp.

PGPR traits of fluorescent *Pseudomonas* strains

Phosphate solubilization: Phosphate solubilization ability was checked on Pikovaskya's agar medium (Hi–Media Ltd., India) which contain insoluble dicalcium phosphate. Overnight grown cultures in 1.5 ml centrifuge tube and with equalized $OD_{600\,nm}$. One loopful of each strain was spotted on Pikovaskya's agar medium and incubated at 30°C for 24 hour and was observed for zone of clearance/colony size.

Antifungal activity: Antifungal activity was checked on potato dextrose agar (Hi–Media Ltd.). Fluorescent *Pseudomonas* strains were grown in King's B broth for 24 hrs. Overnight grown cultures in 1.5 ml centrifuge tube and with equalized $OD_{600\,nm}$ were centrifuged at 6,000 rpm for 5 minute and further washed with normal saline (0.80% NaCl).50 μl of concentrated pellet were inoculated on the four corners of the plate and *R. bataticola* inoculated at the centre of potato dextrose agar plate. The inhibition of *R. bataticola* was observed after 48-72 hours of growth. Percentage fungal inhibition was calculated by the formula. Percentage inhibition=(Radial growth of fungus in absence of inoculants) – (Radial growth of fungus in presence of inoculants/ (Radial growth of fungus in absence of inoculants) × 100.

Siderophore production: For detection of siderophores, overnight grown culture washed with saline and spotted on the Chrome Azurol-S (CAS) agar plates and observed for the colour change from greenish-blue to yellowish orange halo around culture pellet [4]. The halo due to chelation of Fe^{2+} from the CAS-Fe^{2+} conjugate and the diameter of halo zone/colony size was calculated which indirectly represent the ability of siderophore production by isolates. The siderophore production of isolates was compared to the bio-control and siderophore producing standard strain *Pf* CHA0.

HCN production: HCN production by isolates was checked by method of Bakker and Schipper, 1987 on Kings B Medium. King's B agar amended with 4.4 g/l glycine is used in HCN estimation single isolates were streaked in each plate. Whatman no. 1 filter paper disc (9 cm in diameter) was soaked in 0.5% Picric acid in 2% sodium carbonate. Soaked disc was placed in the lid of each petriplate. Petriplates were sealed with parafilm and incubated at 30°C for 4 days. An uninoculated

medium with the soaked filter paper was kept as control for comparison of results.

IAA estimation: IAA estimation was done using Salkowsky method. Overnight grown 100 μl culture was inoculated in 2 ml minimal media amended with 50 μg/ml tryptophan. Incubated at shaking condition for 48 hours at 30°C at 200 rpm. Grown culture was centrifuged at 10,000 g for 15 minutes, 1 ml of supernatant was taken fresh tube and 2 to 3 drops of ortho-phosphoric acid added to the supernatant followed by addition of 2 ml of reagent (1 ml of 0.5 M $FeCl_3$ in 50 ml of 35% $HClO_4$). Samples were incubated for 25 minutes and absorbance was measured at 530 nm. Concentration of IAA was measured against standard graph plot of pure IAA (Hi-media, India) at the range of 10-100 μg/ ml.

Characterization for ACC deaminase/ KMBA pathway: Plate technique using salt minimal medium containing ACC as sole nitrogen source (enrichment technique) was used to characterize the strains for ACC deaminase activity. The composition of salt minimal media containing ACC as sole nitrogen source in g L-1 is as follows, KH_2PO_4, 1.36; Na_2HPO_4, 2.13; $MgSO_4.7H_2O$, 0.2; $CaCl_2.2H_2O$, 0.7; $FeSO_4.7H_2O$, 0.2; $CuSO_4.5H_2O$, 0.04; $MnSO_4.H_2O$, 0.02; $ZnSO_4.7H_2O$, 0.02; H_3BO_3, 0.003; $CoCl_2.6H_2O$, 0.007; $Na_2MoO_4.2H_2O$, 0.004; Substrate ACC, 5 mM; Glucose, 1.0% dissolved in 1000 ml of distilled water.

The presence of KMBA was determined by precipitation with 2,4-dinitrophenylhydrazine according to Primrose [32]. The culture medium was separated and 0.1 ml of 0.1% 2,4-dinitrophenylhydrazine in 2 M HCl was added to 1 ml of culture filtrate and vortexes at room temperature in darkness. Presence of KMBA was confirmed by formation of a yellow precipitate after 30 min.

Effect of nutritional factors on 2,4-DAPG biosynthesis by fluorescent pseudomonad

Quantification of 2,4-DAPG biosynthesis: Each strain was grown in 1/5 diluted 20 ml King's B medium at shaking condition at 130 rpm at 26°C for 72 hours. Culture supernatant was acidified by addition of 1N HCl to make pH equal to 2. 10 ml ethyl acetate was added to supernatant and vigorously vortex for 2 minute and allowed to separate in two layers. Upper ethyl acetate phase was extracted and was allowed to evaporate and antibiotic preparation was dissolved in 1 ml methanol and stored in -20°C for Bioassay for 2,4-DAPG was performed using methicillin resistant *Staphylococcus aureus* 6538 as a sensitive strain and phytopathogen *R. bataticola*. HPLC analysis. A mixture of 30% ACN: 25% Methanol: 45% MQ water was used as mobile phase using C18 reverse phase column (250 x 4.6 mm) and flow rate at 1.0 ml/min at wavelength 272 nm. 2,4-DAPG was quantified using the standard plot of peak area and concentration (10-100 μg/ml).

Effect of carbon sources and Pi level on 2,4-DAPG biosynthesis: Each strain was grown in 1/5 diluted 20 ml King's B medium with the supplementation of phosphate to the final concentrations of 0, 8, 12, 17, 50 and 100 mM. Each strain was grown for 72 hours and OD was observed at 600 nm to monitor cell growth. Ethyl acetate extraction and HPLC for 2,4-DAPG was carried out in similar manner as discussed earlier.

Each strain was inoculated in King's B broth (without glycerol) and supplementation of 1% carbon sources: glucose, sucrose, fructose, mannitol and arabinose. Each strain was grown for 72 hours and OD was observed at 600 nm to monitor cell growth. Ethyl acetate extraction and HPLC for 2,4-DAPG was carried out in the similar manner as discussed earlier.

Plant inoculation study

The seed sterilization and germination was done as described earlier. The germinated seedlings were incubated for 45 minutes in 1.5 ml of overnight grown cultures. Germinated seedlings were inoculated in Murashige and Skoog media supplemented with 0.8% agar as solidifying agent. Plants were allowed to grow for 10 days. At the end of 10 day roots and shoot weight was measured.

Results

Isolation and characterization of fluorescent Pseudomonas strains

The mixtures of rhizosphere bacteria from cotton, sugarcane, groundnut, brinjal, rice, banana and tobacco rhizospheres used to inoculate seedlings of the *Vigna radiata* and enhanced competitive root tip colonizers were enriched as described in the material and methods (Figure 1). At the end of each cycle the colony diversity in terms of morphology, colour and opacity get decreased and the number of fluorescent colony get increased. After the third cycle of enrichment, 12 fluorescent colonies were selected and checked by fluorescent pseudomonad specific biochemical tests and PCR based method (Table 1). Four newly isolated strains G1, G2, G8 and C2 appeared positive for fluorescent pseudomonad specific biochemical (catalase, oxidase, oxidative/fermentative and arginine dihydrolase test) and molecular identification methods (ITS amplification). All strains were gram negative and rod shaped and has showed amplification of 560 bp using fluorescent pseudomonad specific ITS primers (Figure 2).

PGPR traits of fluorescent *Pseudomonas* strains

Strain G1, G8 and C2 has shown higher IAA production than *Pf* CHA0. Strain G1, G2, G8, C2 and *Pf* CHA0 did not showed growth on minimal medium containing ACC as sole nitrogen source but showed growth on L-methionine. Further confirmation for the presence of KMBA pathway in the strains were done by the analysis of precipitate after addition of 2,4-dinitrophenyl hydrazine to the culture supernatant (Figure 3).

All strains showed higher phosphate solubilization than *Pf* CHA0. Strain C2 showed higher siderophore production than *Pf* CHA0 while other strains showed siderophore production similar to *Pf* CHA0. Strain G2, G8 and C2 have shown higher antifungal activity than model bio-control strain *Pf* CHA0. Except C2, strain G1, G2 and G8 showed HCN biosynthesis ability similar to *Pf* CHA0.

Effect of nutritional factors on 2,4-DAPG biosynthesis by fluorescent pseudomonad

Amplification of PhlD and biosynthesis of 2,4-DAPG: All selected strain has shown amplification of *phl D* of 726bp (Figure 4) as previously reported by Raaijmakers, et al. [33]. 2,4-DAPG production by G1, G2, G8 and C2 was significantly high than *Pf* CHA0 (Table 1) and correlated well with bioassay against *S. aureus* and *R. bataticola* (Table 2 and Figure 5).

Effect of carbon sources and Pi level on 2,4-DAPG biosynthesis: Strain G8 has shown remarkably high 2,4-DAPG production in sucrose

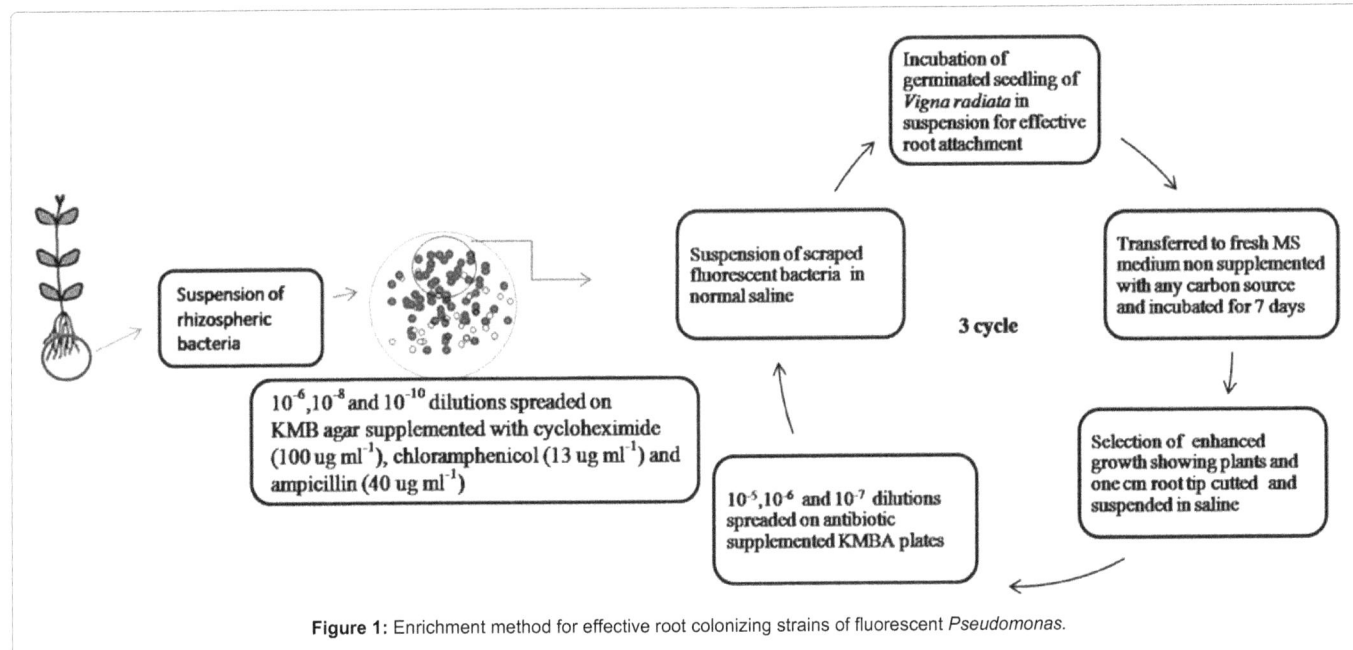

Figure 1: Enrichment method for effective root colonizing strains of fluorescent *Pseudomonas*.

		Pf CHA0	G1	G2	G8	C2
Biochemical methods	Catalase	+	+	+	+	+
	Oxidase	+	+	+	+	+
	Hugh-Leifson's Oxidation/fermentation test	Oxd	Oxd	Oxd	Oxd	Oxd
	Arginine Dihydrolase	+	+	+	+	+
	Gram nature	-ve	-ve	-ve	-ve	-ve
Morphology	Cell shape	Rod	rod	rod	rod	rod
Molecular method	ITS(560bp)Amplification	+	+	+	+	+

Table 1: Identification of fluorescent pseudomonad by biochemical and molecular methods.

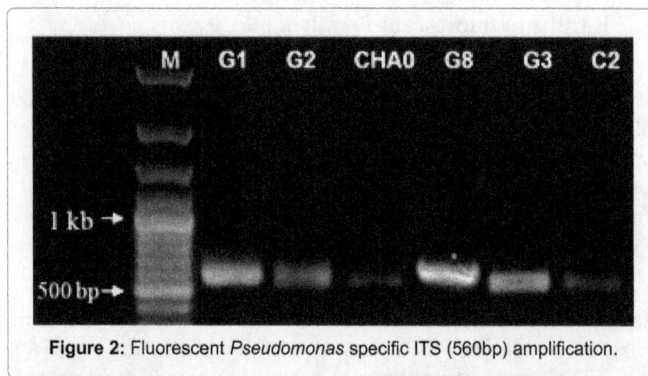

Figure 2: Fluorescent *Pseudomonas* specific ITS (560bp) amplification.

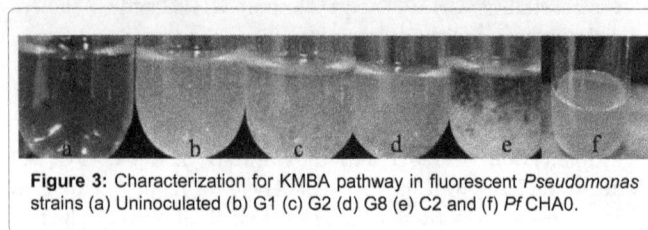

Figure 3: Characterization for KMBA pathway in fluorescent *Pseudomonas* strains (a) Uninoculated (b) G1 (c) G2 (d) G8 (e) C2 and (f) *Pf* CHA0.

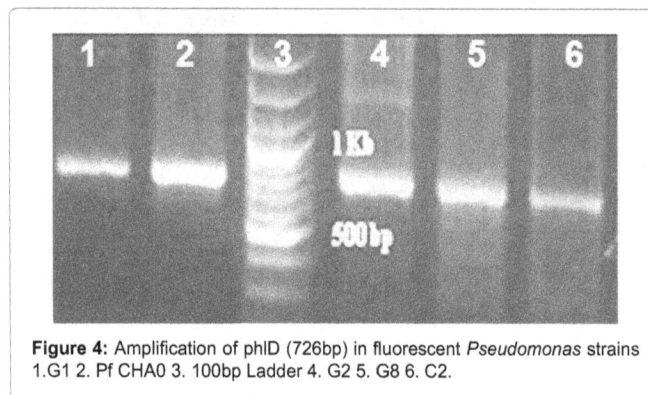

Figure 4: Amplification of phlD (726bp) in fluorescent *Pseudomonas* strains 1.G1 2. Pf CHA0 3. 100bp Ladder 4. G2 5. G8 6. C2.

(Table 3) so G8 could be considered physiology different than other strain including *Pf* CHA0(Figure 6). Strain G1 has also shown good production in sucrose in compare to *Pf* CHA0 so it is close to G8 in the dendogram (Figure 6). Strain G2 has shown good 2,4-DAPG production in presence of glucose as in case of *Pf* CHA0 so it is more close to *Pf* CHAO (Figure 6). Differential bio-control physiology among G1, G2, G8 and C2 was clearly evident after different response of carbon sources (Glucose, Fructose, Sucrose, Arabinose and Mannitol)

on 2,4-DAPG production (Table 3 and Figure 6). The effect of other carbon sources was not much significant and conclusive (Table 3).

As reported previously that inorganic phosphate represses the 2,4-DAPG production and in case of *Pf* CHA0 it get repressed at 10 mM inorganic phosphate level. However strain G2 has resisted the inhibitory effect of Pi on 2,4-DAPG production and has shown good production up to 50 mM supplemented Pi, G1 has shown constant 2,4-DAPG production up to 50 mM (Figure 7). G2, C2, M3 and G8 have shown very much positive effect on plant growth similar to *Pf* CHA0 (Figures 8 and 9).

Discussions

Enrichment method developed in present work was the modification of method invented by Kamilova et al. [15] but in such way that it has yielded only plant beneficial, root colonizing fluorescent pseudomonad. Fluorescent pseudomonads strain G1, G2, G8 and C2 showed the presence of specific biochemical enzymes/pathway (arginine dihydrolase, catalase, oxidase, oxidative respiration) and further confirmed by the Internal Transcribed Spacer (ITS) rRNA coding sequence using specific primers as used by Locatelli et al. [22] which was considered to be more effective rather than the identification by 16S rDNA sequencing. These strains have proved to be efficient PGPR strains as they possess the high IAA production , phosphate solubilization, siderophore production and antifungal activity than model bio-control and PGPR strain *Pf* CHA0. The utilization of L-methionine as a nitrogen source by these strains confirms the presence of KMBA pathway. G1, G2 and G8 showed HCN biosynthesis which has been reported to be effective mechanism of bio-control by fluorescent pseudomonad [20]. Strain G1, G2, G8 and C2 have shown high 2,4-DAPG production and correlated well with bioassay against *S. aureus* and *R. bataticola*.

Quantitative and qualitative differences in the sugar composition of root exudates determine the bio-control mechanism by fluorescent pseudomonad strains in given crop-pathogen systems [8,24]. As earlier reports says that glucose but not glycerol and sucrose has enhanced 2,4-DAPG production in *P. fluorescens* Pf-5, *Pf* CHA0 and other fluorescent *Pseudomonas* strains whereas in *Pf* f113 production of 2,4-DAPG is stimulated by sucrose. Strain G1 and G8 has shown remarkably high 2,4-DAPG production in sucrose so G8 could be considered physiology different than other strain including *Pf* CHA0. Strain G2 has shown good 2,4-DAPG production in presence of glucose similar to *Pf* CHA0. Strain G8 and G1 could be considered as good bio-control strain for the sucrose sufficient plant rhizosphere. Strain G2 and G1 could do good bio-control activity even in phosphate rich soil. As

PGPR Traits		*Pf* CHA0	G1	G2	G8	C2
Phytostimulation	IAA production (ug/ml)	11 ± 1.2	20 ± 0.53	9.3 ± 0.74	14.8 ± 0.75	22 ± 0.73
	Growth on ACC (Sole N-source)	-	-	-	-	-
	Growth on L-methionine (Sole N-source)	++	++	+++	++	+++
	KMBA pathway	+	+	+	+	+
Mineralization	P-Solubilisation (Cz/Cs)	1.3	1.9	2.1	1.8	1.5
	Siderophore activity(Cz/Cs)	2	1.8	1.7	2	3
Antibiosis	Fungal Inhibition (%)	38.4	38.4	50	57.6	41.6
	HCN production	+++	++++	++++	++++	-
	2,4-DAPG production(ng/ml)	740 ± 9.21	2340 ± 87	3350 ±117.5	965 ± 10.27	995 ±25.13
	Bactericidal against *S. aureus*	+++	++++	+++++	+++	+++
	Bactericidal against *S. aureus*	+++	++++	+++++	+++	+++

Table 2: PGPR traits of fluorescent *Pseudomonas* strains.

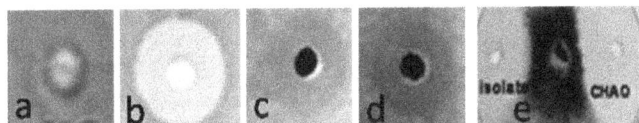

Figure 5: Inhibition of *S. aureus* by 2,4-DAPG extract of (a) Methanol (b) *Pf* CHA0 (c) G1(d) G2 and (e) Inhibition of *R. bataticola*.

	CHA0 (ng/ml)	G1 (ng/ml)	G2 (ng/ml)	G8 (ng/ml)	C2 (ng/ml)
Glucose	859.5 ± 34.95	141 ± 5.05	545 ± 43.5	157.5 ± 17.82	106.5 ± 7.63
Sucrose	203 ± 11.1	572.5 ± 66.4	84 ± 5.75	1328 ± 28.67	162.2 ± 4.2
Mannitol	337 ± 16.85	239 ± 14.95	229 ± 6.32	160.5 ± 21.95	320 ± 8.59
Arabinose	215 ± 4.12	141 ± 15.97	166 ± 14.3	82.5 ± 7.89	233.5 ± 8.67

Table 3: 2,4-DAPG production by fluorescent *Pseudomonas* strains under different carbon sources.

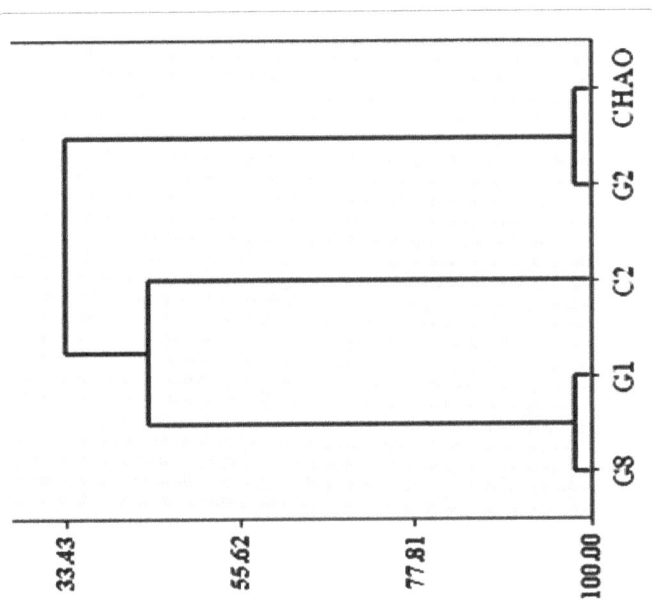

Figure 6: Relatedness based on 2,4-DAPG production under different C-sources.

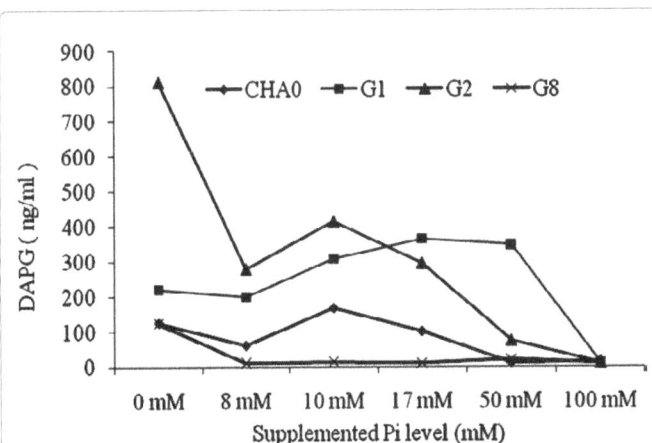

Figure 7: 2,4-DAPG production by fluorescent *Pseudomonas* at different Pi level.

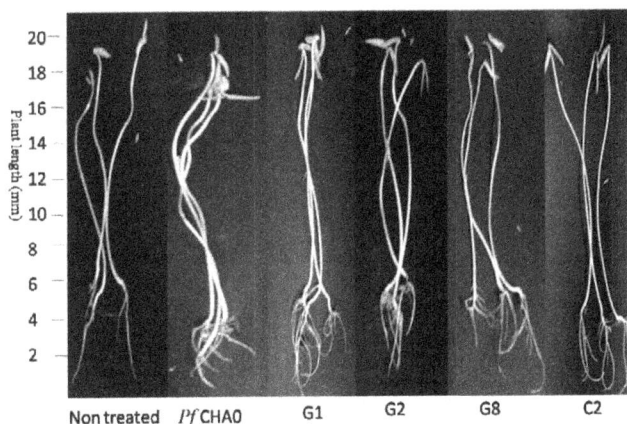

Figure 8: Plant growth promoting effect of isolates growth promoting effect of isolates.

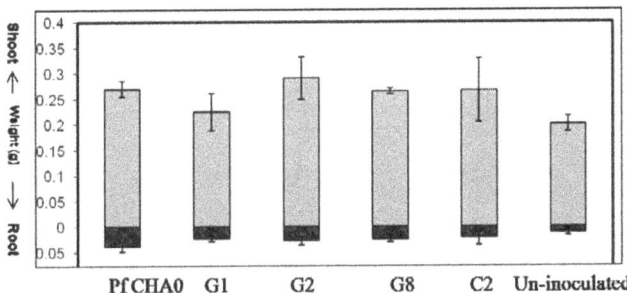

Figure 9: Effect of fluorescent *Pseudomonas* strains on shoot/root weight.

reported previously that inorganic phosphate represses the 2,4-DAPG production and in case of *Pf* CHA0 it get repressed at 10 mM inorganic phosphate level. However strain G2 has resisted the inhibitory effect of Pi on 2,4-DAPG production and has shown good production up to 50 mM supplemented Pi. Differential influence of carbon and mineral sources on 2,4-DAPG production was suggested due to various degrees of adaptation of strains to given rhizospheric nutrient composition.

Acknowledgements

Authors express a gratefully acknowledgement to University Grant Commission for providing Research Fellowship for Meritorious Student (UGC-RFMS) and Gujarat State Biotech Mission (GSBTM), India, for the financial support.

References

1. Haas D, Defago G (2005) Biological control of soil-borne pathogens by fluorescent pseudomonads. Nat Rev Microbiol 3: 307-319.

2. Ojiambo PS, Scherm H (2006) Biological and application-oriented factors influencing plant disease suppression by biological control: a meta-analytical review. Phytopathology 96: 1168-1174.

3. Schippers B, Bakker AW, Bakker PAHM (1987) Interactions of deleterious and benificial rhizosphere microorganisms and the effect of cropping practices. Annu. Rev Phytopathol 25: 339-358.

4. Lugtenberg BJJ, de Weger LA, Bennett JW (1991) Microbial stimulation of plant growth and protection from disease. Curr Opin Biotechnol 2: 457-464.

5. Thomashow LS, Weller DM (1995) Current concepts in the use of introduced bacteria for biological disease control: Mechanisms and antifungal metabolites. In: Stacey G, Keen NT (eds) Plant-Microbe Interactions, vol. 1. Chapman & Hall, New York, USA, pp. 187-235.

6. Vassilev N, Vassileva M, Nikolaeva I (2006) Simultaneous P-solubilizing

and biocontrol activity of microorganisms: potentials and future trends. Appl Microbiol Biotechnol 71: 137-144.

7. Bolwerk A, Lagopodi AL, Wijfjes AH, Lamers GE, Chin-A-Woeng TF, et al. (2003) Interactions in the tomato rhizosphere of twoPseudomonas biocontrol strains with the phytopathgenic fungus Fusarium oxysporum f. sp. radicis-lycopersici. Mol Plant Microbe Interact 16: 983-993.

8. Lemanceau P, Corberand T, Gardan L, Latour X, Laguerre G, et al. (1995) Effect of two plant species flax (Linum usitatissinum L.) and tomato (Lycopersicon esculentum Mill.) of the diversity of soilborne populations of fluorescent pseudomonads. Appl Environ Microbiol 61: 1004 -1012.

9. Ramette A, Moenne-Loccoz Y, Defago G (2003) Prevalence of fluorescent pseudomonads producing antifungal phloroglucinols and/or hydrogen cyanide in soils naturally suppressive or conducive to tobacco black root rot. FEMS Microbiol Ecol 44: 35-43.

10. Shahnahan P, O Sullivan DJ, O Gara F (1992) Isolation of 2,4- Diacetyl phoroglucinol from a Fluorescent Pseudomonas and investigating physiological factors influencing its production. Appl Environ Microbiol 58: 353-38.

11. Locatelli L, Tarnawski S, Hamelin J, Rossi P, Aragno M, et al. (2002) Specific PCR amplification for the genus Pseudomonas targeting the 3' half of 16S rDNA and the whole 16S-23S rDNA spacer. Syst Appl Microbiol 25: 220-227.

12. O'Connell KP, Goodman RM, Handelsman J (1996) Engineering the rhizosphere: expressing a bias. Trends Biotechnol 14: 83-88.

13. Schroth MN, Hancock JG (1982) Disease-suppressive soil and root-colonizing bacteria. Science 216: 1376-1381.

14. Haas D, Keel C (2003) Regulation of antibiotic production in root-colonizing Peudomonas spp. and relevance for biological control of plant disease. Annu Rev Phytopathol 41: 117-153.

15. Kamilova F, Validov S, Azarova T, Mulders I, Lugtenberg B (2005) Enrichment for enhanced competitive plant root tip colonizers selects for a new class of biocontrol bacteria. Environ Microbiol 7: 1809-1817.

16. Haas D, Keel C, Laville J, Maurhofer M, Oberhansli T, et al. (1991) Secondary metabolites of Pseudomonas fluorescens strain CHA0 involved in the suppression of root diseases. Advances in Molecular Genetics of plant-Microbe interactions 1: 450-556

17. Keel C, Défago G (1997) Interactions between beneficial soil bacteria and root pathogens: mechanisms and ecological impact, In: Gange AC, Brown VK (ed.), Multitrophic interactions in terrestrial systems. Blackwell Science, London, England. p. 27-46.

18. Shaharoona B, Arshad M, Khalid A (2007) Differential response of etiolated pea seedlings to inoculation with rhizobacteria capable of utilizing 1-aminocyclopropane-1-carboxylate or L-methionine. J Microbiol 45: 15-20.

19. Picard C, Di Cello F, Ventura M, Fani R, Guckert A (2000) Frequency and biodiversity of 2,4-diacetylphloroglucinol-producing bacteria isolated from the maize rhizosphere at different stages of plant growth. Appl Environ Microbiol 66: 948-955.

20. Juma NG, McGill WB (1986) Decomposition and nutrient cycling in agro-ecosystems. In: Mitchell MJ, Nakas JP (eds) Microfloral and faunal interactions in natural and agro-ecosystems. Martinus-Nijhoff/DR W. Junk Publishers, Boston, Mass, p. 74-137.

21. Jaeger CH 3rd, Lindow SE, Miller W, Clark E, Firestone MK (1999) Mapping of sugar and amino acid availability in soil around roots with bacterial sensors of sucrose and tryptophan Appl Environ Microbiol 65: 2685-2690.

22. Kuiper I, Bloemberg GV, Noreen S, Thomas-Oates JE, Lugtenberg BJJ (2001) Increased uptake of putrescine in the rhizosphere inhibits competitive root colonization by Pseudomonas fluorescens strain WCS365. Mol Plant-Microbe Interact 14: 1096-1104.

23. Nielsen TH, Sørensen D, Tobiasen C, Andersen JB, Christophersen C, et al. (2002) Antibiotic and biosurfactant properties of cyclic lipopeptides produced by fluorescent Pseudomonas spp. from the sugar beet rhizosphere. Appl Environ Microbiol 68: 3416-3423.

24. De Souza JT, Weller DM, Raaijmakers JM (2003) Frequency, Diversity, and Activity of 2,4-Diacetylphloroglucinol-Producing Fluorescent Pseudomonas spp. in Dutch Take-all Decline Soils. Phytopathology 93: 54-63.

25. Keel C, Schnider U, Maurhofer M, Voisard C, Laville J, et al. (1992) Suppression of root diseases by Pseudomonas fluorescens Pseudomonas fluorescens CHA0: importance of the bacterial secondary metabolite 2,4-diacetylphloroglucinol. Mol Plant-Microbe Interact 5: 4-13.

26. Primrose SB, Dilworth MJ (1976) Ethylene production by bacteria. J Gen Microbiol 93: 177-181.

27. Kamei Y, Isnansetyo A (2003) Lysis of methicillin-resistant Staphylococcus aureus by 2,4-diacetylphloroglucinol produced by Pseudomonas sp. AMSN isolated from a marine alga International. Journal of Antimicrobial Agents 21: 71-74.

28. Keel C, Weller DM, Natsch A, Defago G, Cook RJ, et al. (1996) Conservation of the 2,4-diacetylphloroglucinol biosynthesis locus among fluorescent Pseudomonas Pseudomonas strains from diverse geographic locations. Appl Environ Microbiol 62: 552-563.

29. Bangera MG, Thomashow LS (1999) Identification and characterization of a gene cluster for synthesis of the polyketide antibiotic 2,4-diacetylphloroglucinol from Pseudomonas fluorescens Q2-87. J Bacteriol 181: 3155-3163.

30. Landa BB, Mavrodi OV, Raaijmakers JM, McSpadden-Gardener BB, Thomashow LS, et al. (2002) Differential ability of genotypes of 2,4-diacetylphloroglucinol-producing Pseudomonas fluorescens strains to colonize the roots of pea plants. Appl Environ Microbiol 68: 3226-3237.

31. Duffy BK, Défago G (1999) Environmental factors modulating antibiotic and siderophore biosynthesis by Pseudomonas fluorescens biocontrol strains. Appl Environ Microbiol 65: 2429-2438.

32. Raaijmakers JM, Weller DM (1998) Natural plant protection by 2,4-diacetylphloroglucinol-producing Pseudomonas spp. in take-all decline soils. Mol Plant-Microbe Interact 11: 144-152.

33. Raaijmakers JM, Weller DM, Thomashow LS (1997) Frequency of Antibiotic-Producing Pseudomonas spp. in Natural Environments. Appl Environ Microbiol 63: 881-887.

33

Visual Detection of *Curly Top Virus* by the Colorimetric Loop-Mediated Isothermal Amplification

Mohammad Amin Almasi[1], Mehdi Aghapour-ojaghkandi[2]* and Saeedeh Aghaei[2]

[1]Department of Agriculture and Plant Breeding, Faculty of Agriculture, Zanjan University, Zanjan, Iran
[2]Young Researchers and Elite Club, North Tehran Branch, Islamic Azad University Tehran, Iran

Abstract

Curly Top Virus (CTV) in sugar beet, belonging to the family *Geminiviridae*, genius *Curtovirusis* a considerable problem in semiarid areas of the Iran and other semiarid parts of the world. There are several diagnostic methods to detect CTV, But these techniques take a long time for 3 h to tow day, requiring sophisticated tools. The aim of this study, for the first time, was to reduce the time required to detect CTV in sugar beet, using colorimetric loop-mediated isothermal amplification (LAMP) technique requiring only an ordinary water bath or thermoblock.DNA was extracted from infected naturally leaf tissues. Samples were tested for the presence of *Curtovirus* species by PCR and LAMP reactions to amplify replication-associated protein (rep) gene using species primers. LAMP was optimized to amplify CTV DNA under isothermal conditions by incubation at 63°C for 30 min. LAMP products were detected visually using the different dyes. The LAMP amplification products had a ladder-like appearance when electrophorised on an agarose gel and also positive results using the different dyes were a color change. Results confirmed LAMP with different dyes provides a rapid and safe assay for detection of CTV in sugar beet. Since with other molecular methods, equipping laboratories with a thermocycler or expensive detector systems is unavoidable, this assay was found to be a simple, cost-effective molecular method that has the potential to replace other diagnostics in primary laboratories without the need for expensive equipment or specialized techniques. It can also be considered as a reliable alternative viral detection system in further investigations.

Keywords: *Curly top virus*; Colorimetric assay; LAMP reaction; Sugar beet

Introduction

Curly top is one of the most devastating DNA viruses, causing curly top diseases in a wide range of plants. It can be caused by a number of closely related species in the genus *Curtovirus* of the family *Geminiviridae*. These viruses are transmitted by the beet leafhopper (*Circulifertenellus* Baker) and can infect dicotyledonous hosts from 44 families and at least 300 species [1,2]. The viruses responsible for curly top diseases have been separated into three distinct species, based largely on sequence variation in select regions of the curtovirus genome, as well as some differences in host range and disease severity [3]. Curly top disease (CTD) of sugar beet and several crops such as tomato, pepper, bean, cucurbits, tobacco and potato is a known disease affecting these crops for over a century [4,5]. Different symptoms caused by curly top viruses include stunting, leaf curling, vein swelling and yellowing, crumpling, and hyperplasia of the phloem. Early infection causes early death in plants [6,7]. CTD is caused by several curtoviruses, including *beet curly top virus* (BCTV), *Beet Severe Curly Top Virus* (BSCTV), *Beet Mild Curly Top Virus* (BMCTV), *Horseradish Curly Top Virus* (HrCTV), *Spinach Curly Top Virus* (SpCTV) and *Pepper Curly Top Virus* (PepCTV). Furthermore, two divergent curly top viruses, *Beet Curly Top Iran Virus* (BCTIV) and *Turnip Curly Top Virus* (TCTV), have been reported from Iran in the past five years [8-12]. Recombination has also been suggested to have played a part in the evolution of species in this group [13]. BCTV may have arisen by recombination between a whitefly-transmitted geminivirus, which may have donated the complementary genes involved in virus replication, and a leafopper-transmitted geminivirus, which donated the coat protein gene [14].

Curtovirus species possess a monopartite genome with circular single-stranded DNA of approximately 3 kb with three over overlapping virion sense genes (*V1, V2,* and *V3*), four overlapping complementary sense genes (*C1, C2, C3,* and *C4*), and an intergenic region containing the origin of replication [15-18]. During the period from the 1950s to the 1970s a broad range of curly top isolates were described and some were noted to have increased in severity [19]. Some alternative approaches were gradually developed including polymerase chain reaction (PCR), Rolling-circle amplification (RCA), Restriction fragment Length Polymorphism (RFLP) and Chromatin immunoprecipation (ChIP) assays, all of which were unfortunately time-consuming and require expensive or carcinogenic materials to visualize DNA amplification [20-23].

Among various isothermal amplification systems developed over the recent years, the most frequently applied approach seems to be loop-mediated isothermal amplification (LAMP), implemented first by Notomi et al. [24]. Due to its enormous rate of amplification paired with a very high specificity, sensitivity, rapidity and low artifact susceptibility, the method together with its modifications have been strongly recommended for detection of a great number of strains of bacteria as well as viruses worldwide [25]. Briefly, each reaction is carried out by the use of a DNA polymerase with strand displacement activity and set of four different primers: F3 (forward outer primer), B3 (backward outer primer), FIP (forward inner primer), BIP (backward

*Corresponding author: Mehdi Aghapour-Ojaghkandi, Young Researchers and Elite Club, North Tehran Branch, Islamic Azad University Tehran, Iran
E-mail: mehdiagh3@gmail.com

inner primer), designed specifically to recognize six distinct regions on the target sequence in conjunction with two loop primers: LF (loop forward) and LB (loop backward) to accelerate the reaction [24,26]. LAMP assay, alternately, can also amplify nucleic acid under isothermal condition in the range of 60°C-65°C, all turbidity and fluorescent based detections, as well as agarose gel electrophoresis system are applied to visualize suspicious samples [24-26]. LAMP has been shown to be as sensitive and specific as real-time PCR, and several LAMP assays have been developed for the detection of different plant viruses and viroids [27-29]. Here, for the first time, LAMP technique based on visualization system was consequently employed to detection of CTV in sugar beet.

Materials and Methods

Plant samples and extraction of total DNA

Samples were collected from the sugar beet production areas in Iran such as East-Azarbaigan where curly top has been observed. Sugar beet (*Beta vulgaris* L.) leaf samples symptomatic for curly top were collected between 2012 and 2013.

Leaf tissue was sampled by punching out a disk with the cap of a sterile 2 ml microcentrifuge tube and stored at –80°C. Individual tubes containing the frozen tissue were dipped in liquid nitrogen and then pulverized using a Retsch MM 301 mixer mill (Retsch, Inc., Newton, PA) with 5 mm stainless steel beads. DNA was extracted using a DNeasy Plant Mini Isolation kit (Qiagen, Inc., Valencia, CA) following standard protocols suggested by the manufacturer, except DNase and RNase free water (W4502, Sigma-Aldrich, St. Louis, MO) was used in place of Buffer AE in the final steps. The DNA was stored at –20°C.

PCR and LAMP primers

PCR and LAMP assays were developed using specific primers designed based on replication-associated protein (rep) gene (GenBank accession AF379637.1). PCR (Forward and Backward) and LAMP specific primers were designed using the Oligo7 and Primer ExplorerV4 software (specific for LAMP) respectively (Table 1). In addition, Figure 1 shows the position of the PCR and LAMP primers on rep gene. At LAMP assay, six primers recognizing eight distinct regions in the target sequence were used, including Outer primers (F3 and B3), Inner primers (FIP and BIP) and Loop primers (LF and LB). Moreover, each primer was tested for similarities with other sequences available in the GenBank databases (http://blast.ncbi.nlm.nih.gov/Blast.cgi) using the BlastN algorithm.

PCR reaction

PCR were performed in volumes of 25 μl with the componentsin the master mix at the following concentrations: 1×PCR buffer (GeneAmp 10×PCR Gold buffer with 150 mMTris-HCl, pH 8.0, 500 mMKCl; Applied Biosystems, Foster City, CA), 2.5 mM MgCl₂ (Applied

Biosystems), 1.4 mMdNTPs (Applied Biosystems), 0.2 M each of F and B primer, 0.625 UAmpliTaq Gold (Applied Biosystems), and 2 μl DNA. Amplification consisted of 5 min at 95°C followed by 34 cycles 95°C for 1 min, 54°C for 1 min, and 72°C for 1 min. After the last cycle, the reaction was held at 72°C for 5 min and then 4°C. PCR amplification products were visualized with agarose gel electrophoresis, 1.5% in TAE buffer and visualized by staining with ethidium bromide under UV light. To determine specificity of the primers, LAMP and PCR reactions were carried out to BCTV (*Beet Curly Top Virus*) DNA as positive control.

Colorimetric LAMP reaction

In order to select the most appropriate condition, LAMP reaction was conducted according to procedures published previously [24-27]. The amplification reaction was performed at 63°C for 30 min in water bath by mixing 1.6 M each of FIP and BIP primer, 0.2 M each of F3 and B3 primer, 0.8 M each of LF and LB primer, 1.4 mMdNTPs, 0.8M betaine (Sigma–Aldrich, Saint Luis, MO, USA), 8U of BstDNA polymerase (New England Biolabs, Hertfordshire, England, UK) using the supplied buffer (Thermopol buffer containing 2 mM of MgSO₄) and 2 μl DNA. As an additional proof, products from the reaction were also analyzed by agarose gel electrophoresis. Various colorimetric assays including magnesium sulfate (MgSO₄), Calcium chloride (CaCl₂), SYBR® Premix Ex Taq™ II, Hydroxyl naphthol blue (HNB), GeneFinder™, SYBR Green I and Ethidium bromide have been proposed and used in several investigations [30-41]. Here, thus, the validations of positive LAMP reactions were justified by means of these seven staining approaches.

Magnesium sulfate

Like other metal indicators, magnesium sulfate must be added before reaction. At the end of the amplification process, positive reactions were accompanied by a visible darker phase in the tubes in consequence of the formation of magnesium pyrophosphate which can be easily visualized with the naked eye [24,35]. It is noticeable that the turbidity of the positive samples is stable but just for a short time, which should be consequently judged soon after taking out of the samples either from the water bath or the thermal cycler.

Calcium chloride

Use of MgSO₄ in LAMP reaction in order to create turbidity has some drawbacks such as, low stability and concentration. In this research, we used CaCl₂ to address these disadvantages as described previously by Almasi et al. [38]. Thus, 2 mM CaCl₂ was added to LAMP mixture to improve both stability and concentration of turbidity in the reaction.

SYBR® Premix ExTaq™ II

To conquer time-dependent instability of magnesium or calcium

Reaction	Primer	Type	Length of Primer	Sequence(5`-3`)	Length of Product
PCR	F	Forward	24 nt	TGCTCCAATAAGGTGCTTCCAGTG	292 bp
PCR	B	Backward	25 nt	TTTCCTCTGTCCTCATTCACAAACG	
LAMP	F3	Forward outer	18 nt	AAGTAAGTGGGATCTACG	
LAMP	B3	Backward outer	21 nt	GAAATGCAAGAATGGGCTGAT	
LAMP	FIB	Forward inner	47 nt	GGCCCACAATTACATCACAGGGCTTTTGACGTTGTATTCCACTTCAT	Fragment with different sizes
LAMP	BIP	Backward inner	53 nt	CCACATAGTCTTCCCTGTTCTTGATTTTTATTTTGGGGTTGATGCCGCTGCGC	
LAMP	LF	Loop forward	23 nt	TAGATTTTAGCCTTAGAACATAT	
LAMP	LB	Loop backward	21 nt	ACTATGATACTATTATATCTT	

Table 1: Oligonucleotide primers used for PCR and LAMP to detection of CTV.

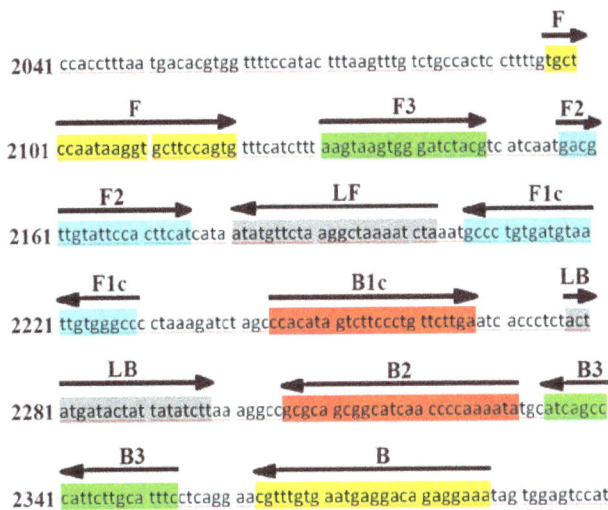

Figure 1: Schematic representation of position and sequence of CTV-specific primer pair used in PCR and LAMP analysis on *rep* gene. F1c plus F2 formed FIP and B1c plus B2 formed BIP.

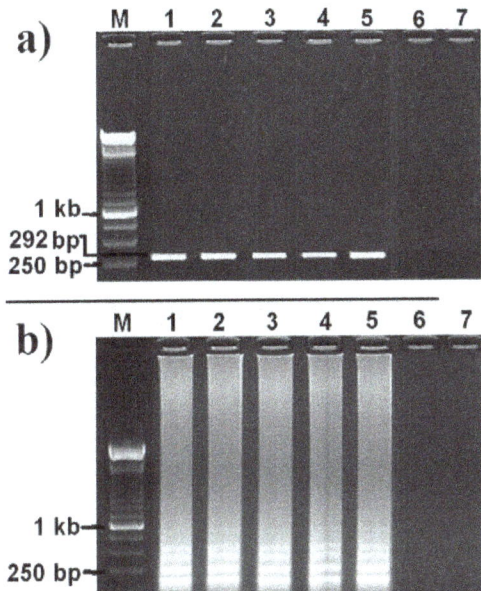

Figure 2: Analysis of products for the detection of CTV (a) PCR products on agarose gel; (b) LAMP products on agarose gel. Left to right: lane M, DNA size marker (1 kb; Fermentas); Lane 1, CTV24; lane 2, CTV36; lane 3, CTV42; lane 4, CTV45; lane 5positive control (BCTV); lane 6,water control; lane 7, negative sample.

pyrophosphate-based detection method, an alternative visual system using SYBR® Premix Ex Taq™ II was employed [34,42]. Hence, 2 µl SYBR® Premix Ex Taq™ II (Perfect Real Time, Takara Bio Co., Ltd., RR081A) was added into each completely finished LAMP reaction containing 25 µl LAMP products.

Hydroxynaphthol blue (HNB)

In this protocol, 1 µl of the hydroxylnaphthol blue dye (3 mM, Lemongreen, Shanghai, China) is mixed prior to amplification; all positive reactions can be easily identified using the naked eye, interestingly with no probable cross contaminations which usually arise from opened tubes after amplification [31-33,38,41].

GeneFinder™

An obvious Green fluorescence pattern was observed to confirm positive LAMP products through visual observation with the naked eye when 1 µl of GeneFinder™, diluted to 1:10 with 6×loading buffer (Biov. Bio. Xiamen, China), was added to each reaction as described previously [30,36,39,41].

SYBR Green I

About 1 µl of SYBR Green I (Invitrogen, Sydney, Australia) diluted to 1:10 with 6×loading buffer was separately added to each reaction as described previously [33,38]. After addition of SYBR Green I, the products can be visualized directly by the naked eye.

Ethidium bromide

Under a UV transilluminator, positive products will be consequently marked if a detectable yellow colour pattern is observed [24]. Ethidium bromide is a polycyclic fluorescent dye that binds to double-stranded DNA molecules by intercalating a planar group between the stacked base pairs of the nucleic acid [34-37]. About 0.5 µg ethidium bromide/ml (Sigma) was added to each tube as designed previously by Tsai et al. [35].

Results

On the whole, 4 out of 83 leaf samples suspicious of having infection with CTV (CTV24, CTV36, CTV42 and CTV45) (4.8%) showed positive responses. When PCR was used for the detection of CTV, band of approximately 292 bp was amplified and visualized under UV light (Figure 2a). The LAMP technique can from amplified products of various sizes, consisting of alternately inverted repeats of the target

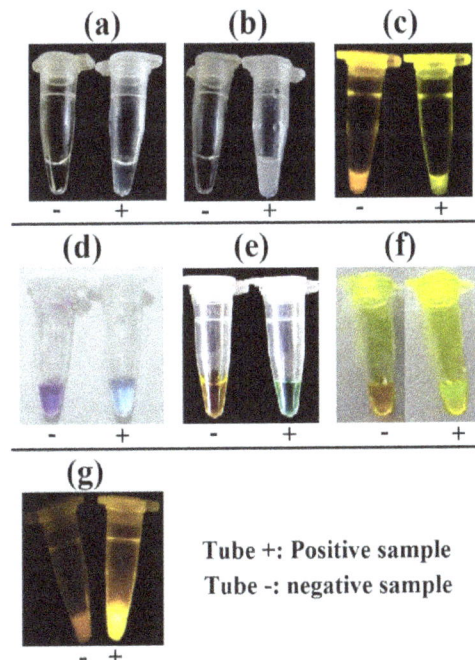

Figure 3: Visual detection of LAMP with different dyes (a) visual detection by magnesium pyrophosphatebased method; (b) visual detection by calcium pyrophosphate-based method; (c) visual detection by SYBR® Premix Ex TaqTM II-based method; (d) visual detection by hydroxynaphthol blue (HNB)-based method; (e) visual detection by GeneFinderTM based method; (f) visual detection by SYBR Green I-based method; (g) visual detection by ethidium bromide-based method.

sequence on the same strand. Therefore, the LAMP primers generated many fragments with different sizes and formed a ladder pattern on an agarose gel (Figure 2b). LAMP amplicons were able to be detected with the naked eye by adding different visual dyes followed by color changing in the solutions. In this regard, all used visual components could successfully make a clear distinction between positive infected samples and negative ones. In the LAMP reaction when DNA polymerizes, producing pyrophosphate ions and interaction between pyrophosphate and magnesium or calcium lead to produce magnesium pyrophosphate or calcium pyrophosphate (white precipitate) which can be observed and detectable as turbidity in positive samples. In contrast, negative control remains transparent without any turbidity (Figure 3a and 3b). The $CaCl_2$ was improved both stability and concentration of turbidity in contrast of $MgSO_4$. Under UV illumination (302 nm), a green color pattern is an identical characteristic of all positive reactions was monitored in this study after was used SYBR® Premix Ex Taq™ II (Figure 3c). When HNB was used for detect, a sky blue color pattern implies the existence of the reference virus, whereas a violet color change is observed when the control(s) are taken into consideration (Figure 3d). An obvious green fluorescence pattern was observed to confirm positive LAMP products through visual observation with the naked eye when GeneFinder™ or SYBR Green I were added to each reaction. Remarkably, concerning negative reaction, the original orange color could be observed (Figures 3e and 3f). Under a UV transilluminator, also a detectable yellow color pattern is observed when ethidium bromide was applied to detection of positive reaction (Figure 3g). As a result, the intensity of the fluorescent emissions moves up in positive tubes, while the reverse is true regarding negative tubes with no amplified fragments. More interestingly, in some cases, just a little fluorescent emission could be observed in negative tubes, presumably arising from the presence of primers and/or DNA templates, leading to an increase to false positive outputs.

Discussion

Curly top virus (CTV) and related viruses (collectively known as curtoviruses) have caused significant problems to irrigated agriculture in Iran. CTV is known to infect a broad range of crop and weed hosts in many plant families. Crop hosts for which natural CTV infection has been reported include sugar beet, tomato, pepper, bean, spinach, and cucurbits [7,22]. Molecular methods including polymerase chain reaction (PCR), Rolling-circle amplification (RCA), Restriction fragment length polymorphism (RFLP) and Chromatin immunoprecipitation (ChIP) assays, are specific and accurate approaches, but most are technically demanding and time-consuming techniques, needing post-process stage or costly detection equipment to visualize amplified DNA [20-22]. Thus, an inexpensive and rapid diagnostic method is required, especially in developing countries with insufficient facilities. The LAMP assay was first applied for detection of yam mosaic virus [27]. In this study, for the first time, a colorimetric LAMP method with the use of different dyes was developed and optimized. In reality, LAMP overall requires just 30 min to accomplish while PCR method 3 h should be served (Table 2). This, in turn, would simplify the detection procedure and result in saving of significant time needing for separating of the amplified products on the gel and the analyzing of the data. Safety regarding a number of detection methods, application of gel electrophoresis systems has emerged as a routine approach with enough potential to observe related amplicons. Just the same, such visual methods not only involve some expensive instruments but also during a period of time, exposure to the UV ray (because it is harmful to the eyes, even watching for a short period would irritate eyes and cause symptoms similar to conjunctivitis) as well as ethidium bromide could accompany a number of serious negative effects on researchers who use these methods [34-43]. More surprisingly, in LAMP, amplified products can be easily visualized by means of different in tube color indicators with no essential requirement of additional staining systems; thus, toxic staining materials would be significantly avoided. Simplicity, cost and user-friendly equipped labs with some molecular instruments as well as trained personnel are prerequisites to perform other assays, all of which are undoubtedly costly. On the contrary, LAMP can be easily accomplished just in a water bath or temperature block with no need of thermocyler and gel electrophoresis [27,44].

According to our results, despite the precise detection of positive LAMP products using all dyes, some were significantly superior when the time of stability, cost and the safety were taken into consideration (Table 3). For instance, regardless of its reasonable color change stability (≥ 2 weeks), when employing ethidium bromide, a UV transilluminator must be available, both of which are toxic, resulting in deleterious consequences on human health and the environment [34,43]. As the

Assay	Total time	Detection method	Safety	Need to UV ray	Need to detect instruments	Cost	User Friendly	Accuracy
PCR	3 hours	Gel Electrophoresis	No	Yes	Yes	High	Low	High
LAMP	30 minutes	Visual+ Gel Electrophoresis	Yes	No and Yes	No and Yes	Low	High	High
Colorimetric LAMP	30 minutes	Visual	Yes	No	No	Low	Very high	High

Table 2: Comparison of PCR, LAMP and colorimetric LAMP assays for detection of CTV.

Visual system	Detection method	Safety	Positive reaction	Negative sample	Need to UV ray	Stability	Cross contamination	Use time
magnesium Sulfate	Visual	Yes	Turbidity	No turbidity	No	5-10 seconds	No	Before the amplification
Calcium chloride	Visual	Yes	Turbidity	No turbidity	No	40-60 seconds	No	Before the amplification
SYBR® Premix Ex Taq™ II	Visual	Yes	Green	Red	Yes	1–3 days	Yes	After the amplification
Hydroxynaphthol blue	Visua	Yes	Sky blue	Violet	No	2-3 weeks	No	Before the amplification
GeneFinder™	Visual	Yes	Green	Red	No	2-3 weeks	No	Before the amplification
SYBR Green I	Visual	Yes	Green	Red	Yes	2-3 weeks	No	Before the amplification
Ethidium bromide	Visual	No	Yellow	No color	Yes	2 weeks	Yes	After the amplification

Table 3: Comparison of colorimetric assays together.

last drawback, the corresponding lab(s) should be equipped with such instruments which are commonly costly. Alternatively, to remedy such problems, the first metal dye called $MgSO_4$ and $CaCl_2$ was utilized, but just an ephemeral color change (i.e. turbidity; no more than a few seconds or minutes) was observed in tubes [24,35]. Even though short stability of the colour change probably cannot be a significant problem as long as just a few numbers of suspicious samples are used, during assessment of a great quantity of infected samples, most of the time, inaccurate results will be accordingly achieved [36-41]. The method, nevertheless, since in one hand is exploited prior to the reaction, leading to a significant reduction on probable contaminations, and on the other there is no need of toxic devices, could be thereby attributed as a reliable visual observation approach. Anyway, to provide a situation to increase time stability, the second fluorescent dye coined SYBR® Premix Ex Taq™ II was employed, leading to a clear green colour pattern. Notably, regardless of observing a significant growth on time stability (about 1–3 days), the method not only requires UV transilluminator but also its stability is more negatively sensitive when amplified reactions are exposed to the daylight. Indeed, since the dye must be added after the amplification so requires opening of the tubes, the occurrence of cross contamination risk will be accordingly enhanced [42]. To avoid such contaminations, using separate rooms can be a solution for LAMP setup and analysis [31]. Interestingly, HNB, SYBR Green I and GeneFinder™ dye-based assays were accompanied by several remarkable advantages compared with other colorimetric-based methods in that of which are mixed prior to amplification, a need to open the assayed samples to add the dye is thereby omitted, and the risk of cross-contamination will be excluded drastically [31-33,42]. Meanwhile, the visual inspection of LAMP products by means of the dyes was seen as advantageous as there was no need for electrophoresis and subsequent staining with carcinogenic ethidium bromide [30]. Lastly, the color brightness and stability of the both HNB, SYBR Green I and GeneFinder™ in the solutions with positive/negative reactions were remained constant after 2 weeks of exposure to ambient light, whereas detection by SYBR® Premix Ex Taq™ II was stable only 1–3 days. For example, at the study of Goto et al. [31], HNB was reported as the best visual system, while the brightness of SYBR green fluorescence and calcein fluorescence was significantly weaker than that of HNB. It is noticeable that since the color presented by HNB was light blue for positive results and dark blue for negative results, which cannot be discriminated precisely [42], so such based detection methods involve a little more attention to provide accurate decision. From the present study it is concluded that colorimetric LAMP could be a reliable and rapid technique that can help at the maximum to this direction.

Acknowledgement

This research was supported financially by Young Researchers and Elite Club, Meybod Branch, Islamic Azad University and desert trier biotechnologist Co. of Meybod.

References

1. Creamer R, Luque Williams M, Howo M (1996) Epidemiology and incidence of beet curly top geminivirus in naturally infected weed hosts. Plant Dis 80: 533-535.

2. Briddon RW, Stenger DC, Bedford ID, Stanley J, Izadpanah K, et al. (1988) Comparison of a beet curly top virus isolate originating from the old world with those from the new world. Eur J Plant Pathol 104: 77–84.

3. Stenger DC (1998) Replication Specificity Elements of the Worland Strain of Beet Curly Top Virus Are Compatible with Those of the CFH Strain But Not Those of the Cal/Logan Strain. Phytopathology 88: 1174-1178.

4. Ke G (1967) Possible incidence of curly top in Iran a new record. Plant Dis 51: 976-977.

5. Bennett CW (1971) The curly top disease of sugarbeet and other plants. Am Phytopathol Soc 7: 81.

6. Briddon RW, Heydarnejad J, Khosrowfar F, Massumi H, Martin DP, et al. (2010) Turnip curly top virus, a highly divergent geminivirus infecting turnip in Iran. Virus Res 152: 169-175.

7. Heydarnejad J, Hosseini Abhari H, BolokYazdi HR, Massumi H (2007) Curly top of cultivated plants and weeds and report of a unique curtovirus from Iran. J Phytopathol 155: 321-325.

8. Yazdi HR, Heydarnejad J, Massumi H (2008) Genome characterization and genetic diversity of beet curly top Iran virus: a geminivirus with a novel nonanucleotide. Virus Genes 36: 539-545.

9. Hormuzdi SG, Bisaro DM (1993) Genetic analysis of beet curly top virus: evidence for three virion sense genes involved in movement and regulation of single- and double-stranded DNA levels. Virology 193: 900-909.

10. Kaffka SR, Wintermantel WM, Lewellen RT (2002) Comparisons of soil and seed applied systemic insecticides to control beet curly top virus in the San Joaquin Valley. J Sugar Beet Res 39: 59-74.

11. Klute KA, Nadler SA, Stenger DC (1996) Horseradish curly top virus is a distinct subgroup II geminivirus species with rep and C4 genes derived from a subgroup III ancestor. J Gen Virol 77 : 1369-1378.

12. Stenger DC, Ostrow KM (1996) Genetic complexity of a beet curly top virus population used to assess sugar beet cultivar response to infection. Phytopathol 86: 929-933.

13. Creamer R, Hubble H, Lewis A (2005) Curtovirus infection in chile pepper in New Mexico. Plant Dis 89: 480-486.

14. Stanley J, Markham PG, Callis RJ, Pinner MS (1986) The nucleotide sequence of an infectious clone of the geminivirus beet curly top virus. EMBO J 5: 1761-1767.

15. Choi IR, Stenger DC (1996) The strain-specific cis-acting element of beet curly top geminivirus DNA replication maps to the directly repeated motif of the ori. Virology 226: 122-126.

16. Soto MJ, Chen LF, Seo YS, Gilbertson RL (2005) Identification of regions of the Beet mild curly top virus (family Geminiviridae) capsid protein involved in systemic infection, virion formation and leafhopper transmission. Virology 341: 257-270.

17. Stanley J, Latham JR, Pinner MS, Bedford I, Markham PG (1992) Mutational analysis of the monopartite geminivirus beet curly top virus. Virology 191: 396-405.

18. Stenger DC, Davis KR, Bisaro DM (1994) Recombinant beet curly top virus genomes exhibit both parental and novel pathogenic phenotypes. Virology 200: 677-685.

19. Duffus JE, Skoyen IO (1977) Relationship of age of plants and resistance to a severe isolate of the beet curly top virus. Phytopathol 67: 151-154.

20. Heydarnejad J, Keyvani N, Razavinejad S, Massumi H, Varsani A (2013) Fulfilling Koch's postulates for beet curly top Iran virus and proposal for consideration of new genus in the family Geminiviridae. Arch Virol 158: 435-443.

21. Demidov V (2005) Rolling-circle amplification (RCA). Encycl Diag Genom Proteom 10: 1175-1179.

22. Johnson L, Cao X, Jacobsen S (2002) Interplay between two epigenetic marks. DNA methylation and histone H3 lysine 9 methylation. Curr Biol 12: 1360-1367.

23. Strausbaugh CA, Wintermantel WM, Gillen AM, Eujayl IA (2008) Curly top survey in the Western United States. Phytopathology 98: 1212-1217.

24. Notomi T, Okayama H, Masubuchi H, Yonekawa T, Watanabe K, et al. (2000) Loop-mediated isothermal amplification of DNA. Nucleic Acids Res 28: E63.

25. Mori Y, Notomi T (2009) Loop-mediated isothermal amplification (LAMP): a rapid, accurate, and cost-effective diagnostic method for infectious diseases. J Infect Chemother 15: 62-69.

26. Nagamine K, Kuzuhara Y, Notomi T (2002) Isolation of single-stranded DNA from loop-mediated isothermal amplification products. Biochem Biophys Res Commun 290: 1195-1198.

27. Fukuta S, Iida T, Mizukami Y, Ishida A, Ueda J, et al. (2003) Detection of Japanese yam mosaic virus by RT-LAMP. Arch Virol 148: 1713-1720.

28. Varga A, James D (2006) Use of reverse transcription loop-mediated isothermal amplification for the detection of Plum pox virus. J Virol Methods 138: 184-190.

29. Nie X (2005) Reverse transcription loop-mediated isothermal amplification of DNA for detection of potato virus Y. Plant Dis 89: 605-610.

30. Ren WC, Wang CM, Cai YY (2009) Loop-mediated isothermal amplification for rapid detection of acute viral necrobiotic virus in scallop Chlamys farreri. Acta Virol 53: 161-167.

31. Goto M, Honda E, Ogura A, Nomoto A, Hanaki K (2009) Colorimetric detection of loop-mediated isothermal amplification reaction by using hydroxy naphthol blue. Biotechniques 46: 167-172.

32. Cardoso TC, Ferrari HF, Bregano LC, Silva-Frade C, Rosa AC, et al. (2010) Visual detection of turkey coronavirus RNA in tissues and feces by reverse-transcription loop-mediated isothermal amplification (RT-LAMP) with hydroxynaphthol blue dye. Mol Cell Probes 24: 415-417.

33. Ma XJ, Shu YL, Nie K, Qin M, Wang DY, et al. (2010) Visual detection of pandemic influenza A H1N1 Virus 2009 by reverse-transcription loop-mediated isothermal amplification with hydroxynaphthol blue dye. J Virol Methods 167: 214-217.

34. Moradi A, Nasiri J, Abdollahi H, Almasi M (2012) Development and evaluation of a loop-mediated isothermal amplification assay for detection of Erwinia amylovora based on chromosomal DNA. Eur J Plant Pathol 133: 609-620.

35. Tsai SM, Chan KW, Hsu WL, Chang TJ, Wong ML, et al. (2009) Development of a loop-mediated isothermal amplification for rapid detection of orf virus. J Virol Methods 157: 200-204.

36. Almasi MA, Erfan Manesh M, Jafary H, Dehabadi SM (2013) Visual detection of Potato Leafroll virus by loop-mediated isothermal amplification of DNA with the GeneFinder™ dye. J Virol Methods 192: 51-54.

37. Almasi MA, Hosseini SMD, Moradi A, Eftekhari Z, Ojaghkandi MA, et al. (2013) Development and application of loop-mediatred Isothermal amplification assay for rapid detection of fusariumoxysporum f. sp. Lycopersici. J Plant Pathol Microbiol.

38. Almasi MA, Jafary H, Moradi A, Zand N, Ojaghkandi MA, et al. (2013) Detection of Coat Protein Gene of the Potato Leafroll Virus by Reverse Transcription Loop-Mediated Isothermal Amplification. J Plant Pathol Microbiol.

39. Almasi MA, Moradi A, Nasiri J, Karami S, Nasiri M (2012) Assessment of performance ability of three diagnostic methods for detection of Potato Leafroll virus (PLRV) using different visualizing systems. Appl Biochem Biotechnol 168: 770-784.

40. Almasi MA, Ojaghkandi MA, Hemmatabadi A, Hamidi F, Aghaei S (2013) Development of Colorimetric Loop-Mediated Isothermal Amplification Assay for Rapid Detection of the Tomato Yellow Leaf Curl Virus. J Plant Pathol Microbiol.

41. Ahmadi S, Almasi MA, Fatehi F, Struik PC, Moradi A (2012) Visual Detection of Potato leafroll virus by One-step Reverse Transcription Loop-Mediated Isothermal Amplification of DNA with Hydroxynaphthol Blue Dye. J phytopathol 161: 120-124.

42. Cheng SJ, Chen ZY, Chu YN, Cui LB, Shi ZY, et al. (2011) Sensitive detection of influenza A (H1N1) virus by isothermal amplification in a single tube. Chin J Anal Chem 39: 335-340.

43. Tomita N, Mori Y, Kanda H, Notomi T (2008) Loop-mediated isothermal amplification (LAMP) of gene sequences and simple visual detection of products. Nat Protoc 3: 877-882.

44. Soliman H, El-Matbouli M (2009) Immunocapture and direct binding loop mediated isothermal amplification simplify molecular diagnosis of Cyprinid herpesvirus-3. J Virol Methods 162: 91-95.

Permissions

All chapters in this book were first published in JPPM, by OMICS International; hereby published with permission under the Creative Commons Attribution License or equivalent. Every chapter published in this book has been scrutinized by our experts. Their significance has been extensively debated. The topics covered herein carry significant findings which will fuel the growth of the discipline. They may even be implemented as practical applications or may be referred to as a beginning point for another development.

The contributors of this book come from diverse backgrounds, making this book a truly international effort. This book will bring forth new frontiers with its revolutionizing research information and detailed analysis of the nascent developments around the world.

We would like to thank all the contributing authors for lending their expertise to make the book truly unique. They have played a crucial role in the development of this book. Without their invaluable contributions this book wouldn't have been possible. They have made vital efforts to compile up to date information on the varied aspects of this subject to make this book a valuable addition to the collection of many professionals and students.

This book was conceptualized with the vision of imparting up-to-date information and advanced data in this field. To ensure the same, a matchless editorial board was set up. Every individual on the board went through rigorous rounds of assessment to prove their worth. After which they invested a large part of their time researching and compiling the most relevant data for our readers.

The editorial board has been involved in producing this book since its inception. They have spent rigorous hours researching and exploring the diverse topics which have resulted in the successful publishing of this book. They have passed on their knowledge of decades through this book. To expedite this challenging task, the publisher supported the team at every step. A small team of assistant editors was also appointed to further simplify the editing procedure and attain best results for the readers.

Apart from the editorial board, the designing team has also invested a significant amount of their time in understanding the subject and creating the most relevant covers. They scrutinized every image to scout for the most suitable representation of the subject and create an appropriate cover for the book.

The publishing team has been an ardent support to the editorial, designing and production team. Their endless efforts to recruit the best for this project, has resulted in the accomplishment of this book. They are a veteran in the field of academics and their pool of knowledge is as vast as their experience in printing. Their expertise and guidance has proved useful at every step. Their uncompromising quality standards have made this book an exceptional effort. Their encouragement from time to time has been an inspiration for everyone.

The publisher and the editorial board hope that this book will prove to be a valuable piece of knowledge for researchers, students, practitioners and scholars across the globe.

List of Contributors

Khadija Khataby and Moulay Mustapha Ennaji
Laboratory of Virology, Microbiology, Quality and Biotechnologies/Ecotoxicology and Biodiversity, Faculty of Science and Techniques Mohammedia, University Hassan II of Casablanca, Morocco

Saaid Amzazi
Laboratory of Biochemistry and Immunology, Faculty of Sciences, University of Mohammed V, Rabat, Morocco

Amal Souiri
Laboratory of Virology, Microbiology, Quality and Biotechnologies/Ecotoxicology and Biodiversity, Faculty of Science and Techniques Mohammedia, University Hassan II of Casablanca, Morocco
Laboratory of Biochemistry and Immunology, Faculty of Sciences, University of Mohammed V, Rabat, Morocco
Laboratory of Sanitary Control, Control Unit of Plants, Domaines Agricoles Maâmora, Salé, Morocco

Mustapha Zemzami and Hayat Laatiris
Laboratory of Sanitary Control, Control Unit of Plants, Domaines Agricoles Maâmora, Salé, Morocco

Pallavi JK and Usha Rao I
Department of Botany, University of Delhi-110007, New Delhi, India

Anupam Singh and Prabhu KV
National Phytotron Facility, Indian Agricultural Research Institute, New Delhi-110012, India

Mohamed E. Selim
Agricultural Botany Department, Faculty of Agriculture, Menoufiya University, Egypt

Mohammed Amin, Negeri Mulugeta and Thangavel Selvaraj
Department of Plant Science and Horticulture, College of Agriculture and Veterinary Science, Ambo University, Ethiopia

Vishnu Sukumari Nath, Shyni Basheer, Muthulekshmi Lajapathy Jeeva and Syamala Swayamvaran Veena
Division of Crop Protection, Central Tuber Crops Research Institute, Thiruvananthapuram, Kerala, India

Ahmed IS Ahmed
Plant Pathology Unit, Department of Plant Protection, Desert Research Center, Cairo, Egypt

Mekonnen T and Haileselassie T
Addis Ababa University, Institute of Biotechnology, Addis Ababa University, Ethiopia

Tesfaye K
Ethiopian Biotechnology Institute, Ethiopia

Hina Ali, Syed Sarwar Alam and Nayyer Iqbal
Nuclear Institute for Agriculture and Biology, Faisalabad, Pakistan

Jake R. Erickson and Kshitij Shrestha
Department of Biological Sciences, Idaho State University, 650 Memorial Dr. Gale Life Sciences Building, Pocatello, ID 83209, USA

Peter Rafael Ferrer, Shannon Piele, Caleb Fiedor and James Titius
Department of Biology, Fresno Pacific University, 1717 S Chestnut Ave, Fresno, CA 93702, USA

Moytri Roy Chowdhury
Department of Biological Sciences, Idaho State University, 650 Memorial Dr. Gale Life Sciences Building, Pocatello, ID 83209, USA
Department of Biology, Fresno Pacific University, 1717 S Chestnut Ave, Fresno, CA 93702, USA

Brian Foley
Los Alamos National Laboratory, Santa Fe, NM 87545, USA

Anteneh Ademe
Sekota Dryland Agricultural Research Center, Sekota, Ethiopia

Amare Ayalew and Kebede Woldetsadik
Department of Plant Sciences, Haramaya University, Haramaya, Ethiopia

Michelli de Souza dos Santos, Kavamura VN, Reynaldo ÉF and Souza DT
Laboratory of Environmental Microbiology, Brazilian Agricultural Research Corporation, EMBRAPA Environment, SP 340, Km 127.5, 13820-000, Jaguariúna, SP, Brazil

da Silva EHFM
College of Agriculture "Luiz de Queiroz", University of São Paulo, Av. Pádua Dias, 11, 13418-900, Piracicaba, SP, Brazil

May A
Laboratory of Plant Physiology, Brazilian Agricultural Research Corporation, EMBRAPA Maize and Sorghum, MG 424, Km 45, 35701-970, Sete Lagoas, MG, Brazil

Pallavi JK and Anupam Singh
National Phytotron Facility, Indian Agricultural Research Institute, New Delhi-110012, India

Prabhu KV
Joint Director (Research), Directorate, Indian Agricultural Research Institute, India

Temuujin U
Department of Biotechnology and Breeding, Mongolian University of Life science, Ulaanbaatar 17024, Mongolia

Kang HW
Department of Horticulture, Hankyong National University, Ansung 17579, Korea

Institute of Genetic Engineering, Hankyong National University, Ansung 17579, Korea

Xiangyang Shi
Department of Plant Science, University of California, Davis, CA 95616, USA

Crop Diseases, Pests & Genetics Research Unit, San Joaquin Valley Agricultural Sciences Center, USDA-ARS, Parlier, CA 93648, USA

Hong Lin
Crop Diseases, Pests & Genetics Research Unit, San Joaquin Valley Agricultural Sciences Center, USDA-ARS, Parlier, CA 93648, USA

Chavan NP, Pandey R, Nawani N and Tandon GD
Dr. D.Y. Patil Biotechnology and Bioinformatics Institute, Dr. D.Y. Patil Vidyapeeth, Pune, India

Khetmalas MB
Rajiv Gandhi Institute of Information Technology and Biotechnology, Bharati Vidyapeeth Deemed University, Pune, India

Rabab Sanoubar
Department of Horticulture, Agriculture Faculty, Damascus University, Syria

Astrid Bauer and Luitgardis Seigner
Iinstitute of Plant protection (Bayerische Landesanstalt für Landwirtschaft Institut für Pflanzenschutz, LfL), Freising, Germany

Ivone M Martins and Altino Choupina
Department of Biology and Biotechnology, Polytechnic Institute of Bragança, Apartado 1172, 5301-854 Bragança, Portugal
Mountain Research Center, Polytechnic Institute of Bragança, Apartado 1172, 5301-854 Bragança, Portugal

M Carmen López and Angél Dominguez
Department of Microbiology and Genetics, CIETUS-IBSAL, University of Salamanca / CSIC, Plaza de Drs. Queen s/n, 37007 Salamanca, Spain

Abd El-Moneim MR Afify and Ghada M Ibrahim
Department of Biochemistry, Faculty of Agriculture, Cairo University, Giza, Egypt

Mohamed A Abo-El-Seoud and Bassam W Kassem
Department of Plant Research, Nuclear Research Center, Egyptian Atomic Energy Authority, Abu-Zabal 13759, Egypt

Dinesh Singh
Division of Plant Pathology, Indian Agricultural Research Institute, New Delhi-110012 India

Richa Raghuwanshi
Department of Botany, Mahila Mahavidyalya, Banaras Hindu University, Varanasi-221005, U.P, India

Priyanka Singh Rathaur
Division of Plant Pathology, Indian Agricultural Research Institute, New Delhi-110012 India
Department of Botany, Mahila Mahavidyalya, Banaras Hindu University, Varanasi-221005, U.P, India

Yadava DK
Division of Genetics, Indian Agricultural Research Institute, New Delhi-110012, India

Yogesh Ruwali, Lalan Kumar and JS Verma
G.B. Pant University of agriculture and technology, Pantnagar, U.S.Nagar, Uttarakhand, India

Tatiana A Valueva, Natalia N Kudryavtseva, Alexis V Sofyin and Boris Ts Zaitchik
A.N. Bach Institute of Biochemistry of the Russian Academy of Sciences, Leninsky prospect 33, Moscow 119071, Russia

Marina A Pobedinskaya and Lyudmila Yu Kokaeva
Department of Mycology, M.V. Lomonosov Moscow State University, Moscow 119991, Russia

Sergey N Elansky
Department of Mycology, M.V. Lomonosov Moscow State University, Moscow 119991, Russia

A.G.Lorkh Potato Research Institute of the Russian Academy of Sciences, Moscow region, 140051, Kraskovo-1, Lorkh Street 23, Russia

Swatilekha Mohanta, Sial P and Swain PK
Department of Nematology, College of Agriculture, OUAT, Regional Research Technology Transfer Station, Pottangi, Bhubaneswar-751003, Odisha, India

Rout GR
Department of Biotechnology, College of Agriculture, OUAT, Bhubaneswar-751003, India

Aseel DG and Hafez EE
Department of Plant Protection and Biomolecular Diagnosis, Arid Lands Cultivation Research Institute (ALCRI), Egypt

Jarred Yasuhara-Bell
Department of Molecular Biosciences and Bioengineering, College of Tropical Agriculture and Human Resources, University of Hawai'i at Mānoa, 3190 Maile Way, St. John Room 315, Honolulu, HI 96822, USA

Caleb Ayin, Anne M. Alvarez, April Hatada and Yonghoon Yoo
Department of Plant and Environmental Protection Sciences, College of Tropical Agriculture and Human Resources, University of Hawai'i at Mānoa, 3190 Maile Way, St. John Room 315, Honolulu, HI 96822, USA

Robert L. Schlub
Cooperative Extension Service, University of Guam, Agriculture and Life Sciences Building Room 105E, Mangilao, Guam 96923, USA

Kangmin Kim, Hee-Sun Kook, Ye-Jin Jang, Seralathan Kamala-Kannan, Jong-Chan Chae and Kui-Jae Lee
Division of Biotechnology, Advanced Institute of Environment and Bioscience, Chonbuk National University, Iksan 570-752, Korea

Wang-Hyu Lee
Department of Agricultural Biology, College of Agriculture and Life Sciences, Chonbuk National University, Jeonju 561-756, Korea

Hoda MA Waziri
Plant Virus and Phytoplasma Research Department, Plant Pathology Research Institute, Agricultural Research Center, Cairo, Egypt

Adam OJ and Midega CAO
International Centre of Insect Physiology and Ecology (icipe), Nairobi, Kenya

Runo S and Khan ZR
Kenyatta University, Nairobi, Kenya

Yasser El-Halmouch
Botany Department, Faculty of Science, Damanhour University, Damanhour 22511, Egypt
Biotechnology Department, Faculty of Science, Taif University, Taif 21974, Kingdom of Saudi Arabia

Ahlam Mehesen
Soils, Water and Environment Research Institute, Agriculture Research Center, Giza 12619, Egypt

Abd El-Raheem Ramadan El-Shanshoury
Botany Department, Faculty of Science, Tanta University, Tanta 31527, Egypt
Biotechnology Department, Faculty of Science, Taif University, Taif 21974, Kingdom of Saudi Arabia

Hassan Momeni
Department of Plant Pathology, Iranian Research Institute of Plant Protection (IRIPP), Tehran, Iran

Fahimeh Nazari
Department of Plant Pathology, University of Tarbiat Modarres, Tehran, Iran

Mohammad Shahid, Mukesh Srivastava, Antima Sharma, Vipul Kumar, Sonika Pandey and Anuradha Singh
Biocontrol Laboratory, Department of Plant Pathology, Chandra Shekhar Azad University of Agriculture and Technology, Kanpur, 208002, India

James O Obuya and Anthony Ananga
Center for Viticulture and Small Fruit Research, Florida A&M University, 6505 Mahan Drive, Tallahassee, FL 32317, USA

Gary D. Franc
Plant Sciences Department-3354, University of Wyoming, 1000 E. University Ave, Laramie, WY 82071, USA

Sumant Chaubey, Malini Kotak and Archana G
Department of Microbiology & Biotechnology Centre, Faculty of Science, The Maharaja Sayajirao University of Baroda, Vadodara-390 002, Gujarat, India

Mohammad Amin Almasi
Department of Agriculture and Plant Breeding, Faculty of Agriculture, Zanjan University, Zanjan, Iran

Mehdi Aghapour-ojaghkandi and Saeedeh Aghaei
Young Researchers and Elite Club, North Tehran Branch, Islamic Azad University Tehran, Iran

Index